SOILS
and
HUMAN HEALTH

SOILS
and
HUMAN HEALTH

Edited by
Eric C. Brevik • Lynn C. Burgess

CRC Press
Taylor & Francis Group
Boca Raton   London   New York

CRC Press is an imprint of the
Taylor & Francis Group, an **informa** business

CRC Press
Taylor & Francis Group
6000 Broken Sound Parkway NW, Suite 300
Boca Raton, FL 33487-2742

First issued in paperback 2016

© 2013 by Taylor & Francis Group, LLC
CRC Press is an imprint of Taylor & Francis Group, an Informa business

No claim to original U.S. Government works

ISBN 13: 978-1-138-19931-6 (pbk)
ISBN 13: 978-1-4398-4454-0 (hbk)

Library of Congress Cataloging-in-Publication Data

Soils and human health / editors: Eric C. Brevik, Lynn C. Burgess.
    p. cm.
  Includes bibliographical references and index.
  ISBN 978-1-4398-4454-0 (hardcover : alk. paper)
  1. Soils. 2. Soils--Health aspects. 3. Soil science. 4. Soils--Environmental aspects. 5. Soils and climate. I. Brevik, Eric C. II. Burgess, Lynn C.

S591.S7575 2013
635.9'55--dc23   2012034999

Visit the Taylor & Francis Web site at
http://www.taylorandfrancis.com

and the CRC Press Web site at
http://www.crcpress.com

I dedicate this book to my wife, Corinne, and my children, David, Joshua, and Kathryn. Without their loving support and understanding, it would not have been possible to find the time needed to complete this project.

Eric C. Brevik

There is one person who has helped me get to the point that I could write and edit a book like this, and that person is my wife, Cindy. If not for her, I never would have returned to college twice and completed my degrees. She struggled through the poor economic conditions of raising a family and working while her husband was a student. Cindy taught me a rudimentary understanding of soil science that helped me in editing this book. Lastly, I would like to thank Dr. Eric Brevik who thought that I could assist him in this project and engaged me in something I never would have attempted without his guidance.

Lynn C. Burgess

Contents

SECTION I   Topic Overview

SECTION II   Human Health as It Relates to Materials Found in Soil

SECTION III   Human Use of and Interactions with Soil

SECTION IV   Food Security and Climate Change

Preface

Many different fields are interested in soils and human health, including but not limited to soil science, geology, geography, anthropology, biology, agronomy, sociology, public health, and medicine. Individuals from many of these fields were solicited as both authors and reviewers of the chapters in this book. Due to the wide-ranging target audience, it was necessary for chapter authors to communicate useful and accurate information without getting too deep into the professional jargon of any particular field. Therefore, readers of this book will likely find that the chapters do not explore their specialty areas as deeply as a publication that was written specifically for that specialty would. However, our goal was to make this publication accessible and valuable to these disparate fields of study. It is our hope that the approach taken will provide useful information to all target fields while also being understandable by a professional in any of those fields. It is up to you as readers and users of this book to decide whether or not this effort was successful.

Another major goal was to elucidate the current state of knowledge on soils and human health and to identify topics where more work is needed. The study of soils and human health is very complex. Traditional scientific approaches that isolate a single variable and investigate that variable do not work well in the study of soils and human health because many of the issues that affect human health involve complex, synergistic relationships. To truly understand soils and human health, we have to gain an understanding of interactions between multiple variables. We also have to gain an understanding not only of the properties of soils and processes occurring in soils, but also of relevant properties and processes outside of soils that influence what occurs within soils. And, of course, we have to understand how soil exposure in its many potential forms and ways influences the human organism.

Going forward, the diverse array of fields with an interest in soils and human health need to have increased cross-disciplinary communication and cooperation if we are going to adequately address soil and human health issues. Complex interdisciplinary research teams will be needed with expertise in the relevant areas and the ability for the team members to communicate effectively with one another at a professional level. In a scientific world, where the trend has often been ever-increasing specialization, it has become more difficult to communicate information across disciplinary boundaries. In some cases it has even become difficult to communicate with other experts within a given discipline but in separate subfields. The interdisciplinary nature of soils and human health studies presents a significant challenge going forward.

Eric C. Brevik and Lynn C. Burgess
Dickinson, North Dakota

Editors

Eric C. Brevik is a professor of geology and soils at Dickinson State University, North Dakota, United States. Dr. Brevik earned his BS and MA degrees in geology from the University of North Dakota and his PhD in soil science (soil morphology and genesis) from Iowa State University. He has taught courses in soil science and geology at Valdosta State University (Georgia, United States) and Dickinson State University since 2001. His research interests include carbon sequestration by soil, soil health and productivity, soils and society, and the integration of geological and soils information. He is active professionally, having published over 160 peer-reviewed articles, abstracts, and other publications. He is also active both in the United States and internationally in researching the historical and sociological aspects of soil science. Dr. Brevik served as the vice chair and then the chair for the Soil Science Society of America's (SSSA's) Council on the History, Philosophy, and Sociology of Soil Science from 2004 to 2006 and currently serves as the SSSA historian. He has organized and chaired several sessions at meetings of professional groups including SSSA and the European Geosciences Union, including a session on soils and human health at the 2006 SSSA's meeting in Indianapolis, Indiana, United States.

Lynn C. Burgess is a professor of biology at Dickinson State University, North Dakota. Dr. Burgess earned his BS degree in zoology from Utah State University, his MS degree in biology from Eastern Washington University, and his PhD in toxicology from Utah State University. Before working in academia, he was an air-force officer and pilot, and business owner, and managed both milk-processing and feed-production plants. He has taught courses in biology and environmental health at Dickinson State University since 1999. He is the director of the university's Environmental Health Program, which is nationally accredited. He is also a member of the National Environmental Health Science and Protection Accreditation Council, and he is the current president of the board of directors of the Association for Environmental Health Academic Programs. His research interests include the cancer preventative properties of natural compounds, the toxicological effect of heavy metals on gap junctional communications, and the angiogenesis involved in the exuberant granulation tissue in equines. His research is funded by the NIH specifically for undergraduate students. Dr. Burgess is involved in the Society of Toxicology and their subcommittee on undergraduate education, and in the National Environmental Health Association.

Contributors

Winfried E.H. Blum
Institute of Soil Research
University of Natural Resources and Life
    Sciences (BOKU) Vienna
Vienna, Austria

Eric C. Brevik
Departments of Natural Sciences, and
    Agriculture and Technical Studies
Dickinson State University
Dickinson, North Dakota

Lynn C. Burgess
Department of Natural Sciences
Dickinson State University
Dickinson, North Dakota

C. Lee Burras
Department of Agronomy
Iowa State University
Ames, Iowa

Lorna Butler
Department of Anthropology
Iowa State University
Ames, Iowa

Cynthia A. Cambardella
USDA-ARS National Laboratory for
    Agriculture and the Environment
Ames, Iowa

Cheryl Carmona
Department of Environmental Science
    and Technology
University of Maryland
Baltimore, Maryland

Patrick M. Carr
Dickinson Research and Extension Center
North Dakota State University
Dickinson, North Dakota

Pattie L. Carr
Departments of Health and Physical
    Education, and Fine and Performing Arts
Dickinson State University
Dickinson, North Dakota

F. Daniel Cring
Department of Anthropology and Sociology
University of Louisiana-Lafayette
Lafayette, Louisiana

Kathleen Delate
Departments of Agronomy and Horticulture
Iowa State University
Ames, Iowa

Joseph R. Heckman
Department of Plant Biology and Pathology
Rutgers, The State University of New Jersey
New Brunswick, New Jersey

Martin F. Helmke
Department of Geology and Astronomy
West Chester University of Pennsylvania
West Chester, Pennsylvania

Jacques M. Henry
Department of Anthropology and Sociology
University of Louisiana-Lafayette
Lafayette, Louisiana

Rebecca Kanter
Center for Human Nutrition, International
    Health
Bloomberg School of Public Health
The Johns Hopkins University
Baltimore, Maryland

and

The Integral Center for the Prevention of
    Chronic Diseases
Institute of Nutrition of Central America and
    Panama—INCAP
Guatemala City, Guatemala

Anna S. Knox
Environmental Analysis Section
Savannah River National Laboratory
Aiken, South Carolina

Wendy W. Kuhne
Environmental Analysis Section and National
    Center for Radioecology
Savannah River National Laboratory
Aiken, South Carolina

Miroslav Kutílek
Soil Science and Soil Physics
Prague, Czech Republic

Russell L. Losco
Lanchester Soil Consultants, Inc.
West Grove, Pennsylvania

Thomas E. Loynachan
Department of Agronomy
Iowa State University
Ames, Iowa

Monday Mbila
Department of Biological and Environmental
    Sciences
Alabama A&M University
Normal, Alabama

Richard Morgan
Geography Department
University of Otago
Dunedin, New Zealand

Roni A. Neff
Center for a Livable Future
Bloomberg School of Public Health
The Johns Hopkins University
Baltimore, Maryland

Donald R. Nielsen
Department of Land, Air, and Water Resources
University of California, Davis
Davis, California

Stephen Nortcliff
Department of Geography and Environmental
    Science
University of Reading
Reading, United Kingdom

Mary Nyasimi
Department of Agronomy
Iowa State University
Ames, Iowa

Charles E. Turick
Environmental Biotechnology Section
Savannah River National Laboratory
Aiken, South Carolina

Xin Zhao
Horticultural Sciences Department
University of Florida
Gainesville, Florida

Reviewers

We thank the following reviewers who contributed their time and expertise to the improvement of this project.

Peter W. Abrahams
Institute of Geography and Earth Sciences
Aberystwyth University
Ceredigion, Wales

John Bluemle
North Dakota Geological Survey (retired)
Bismarck, North Dakota

Christian Feller
Functional Ecology and Biogeochemistry of
    Soils and Agroecosystems
Research Institute for Development
Montpellier, France

Thomas E. Fenton
Agronomy Department
Iowa State University
Ames, Iowa

Michael Fletcher
Missouri Southern State University
Joplin, Missouri

Alfred Hartemink
Department of Soil Science
University of Wisconsin-Madison
Madison, Wisconsin

Michael T. Homsher
Environmental, Safety, and Occupational
    Health Management
The University of Findlay
Findlay, Ohio

Rupert Hough
The James Hutton Institute
Aberdeen, United Kingdom

Robert K. Hubbard
Southeast Watershed Research Laboratory
USDA-ARS
Tifton, Georgia

Samuel J. Indorante
USDA-NRCS
Carbondale, Illinois

Anna Jeng
College of Health Sciences
Old Dominion University
Norfolk, Virginia

Michael B. Jenkins
J. Phil Campbell Sr Natural Resource
    Conservation Center
USDA-ARS
Watkinsville, Georgia

M.B. Kirkham
Department of Agronomy
Kansas State University
Manhattan, Kansas

Frederick Kirschenmann
Leopold Center for Sustainable Agriculture
Iowa State University
Ames, Iowa

and

Stone Barns Center for Food and Agriculture
Pocantico Hills, New York

Robin C. Leonard
Department of Health
West Chester University of Pennsylvania
West Chester, Pennsylvania

Suzette Morman
U.S. Geological Survey
Denver, Colorado

Elica M. Moss
Department of Biological and Environmental
    Sciences
Alabama A&M University
Huntsville, Alabama

David Pimentel
Departments of Entomology, and Ecology
    and Evolutionary Biology
Cornell University
Ithaca, New York

Jacob Prater
Agronomy Department
Iowa State University
Ames, Iowa

Susan E. Samson-Liebig
USDA-NRCS
Bismarck, North Dakota

David B. Sing
Department of Human Genetics
University of Michigan
Ann Arbor, Michigan

Balwant Singh
Faculty of Agriculture, Food and Natural
    Resources
The University of Sydney
Sydney, Australia

Dan H. Yaalon
Hebrew University of Jerusalem
Jerusalem, Israel

Gregory Zimmerman
Department of Biology
Lake Superior State University
Sault Ste. Marie, Michigan

In addition to the named reviewers, we wish to thank those who provided reviews for this project but requested that they remain anonymous.

Section I

Topic Overview

1 An Introduction to Soil Science Basics

Eric C. Brevik

CONTENTS

1.1 INTRODUCTION

This book is intended to inform professionals interested in human health about the many interactions between soils and human health. However, many in the target audience may not have a background in soil science. Readers who do not have a soil science background should find this chapter useful in explaining the soil science terms and concepts discussed in later chapters. There is also a list of additional readings at the end of this chapter that can deepen your knowledge of soil science if you require a more detailed background than is provided here.

1.2 SOIL FORMATION

Five factors combine to create the properties of any given soil. Those factors are (1) parent material, (2) topography, (3) climate, (4) organisms, and (5) time. Different combinations of these five factors lead to the creation of soils with different properties. It is also important to note that these factors are not completely independent of one another. For example, climate will influence the organisms that we find at any given location. Topography and parent material can also influence organisms, and there are multiple other influences that are possible. Therefore, the interactions between the soil-forming factors can become quite complex.

Parent material refers to the material from which the soil is formed. That material can be weathered rock, called residuum, or thick accumulations of organic material such as plant debris. But the most common parent materials are sediments deposited by various geologic processes. Parent materials are summarized in Table 1.1. Parent materials are important because they determine many of the physical and chemical characteristics of a soil. For example, calcium is an important plant nutrient, but a soil is likely to be deficient in calcium if the parent materials do not contain minerals that

TABLE 1.1
Classification of Soil Parent Materials

Parent Material	Mode of Creation
Aeolian	Sediments that were transported (moved) and deposited (left behind) by wind. A commonly known example would be sand in sand dunes.
Alluvium	Sediments that were transported and deposited by flowing water (rivers or streams).
Colluvium	Sediments that were transported and deposited by gravity, usually found at the base of high spots such as mountains or hills.
Lacustrine	Sediments that were deposited in lakes.
Marine	Sediments that were deposited in oceans.
Organic	Created when organic materials, primarily plant debris, accumulate to thicknesses that can be several meters.
Outwash	Sediments that were deposited by flowing water (rivers or streams) flowing off melting glaciers. The glacial source typically gives these sediments different properties from alluvium derived from nonglacial sources.
Residuum	Forms when solid rocks weather (break down) in place, leaving behind loose material.
Till	Sediments that were transported and deposited by glacial ice.

include calcium in their chemical composition. Likewise, the proportions of sand, silt, and clay in a soil are determined largely by the sand, silt, and clay contents of the parent materials.

Topography refers to the slope, aspect, and landscape position of a given soil. The steeper a slope is, the greater erosion tends to be, which limits the depth of the soils on those steep slopes (Figure 1.1). Gentler slopes allow more water infiltration and less runoff, leading to more developed soils. *Aspect* refers to the direction the slope faces, which influences the amount of solar energy that hits the slope. South-facing slopes in the northern hemisphere get more incoming solar energy than north-facing slopes, thus warming and drying soils on the southern slopes relative to the northern slopes (Figure 1.2). The same relationship holds in the southern hemisphere, with north-facing slopes getting more solar energy than south-facing slopes. *Landscape position* refers to where, on a slope, a soil is formed: at the top, along the side, or at the bottom. Landscape position is important in determining whether a soil will be subjected to slight erosion, intense erosion, slight deposition, or intense deposition. The processes influenced by slope, aspect, and landscape position influence the final soil formed at any given location.

Climate primarily refers to precipitation and temperature. Figure 1.3 shows examples of temperature and precipitation data for the cities of Dickinson (North Dakota), Bangor (Maine), and San Diego (California), United States, and demonstrates how climate variables can be different from place to place. The average monthly temperatures in Dickinson and Bangor are very similar, especially from February through October. The precipitation in these two cities, on the other hand, is very different. Dickinson averages about 41.5 cm of precipitation per year, while Bangor averages 100.5 cm. Therefore, from a soil-formation perspective, the climate is different. But climate does not refer only to the amounts of precipitation and the average temperature. Climate includes distribution and extremes of precipitation and temperature throughout the year as well. Notice in Figure 1.3 that the temperature does not vary much from month to month in San Diego, but is highly variable from month to month in Dickinson and Bangor. Dickinson's precipitation is highly variable by season: 57.9% of its average annual precipitation is in spring and early summer (April through July), with 20.2% in June but only 2.1% in December. Bangor, on the other hand, has a much more evenly distributed precipitation pattern, with no more than 9.3% and no less than 6.4% of its average annual precipitation coming in any given month. By contrast, San Diego usually gets most of its precipitation in the winter months, with little precipitation falling from May through September. Temperature can vary from place to place in a similar manner to these

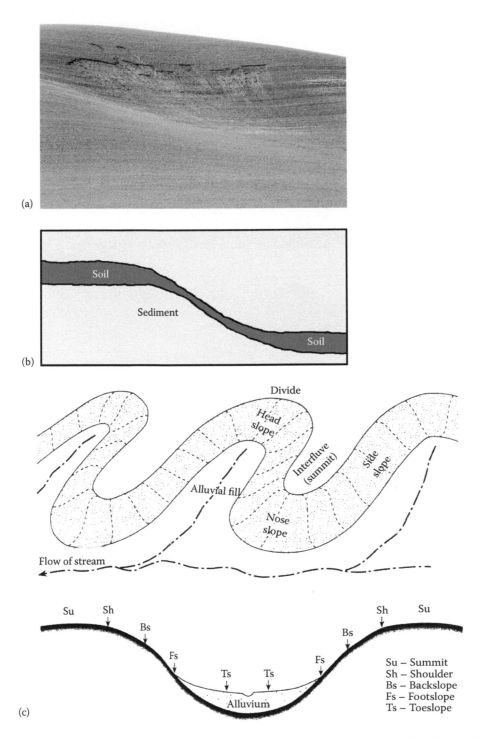

FIGURE 1.1 (a) The steepest portions of slopes often see the greatest erosion rates, limiting the depth of the soil on steep slopes. The photo shows accelerated erosion along the hillside (Photo by Tim McCabe, USDA NRCS). (b) The diagram shows a profile view with thinner soils along the slope. (c) Landscape models help us understand and communicate where on the landscape a soil is forming. These two figures are examples of three-dimensional (top) and two-dimensional (bottom) landscape models developed by Robert Ruhe that are widely used in soil science. (From Schaetzl, R. and Anderson, S., *Soils: Genesis and Geomorphology*, Cambridge University Press, New York, 2005.)

(a)

(b)

FIGURE 1.2   Demonstrations of the importance of aspect on slope microclimates. (a) A view of the Rocky Mountains in Colorado, United States, in November, 2005. Note that the north-facing slopes have snow cover, but the south-facing slopes are bare. This is because the south-facing slopes receive more solar energy and are therefore warmer, melting the snow. (b) A hill in the badlands of North Dakota, United States. The trees on the left side of the photo are growing on the north-facing slope, where soil water is adequate to support vegetative growth. The nearly unvegetated slopes on the right and at the top of the picture are south-facing slopes. High inputs of solar energy warm and dry the soils, leading to low soil water content and sparse vegetative growth. (Photos by Eric Brevik.)

precipitation examples. So the total amount of precipitation, the distribution of that precipitation, the average annual temperatures, and the distribution of temperature all come together to create the climate variable in soil formation.

Organisms refer to plants and animals in the soil at both a microscale and a macroscale. Important microorganisms include various algae, bacteria, and protozoa, while important macroorganisms include earthworms, termites, and ants. Organisms are important in soil formation because they add organic materials to the soil; facilitate weathering (the breakdown) of soil solids, which releases nutrients; and mix materials throughout the soil profile through burrowing activities. Different ecosystems lead to different organism interactions and ultimately to different soils (Figure 1.4). Organisms are also important in nutrient cycling and aggregate stability, topics that will be discussed in more detail later in the chapter. While it is possible and tempting to place

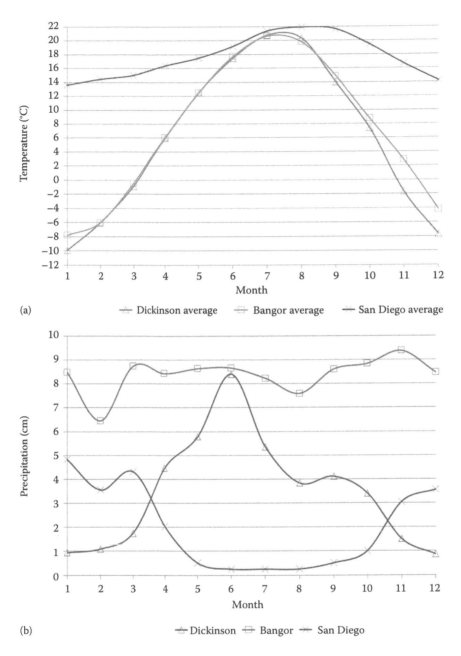

FIGURE 1.3 Long-term average monthly temperatures (a) and precipitation (b) in the cities of Dickinson, Bangor, and San Diego. Months are from January (1) to December (12). (Data for Dickinson is from http://www.crh.noaa.gov/bis/climate_extremes.php; Bangor is from http://www.erh.noaa.gov/er/gyx/climo/ME_STATS_NEW.htm; San Diego is from http://www.wrh.noaa.gov/sgx/climate/san-san.htm.)

humans into the organism category as well, as we are organisms, some scientists feel that the organism state factor does not adequately account for humans, as our activities have the ability to modify the other state factors (Jenny, 1980).

The *time factor* refers to how long a soil has had to form. In general, the longer a given point in the landscape has been stable, the deeper and more well developed the soil at that point will be (Table 1.2). Horizons (layers) in the soil start to form as a soil develops, with O or A horizons

FIGURE 1.4 The soil ecosystem under the trees (*background*) will be different from the soil ecosystem under the grasses and flowers (*foreground*), leading to differences in the soils formed even though the landscape position, climate, parent material, and time factors for the two locations are the same. (Photo by Gary Kramer, USDA NRCS.)

typically forming first, followed by E and B horizons (Table 1.3). The soil-forming factors all come together at any given point to form a soil with a sequence of horizons that reflect the influence of those factors. A vertical exposure of this sequence of horizons is known as a *soil profile* (Figure 1.5), with each horizon differing chemically and/or physically from the other horizons in the profile.

TABLE 1.2
Comparisons of Some Soil Depths Reached over Given Periods of Soil Formation

Horizon or Profile	Years of Soil Formation	Soil Depth (cm)
Hardening of tropical clay surface to laterite following deforestation	35	15
Grassy area, Georgia borrow pit	41	25
Entisol on volcanic ash	45	35
A1 horizon, Vertic Argiaquoll (Oregon)	133	15
A1 horizon, Hapludoll (Iowa)	400	33
Histosol (Wisconsin)	3000	200

Source: Modified from Dixon-Coppage, T.L., Davis, G.L., Couch, T., Brevik, E.C., Barineau, C.I., and Vincent, P.C., *Soil and Crop Science Society of Florida Proceedings*, 64, 8–15, 2005.

Note: The trend of increasing depth with increasing time of soil formation.

TABLE 1.3

Soil Horizons Shown in Their Common Order from the Surface Down

Horizon	Characteristics
O	An organic layer, frequently greater than 35% organic material, although it can be as low as 25% given certain other conditions. Can be very thin, such as a layer of leaves on the forest floor, or several meters thick, such as the accumulation of organic material in a bog.
A	An organic-rich mineral horizon, usually less than 5% organics, although it can contain as much as 35% organic material given certain conditions.
E	A light colored mineral horizon, characterized by intense leaching and the loss of clay (eluviation) to lower horizons. Common beneath forest vegetation.
B	A zone marked by a significant accumulation due to soil-forming processes. These accumulations can be silicate clays (Bt horizon), pigmenting minerals marked by a color change relative to the horizons above and below (Bw horizon), or a number of other materials. The lower case letter behind the upper case B tells a soil scientist what has accumulated to create the B horizon.
C	Slightly altered parent material, represents the geologic sediment that was present prior to soil formation.
R	A layer of consolidated rock within the soil profile.

Source: Brevik, E.C., *Journal of Geoscience Education,* 48, 19–23, 2000.

Note: It is important to note that not all of these horizons are found in the typical soil, but various combinations of these horizons are found in all soils.

FIGURE 1.5 Examples of soil profiles formed under different combinations of soil-forming factors. The photo on the *left* shows the typical O-E-B-C sequence (the C horizon is in the bottom of the hole by the shovel blade) of a soil called a Spodosol with shovel for scale. Spodosols form in sandy parent materials under a pine forest ecosystem in humid climates. The photo on the *right* shows the A and upper B horizons of a soil called a Mollisol, with maize stalks for scale. Mollisols are typically found in subhumid to semiarid grassland ecosystems. (Left photo by Eric Brevik and right photo by Lynn Betts, USDA NRCS.)

1.3 SOIL PHYSICAL PROPERTIES

Soil physical properties can be directly observed through sight or feel. The physical properties of a soil influence such aspects as water, air, and organism movement through the soil, water storage in the soil, and the ease with which roots can penetrate the soil. These in turn influence chemical and biological processes that occur in the soil. Important soil physical properties include composition, structure, pore space, bulk density, water-holding capacity, and color.

Soils are composed of four basic materials: (1) minerals, (2) organic matter, (3) water, and (4) air. The "ideal" soil is about 45% mineral material and 5% organic matter, the solid materials in soil, and 50% pore space by volume. About half of that pore space would ideally be filled with water and half with air, making our "ideal" soil 25% water and 25% air (Figure 1.6). This balance provides enough mineral material to act as an anchor for plant roots and to supply nutrients as the minerals weather; enough organic matter to promote important soil physical, chemical, and biological properties and processes; and adequate but not excessive water and air to support important biological activity. In actuality, real-world soils rarely achieve this "ideal" balance. Soil water and air levels fluctuate constantly throughout the year, and organic matter levels fluctuate over time periods of years to decades, particularly in agricultural soils, in response to human soil management.

Texture refers to the proportions of sand-, silt-, and clay-sized particles in a soil. Soil texture is important in determining properties such as soil fertility, water and air movement, and water storage in soils, and texture is a property that does not readily change. For these reasons, particle size analysis is used to determine soil textural classes and is one of the most commonly used analyses in soil science. Textures are assigned names such as loam, silt loam, and sandy clay based on the sand/silt/clay proportions (Figure 1.7). Note in Figure 1.7 that the zones representing sand or silt textures are quite small, while the zone representing clay texture is large. This is because clays have a large surface area per volume due to their small size. Chemical reactions happen on surfaces, so this large surface area per volume means clays have more influence on soil properties per unit volume than sands or silts. Particles larger than sand (gravel) are not considered when calculating soil texture because gravels have a very low surface area per volume. It is also important to know that there are several different systems that place mineral particles into sand, silt, and clay size classes, and these different systems use different size ranges to assign classes (Table 1.4). Therefore, it is important to know which system is being used to classify particle size when being given sand/silt/clay percentages.

Sand, silt, and clay particles are not usually found on their own in the soil. Instead, they tend to be grouped together into aggregates called peds (Figure 1.8). Decomposed organic matter; plant roots and fungal hyphae; electrostatic attraction between negatively charged clay particles and cations

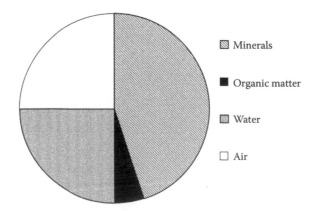

FIGURE 1.6   The basic composition of an "ideal" soil.

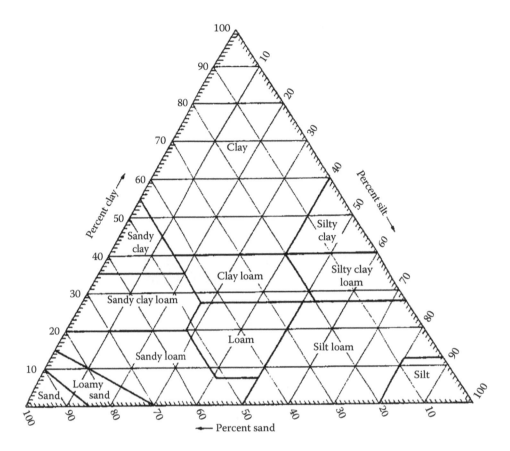

FIGURE 1.7   The textural triangle is used to assign names to soil textures after the percentages of sand, silt, and clay have been determined.

found in the soil; and minerals such as iron oxides, calcium carbonates, and silica can all serve as "glues" to help bind individual particles into aggregates. These aggregates or peds tend to take on characteristic shapes, referred to as *structure* (Figure 1.9). Aggregates are important in the soil because the spaces between the aggregates create large openings called macropores, which serve as pathways for the movement of air, water, and organisms, including plant roots, through the soil. Soils that have lost their structure, often due to a loss of organic matter or compaction, also tend to

---

TABLE 1.4

Examples of the Different Size Ranges Used by Some Common Systems for Classifying Mineral Particles

	Size Range (mm) by System				
Particle Size Name	U.S. Department of Agriculture	International Society of Soil Science	American Association of State Highway and Transportation Officials	Udden-Wentworth Scale	Unified System
Gravel	>2.0	>2.0	>2.0	>2.0	>5.0
Sand (total)	0.05–2.0	0.02–2.0	0.074–2.0	0.0625–2.0	0.074–5.0
Silt (total)	0.002–0.05	0.002–0.02	0.005–0.074	0.0039–0.0625	<0.074 (silt and
Clay (total)	<0.002	<0.002	<0.005	<0.0039	clay are combined)

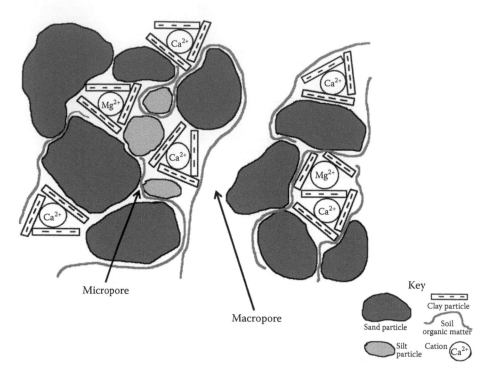

FIGURE 1.8 Diagram showing two aggregates, or peds, side by side in the soil. The aggregates are composed of individual sand, silt, and clay particles that are being held together by soil organic matter and electrostatic attraction between the negatively charged clay particles and cations. The large pore, or opening, between the aggregates is a macropore. Macropores serve as pathways for air, water, and many organisms to move through the soil as well as places for the penetration of plant roots. The small pores within the aggregates are micropores that serve to store water, some of which can be accessed by plants during dry periods.

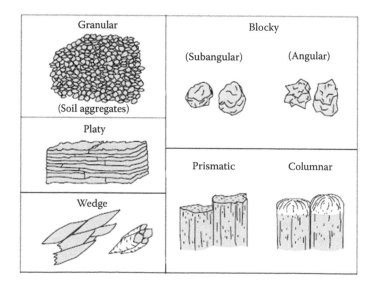

FIGURE 1.9 Common ped shapes. (Modified from Schoeneberger, P.J., Wysocki, D.A., Benham, E.C., and Broderson, W.D. (eds), *Field Book for Describing and Sampling Soils*, Natural Resources Conservation Service, National Soil Survey Center, Lincoln, NE, 2002.)

lose their macropores. However, the small openings, or micropores, are important too. Micropores serve as a storage point for water in the soil, so soils with adequate micropores tend to be more drought-resistant than soils that lack abundant micropores. In a well-aggregated soil, micropores are found within the soil aggregates. Therefore, because a good mix of both micropores and macropores is important in a soil, good soil structure is an important soil property. In an "ideal" soil, about 50% of the soil is solid material and 50% is pore space (Figure 1.6).

Bulk density is the mass per unit volume of oven-dry soil. Oven drying is important to drive off the water in the soil, as water content varies tremendously in any given soil over time. Therefore, water weight is not representative of the "typical" condition of the soil at any given time. The mass of the soil includes both the solid part of the soil and the pore space. Therefore, for any given soil, increased bulk density results in decreased pore space. Bulk density can be increased by compaction of the soil or by a loss of aggregation due to overtilling (plowing) of the soil or a decrease in organic matter content, among other causes. Macropores are the primary type of pore lost when bulk density increases. Because of this, the soil tends to become less well aerated and root growth becomes restricted, limiting plant growth, when bulk densities are too high (Figure 1.10 and Table 1.5). This in turn has a negative feedback on the soil ecosystem and nutrient cycling. Well-aggregated soils, on the other hand, have lower bulk densities than the same soils with poor aggregation.

Water-holding capacity refers to how much plant-available water a soil can store following a soaking rainfall. Soils with higher water-holding capacities are better able to supply plants growing in them with water through dry periods of short to moderate length, thus increasing plant production. Therefore, good water-holding capacity is highly desirable in a soil. Table 1.6 shows typical water-holding capacities for soils of various textures. Note that, as a general rule, water-holding capacity increases as the soils become finer textured, reaching a maximum value with the loam-textured soils. This is because sandy-textured soils tend to have a high percentage of macropores with few micropores, resulting in low water storage capacity. Moving from the loam textures into the clay textures, there is a decrease in water-holding capacity. Clay-textured soils will actually store more total water than loam-textured soils, but due to a high percentage of micropores that hold very tightly to water, much of that water is not plant-available. Loam-textured soils tend to have a good balance between macropores and micropores, resulting in the highest levels of stored plant-available water. However, it must be understood that soil texture is not the only factor that determines how much plant-available water a soil can store. Structure, organic matter content,

FIGURE 1.10 A section of an old wagon trail (line through the center of the photo) that ran through Iowa, United States, in the late 1800s shows the effects of soil compaction on vegetative growth. Note that the grass in the trail, where soils are still compacted, is considerably shorter than to the sides of the trail. Vegetation clippings showed that plant production off the trail was significantly higher than plant production on the trail (Brevik et al., 2002). (Photo by Eric Brevik.)

TABLE 1.5
Typical Root-Limiting Bulk Densities by Soil Texture

Texture	Root-Limiting Bulk Density (g/cm³)
Sand	1.80
Fine sand	1.75
Sandy loam	1.70
Fine sandy loam	1.65
Loam	1.55
Silt loam	1.45
Clay loam	1.50
Clay	1.40

Source: Data from Coder, K.D., *Soil Constraints on Root Growth*, University of Georgia Cooperative Extension Service Forest Resources Publication FOR98-10, Athens, GA, 1998.

Note: Root growth is limited when bulk density values are at or above the values given. Coarser soils tend to have higher root-limiting bulk densities because in poorly structured soils the coarse particles create more macropores than the smaller particles of fine-textured soils do.

and the type of clay minerals present can also make a difference, such that water-holding capacity values can vary for any given texture by as much as ±25% (Troeh, Hobbs, and Donahue, 2004). Within textural classes, well-aggregated soils and soils with high organic matter contents will have higher water-holding capacities than soils that are poorly aggregated or that have relatively less organic matter.

Color is probably one of the most obvious physical properties of soil. Anyone who has gone into the backyard and dug a hole to plant a tree, install fence posts, put in a foundation for a building, etc., has probably noticed that you do not have to dig very deep in most cases before the color of the soil starts to change. If you dig deep enough, you'll go through several different layers with varying colors (Figure 1.5). Most soil colors come from organic matter or iron oxides and provide valuable information about the conditions under which a soil formed (Table 1.7). While color is a physical property, it provides information about the chemical conditions driving soil formation.

TABLE 1.6
Typical Water-Holding Capacities for Soils of Various Textures

Soil Texture	Plant-Available Water (cm of water per 30 cm of soil depth)
Sand	0.6–2.5
Loamy sand	2.8–3.0
Sandy loam	3.2–3.6
Silt loam	5.1–6.4
Silty clay loam	4.6–5.1
Silty clay	3.8–4.3
Clay	3.0–3.8

Source: Data from Ball, J., *Ag News and Views*, 2001. http://www.noble.org/Ag/Soils/SoilWaterRelationships/Index.htm.

TABLE 1.7

Some Common Soil Colors, What Causes Them, and What They Tell Us about the Soil

Color	Cause (Mineral Name)	Indicates
Iron Oxide Pigmenting Agents		
Yellow	Goethite	Goethite forms in cool temperatures, or in moist soils in warm settings that are not water saturated. Goethite often occurs in association with hematite.
Red	Hematite	Hematite forms in high-temperature soils that are well oxygenated.
Brown	Maghemite	Maghemite is formed when other iron oxides are transformed by heating in association with organic matter, such as in forest fires.
Orange	Lepidocrocite	Lepidocrocite forms in a local zone of oxidation in soils that are otherwise reduced, such as around the roots of wetland plants.
Gray-green	Hydromagnetite	Hydromagnetite forms in soils that are frequently saturated, creating reducing conditions.
Other Pigmenting Agents		
Black or dark colors near the surface	Coatings of organic matter on soil particles	Conditions are favorable for the accumulation of organic matter in surface horizons. This may indicate wet conditions or high levels of organic matter additions.
Black in the subsurface	Organic matter or manganese oxides	*Organic matter*: intense leaching is moving organic matter from surface horizons into the subsurface. *Manganese oxides*: form in wet soils that undergo frequent alterations between oxidizing and reducing conditions. Manganese oxides often occur as concretions (small hard nodules).
White	Carbonates or other salts	The soils formed in an environment that is dry enough that there was not enough water to leach the salts from the soil.

1.4 SOIL CHEMICAL PROPERTIES

The term "clay" means two different things in soil science. On the one hand, it refers to a size range (Table 1.4). Any mineral particle that is less than the maximum size range, as determined by the textural classification system being used, can be considered clay regardless of its chemical composition. On the other hand, the term "clay" can also refer to a group of minerals. While clay minerals do tend to be small, chemical composition is the important factor under this definition. Clay minerals are important components of the soil system. Chemical reactions occur on surfaces, and some clay minerals have as much as 820 m^2 of surface area per gram of clay. Sand and silt grains have much smaller surface areas, as shown in Table 1.8. Therefore, clays are the dominant mineral

TABLE 1.8

Surface Areas for Different Particle Sizes Found in Soil

Particle Size Name	Particle Size (mm)	Surface Area (m^2/g)
Coarse sand	1.0–0.5	0.0023
Fine sand	0.25–0.10	0.0091
Silt	0.05–0.002	0.0454
Clay	<0.002	10–820

Source: Data from Pittenger, D.R., *California Master Gardener Handbook*, University of California Agriculture and Natural Resources Publication 3382, Oakland, CA, 2002; Brady, N.C. and Weil, R.R., *Nature and Properties of Soils*, Prentice Hall, Upper Saddle River, NJ, 2008.

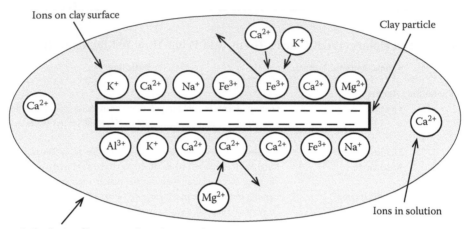

FIGURE 1.11 The concept of cation exchange on clay surfaces. Clays have a net negative charge, which attracts cations to them. Cations can be exchanged (note the Mg^{2+} exchanging for the Ca^{2+} and the Ca^{2+} and K^+ exchanging for the Fe^{3+}) between the soil solution (water plus dissolved ions) and the clay surface on a charge-for-charge basis.

components of soil participating in chemical reactions. Clays are also important in attracting and holding water in the soil.

One of the important soil properties that clays contribute to is *cation exchange capacity* (CEC), or the ability for negatively charged surfaces in the soil to attract and exchange positively charged ions (cations) (Figure 1.11). Humus, or decayed organic matter, is also an important contributor to CEC. Clays and humus both fall into a group of very small soil particles called colloids. CEC is commonly expressed as centimoles of charge per kilogram (cmol/kg) of soil (Table 1.9). Many of the cations held on exchange sites are plant nutrients, so soils with high CEC tend to be more fertile than soils with low CEC.

Soil pH is important because pH controls the availability of soil nutrients utilized by plants, and the diversity and abundance of soil organisms. There is no single soil pH at which all

TABLE 1.9

Total Surface Area and Net Negative Charge of Common Soil Colloids

Colloid	Total Surface Area (m²/g)	Net Negative Charge (cmol/kg)
Kaolinite	10–30	2–5
Chlorite	70–100	15–40
Illite	70–100	15–40
Smectite	650–800	80–120
Vermiculite	670–820	100–180
Humus	20–800	100–550

Source: Data from Brady, N.C. and Weil, R.R., *Nature and Properties of Soils*, Prentice Hall, Upper Saddle River, NJ, 2008.

Note: Net negative charge (and thus cation exchange capacity) tends to increase with increasing surface area. The first five colloids are clay minerals.

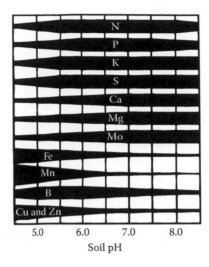

FIGURE 1.12   The availability of selected soil nutrients at various pH values.

nutrients are most available, but the widest range of nutrients are available between pH values of about 5.5 and 7.5 (Figure 1.12). The widest array of crops will grow well and the soil ecosystem is at its most diverse and abundant in this relatively neutral pH range. Many agricultural chemicals are also designed to provide maximum performance in this near-neutral pH range. When soils become too strongly acidic or basic, nutrient deficiencies and toxicities can become an issue and the diversity of the soil ecosystem also declines. Soils tend to behave as buffered systems in regards to pH changes and, generally speaking, higher CEC means greater buffering capacity. This helps to prevent drastic changes in pH that could be harmful to soil organisms. However, soil pH can be changed over time with sufficient inputs of acids or bases. A good example of this is acid rain, which has been shown to lower pH even in well-buffered soils over a period of decades.

Aeration refers to the ability of gases to move in and out of the soil, and is influenced by soil water content, the number and connectivity of macropores, and the rate of oxygen consumption by soil organisms. Soils that have adequate oxygen in them are referred to as *well aerated*, while soils that are deficient in oxygen are referred to as *anaerobic*. The aeration state of a soil has profound influences on soil chemical and biological processes. Organic matter decomposes more rapidly in well-aerated soils than in anaerobic soils; for that reason thick O horizons are typically found in wet (anaerobic) settings. The color of a soil can be used to determine aeration status. Gray-green iron oxides and black manganese oxide nodules typically form in anaerobic soils, while yellow, brown, and red iron oxides form in well-aerated soils (Table 1.7). Well-aerated soils also tend to have larger and more diverse biological populations than anaerobic soils.

1.5   SOIL BIOLOGY

Most scientists believe there are more species in existence below the soil surface than above it. Soil organisms include microorganisms and macroorganisms, and are very important to soil physical and chemical processes that influence soil fertility and productivity. Some important soil organisms and the roles they play are shown in Figure 1.13 and Table 1.10. While some soil organisms are plant or human pathogens, most are beneficial or benign.

Several small vertebrates live in the soil, including gophers, mice, moles, and prairie dogs. The burrowing action of these organisms creates large macropores and redistributes soil material from

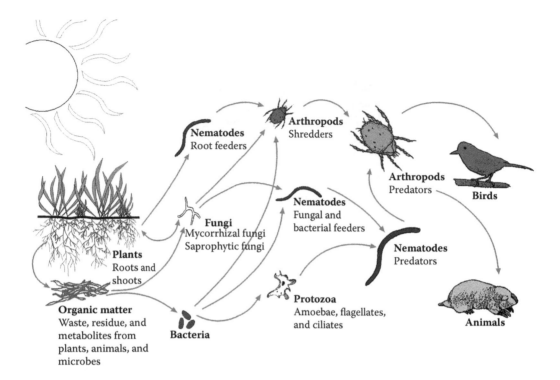

FIGURE 1.13 The various organisms that live in soil are interconnected in an intricate food web. This diagram shows some of the more common links in this web. (Courtesy of Tugel, A., Lewandowski, A., and Happe-von Arb, D. (eds), *Soil Biology Primer*, Soil and Water Conservation Society, Ankeny, IA, 2000.)

deeper reaches of the soil to the surface. They also shred organic matter into small pieces that smaller organisms can more efficiently use. Although they are often the most noticeable of soil organisms, smaller invertebrates and microorganisms are believed to be more important for most soil processes.

Important macroorganisms in the soil include earthworms, termites, and ants. Earthworms are the most important macroanimals in temperate soils, while termites are important in tropical soils and perform many of the same functions that earthworms do. Ants are found in a number of different environments. Earthworms, termites, and ants "till" the soil through their burrowing action, creating macropores. The borrowing activity can redistribute large quantities of soil (Figure 1.14). These macroorganisms also move organic materials from the soil surface into the soil as they move it into their burrows and break the organic matter down into smaller pieces that can be more readily decomposed by smaller organisms. Their feces serve as rich sources of soil nutrients and can be important in the formation of soil structure; much of the granular structure in temperate grassland soils is earthworm feces.

Termites and ants are members of the arthropod group, which includes many important soil predators such as spiders, beetles, centipedes, and mites. Small soil arthropods are often referred to as microarthropods. The most common microarthropods in the soil are mites and springtails, but they also include organisms such as rootworms and symphylans. Arthropods can be shredders, predators, herbivores, fungal feeders, or root feeders. The root feeders can cause significant damage to crops, but are less common than the other types of arthropods. The shredders, predators, and fungal feeders can all be beneficial to crop growth. The shredders participate in nutrient release through organic matter decomposition, and the predators and fungal feeders help control crop pests.

TABLE 1.10

Functions Carried Out by Soil Organisms

Type of Organism by Activity	Organism Groups	Major Functions
Photosynthesizers	Higher plant roots Algae Bacteria	Use solar energy to fix atmospheric CO_2 and add biomass to the soil system.
Decomposers	Actinomycetes Bacteria Fungi	Break down organic residues, releasing nutrients that are available to other soil organisms and producing compounds that bind soil particles into aggregates. Fungal hyphae also wrap around soil particles, creating aggregates. Decomposers compete with and inhibit pathogens.
Mutualists	Actinomycetes Bacteria Fungi	Enhance plant growth by protecting roots from pathogens, fixing atmospheric nitrogen (some actinomycetes and bacteria), and forming symbiotic associations with roots (mycorrhizal fungi).
Pathogens	Bacteria Fungi	Promote disease by consuming roots and other plant parts.
Parasites	Nematodes Arthropods	Parasitize nematodes and insects, including pathogens.
Root feeders	Nematodes Arthropods	Can cause significant crop yield losses through the consumption of plant roots.
Bacterial feeders	Protozoa Nematodes	Release plant-available nutrients, control pests and pathogens, and regulate bacterial populations.
Fungal feeders	Nematodes Arthropods Protozoa	Release plant-available nutrients, control pests and pathogens, and regulate fungal populations.
Shredders	Earthworms Arthropods	Break down plant residues and enhance soil structure through the production of fecal pellets.
Burrowers/Mixers	Earthworms Arthropods Small vertebrates	Mix the soil horizons they burrow through, redistribute nutrients, incorporate organic matter into the soil, create macropores, and enhance soil structure.
Higher-level predators	Nematodes Arthropods Small vertebrates	Control the populations of other soil organisms.

Source: Modified from Tugel, A., Lewandowski, A., and Happe-von Arb, D. (eds), *Soil Biology Primer*, Soil and Water Conservation Society, Ankeny, IA, 2000.

Tens of thousands of species of fungi have been identified in soil. While found in a wide range of soil environments, fungi tend to be the dominant organisms in forest soils or when soils are acidic or sandy. Fungi are important in the decomposition of organic materials to form humus (Figure 1.15) and thus are also important in the formation of soil aggregates. Mycorrhizae are long, thread-like fungi that form a symbiotic relationship with the roots of higher plants. The mycorrhizae derive needed materials from the plant, but in turn act as a sort of root themselves to help the plant more efficiently extract nutrients from the soil. Studies show that most higher plants grow better with mycorrhizal infections of their root system. Mycorrhizae can even form links between two higher plants, transferring nutrients between them.

There is a great abundance and diversity of microorganisms in the soil, including actinomycetes, algae, bacteria, nematodes, and protozoa. Actinomycetes are bacteria-like organisms that live on decaying organic matter, breaking it down and adding plant-available nutrients to the soil. They give soil its characteristic "earthy" aroma. Soil algae are only found near the soil surface, as they require

FIGURE 1.14   A termite mound in Africa with a person standing beside it for scale. All the material in the mound has been dug out of the termite tunnels and moved to the soil surface, a process that results in significant mixing of soil materials and cycling of nutrients. (Courtesy of Jeffrey Homburg.)

access to sunlight for photosynthesis. Soil algae secrete polysaccharides that have a favorable effect on soil aggregation and are also important in desert environments where they form microbiotic crusts on the soil surface with cyanobacteria, fungi, and other microorganisms. Bacteria are the most diverse group of soil organisms; they occur in all soils, under all conditions, and in all climates. Bacteria participate in virtually all soil organic transactions, including those that promote soil aggregation, making their presence very important to a healthy soil ecosystem. Nematodes are unsegmented roundworms that live in soil water. Most nematodes are either predators of other microorganisms or plant parasites. The plant parasite nematodes can cause serious problems in crop production, but managing agricultural soils to promote conditions that favor predator nematodes can help keep plant parasites in check. Protozoa are mobile, single-celled creatures that include amoebas, ciliates, and flagellates. They release plant-available nitrogen into the soil as they feed and help regulate bacteria populations in the soil through predation.

FIGURE 1.15   The fungi growing on the end of this log are slowly breaking the log down. Eventually, some of the organic materials from the log will be incorporated into the soil the log is laying on. (Photo by Eric Brevik.)

Some actinomycetes, bacteria, and soil algae can take atmospheric nitrogen and fix it in the soil in plant-available forms. Some soil bacteria also fix atmospheric nitrogen in a symbiotic relationship with the roots of legumes. Soil fungi and actinomycetes produce antibiotics as a means of gaining a competitive advantage over other organisms, and several of those antibiotics have been isolated and are now used in medical applications.

The roots of higher plants are also part of the soil ecosystem. The activity of plant roots has several significant effects on soil physical and chemical properties. Living roots secrete exudates that include amino acids, organic acids, carbohydrates, sugars, vitamins, mucilage, and proteins into a zone around the roots about 5 mm across that is called the rhizosphere. These root exudates support soil organisms, making the rhizosphere the most biologically active part of the soil, and can aid in the aggregation of soils. Roots lose parts of the root system that die, decompose, and become a part of the surrounding soil. Those root parts and the exudates release plant nutrients into the soil as they decompose, and the pathways occupied by roots prior to decomposition become macropores.

Greater diversity in soil organisms corresponds to a healthier and more vibrant soil ecosystem, which is also beneficial to plant growth, including crop production. With increased diversity the decomposition of organic matter, and thus the release of nutrients into the soil system, takes place more efficiently because no single organism is responsible for all the various steps and processes involved in decomposition. Greater diversity also insures production of sufficient quantities of beneficial organic substances that are introduced to the soil system, and diversity provides competition for plant pests and pathogens, helping to reduce the problems they cause. In many cases, degraded unhealthy soil ecosystems favor plant pests and pathogens over their competitors.

1.6   NUTRIENT CYCLING

One of the most important processes that occur in soil is nutrient cycling. The fundamental concept behind nutrient cycling is that nutrients are taken up and utilized by organisms during their life cycle. When the organism or part of the organism dies, its nutrients are returned to the natural environment. Within any given nutrient cycle there are a number of pools that the nutrient can move between. An example of the carbon cycle is given in Figure 1.16. Other nutrient cycles that are important in the soil include the nitrogen, phosphorus, potassium, calcium, magnesium, sulfur, iron, copper, manganese, zinc, boron, chorine, molybdenum, and nickel cycles. These 14 nutrients are all essential plant nutrients that are supplied by the soil.

Much of the nutrient cycling that takes place in the soil is based on the decomposition of organic matter. The 14 nutrients mentioned in the previous paragraph are all present in organism tissues, as are carbon, oxygen, hydrogen, and a number of other elements. While a cycle has no beginning or end, for the sake of example we can start with a plant growing in the soil. When that plant dies, organisms such as small vertebrates, arthropods, insects, and earthworms consume the plant tissues and incorporate some of the nutrients found in the plant's tissues into their own body mass. Tissues not utilized by this first round of organismal consumption are utilized by soil bacteria, fungi, and actinomycetes, transforming the decomposed plant tissues into soil humus and releasing plant-available nutrients into the soil. A growing plant can then take up those nutrients for use in its own tissues, resetting the entire cycle. The same decomposition process and nutrient cycling also takes place when vertebrates, earthworms, bacteria, etc. die and their bodily tissues are processed by other soil organisms. In this fashion, the same atom of nitrogen, calcium, iron, etc. may be used repeatedly by various organisms growing in the same ecosystem at different times. A good example of nutrient cycling is found in tropical rainforests. Most of the nutrients in tropical rainforest ecosystems are actually stored in the living biomass as opposed to the soils. The rainforest ecosystems can support vigorous plant growth despite soils that are typically nutrient-poor, because the overall ecosystem is highly efficient at cycling nutrients. Dead organismal tissues introduced into the soil system are rapidly decomposed and their nutrients distributed to various soil organisms, including the plants for which they are so well known.

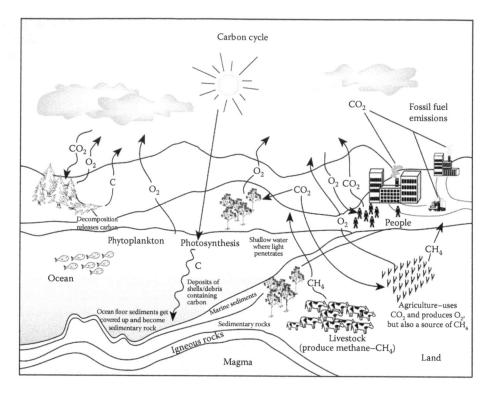

FIGURE 1.16 A diagram showing the parts of the global carbon cycle. Carbon is one of the fundamental building blocks of life and of soil organic matter. It is important to note that carbon is supplied to plants from the atmosphere, not from soil, and therefore is not a soil-derived plant nutrient, but soil is an important component of the carbon cycle. (Courtesy of NASA.)

New nutrients can be added to the soil system at any given location through processes such as the weathering (breakdown) of soil minerals, fixation of atmospheric nitrogen, deposition of new materials on the soil surface, or fertilization in the case of human-managed soils. Nutrients can also be lost from the soil system through processes such as erosion, leaching through the soil profile, or the transformation of nutrients such as nitrogen and carbon into gas form and outgassing. So while it is possible for a significant amount of nutrient cycling to take place within the soil system, it is also important to realize that soils are not a closed system in regard to nutrient cycling. Soils are simply one pool in a much larger global cycle (i.e., Figure 1.16).

1.7   ORGANIC MATTER

Organic matter (or humus) plays many significant roles in the soil system. It is important in promoting and maintaining physical properties such as aggregation, water-holding capacity, and low bulk density. Organic matter is an important contributor to cation exchange capacity (Table 1.9) and is the fundamental source of energy and nutrients for soil organisms. Most natural soils have adequate organic matter to develop and sustain the physical, chemical, and biological processes needed for a healthy soil ecosystem. As previously discussed and seen in Figure 1.16, soil organic matter is also an important part of the global carbon cycle. But the loss of organic matter in human-managed soils, such as agricultural fields, can become a major problem. These losses take place through processes such as accelerated rates of soil organic matter decomposition, decreased additions of organic matter in agricultural systems as compared to natural systems, and accelerated erosion in fields with exposed, unprotected soil.

FIGURE 1.17 The tissues of these maize plants (a) are potential organic matter additions to the soil, but practices such as silage (b) that remove most of the plant from the field reduce the plant tissues returned to the soil for nutrient cycling. Practices that leave significant residue behind, such as a thick cover of maize residues (c), are beneficial to soil properties. ([a, b] Courtesy of USDA NRCS; [c] photo by Gene Alexander, USDA NRCS.)

The primary sources of soil organic matter are plant tissues, animal tissues, and wastes (i.e., manure). Of these three, plant tissue additions far outweigh the animal tissue and waste additions in most natural systems. In many modern agricultural systems, the plant tissues returned to the soil have been significantly reduced. Harvest removes the grain, fruit, vegetable, or fiber being grown for economic gain from the field. Removing straw for animal bedding or silage for feed removes the majority of the biomass produced in a field over the course of a year, further reducing the potential organic returns to the soil. The same is true when fallen branches are removed from an orchard or vines from a vegetable field for aesthetic purposes. For these reasons, soil scientists and many other agronomists have long been advocates of leaving as many crop residues as possible on a field following harvest (Figure 1.17). For the same reason, many soil scientists are concerned about the push to convert crop residues into biofuels, fearing that such a move will rob the soils in those fields of critically needed organic matter.

1.8   SOIL VS. SEDIMENT

One common point of confusion for people not familiar with soils is the distinction between soils and geologic sediments found at or near Earth's surface. To demonstrate the difference between these materials, it is useful to look at and compare their respective definitions. Soil can be defined as,

"1. A dynamic natural body composed of mineral and organic solids, gases, liquids, and living organisms which can serve as a medium for plant growth" or "2. The collection of natural bodies occupying parts of the Earth's surface that is capable of supporting plant growth and that has properties resulting from the integrated effects of climate and living organisms acting upon parent material, as conditioned by topography, over periods of time." (Brady and Weil, 2008)

FIGURE 1.18 (a) Soil layers (horizons) are usually parallel to the surface. (b) Sedimentary layers (beds, pointed at by the white arrows) are often angled relative to the surface. (Photos by Eric Brevik.)

In contrast, sediment can be defined as:

"1. Solid material that has settled down from a state of suspension in a liquid" or "2. ... solid fragmented material transported and deposited by wind, water, or ice, chemically precipitated from solution, or secreted by organisms, that forms in layers in loose unconsolidated form, e.g. sand, mud, till." (Bates and Jackson, 1984)

Note that there is no mention of organic materials, the support of plant growth, or the alteration of properties over time in the sediment definition. Most soils form in sediment and are therefore altered sediments. One of the first alterations often seen as a sediment transforms into soil is the addition of organic materials to the sediment. Both soils and sediments also tend to be layered. However, layers in the soil (called horizons) are formed by the alterations that convert sediment into soil and tend to be roughly parallel to the land surface. Layers in sediment (called beds) are created by the processes that deposited, or left behind, the sediment, and may be parallel to the land surface but frequently are not (Figure 1.18). Being able to differentiate between soils and sediments is an important skill for soil scientists, geologists, geographers, and others who work with near-surface materials.

1.9 SOIL CLASSIFICATION

Unfortunately, the soil science world is not as unified as the biological world when it comes to classification. Several countries have their own systems of soil classification, including Australia, Brazil, Canada, France, New Zealand, the United States, and others. In addition, the United Nations Food and Agriculture Organization (FAO) developed a system that was intended to "serve as a common denominator" for the various soil classification systems of the world (FAO, 1974). The FAO system has since been converted to the World Reference Base for Soil Resources (WRB). All these systems can lead to some confusion when it comes to sharing information between countries, but they are all intended to communicate meaningful information about the soil being classified. Exactly what is meaningful depends to some extent on the intended use of the classification system and the opinions of those who formed it. In some cases, that information is the degree of evolution of the soil profile, sometimes it is the properties of the soil, and at other times the climate and vegetation differences are emphasized (Buol et al., 1997).

The U.S. system developed by the U.S. Department of Agriculture is one of the most widely used classification systems (Krasilnikov and Arnold, 2009) and will be utilized here in a very basic way as an example of how these systems tend to work. Most systems have multiple levels; the USDA Soil Taxonomy has six levels of classification. Starting with the most general and proceeding to the most specific levels, they are order, suborder, great group, subgroup, family, and series. Increasingly detailed information is provided about the soil proceeding from the order level to the series level.

The term Kandiudox is an example of a USDA Soil Taxonomy name at the great group level. The "ox" at the end signifies that this soil is in the Oxisol order, the "ud" indicates it formed in a humid climate, and the "kandi" indicates there is an accumulation of low-activity clays (meaning clays with a low cation exchange capacity, or CEC) in the subsurface. By using soil taxonomic names, a large amount of information about any given soil can be communicated in a relatively short word or sequence of words.

The USDA Soil Taxonomy is designed to be flexible and there is a process to add to the system as needed; updates are made to the USDA Soil Taxonomy every 2 years. As one might expect, there are fewer orders than any other level, with increasing numbers as one proceeds through the system. At present there are 12 orders, and over 20,000 series have been identified in the United States alone. A basic description of the 12 orders in the USDA Soil Taxonomy and some of the information they convey about the soils are given in Table 1.11.

Both the USDA Soil Taxonomy and WRB soil classification systems are used by authors in the chapters found in this book. Table 1.12 gives a basic conversion between the two systems, although it is important to note that the conversions given in Table 1.12 are not perfect 1:1 correlations.

TABLE 1.11
The 12 Soil Orders Currently Recognized in the USDA Soil Taxonomy System

Order	Basic Description	Controlling Factors[a]
Alfisols	Soils of humid and subhumid climates with a subsurface horizon of clay accumulation, not strongly leached, common in forested areas	Climate, organisms
Andisols	Soils that formed in volcanic ash and contain aluminum-organic compounds	Parent material
Aridisols	Soils formed in dry climates, low in organic matter, and often having subsurface horizons with salt accumulations	Climate
Entisols	Soils lacking subsurface horizons because the parent material accumulated recently or because of constant erosion, common on floodplains and in mountains and badland areas	Time, topography
Gelisols	Weakly weathered soils formed in areas that contain permafrost within the soil profile	Climate
Histosols	Soils with a thick upper layer very rich in organic matter (>25%) and containing relatively little mineral material	Topography
Inceptisols	Soils with weakly developed subsurface horizons, little or no subsoil clay accumulation because the soil is young or the climate does not promote rapid weathering	Time, climate
Mollisols	Mineral soils of semiarid and subhumid midlatitude grasslands that have a dark organic-rich A-horizon and are not strongly leached	Climate, organisms
Oxisols	Very old, extremely leached and weathered soils with a subsurface accumulation of Fe and Al oxides, commonly found in humid tropical environments	Climate, time
Spodosols	Soils formed in cold, moist climates that have a well-developed B-horizon with accumulation Al and Fe oxides, form under pine vegetation in sandy parent material	Parent material, organisms, climate
Ultisols	Soils with a subsurface horizon of clay accumulation, strongly leached (but not as strongly as Oxisols), commonly found in humid tropical and subtropical climates	Climate, time, organisms
Vertisols	Soils that develop deep, wide cracks when dry (shrink and swell) due to high clay content (>35%) and are not strongly leached	Parent material

Source: Brevik, E.C., *Journal of Geoscience Education*, 50, 539–543, 2002.

[a] All five soil-forming factors combine to create any given soil, but this column lists the factor(s) that are often most dominant in forming soils within the given order.

TABLE 1.12

Conversion between the Current WRB and USDA Soil Taxonomy Classification Systems

WRB	USDA	Basic Soil Properties
Acrisols	Udults, Ustults, Xerults	Ultisols formed in humid climates, semiarid climates, and climates with moist winters and dry summers, respectively.
Albeluvisols	Glossudalfs	Alfisols that formed in a humid climate and have a light colored subsurface horizon.
Alisols	Udults, Ustults, Xerults	Ultisols formed in humid climates, semiarid climates, and climates with moist winters and dry summers, respectively.
Andosols	Andisols	See Table 11.
Anthrosols	Arents	Entisols that are heavily influenced/mixed by human activity such as plowing, mining, etc.
Arenosols	Psamments	A suborder of the Entisol order with a sandy texture.
Calcisols	Calcids, Ustepts, Xerepts	Aridisols with high calcium carbonate content in a subsurface horizon, Inceptisols that formed in a semiarid environment, and Inceptisols that formed in an area with moist winters and dry summers, respectively.
Cambisols	Cambids, Udepts, Ustepts, Xerepts	The first is an Aridisol with a low degree of soil development. The last three are Inceptisols formed in humid, semiarid, and moist winter/dry summer climates, respectively.
Chernozems	Cryolls, Udolls, Ustolls, Xerolls	Mollisols formed in cold climates, humid climates, semiarid climates, and areas with moist winters and dry summers, respectively.
Cryosols	Gelisols	See Table 1.11.
Durisols	Durids	Aridisols with a subsurface horizon that is cemented by silica, this subsurface horizon can be very hard.
Ferralsols	Oxisols	See Table 1.11.
Fluvisols	Fluvents	Entisols along streams that frequently flood.
Gleysols	Aquepts and Aquents	Suborders of the Inceptisol and Entisol orders that formed in very wet, often saturated conditions.
Gypsisols	Gypsids	Aridisols with a high gypsum content in a subsurface horizon.
Histosols	Histosols	See Table 1.11
Kastanozems	Udolls, Ustolls, Xerolls	Mollisols that formed in humid, semiarid, and moist winter/dry summer climates, respectively.
Leptosols	Rendolls	Mollisols formed in high calcium carbonate parent materials, usually in a humid, temperate climate.
Lixisols	Kanhaplustalfs, Kanhapludalfs	Alfisols that formed in semiarid and humid climates, respectively, that have subsurface accumulations of low-activity (low CEC) clays.
Luvisols	Udalfs, Argids	Alfisols that formed in a humid climate, Aridisols with a subsurface accumulation of high activity (high CEC) clays.
Nitisols	Kandiudox, Ultisols	Oxisols formed in a humid climate with a subsurface accumulation of low-activity (low CEC) clays, Ultisols.
Phaeozems	Cryolls, Udolls, Ustolls	Mollisols formed in cold climates, humid climates, and semiarid climates, respectively.
Planosols	Kandiaqualfs	Alfisols formed in very wet conditions with a subsurface accumulation of low-activity (low CEC) clays.
Plinthosols	Humults, Udults	Ultisols with high organic carbon content in the A and B horizons and that formed in a humid climate, respectively.
Podzols	Spodosols	See Table 1.11.
Regosols	Orthents	Entisols with very shallow soil development.
Solonchaks	Salorthids	Aridisols that are shallow and have a high salt content.
Solonetz	Natudalfs, Natrargids, Natrixeralfs	Alfisols formed in a humid climate with accumulations of sodium salts, Aridisols with an accumulation of high-activity (high CEC) clays and sodium salts, and Alfisols formed in a climate with moist winters, dry summers, and having an accumulation of sodium salts, respectively.

TABLE 1.12 (Continued)
Conversion between the Current WRB and USDA Soil Taxonomy Classification Systems

WRB	USDA	Basic Soil Properties
Stagnosols	Kandiaqualfs	Alfisols formed in a wet environment with a subsurface accumulation of low-activity (low CEC) clays.
Technosols	Arents	Entisols that are heavily influenced/mixed by human activity such as plowing, mining, etc.
Umbrisols	Inceptisols	See Table 1.11.
Vertisols	Vertisols	See Table 1.11.

Source: Conversions are from Krasilnikov and Arnold (2009) and Deckers et al. (2003).

Note: It is important to note that these conversions are only rough approximations, and do not represent exact equivalents. This table is only provided so that general soil properties can be understood when WRB taxonomic names are used in certain chapters in this book.

1.10 SOIL HEALTH/QUALITY

Soil health and soil quality are essentially the same idea, with the term "soil health" more frequently used by farmers and "soil quality" more frequently used by academic researchers (Magdoff and van Es, 2009). A healthy soil is said to be one that has the capacity to function, within ecosystem and land use boundaries, to sustain biological productivity, maintain environmental quality, and promote plant and animal health (Doran et al., 1994). There are several physical, chemical, and biological properties a soil needs to have in order to be considered healthy. While the exact specifications differ from soil to soil depending on the environment the soil is in and the use it is being put to, a healthy soil does need adequate organic matter, good structure, and a diverse mixture of microorganisms and macroorganisms (Brevik, 2009). Therefore, the physical, chemical, and biological properties previously discussed are all important to the ability of a soil to carry out its various important functions. This is an oversimplification of a fairly complex idea, but the overriding theme is the establishment of the soil as a medium that can do a good job of whatever function it is being put to. In natural systems soils often reach this point by achieving equilibrium with the surrounding environment; in a human-managed system it is up to the caretaker to provide the inputs needed to achieve a state of soil health (Wolf and Snyder, 2003). Healthy soils are also very important to human health. Degraded soils reduce crop yields and produce crops with poor nutritional value, leading to malnutrition in the people who depend on those soils to produce their food. Healthy soils also lead to reduced erosion and better air and water quality.

1.11 ADDITIONAL READINGS

This chapter is a brief introduction to important soil properties and processes you will encounter as you make your way through this book. If you wish to expand on your soil science background, the following books are likely to be helpful.

Brady, N.C. and R.R. Weil. 2008. *Nature and Properties of Soils*, 14th edn. Prentice Hall, Upper Saddle River, NJ.

Eash, N., C.J. Green, A. Ravzi, and W.F. Bennett. 2008. *Soil Science Simplified*, 5th edn. Wiley-Blackwell, Hoboken, NJ.

Kohnke, H. and D.P. Franzmeier. 1994. *Soil Science Simplified*, 4th edn. Waveland Press, Long Grove, IL.

Plaster, E.J. 2009. *Soil Science and Management*, 5th edn. Delmar Cengage Learning, Clifton Park, NY.

Singer, M.J. and D.N. Munns. 2006. *Soils: An Introduction*, 6th edn. Prentice Hall, Upper Saddle River, NJ.

REFERENCES

Ball, J. 2001. Soil and water relationships. *Ag News and Views*. http://www.noble.org/Ag/Soils/SoilWaterRelationships/Index.htm. Accessed 16 September 2011 (verified 4 January 2012).

Bates, R.L. and J.A. Jackson. 1984. *Dictionary of Geological Terms*, 3rd edn. Prepared by the American Geological Institute. Anchor Books, New York.

Brady, N.C. and R.R. Weil. 2008. *Nature and Properties of Soils*, 14th edn. Prentice Hall, Upper Saddle River, NJ.

Brevik, E.C. 2000. The value of soils courses to the geology student. *Journal of Geoscience Education* 48: 19–23.

Brevik, E.C. 2002. Problems and suggestions related to soil classification as presented in introduction to physical geology textbooks. *Journal of Geoscience Education* 50(5): 539–543.

Brevik, E.C. 2009. Soil health and productivity. *In* W. Verheye (ed.), *Soils, Plant Growth and Crop Production*. Encyclopedia of Life Support Systems (EOLSS), developed under the Auspices of the UNESCO, EOLSS Publishers, Oxford. http://www.eolss.net (verified 4 January 2012).

Brevik, E., T. Fenton, and L. Moran. 2002. Effect of soil compaction on organic carbon amounts and distribution, south-central Iowa. *Environmental Pollution* 116: S137–S141.

Buol, S.W., F.D. Hole, R.J. McCracken, and R.J. Southard. 1997. *Soil Genesis and Classification*, 4th edn. Iowa State University Press, Ames, IA.

Coder, K.D. 1998. *Soil Constraints on Root Growth*. University of Georgia Cooperative Extension Service Forest Resources Publication FOR98-10, Athens, GA.

Deckers, J., F. Nachtergaele, and O. Spaargaren. 2003. Tropical soils in the classification systems of USDA, FAO and WRB. www.fao.org/ag/agl/agll/wrb/doc/KAOWDeckerscorr280203.doc (verified 11 February 2012).

Dixon-Coppage, T.L., G.L. Davis, T. Couch, E.C. Brevik, C.I. Barineau, and P.C. Vincent. 2005. A forty-year record of carbon sequestration in an abandoned borrow-pit, Lowndes County, GA. *Soil and Crop Science Society of Florida Proceedings* 64: 8–15.

Doran, J.W., D.C. Coleman, D.F. Bezdicek, and B.A. Stewart (eds). 1994. *Defining Soil Quality for a Sustainable Environment*. Special Publication #34, American Society of Agronomy, Madison, WI.

FAO. 1974. *Soil Map of the World*. Vols 1–10. Food and Agriculture Organization of the United Nations and UNESCO, Paris.

Jenny, H. 1980. *The Soil Resource: Origin and Behavior*. Ecological Studies, Vol. 37. Springer-Verlag, New York.

Krasilnikov, P. and R. Arnold. 2009. The United States Soil Taxonomy. *In* P. Krasilnikov, J.-J.I. Martí, R. Arnold, and S. Shoba (eds), *Soil Terminology, Correlation, and Classification*, pp. 75–95. Earthscan, London.

Magdoff, F. and H. van Es. 2009. *Building Soils for Better Crops: Sustainable Soil Management*, 3rd edn. Sustainable Agriculture Network Handbook Series #10. Sustainable Agriculture Publications, Waldorf, MD.

Pittenger, D.R. 2002. *California Master Gardener Handbook*. University of California Agriculture and Natural Resources Publication 3382, Oakland, CA.

Schaetzl, R. and S. Anderson. 2005. *Soils: Genesis and Geomorphology*. Cambridge University Press, New York.

Schoeneberger, P.J., D.A. Wysocki, E.C. Benham, and W.D. Broderson (eds). 2002. *Field Book for Describing and Sampling Soils*, Version 2.0. Natural Resources Conservation Service, National Soil Survey Center, Lincoln, NE.

Troeh, F.R., J.A. Hobbs, and R.L. Donahue. 2004. *Soil and Water Conservation for Productivity and Environmental Protection*, 4th edn. Prentice Hall, Upper Saddle River, NJ.

Tugel, A., A. Lewandowski, and D. Happe-von Arb (eds). 2000. *Soil Biology Primer*, revised edition. Soil and Water Conservation Society, Ankeny, IA.

Wolf, B. and G.H. Snyder. 2003. *Sustainable Soils: The Place of Organic Matter in Sustaining Soils and Their Productivity*. Haworth Press, New York.

2 Soils and Human Health
An Overview

Eric C. Brevik

CONTENTS

2.1 INTRODUCTION

Many things are likely to come to mind when people think about their health, such as an active exercise program, wise food choices, good medical care, and proper sanitation. Few people recognize the connection between soils and human health, even though they are actually very important to health. Soils influence health through the nutrients taken up by plants and the animals that eat those plants, nutrients that are needed for adequate nutrition for growth and development. Soils can also harm human health in three major ways: (i) toxic levels of substances or disease-causing organisms may enter the human food chain from the soil, (ii) humans can encounter pathogenic organisms through direct contact with the soil or inhaling dust from the soil, and (iii) degraded soils produce nutrient-deficient foods, leading to malnutrition. Therefore, soils form an integral link in the holistic view of human health. In this book, soils and their influence on human health are viewed from a broad perspective, including both direct influences of soils on health and indirect influences through things such as occupational exposure to soil amendments.

2.2   CONCEPT OF HUMAN HEALTH

Health was defined as "a state of complete physical, mental and social well-being and not merely the absence of disease or infirmity" by the World Health Assembly in 1946 (Grad, 2002). Note that this definition includes three primary aspects of health: (i) physical, (ii) mental, and (iii) social.

Physical fitness is achieved through proper nutrition in the daily diet and regular exercise. Mental fitness is achieved through emotional and psychological well-being and is also partially dependent on proper nutrition, and social fitness is achieved through the ability to operate comfortably within the expectations of the society the individual lives in. These components of health can also be seen within the Food and Agriculture Organization (FAO)'s definition of food security (FAO, 2003), demonstrating the interrelationship between food security and human health. Given that soils are integral to food security, soils are also integral to human health (Ljung et al., 2009; Pimentel, 2006; Abrahams, 2002; Cakmak, 2001).

2.3   PROMOTION OF HUMAN HEALTH THROUGH SOILS

There are 14 elements that are essential for plant growth that come from the soil (Havlin et al., 2005) and many of these elements, such as calcium (Ca), iron (Fe), potassium (K), phosphorus (P), and others, are also essential for human health (Leitzmann, 2009; Combs, 2005; Klasing et al., 2005). Essential soil elements that end up in the human diet are primarily supplied through food from either plants (that took the elements up from the soil during growth) or animal products (after the animal obtained those essential elements from plants through soils) (Klasing et al., 2005; Abrahams, 2002; Voisin, 1959).

Because plants depend on the soil for their nutritional needs and all higher animals, including humans, depend directly or indirectly on plants for their nutrition, plants form the base of the food chain and, consequently, a major portion of the nutrients needed for human health originate with the soil. This section will take a closer look at the elemental content of the soil and at some of the ways soil nutrients are taken up and influence human health.

2.3.1   SOIL ELEMENTS NECESSARY FOR HUMAN HEALTH

The 14 elements in the soil that are essential for plant growth are nitrogen (N), calcium (Ca), phosphorus (P), magnesium (Mg), potassium (K), sulfur (S), iron (Fe), copper (Cu), manganese (Mn), zinc (Zn), boron (B), chlorine (Cl), molybdenum (Mo), and nickel (Ni) (Havlin et al., 2005). There are additional elements that are needed by some but not all plants such as cobalt (Co), bromine (Br), vanadium (V), silicon (Si), and sodium (Na) (Havlin et al., 2005). In addition to these soil elements, hydrogen (H), oxygen (O), and carbon (C) are also essential for plant growth but are obtained from air and water (Havlin et al., 2005). Most of these elements are also essential for human health.

Eleven elements comprise 99.9% of the atoms found in the human body, subdivided into major and minor elements. Four major elements, H, O, C, and N, make up about 99% of the atoms in the body; seven minor elements, Na, K, Ca, Mg, P, S, and Cl, make up an additional 0.9% of the atoms in the body (Combs, 2005). In addition to these major and minor elements, there are approximately 18 additional elements considered essential in small amounts to maintain human life, although the exact number and identity of these elements are not universally agreed on by human health experts (Leitzmann, 2009; Combs, 2005; Klasing et al., 2005; Deckers and Steinnes, 2004; Abrahams, 2002). These 18 additional elements, known as trace elements, include lithium (Li), vanadium (V), chromium (Cr), manganese (Mn), iron (Fe), cobalt (Co), nickel (Ni), copper (Cu), zinc (Zn), tungsten (W), molybdenum (Mo), silicon (Si), selenium (Se), fluorine (F), iodine (I), arsenic (As), bromine (Br), and tin (Sn) (Combs, 2005).

Note that of the approximately 29 elements considered essential for human life, 13 are essential plant nutrients obtained from the soil and another 5 are elements obtained from the soil that are needed by some, but not all, plants. Although the elements Cr, W, Se, F, I, As, and Sn are not considered essential for plant health, these elements are also found in trace amounts in plants that grow in soils containing them. Therefore, soils that provide a healthy, nutrient-rich growth medium for plants will result in plant tissues that contain many of the elements required for human life. In fact, most of the elements necessary for human life are obtained from either plant or animal tissues (Leitzmann, 2009; Shetty, 2009; Buol, 2008; Combs, 2005). Plant tissues are among the most important sources of Ca, P, Mg, K, Cu, Zn, Se, Mn, and Mo in the human diet (Table 2.1) and these elements are obtained by plants from the soil.

2.3.2 HEALTH AND NUTRIENT IMBALANCES IN SOIL

There are several adverse health effects that can arise from nutrient deficiencies. Iron deficiency is probably the most common example (WHO, 2007a; Sanchez and Swaminathan, 2005; Deckers and Steinnes, 2004), with about 2 billion people estimated to be anemic due to Fe deficiency (WHO, 2007a). Iron-deficient soils can lead to low Fe content in plant products and deficiency in the humans who eat them, but low Fe in soils is rarely a problem except in arid regions (Combs, 2005; Deckers and Steinnes, 2004). Blood loss to parasites such as hookworms, a disease-causing organism associated with the soil, is another major cause of Fe deficiency (Deckers and Steinnes, 2004). The best source of Fe is meat, especially red meat. High reliance on cereal-based diets low in meat, including cereal-based weaning foods for infants, can contribute to Fe deficiency.

Another soil-related form of malnutrition is iodine deficiency, which leads to goiter (abnormal enlargement of the thyroid gland), severe cognitive and neuromotor deficiencies, and other neuropsychological disorders (Fuge, 2005a; Deckers and Steinnes, 2004). Iodine deficiency is the single most important preventable cause of brain damage and the World Health Organization has made the elimination of iodine deficiency disorders a priority (WHO, 2007b). Regions known to have soils deficient in iodine are mainly located in the high-altitude interior of continents (Combs, 2005; Fuge, 2005a; Deckers and Steinnes, 2004), although iodine deficiency has been eliminated in many developed countries by introducing iodine supplements to foods such as salt and bread (WHO, 2007b, Fuge, 2005a). Most iodine deficiency problems today are found in developing

TABLE 2.1
Some Important Plant-Tissue Sources of Elements Essential to Human Life

Element	Important Sources
Ca	Kale, collards, mustard greens, broccoli
Cu	Beans, peas, lentils, whole grains, nuts, peanuts, mushrooms, chocolate
I	Vegetables, cereals, fruit
K	Fruits, cereals, vegetables, beans, peas, lentils
Mg	Seeds, nuts, beans, peas, lentils, whole grains, dark green vegetables
Mn	Whole grains, beans, peas, lentils, nuts, tea
Mo	Beans, peas, lentils, dark green leafy vegetables
P	Nuts, beans, peas, lentils, grains
Se	Grain products, nuts, garlic, broccoli (if grown in high-Se soils)
Zn	Nuts, whole grains, beans, peas, lentils

Source: Based on *Essentials of Medical Geology,* Combs Jr, G.F., Geological impacts on nutrition, pp. 161–177, Copyright 2005, from Elsevier.

countries, particularly in South America, Africa, and Southeast Asia (Fuge, 2005a; Deckers and Steinnes, 2004).

A low level of selenium (Se) in soils used to grow grains and other food crops is associated with a Se-deficient diet. Inadequate intake of Se can cause Keshan disease, a heart disease that likely has a viral component as well, and Kashin-Beck disease, a disease which results in chronic disabling degenerative osteoarthrosis that may form through a combination of Se and iodine deficiency (Fordyce, 2005; Hartikainen, 2005; Deckers and Steinnes, 2004). Insufficient Se levels may also be associated with cancer (Combs, 2005; Hartikainen, 2005), stunting of growth, and immune system and reproductive problems (Fordyce, 2005). Places that experience Se deficiency problems in the soil include the mountainous belt of northeastern China to the Tibetan Plateau, parts of Africa, and the Pacific Northwest, Great Lakes region, and East Coast in the United States (Combs, 2005). Of all nutrients, Se has one of the narrowest ranges between deficiency and toxicity, so attempts to correct Se deficiency must be carefully regulated (Fordyce, 2005).

Zinc deficiency causes stunted growth, anorexia, skin lesions, diarrhea, and impaired immune and cognitive functions (Combs, 2005; Abrahams, 2002). Some foods such as whole grains can be rich in Zn, but low in bioavailable Zn, which can then lead to Zn shortage in the human organism (Abrahams, 2002). Zinc-deficient soils are widespread and include about half the world's soils, but calcareous soils and leached, acidic soils are most likely to be Zn deficient (Combs, 2005; Abrahams, 2002).

It is also important to note that several elements normally required in small amounts for human health can become toxic when present in high amounts, such as As, Cd, Fe, Zn, Cu, Cr, Se, and others (Abrahams, 2002). Therefore, it is important not to overcorrect soil nutrient deficiencies. High levels of cadmium (Cd) in the soil, for example, can lead to high levels of Cd in plant tissues and Cd toxicity problems for people who consume plant materials raised in those soils. Itai-itai disease in Japan is a classic example of the problems created by high Cd levels (Nordberg and Cherian, 2005; Abrahams, 2002; Oliver, 1997). Mining in the Toyama Prefecture of Japan released Cd into the Jinzu River. The contaminated river water was used for rice irrigation, leading to high Cd levels in the rice and accumulation of Cd in people who ate the contaminated rice and subsequently developed itai-itai disease, characterized by weak, brittle bones, pain in the legs and spine, coughing, anemia, and kidney failure. It is important to note that the level of Cd in the Japanese rice was influenced by soil aeration status (Oliver, 1997), the important point being the relationship between soil Cd levels and Cd in the rice is more complex than just the soil Cd levels. As another example, excessive Se intake can result in brittle hair and nails, hair and nail loss, skin rash, weakness, excessive tooth decay, lack of mental alertness, and diarrhea (Combs, 2005; Fordyce, 2005).

Another important point that must be understood is that elements tend to have synergistic or antagonistic effects in the human body, so it is inadequate to discuss only single-element studies. Cadmium provides a good example of this. When Fe and Zn are deficient in the diet, Cd retention is increased by 15 times compared to individuals who have adequate Fe and Zn intake (Davies et al., 2005). Rice grain is deficient in Fe and Zn relative to human dietary needs (Davies et al., 2005), and the main component of the diet in the region of Japan affected by itai-itai disease was rice supplemented with small amounts of beans and vegetables (Oliver, 1997). By contrast, residents of the village of Shipham, in England, also live with high Cd soils but do not appear to suffer any adverse health effects (Davies et al., 2005). The people who suffered from itai-itai in Japan were likely made more susceptible to it than some other populations by their Fe- and Zn-deficient diet. A more in-depth discussion of this issue is provided by Morgan (Chapter 3, this volume).

2.3.3 Soil Chemical Factors That Influence Nutrient Availability

Soil nutrient deficiencies may occur for a number of reasons. The most obvious reason is that there is simply not enough of the nutrient in the soil to lead to healthy plant growth. However, there are also situations where the nutrients are present in plentiful amounts, but unavailable to plants. For example, Zn deficiency commonly occurs in calcareous soils (Deckers and Steinnes, 2004) and P is easily bound by Ca, Al, or Fe depending on the soil pH (Brady and Weil, 2008). In both cases, it is possible to have ample amounts of Zn or P in the soil for nutritional needs, but inadequate amounts of Zn or P to be taken up by plants due to chemical reactions occurring within the soil that bind these elements. Soil pH also influences nutrient and toxic element availability. Acidic soil pH levels tend to make Fe, Al, Mn, and heavy metals such as lead (Pb), cadmium (Cd), and Ni more available and nutrients such as iodine and Se less available (Oliver, 1997).

Another way that nutrient deficiencies may occur is through antagonism, a process by which ions with the same valence will reduce the uptake of another ion. Examples of antagonism include arsenic (As) antagonizing P and strontium (Sr) antagonizing Ca (Hurd-Karrer, 1939). Some apple orchards in the United States have problems with As antagonizing P uptake because As-containing pesticides were used in the past and As accumulated in the soil (Manuel, 2000; Walsh et al., 1977). There is also concern that Sr released during the Chernobyl nuclear disaster in 1986 could antagonize Ca uptake (Ramberg, 1986), especially in fields in northern Europe where Chernobyl fallout was highest (Balonov et al., 1999). Zinc antagonisms are possible with Ca, Fe, Cu, and Ni (Deckers and Steinnes, 2004; Oliver, 1997). Because there are several different ways that soil nutrient deficiencies may occur, it is important to determine (i) what kind of and (ii) why a deficiency is occurring when nutrient deficiency is an issue.

2.4   ANIMAL PRODUCTS AND SOIL NUTRIENT STATUS

Up to this point, most of the discussion has focused on the relationship between nutrients in plants and the nutrient status of the soil they grow in. However, the nutrient status of the soil also impacts the nutritional quality of meat, milk, and other animal products produced for human consumption (Jones, 2005; Klasing et al., 2005). This derives from the fact that the feed for animals, whether it is grass, cereals, or other plant materials, is grown in the soil. Just as with plants, the nutritional content of these animal products in turn influences the general health of the people who consume them. Some minerals, such as Cd, Pb, Se, and Hg, can accumulate in animal products at levels that are not detrimental to animal health but are detrimental to human health if those animal products are consumed (Klasing et al., 2005). Table 2.2 shows some of the most important animal nutrient sources in the human diet.

TABLE 2.2
Some Important Animal-Product Sources of Elements Essential to Human Life

Element	Important Sources
Ca	Dairy products
Cl	Dairy products, meats, eggs
Cu	Organ meats
Fe	Meats, especially red meat
K	Dairy products, meats
Mo	Organ meats
Na	Dairy products, meats, eggs
P	Meats, eggs, dairy products
Se	Meats from Se-fed livestock
Zn	Meats, organ meats

Source: Based on *Essentials of Medical Geology,* Combs Jr, G.F., Geological impacts on nutrition, pp. 161–177, Copyright 2005, from Elsevier.

Nutrient deficiencies are not usually a problem in domestic animals raised in the intensive animal husbandry systems commonly used in North America, Europe, and other parts of the developed world. The nutrient content in the feed supplied to these animals is usually closely controlled and deficiencies compensated for. However, deficiencies in feed are still common in many developing countries where farmers rely on locally produced feeds and do not have access to feed supplements (Jones, 2005).

Deficiencies can also be an issue in organic farming systems in developed countries due to the purposeful avoidance of feed supplements, but careful selection of a variety of feed plants for these organic operations can often overcome the lack of additional nutrients from feed supplements (Jones, 2005). Any nutrient that is deficient in an animal's diet will also be deficient in products obtained from that animal for human consumption.

2.5 TOXICITY ISSUES RELATED TO SOILS

In addition to providing elements at levels that are essential for human health, soils can also provide elements such as Pb, Cd, and As, as well as radioactive elements such as uranium (U), radium (Ra), and radon (Rn), at levels that are detrimental to human health. The soil is also a source of several organic compounds, introduced primarily by industrial and agricultural functions, that are toxic to humans when exposure occurs at high enough levels. In some cases these organic compounds were purposefully applied to crops or directly to the soil as pesticides. In other cases, the organic compounds were inadvertently introduced to the soil. These materials can enter the body from food sources or directly from the soil due to ingestion or inhalation. The main direct health effect of inhaled dust is irritation of the respiratory passages and diseases such as lung cancer (Derbyshire, 2005; Abrahams, 2002; Oliver, 1997). However, airborne dust can carry additional materials such as pathogens, harmful gases, organic chemicals, heavy metals, and radioactive materials with it that can cause other health problems (Bartoš et al., 2009; Derbyshire, 2005; Abrahams, 2002; Oliver, 1997). Airborne dust containing such toxins may carry these materials into the lungs, where they can enter the bloodstream. Cultivation for agricultural production and deflation (wind erosion) can introduce such dusts into the atmosphere (Abrahams, 2005).

2.5.1 HEALTH EFFECTS FROM EXPOSURE TO HEAVY METALS IN SOIL

Heavy metals are metallic elements that have densities greater than 4500 kg/m^3. Heavy metals cannot be degraded into nontoxic forms, but it is possible to create insoluble forms that are not biologically available (Baird and Cann, 2005). Heavy metals originate naturally from the weathering of rocks (Fuge, 2005b; Garrett, 2005), but have also been introduced to soils through human activity (Akabzaa and Yidana, 2011; Taylor et al., 2010; Fuge, 2005b; Mbila and Thompson, 2004; Senesi et al., 1999; Oliver, 1997). Heavy metals may occur as a by-product of mining ores and are therefore present in mine spoils and in the immediate surroundings of metal-processing plants (Akabzaa and Yidana, 2011; Meza-Figueroa et al., 2009; Fuge, 2005b). E-wastes, or those associated with electronic appliances such as computers and mobile phones, are also becoming an increasing source of heavy metals such as Pb, Sb, Hg, Cd, and Ni in the soil (Ha et al., 2009; Robinson, 2009). Urban soils are particularly susceptible to significant accumulations of heavy metals (Filippelli and Laidlaw, 2010; Handschumacher and Schwartz, 2010). However, not all anthropogenic sources are what we would traditionally view as industrial. Fertilizers have been a source of heavy metal additions to soils (Chen et al., 2009; Cupit et al., 2002; Senesi et al., 1999; Troeh and Thompson, 1993). Heavy metals have also been used in chicken feed (As) and swine feed (Cu, Zn) to promote growth and control disease (Bradford et al., 2008). These metals can end up in the soil if the manures produced are spread on fields. Heavy metal contents in agricultural soils have increased significantly in industrialized countries over the past century (Senesi et al., 1999).

Transport of heavy metals from one place to another most commonly occurs through the atmosphere as metal-containing gases or when the metals are suspended on particles such as dust (Baird and Cann, 2005; Troeh and Thompson, 1993). Many heavy metals in the atmosphere are linked to the burning of fossil fuels or industrial waste products (Senesi et al., 1999). Surface runoff and river sediments are another facet of heavy metal transport (Cech, 2010; Albering et al., 1999). The ultimate sink for heavy metals, however, is in soils and sediments (Baird and Cann, 2005).

One of the big health concerns related to soils in North America is airborne dust from Africa. Clouds of dust from the Sahara and Sahel deserts follow the trade winds across the Atlantic Ocean (Sing and Sing, 2010; Derbyshire, 2005). African dust has been documented in the Caribbean and all along the United States' east coast from Florida to Maine, and the amount of African dust making its way across the Atlantic is increasing as desertification problems in Africa become more pronounced (Derbyshire, 2005; Prospero and Lamb, 2003). Influxes of African dust have been linked to elevated levels of elements such as mercury (Hg), Se, and Pb (Garrison et al., 2003).

There are several heavy metals that cause concerns from a human health perspective. These include Hg, Pb, Cd, chromium (Cr), Fe, Sn, and As (Baird and Cann, 2005; Sparks, 2003; Oliver, 1997). Arsenic is actually a metalloid, but is commonly grouped with the heavy metals for the purpose of human health discussions. The metals listed here are not usually toxic in their free element form, but are often toxic in their ion forms and most of them are also highly toxic when bound to organic molecules (Baird and Cann, 2005). Human exposure to these metals can occur through routes such as inhalation of contaminated soil dust, consumption of crops grown in contaminated soils (Handschumacher and Schwartz, 2010), or purposeful or incidental consumption of contaminated soil (Abrahams, 2005).

Heavy metals affect human health by bonding with sulfhydryl groups on enzymes so that the enzyme cannot function normally (Baird and Cann, 2005). Damage to the central nervous system is common, leading to problems with systems such as coordination, eyesight, the sense of touch, and lowering of IQ; bone deterioration, gastrointestinal damage, hypertension, organ damage, and increased cancer rates can also occur with heavy metal exposure (Baird and Cann, 2005; Deckers and Steinnes, 2004; Sparks, 2003; Oliver, 1997). Worldwide, Pb is the most problematic heavy metal due to the extensive variety of lead sources leading to widespread exposure problems (Baird and Cann, 2005). A list of health problems associated with selected heavy metals and some of the common sources of those metals are given in Table 2.3.

Soil pH and drainage are important considerations when dealing with heavy metal contaminated soils. Maintaining the soil pH at about 7.0 will reduce the mobility of heavy metals, making them less available for plant uptake. The abundance of alkaline materials such as concrete and cinder blocks used during construction often increases the pH of urban soils, limiting metal mobility (Handschumacher and Schwartz, 2010). With the exception of Cr, draining wet soils will also decrease heavy metal mobility. Applications of phosphate fertilizers will reduce the availability of most metal cations due to the formation of P-metal complexes, but P-fertilization makes As more available in the soil (Brady and Weil, 2008). Heavy metal issues are discussed in more detail in Morgan (Chapter 3, this volume).

2.5.2 Health Effects from Exposure to Organic Pollutants in Soil

The main concern with organic chemicals comes from materials known as persistent organic pollutants (POPs). These are organic chemicals that resist decomposition in the environment or that bioaccumulate through the food web and therefore pose a risk of causing adverse effects to human health and the environment (Lee et al., 2003). Some common organic chemicals of concern include organochlorines, organophosphates, carbamates, chloroacetamides, glyphosate, and phenoxy herbicides. Examples of common organic chemicals are given in Table 2.4. Soils and human health issues arise regarding organic chemicals due to their widespread use

TABLE 2.3

Common Sources of and Health Problems Associated with Selected Heavy Metals

Heavy Metal	Anthropogenic Sources	Health Problems
Hg	Electrical switches, fluorescent light bulbs, mercury lamps, batteries, thermometers, dental fillings, burning of coal and fuel oil, medical wastes, pesticides, mining	Central nervous system damage, coordination difficulties, eyesight problems, problems with the sense of touch, liver, heart, and kidney damage
Pb	Batteries, solder, ammunition, pigments, ceramic glaze, hair coloring, fishing equipment, leaded gasoline, mining, plumbing, burning of coal	Neurological impacts, lowers IQ and attention spans, impaired hand-eye coordination, encephalopathy, deterioration of bones, hypertension
Cd	Zinc smelting, burning coal or Cd-containing garbage, rechargeable batteries, pigments, TVs, solar cells, steel, phosphorus fertilizer, metal plating, water pipes	Liver and kidney damage, carcinogenic, low bone density
As	Pesticides, mining and smelting of gold, lead, copper, and nickel, iron and steel production, burning of coal, wood preservatives	Gastrointestinal damage, skin damage, carcinogenic, heart, neurologic, and liver damage
Cr	Electroplating, corrosion protection, leather tanning, wood preservative, cooling-tower water additive	Carcinogenic, gastrointestinal disorders, hemorrhagic diathesis, convulsions

Source: Based on Baird, C. and Cann, M., *Environmental Chemistry*, W.H. Freeman, New York, NY, 2005; Sparks, D.L., *Environmental Soil Chemistry*, Academic Press, Amsterdam, The Netherlands, 2003; http://www.cdc.gov/.

as pesticides in agricultural situations and for lawns and households, and through their accumulation in landfills or other disposal sites due to inadequate disposal practices (Brady and Weil, 2008; Vega et al., 2007). E-wastes are also a new source of POPs such as polychlorinated biphenyls (PCBs), and burning of e-wastes can generate other POPs such as dioxins and furans (Robinson, 2009). Common routes of exposure to organic chemicals include dermal contact with soil and soil ingestion (Peterson et al., 2006).

The organochlorines are an extensive group of organic chemicals that were once widely used in a variety of applications, including as pesticides for insects, weeds, and fungi and as industrial chemicals. Organochlorines are also soluble in fatty tissues, and tend to bioconcentrate and biomagnify. Many organochlorines have been shown to cause cancer and birth defects, and have other toxic effects in higher animals, and for this reason many of them have been banned in developed countries (Brady and Weil, 2008; Baird and Cann, 2005; Fuge, 2005b; Grassman et al., 1998). A number of developing countries are allowed to use them until exhaustion of stocks, but this situation is rapidly changing. Large-scale agroindustrial projects that rely on World Bank, International Finance Corporation, or EU funds are no longer allowed to use them.

Organophosphates do not bioaccumulate, but they are more toxic to humans and other mammals than organochlorines (Brady and Weil, 2008). Organophosphates can cause health problems including leukemia and brain cancer in children (Baird and Cann, 2005; Harnly et al., 2005). For this reason, the World Health Organization (WHO) and the U.S. Environmental Protection Agency (EPA) are currently reevaluating the use of organophosphates. The World Bank and FAO are carrying out similar investigations worldwide. Parathion, an extremely toxic organophosphate, is considered responsible for more agricultural workers' deaths than any other pesticide and is currently banned internationally (Baird and Cann, 2005).

As with organophosphates, carbamates are less environmentally persistent than organochlorines but can be much more toxic to humans and other mammals. The carbamate carbaryl is widely used

TABLE 2.4

Some Commonly Used Organic Chemical Groups and Examples of Some of the Specific Chemicals Found within Those Groups

Chemical Group	Examples	Notes
Organochlorines[a]	*para*-Dichlorodiphenyltrichloroethane (DDT)	Used in mosquito control
	Aldrin	Used in termite control
	Dieldrin	Protects crops from insect pests
	Endrin	Used in rodent control
	Chlordane	Used in termite control
	Heptachlor	Controls soil insects
	Hexachlorobenzene	Used in fungus control
	Mirex	Used in ant, termite control
	Toxaphene	Used in tick, mite control
	Polychlorinated biphenyls (PCBs)	Used in industrial applications
	Dioxins	Group of related compounds, by-product of manufacturing
	Furans	Created by heating of PCBs in the presence of oxygen
Organophosphates	Dichlorvos	Used in fly and flea control
	Parathion	Highly toxic to humans, banned in many Western countries
	Diazinon	Insect control in homes, lawns
	Chlorpyrifos	Insect control in homes
	Malathion	Toxic to insects, fairly low toxicity to mammals
Carbamates	Carbaryl	Insect control on lawns, highly toxic to honey bees
	Aldicarb	Highly toxic to humans
Chloroacetamides	Alachlor	Weed control, carcinogen in animals
	Metolachlor	Weed control, suspected carcinogen in animals
	Acetochlor	Weed control
Glyphosate		Commercial name is Roundup
Phenoxy herbicides	2,4-D	Broadleaf weed control
	2,4,5-T	Used to clear brush along roads and powerlines

Source: Based on Baird, C. and Cann, M., *Environmental Chemistry*, W.H. Freeman, New York, NY, 2005.

[a] All of the organochlorines in this list are banned by the U.S. Environmental Protection Agency due to adverse environmental impacts but may still be used in countries other than the United States.

on lawns and gardens as an insecticide and has low mammal toxicity, but is very toxic to honey bees and other beneficial insects (Brady and Weil, 2008; Baird and Cann, 2005).

Herbicide use in the United States exceeds the use of all other pesticides combined (Brady and Weil, 2008). Chloroacetamides, glyphosate, and phenoxy herbicides are applied to control weeds and tend to be less toxic to humans than the organic chemicals used to control animal pests. The most heavily used herbicide in the United States is atrazine, which is used to kill weeds in maize. Exposure to high rates of atrazine, particularly during application, may be linked to cancer and birth defects (Baird and Cann, 2005). Alachlor and metolachlor, two chloroacetamides, are carcinogenic and a suspected carcinogen, respectively, in animals, and all the commonly used chloroacetamides are significant risks for groundwater contamination when they leach through soil. Phenoxy herbicides have been linked to non-Hodgkin's lymphoma and other cancers (Baird and Cann, 2005; Mannetje et al., 2005).

An estimated 2 million people are poisoned worldwide each year by pesticides with an estimated 10,000–40,000 deaths annually; approximately 75% of those deaths occur in developing countries due to lack of experience in using these products and to inferior medical care for exposed individuals (Horrigan et al., 2002). Within the United States, it is estimated that about 20,000 people receive medical treatment for pesticide exposure each year with approximately 30 deaths (Baird and Cann, 2005).

Pesticide pollution is a major problem in the world's water supplies. Researchers in Switzerland have found atrazine, alachlor, and other pesticides in rainwater, presumably evaporated from farm fields, and pesticides such as toxaphene have been found in countries where it is no longer used due to atmospheric transportation of the pesticide (Baird and Cann, 2005). In the United States, over 90% of stream water samples and about 45% of groundwater samples analyzed contain detectible levels of at least one pesticide (Gilliom et al., 2007).

Other organic materials found in soils that may represent human health hazards include environmental estrogens and antibiotic residues. Environmental estrogens are substances that interfere with the endocrine system. While the exact impacts of environmental estrogens are not yet clear, suspected influences include infertility and increased cancer rates in reproductive organs (Baird and Cann, 2005; Safe, 2000). Potential sources of environmental estrogens include veterinary pharmaceuticals, agricultural animal wastes (Bradford et al., 2008), some organochlorine chemicals, detergents, spermicides, paints, and plastics (Baird and Cann, 2005). In the case of antibiotic residues, the possible development of antibiotic-resistant bacteria is a concern, something that could lead to diseases that are more difficult to treat (Chee-Sanford et al., 2009; Bradford et al., 2008; Brady and Weil, 2008), but again more research is needed in this area. Potential sources of antibiotic residues in the soil include animal manures and sewage sludge. Organic pollutants in soil are discussed in more detail in Burgess (Chapter 4, this volume).

2.5.3 Toxic Materials in Fertilizers

Many people probably view fertilizers simply as a source of nutrients for plant growth. However, few fertilizers are "pure," providing only the nutrient of interest. Toxic materials, particularly heavy metals, are found in many fertilizers. Given multiple applications over time, these toxic materials can build up in the soil to the point of causing problems (Keller et al., 2001).

Many trace elements that fall within the heavy metals category have been documented in phosphate and nitrate fertilizers, including As, Cd, Cr, Hg, Ni, Pb, and Zn (Fuge, 2005b). Increases in the soil content of heavy metals such as As, Cd, Hg, and Pb have been documented in association with the application of P and micronutrient fertilizers (Bourennane et al., 2010; Cakmak et al., 2010; Chen et al., 2008). Composted municipal solid wastes (Zheljazkov and Warman, 2004), animal manures (Bourennane et al., 2010; Bradford et al., 2008), and sewage sludges (Bourennane et al., 2010; Mbila et al., 2001) used as soil amendments can also be sources of heavy metals. Simply documenting an increase in heavy metals does not necessarily mean there is a problem; soils naturally contain heavy metals (Garrett, 2005) and the metals only become a problem in the food chain if soil chemical characteristics favor metal mobilization and bioavailability. However, the fact that metals can accumulate from these nutrient sources means long-term monitoring and care is needed when they are used (Cakmak et al., 2010; Zheljazkov and Warman, 2004).

The filler content of fertilizers is not always regulated, which has led to problems with some companies that have used fertilizers as an inexpensive way to dispose of hazardous wastes. In the United States, the most famous example may be from the state of Washington. Concerns about contaminated fertilizer in the Columbia Basin of Washington started in the 1980s. In 1997, a series of articles in the *Seattle Times* detailed how hazardous wastes were being mixed into fertilizers to allow companies to dispose of those wastes without following established state and federal laws governing hazardous waste disposal (Stevens and Pan, 2004). This was made possible by the fact that neither the state of Washington nor the U.S Federal Government had laws in place regulating hazardous material content in fertilizers. In 1998 the state of Washington passed new fertilizer regulations setting standards for nine metals sometimes found in commercial fertilizers (Norman, 2004). The state of Minnesota also passed a law regulating the amount of arsenic in fertilizers in 2003, again in response to a contamination issue with fertilizers (Rosen, 2004). Examples such as these serve as reminders of how important it is to provide adequate protection to the soil we grow our food in and regulation of the materials we put into that soil.

2.5.4 RADIOACTIVE MATERIALS IN SOILS

Soils can also be a reservoir of radioactive elements introduced through both natural (Mehra and Singh, 2011; Osoro et al., 2011; Elstner et al., 1987; Beck et al., 1986) and anthropogenic (Hossain et al., 2010; Balonov et al., 1999; Elstner et al., 1987) sources. About 90% of human radiation exposure worldwide is from natural sources, but anthropogenic exposure sources can be significant in select locations (Hossain et al., 2010).

Radon is a major source of natural radiation exposure, representing about half the natural radiation dose to humans (Appleton, 2007; Yamada, 2003). The ultimate source of radon is the radioactive decay of uranium found in rocks, with granites, felsic metamorphic rocks, organic-rich shales, and phosphatic rocks particularly associated with high uranium content (Appleton, 2007; Owen, 1993). However, most of the radon formed from decay of uranium locked up in these rocks remains trapped within the mineral grains (Appleton, 2007). Soils formed from parent materials high in uranium will also contain uranium, and significant amounts of the radon formed in these soils ends up in soil pore spaces where it can migrate through the soil via diffusion through soil gases or with water moving through the soil (Appleton, 2007). When radon moves through the soil or degasses from radon-containing water sources it can accumulate in enclosed spaces such as basements, cave dwellings, mines, and other enclosed structures in concentrations that negatively impact human health (Appleton, 2007; Yamada, 2003; Owen, 1993). Lung cancer is the primary health concern that arises from radon exposure (Tracy et al., 2006). Radon exposure in conjunction with smoking cigarettes is a particularly lethal combination, with about 90% of the deaths attributable to residential radon occurring in smokers (Tracy et al., 2006).

Other than radon, radionuclides of significant environmental concern include isotopes of cesium, cobalt, curium, neptunium, strontium, plutonium, uranium, technetium, tritium, thorium, americium, radium, and iodine (Cygan et al., 2007). Background radiation levels are fairly constant throughout most of the world (Magill and Galy, 2005), but isolated areas of higher background radiation do occur in regions where high concentrations of minerals, such as monazite, that contain radioactive elements are found and in areas containing large volumes of concentrated plant biomass (i.e., coal) (Turick et al., Chapter 6, this volume). Areas of particularly high background radiation include Ramsar in Iran, Guarapari in Brazil, Karala in India, and Yangjiang in China (Magill and Galy, 2005). Important anthropogenic sources of radioactive materials in the environment include nuclear weapons manufacture and testing, accidental release from nuclear facilities such as Chernobyl and Fukushima, the burning of coal, smelting of nonferrous metals, mining activities, and medical wastes (Turick et al., Chapter 6, this volume; Hu et al., 2010). Once in the environment, exposure to radioactive elements in the soil can be direct, through contact with the soil, or indirect, through consumption of food products grown in the contaminated soils (Hossain et al., 2010). The most common health risks from environmental exposure to radioactive materials include various forms of cancer and genetic mutations (Turick et al., Chapter 6, this volume; Magill and Galy, 2005; Cheever, 2002). The level of risk and health problems encountered are highly dependent on both the dose received and the amount of time over which exposure occurred; average background radiation levels rarely cause health problems (Cheever, 2002). Radioactive issues in the soil are discussed in more detail in Turick et al. (Chapter 6, this volume).

2.6 PATHOGENS IN SOIL

The potential adverse health effects from soil exposure are not limited to chemical content issues. The soil is home to a vast array of organisms; in fact, many scientists believe there are more species of organisms below the soil surface than above it. A single gram of healthy soil may contain more than 1 billion microorganisms (Brady and Weil, 2008; Bultman et al., 2005). Most of the organisms found in soil are not harmful to humans, but soil serves as the home for a number of organisms that can cause human diseases. Conditions that promote an abundance of disease-causing organisms are detrimental to long-term human health, while conditions that reduce the numbers of such organisms promote human health. Soil organisms also influence crop production (Wolf and Snyder, 2003), representing another way they influence human health.

One of the biggest concerns with long-distance transport of dust involves the movement of disease-causing organisms (Polymenakou et al., 2008; Abrahams, 2002). Spore-forming bacteria, viruses, and fungi are able to hitch rides on dust particles over long distances (Sing and Sing, 2010), causing allergy problems and disease in humans; crop pathogens have also been documented (Polymenakou et al., 2008; Abrahams, 2002; Taylor, 2002). Ultraviolet light would normally kill microbes transported through the stratosphere on dust, but modern dust levels might be so thick that it provides protection from ultraviolet radiation allowing some viable organisms to travel long distances, such as across the Atlantic from Africa (Taylor, 2002).

2.6.1 CLASSIFICATION OF SOIL PATHOGENS

A common classification used for disease-causing organisms in the soil is one that classifies them based on the time they spend in the soil. Some pathogens spend their entire life cycle in the soil; others require the soil for part of their life cycle but spend other parts of their life cycle elsewhere. Some pathogens may naturally occur in the soil, but do not actually require the soil to complete their life cycle, while still others may be introduced to the soil through anthropogenic activity but do not occur in the soil naturally. With these various residencies in mind, the following classification system was developed (Bultman et al., 2005):

- *Permanent*: The organism is a permanent inhabitant of the soil that is capable of completing its entire life cycle in the soil.
- *Periodic*: The organism requires part of its life cycle to be completed in the soil.
- *Transient*: The organism may be found in the soil naturally, but the soil is not required to complete its life cycle.
- *Incidental*: The organisms is introduced into the soil through anthropogenic activities.

This system is certainly not the only way to classify pathogens found in the soil, but it is the system that will be followed in this paper. Pepper et al. (2009), for example, present a similar but slightly different classification.

2.6.2 INFECTION GATEWAYS

An infectious organism is only capable of causing disease if it is able to enter the body. There are three common ways that soil organisms gain access to the human body: (i) ingestion, (ii) respiration, and (iii) skin penetration (Bultman et al., 2005; Abrahams, 2002; Oliver, 1997).

Ingestion of soil containing disease-causing organisms can occur incidentally, such as when soil particles cling to poorly washed foods like fruits and vegetables, or on purpose in cultures that practice geophagy. Ingestion of soil organisms can also occur indirectly, when animals ingest the organisms, the organisms end up in the animal's muscles, and the animal is then slaughtered and eaten. In such cases, complete cooking of the meat before ingestion will kill the disease-causing organism. Ingestion is probably the most common way that humans are infected by soil pathogens (Bultman et al., 2005).

Respiration involves inhaling airborne soil organisms. In some cases these soil organisms are attached to dust particles that were originally soil solids; in other cases the organism itself becomes airborne. Dry conditions, lack of vegetative cover, and soil disturbance such as during tillage or construction activity are factors that promote airborne dust production.

Most soil microorganisms are not capable of penetrating healthy, intact human skin. However, several organisms are capable of penetrating the skin if it has been compromised by a cut or a crack, and a small number may penetrate healthy skin (Bultman et al., 2005).

2.6.3 ORGANISM GROUPS

There are several groups of soil organisms that can cause human diseases. While a comprehensive discussion of these organisms would be too extensive to cover in this chapter, some common examples will be given and briefly discussed. Information on the incidence of soil organism related diseases is given in Loynachan (Chapter 5, this volume) and Brevik (2009).

Helminths are parasites that may inhabit the human intestines, lymph system, or in some cases other tissues (Figure 2.1). Common helminths include permanent, periodic, and incidental forms. Human diseases caused by helminths are zoonotic diseases that require a nonanimal development site or reservoir for transmission, with the soil being a common development site. Some soils contain more than 10,000,000 helminths/m^2 of soil to a depth of 10 cm. The helminths important in human disease are subdivided into the nematodes, the trematodes, and the cestodes (Bultman et al., 2005). Billions of people are infected with nematodes worldwide each year, with an estimated 130,000 deaths annually (Bultman et al., 2005). Infection generally occurs through ingestion or skin penetration, and in most cases involves infection of the intestines. Trematodes cause severe infections involving the lungs, bladder, blood, liver, and gastrointestinal tract. Cestodes cause parasitic tapeworm infections in the intestines. Common soil helminth pathogens include hookworms (*Ancylostoma duodenale* and *Necator americanus*), roundworms (*Ascaris lumbricoides*, *Strongyloides stercoralis*, *Toxocara canis*, and *Toxocara cati*), and tapeworms (*Taenia saginata* and *Taenia solium*) (Brevik, 2009).

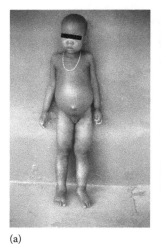
(a)
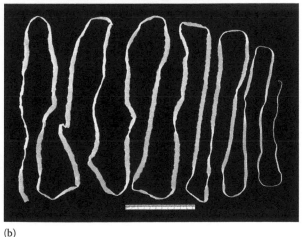
(b)

FIGURE 2.1 (a) A child with hookworm showing visible signs of edema, who was also diagnosed with anemia. (b) An adult *Taenia saginata* tapeworm with a 30 cm ruler at bottom center for scale. (Courtesy of Centers for Disease Control and Prevention, images #5243 and #5260.)

Protozoa are single-celled eukaryotic organisms. Most protozoa found in soil feed on bacteria and algae, but some cause disease in humans (Brady and Weil, 2008) (Figures 2.2 and 2.3). Archeozoa are similar to protozoa, but they lack mitochondria. Most of the health issues caused by protozoa and archeozoa involve the gastrointestinal tract and include problems such as diarrhea, cramps, and abdominal pain. Associated diseases include amebiasis, cryptosporidiosis, giardiasis, and toxoplasmosis, and most pathogenic soil protozoa and archeozoa are transient or incidental (Brevik, 2009).

Approximately 300 species of soil fungi are known to cause disease in humans out of more than 100,000 total fungi species (Bultman et al., 2005). Most fungi are saprophytes that absorb nutrients by aiding in the decomposition of dead organisms. As a group, fungi thrive in the upper 15 cm of moist, acidic soils, although fungi are also found in other soils (Brady and Weil, 2008). Common diseases caused by soil fungi include valley fever (coccidioidomycosis), histoplasmosis, and ringworm (tinea corporis) (Figure 2.4). Coccidioidomycosis is caused by a fungus that lives in the soil called *Coccidioides immitus* (Bultman et al., 2005; Abrahams, 2002). Any activity that stirs up the soil, including dust storms, dry conditions, unvegetated fields, and earthquakes can cause fungal spores to become airborne where they can be inhaled. Valley fever is a serious disease that can be fatal (Bultman et al., 2005). Between 50,000 and 100,000 people develop symptoms of valley fever each year in the United States, with about 35,000 of those cases occurring in California alone (Brevik, 2009; Abrahams, 2002) (Figure 2.5). Epidemics of valley fever occurred in California in

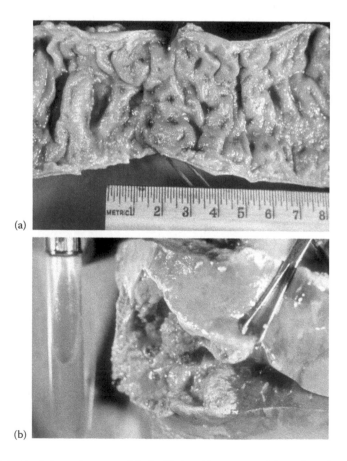

(a)

(b)

FIGURE 2.2 (a) Intestinal ulcers due to amebiasis. (b) Amebic abscess of liver; the tube on the left contains abscess pus. (Courtesy of Centers for Disease Control and Prevention, images #361 and #362.)

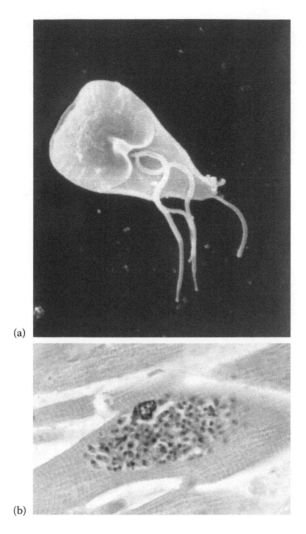

(a)

(b)

FIGURE 2.3 (a) Scanning electron micrograph of the *Giardia lamblia* protozoan parasite. (b) Toxoplasmosis of the heart in an AIDS victim; numerous tachyzoites of *Toxoplasma gondii* are visible within a pseudocyst in a myocyte. (Courtesy of Centers for Disease Control and Prevention, images #8698 and #966.)

the 1990s during a prolonged drought when there was significant windblown dust in the air and following the 1994 Northridge earthquake (Abrahams, 2002). Most pathogenic soil fungi are permanent soil residents (Brevik, 2009).

Bacteria are the most abundant organisms in the soil, with up to 1 trillion in a single gram of soil. Bacteria perform a wide range of ecological functions (Brady and Weil, 2008). They are most numerous in warm, moist soils, but can be found in any soil on Earth. Actinomycetes are similar to bacteria but have a filamentous growth pattern and are often abundant in the soil (Brady and Weil, 2008). Most actinomycetes do not cause disease in humans, but a few pathogenic actinomycetes are known to exist (Bultman et al., 2005). Diseases that can be caused by soil bacteria and actinomycetes include anthrax, tetanus (lockjaw), botulism, gas gangrene, salmonellosis, and Rocky Mountain spotted fever and other tick fevers (Brevik, 2009). Meningitis can be caused by the bacterium *Neisseria meningitides* carried on airborne soil particles; some ocular infections can also be caused by airborne soil (Handschumacher and Schwartz, 2010). Examples of diseases caused

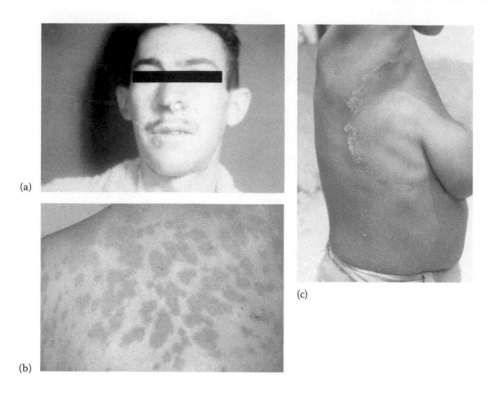

(a)

(b)

(c)

FIGURE 2.4 Paracoccidioidomycosis lesions on the face of a patient (a) and erythema nodosum lesions on skin of the back (b) due to coccidioidomycosis. Ringworm on the skin of the right axilla and flank due to *Trichophyton rubrum* (c). (Courtesy of Centers for Disease Control and Prevention, images #4027, #482, and #2909.)

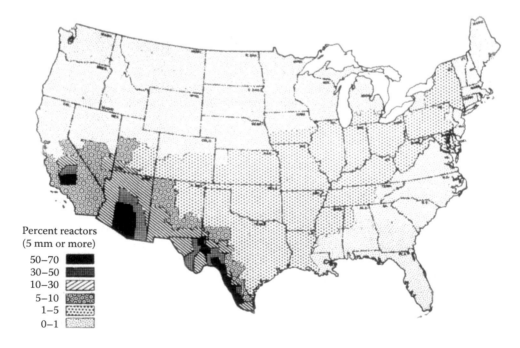

Percent reactors
(5 mm or more)

50–70
30–50
10–30
5–10
1–5
0–1

FIGURE 2.5 Geographic variation in the prevalence of coccidioidin sensitivity in young adults in the United States. (Courtesy of Centers for Disease Control and Prevention, image #474.)

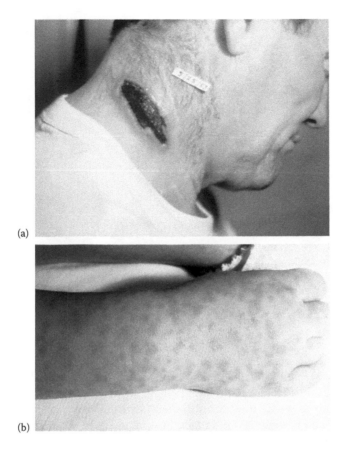

(a)

(b)

FIGURE 2.6 (a) Anthrax lesion on the neck. (b) Right hand and wrist displaying the characteristic rash of Rocky Mountain spotted fever. (Courtesy of Centers for Disease Control and Prevention, images #1934 and #1962.)

by soil bacteria are given in Figure 2.6. Bacteria span the range from permanent to incidental soil residents.

The soil is not a natural reservoir for viruses; however, viruses are known to survive in the soil environment (Pepper et al., 2009; Bultman et al., 2005). Viruses that cause human diseases are usually introduced to the soil with human wastes, such as through septic systems or sewage sludge (Hamilton et al., 2007; Bultman et al., 2005; Fetter, 2001). Viruses usually survive best in cool, wet soils with neutral pH and low microbial activity (Bultman et al., 2005). The presence of clays in the soil also tends to prolong viruses' ability to survive. Viruses that cause conjunctivitis, gastroenteritis, hepatitis, polio, aseptic meningitis, and small pox have all been found in soil, in each case as an incidental organism (Hamilton et al., 2007; Bultman et al., 2005). Hantavirus is one of the viral diseases that may spend part of their life cycle in the soil, although soil is not essential to hantaviruses' life cycle. Hantavirus is excreted in the urine and feces of rodents and may then later be inhaled as airborne dust through disturbed feces or soil (Bultman et al., 2005). The early symptoms of hantavirus pulmonary syndrome (HPS) include fever, headaches, muscle aches, stomach problems, dizziness, and chills. Advanced symptoms include fluid in the lungs and shortness of breath, which can lead to death (Bultman et al., 2005). Hantavirus occurs worldwide (Bultman et al., 2005), with most cases of HPS in the United States occurring in the dusty arid and semiarid western states (Brevik, 2009) (Figure 2.7). Soil pathogens are discussed in more detail in Loynachan (Chapter 5, this volume).

FIGURE 2.7 HPS cases in the United States, January 2007. (Courtesy of Centers for Disease Control and Prevention, image #90.)

2.7 GEOPHAGY

Geophagy is the deliberate ingestion of soil, and is practiced by many animals and humans (Young et al., 2011; Abrahams, 2005; Henry and Kwong, 2003; Oliver, 1997). While geophagy is not common in the urban developed world, there are still many societies where geophagy is common (Sing and Sing, 2010; Henry and Kwong, 2003). In fact, in some societies where geophagy is practiced it is possible to purchase soil for consumption in the market places (Sing and Sing, 2010; Abrahams, 2005). Geophagy offers several potential benefits to those who practice it, but there is the potential for serious health problems as well.

Geophagy has been documented all over the world, in both developed and developing countries (Abrahams, 2005; Henry and Kwong, 2003; Oliver, 1997). The most frequently cited benefit of geophagy is using the soil as a source of mineral nutrients, particularly Ca, Fe, Cu, and Mn (Abrahams et al., 2006), but it is also practiced as a way to appease hunger, as a medicine, and as a way to detoxify foods (Sing and Sing, 2010; Abrahams, 2005; Oliver, 1997). Young et al. (2011) argue that human geophagy is best explained as a means of protection from the intake of chemicals, pathogens, and parasites. Medical uses of ingested soil material that have been shown effective include the treatment of some types of poisoning and the treatment of gastrointestinal disorders such as stomach aches, acid indigestion, nausea, and diarrhea (Abrahams, 2005). Geophagy is most often associated with people of low socioeconomic standing, children, and pregnant women (Abrahams, 2005; Henry and Kwong, 2003; Oliver, 1997).

Despite the potential advantages of geophagy, there are a number of negative effects as well. Ingestion of clays with high cation exchange capacity, such as smectites, can cause deficiency symptoms of nutrients such as Fe, Zn, and K (Abrahams, 2005; Oliver, 1997). Geophagy has also been associated with heavy metal toxicity (Sing and Sing, 2010; Abrahams, 2005; Calabrese et al., 1997), iodine deficiency disorders, and infection by a number of parasitic soil organisms (Sing and Sing, 2010; Hough, 2007). Toxic substances such as pesticides may also be found in soil (Boxall et al., 2009). Liver disorders may be associated with soil bacteria and fungi ingested through geophagy, minerals in the soil can cause excessive tooth wear, and accumulation of soil in the gastrointestinal tract can cause constipation, abdominal pain, reduced absorption of food, and obstruction and perforation of the colon (Abrahams, 2005). Humus in the soil may chelate with iron, creating anemia and other malnutrition problems (Hough, 2007; Henry and Kwong, 2003; Oliver, 1997).

Involuntary ingestion of soil occurs in all societies (Abrahams, 2005; Calabrese et al., 1997). Foods often have soil particles adhered to them, which can introduce contaminants during consumption (Finster et al., 2004; Motarjemi et al., 1994). Another route of involuntary ingestion involves dust particles in the air that are inhaled, at which point most of the dust is trapped and carried into the esophagus and finally the gastrointestinal tract. Children are particularly prone to ingesting soil from their hands when they put their hands into their mouths (Abrahams, 2005). There are some potential benefits to the involuntary ingestion of soil, such as the intake of Fe-rich dust on foods, but most studies have focused on the potentially negative impacts of soil ingestion. Geophagy is discussed in more detail in Henry and Cring (Chapter 8, this volume).

2.8   WATER

Soil is an important part of the hydrologic cycle. As water comes into contact with soil, it has the opportunity to pick up toxic materials, including heavy metals, organic pollutants, and soil pathogens. These materials and organisms then have the potential to end up in potable surface-water or groundwater sources and cause adverse effects on human health (Cech, 2010; Kresic, 2009). Humans also purposely introduce contaminated water into the soil system with the intent of allowing natural soil processes to clean that water for us. Examples of this include septic systems, the use of artificial wetland systems for sewage processing, and land application of manure (Cech, 2010; Kresic, 2009; Baird and Cann, 2005; Fetter, 2001). Using the soil system to clean such contaminated water sources works well if the system is properly designed, sited, and not overwhelmed with an overload of contaminants, but the potential for contamination is a constant concern (Cech, 2010; Kresic, 2009; Brady and Weil, 2008; Hamilton et al., 2007; Fetter, 2001). Contaminated water can also be introduced to the soil system by accident via sewage spills, breaching or leaching of manure pits or lagoons, or the washing of surface-applied manure into streams by rainwater (Owens et al., 2011; Brady and Weil, 2008; Baird and Cann, 2005). Soil and water issues are discussed in more detail by Helmke and Losco (Chapter 7, this volume) and Neff et al. (Chapter 13, this volume).

2.9   MEDICINES FROM SOIL

The most significant medical use of soils is the isolation of antibiotics from soil organisms; in fact, most of our current clinically relevant antibiotics come from soil actinomycetes (Pepper et al., 2009; D'Costa et al., 2006). Many soil organisms make antibacterial molecules, including actinomycetes and fungi (Pepper et al., 2009; Wolf and Snyder, 2003). Antibiotics are used to fight bacterial infections in humans, but soil microorganisms create them in an effort to gain a competitive advantage in the soil ecosystem (Brady and Weil, 2008; Ratnadass et al., 2006; Wolf and Snyder, 2003). Antibiotic compounds were first isolated from soil actinomycetes in 1940 by a team lead by Selman Waksman, a soil biochemist at Rutgers University; Dr. Waksman was awarded the Nobel Prize in Physiology or Medicine in 1952 for his work (Ginsberg, 2011). Antibiotic groups derived from actinomycetes include the aminoglycosides, glycopeptides, and tetracyclines. Antibiotics derived from soil fungi include the cephalosporin group. In all, about 78% of antibacterial agents approved between 1983 and 1994 had their origins in the soil (Pepper et al., 2009). But the importance of soil as a source of medicines doesn't stop with antibiotics. About 40% of prescription drugs have their origin in the soil, as did about 60% of all newly approved drugs between 1989 and 1995 (Pepper et al., 2009). About 60% of new cancer drugs approved between 1983 and 1994 had a soil origin (Pepper et al., 2009). In short, soils are helping us combat numerous diseases on many fronts.

The clay mineral kaolin, which is found in some soils, has been widely used as a digestive aid. The original formula for the anti-diarrheal medicine Kaopectate was a combination of kaolin and pectin (Allport, 2002; Tobey and Covington, 1975). The formula was changed in the early 1990s when the kaolin–pectin mix was replaced by attapulgite, another soil mineral. The attapulgite-based formula of Kaopectate was replaced with a new formula in the United States in 2003 due to changes in U.S. law (AVMA, 2003). Outside of North America, kaolin products are still used to treat diarrhea and kaolin tablets can still be purchased for medicinal use (Sing and Sing, 2010; Abrahams, 2005). Other medical uses of kaolin include treatment of diaper rash and as an emollient and drying agent in treating poison ivy, poison oak, and poison sumac cases (AJN, 1989). On the dental health side, kaolin is used in some toothpaste formulas (EPA, 1999). Fuller's earth and bentonite both contain large quantities of the clay mineral montmorillonite, which can be found in soils and is used in developed nations to treat poisoning by the herbicides paraquat and diquat (Abrahams, 2005; Clark, 1971). Clays are also widely used by the pharmaceutical and cosmetics industries in products designed to prevent wrinkles and skin ageing (Handschumacher and Schwartz, 2010). Medicines from soils and medical uses of soil materials are further discussed in Mbila (Chapter 9, this volume).

2.10 INFLUENCE OF SOIL CONSERVATION ON HUMAN HEALTH

Approximately 5 billion hectares (ha) of land are currently considered degraded; this represents about 43% of the vegetated land on Earth (Brady and Weil, 2008). Of these 5 billion ha, soil degradation is considered to be the primary problem on about 2 billion ha (Brady and Weil, 2008). Soil degradation is manifested in several changes to the soil, including loss of organic matter, loss of soil nutrients, reduced cation exchange capacity, lowered water-holding capacity, and loss of soil structure. These changes reduce the ability of a soil to support plant and animal life, and thus, to provide humans with food (Figure 2.8). The changes also increase water runoff, which increases overland erosion and exacerbates future soil degradation. About 80% of the world's agricultural land is moderately to severely eroded and another 10% slightly eroded (Pimentel, 2006). Each year, approximately 10 million ha of additional cropland are lost to soil erosion, further reducing the amount of land available for food and fiber production (Pimentel, 2006). In all, humans have abandoned a total of approximately 2 billion ha of arable land since the advent of agriculture, land that is now capable of marginal productivity at best (Pimentel, 2006).

At the same time, the human population is rapidly increasing, and with the increase in population comes a corresponding increase in food needs. In 1985 the world's population was approximately 4.8 billion. By 2008 the population had increased to 6.8 billion, and by 2050 it is expected to reach

FIGURE 2.8 Topsoil loss along the top of this hill has compromised its ability to support crops as evidenced by the visibly reduced growth. These kinds of productivity losses can jeopardize food security if they occur in enough places. (Photo by Gene Alexander, USDA NRCS.)

9 billion. Feeding all these new people as land is degraded and prime farmland is lost to urbanization represents a serious challenge (Lal, 2010). Meanwhile, the huge increases in crop yields realized through improved hybrids during the Green Revolution have tailed off. World cereal grain production per capita decreased steadily from the early 1980s through 2000 (Pimentel, 2006). All this means that maintaining, and even improving and expanding, good croplands is even more important for future food security, which puts a premium on soil conservation. Some of the impacts of soils, soil management, and land use decisions on societies are further explored in Burras et al. (Chapter 10, this volume), Carr et al. (Chapter 12, this volume), and Heckman (Chapter 11, this volume).

2.11 FOOD SECURITY

The concept of food security has several facets. These include an appropriate volume of stable food supplies, access to available supplies, food safety, nutritional balance, and social or cultural food preferences. This concept of food security has developed over several decades, starting in the 1970s and being constantly and steadily refined throughout the 2000s. By 2001, the FAO definition of food security had been refined to:

> Food security [is] a situation that exists when all people, at all times, have physical, social and economic access to sufficient, safe and nutritious food that meets their dietary needs and food preferences for an active and healthy life. (FAO, 2003)

After reviewing some of the many soils-related issues that have an impact on human health outlined in the previous sections of this chapter, it is obvious that there are several that influence the physical and economic access to adequate, nutritious, and safe food that meets dietary needs and food preferences and allows for an active and healthy lifestyle. Healthy soils containing adequate amounts of the nutrients essential to humans are required to produce sufficient amounts of nutritious plant and animal products. While this alone will not assure human health, it is an important step. In addition, activities that contaminate soils with heavy metals, organic chemicals, or human pathogens can lead to an unsafe food supply by introducing unhealthy levels of these substances into the foods that we eat. Clearly, the health of our soils is closely tied to the overall health of the human race.

There are few alternative places to turn to for new food sources as most places not already utilized for agriculture are too steep, too cold, too wet, or too dry for agricultural use. The oceans do not offer a pathway to increased food production either. Even though only 2.4% of the world's food supplies (by calorie intake) come from oceans and other aquatic sources (Table 2.5), overfishing is already a major problem in many of the world's prime fishing grounds, including both marine and freshwater fisheries (Allan et al., 2005; Jackson et al., 2001). Therefore, to insure an adequate food supply, it is absolutely essential that we reduce soil erosion and other forms of soil degradation and start managing our soils in a sustainable fashion.

As can be seen in Table 2.5, food security through proper soil management is not limited to plant production. Animal products including various meats, milk, and eggs are major sources of nutrition in the human diet as well (Tables 2.2 and 2.5), and adequate production of animal products also requires soils with the proper nutrients to allow for adequate and healthy forage and feed production without introducing toxic levels of contaminants into these products.

It has been estimated that 0.5 ha of arable land per capita is required to provide a healthy, diverse diet (Pimentel, 2004). With the current world population and the amount of arable land available there is only about 0.25 ha per capita, and that amount is expected to decrease given current rates of population growth and agricultural land loss (Pimentel, 2004). Land in some countries with suitable climates can provide two or more crops per year, meaning more people can be fed per hectare of land. However, rapidly expanding populations reduce the proportion of the population that these

TABLE 2.5
Daily Per Capita Food Intake as a Worldwide Average, 2001–2003

Food Source	Calories[a]	Percent of Calories
Rice	557	25.5
Wheat	521	23.9
Maize	147	6.7
Sorghum	33	1.5
Potatoes	60	2.7
Cassava	42	1.9
Sugar	202	9.3
Soybean oil	87	4.0
Palm oil	50	2.3
Milk	122	5.6
Animal fats (raw and butter)	62	2.8
Eggs	33	1.5
Meat (pig)	117	5.4
Meat (poultry)	46	2.1
Meat (bovine)	40	1.8
Meat (sheep and goats)	11	0.5
Fish and other aquatic products[b]	52	2.4
Total	2182	

Source: Data from Brevik (2009) based on information from http://faostat.fao.org/Portals/_Faostat/
documents/pdf/world.pdf and http://www.fao.org/fishery/statistics/global-consumption/en.

[a] Aquatic products data from 2003. All other data from 2001 to 2003.

[b] Includes both marine and freshwater products.

sequential cropping systems can feed per hectare as well, and not all arable land is located in climates where sequential cropping techniques can be used to increase overall production. Despite this, soil erosion has increased throughout much of the world over the last 80 years (Brady and Weil, 2008). Soil conservation needs to be given a high priority by governments. Clearly, enhanced care of our soil resources will be critical to food security and human health in the future. Food security issues are further explored in Blum and Nortcliff (Chapter 14, this volume) and Brevik (Chapter 16, this volume).

2.12 CLIMATE CHANGE AND HUMAN HEALTH

The most recent report of the Intergovernmental Panel on Climate Change (IPCC) indicates that the average global temperature will probably rise by between 1.1°C and 6.4°C by 2090–2099 as compared to 1980–1999 temperatures, with the most likely rise being between 1.8°C and 4.0°C (IPCC, 2007). The exact effects that changes in global temperatures and atmospheric CO_2 levels will have on soils are not well understood. It is highly likely that these changes will lead to alterations in important soil cycles such as the carbon and nitrogen cycles (Wan et al., 2011; Reich et al., 2006; Gorissen et al., 2004; Hungate et al., 2003). Climate change is also expected to alter soil water dynamics (Hatfield, 2011; Kirkham, 2011; Kang et al., 2009) and the soil ecosystem (Kardol et al., 2011; Lipson et al., 2006; Briones et al., 1997), and increase erosion (Ravi et al., 2010; Zhang et al., 2004). Given that these soil cycles and systems are all important to plant growth, it can be anticipated that climate change will influence the agricultural production of food and, thus, global food security and human health. As a general trend, food production is expected to increase at mid- to high latitudes and to decrease near the equator, but there is significant uncertainty in these projections (Parry et al., 2005; Olesen and Bindi, 2002). It is also anticipated that soils could negatively influence human health in a changing climate based on increased dust generation and human exposure to dust-related health issues (Ravi et al., 2010; Boxall et al., 2009). Asthma cases in the United States more than doubled between 1980 and 2000 (Redd, 2002), and asthma rates have also increased in the Caribbean (Taylor, 2002). African dust in North America, which has been increasing as desertification problems in Africa have increased, has been tentatively linked to increased asthma, but much additional work is needed to investigate this possibility (Abrahams, 2002; Taylor, 2002). It is important that we learn more about how climate change affects soils so that we can also better understand the potential implications for human health. The impact of past climate changes on human society are discussed in Kutílek and Nielsen (Chapter 15, this volume), while possible future implications of climate change are discussed in Brevik (Chapter 16, this volume).

2.13 CONCLUSIONS

Soils influence human health in a number of ways, including the supply of nutrients, presence of toxic levels of various materials, and as a reservoir for disease-causing organisms. Wind-blown dust from soils can also impact human health, and medicines derived from soil compounds and organisms help enhance human health. Throughout this book, the topics covered here will be addressed in greater depth than the coverage given in this chapter.

Despite the obvious connections between soils and human health, there has been less research in this area than in many other fields of scientific and medical study. The subfield of medical geology is now recognized as one of the exciting new areas of geological research. There are also scientists in the soils community conducting research on soils and human health; however, at present this is not a formally established subfield of soil science. Increasing research in soils and human health is essential to protect and enhance health.

That being said, this chapter should not leave the reader with the impression that nothing has been done to specifically deepen our knowledge of soils and human health. Some directed efforts in this area have been made. One example is the USDA-Agricultural Research Service (ARS) Plant, Soil and Nutrition Research Unit (PSNRU) on the Cornell University campus in the United States. The PSNRU was created specifically to conduct research in the area of soils and human health, and although its mission has expanded since its founding, soils and human health is still a major research area at the PSNRU. The reader is also referred to the references section of this paper for many additional readings on soils and human health. However, more efforts in this all-important field of study are needed.

REFERENCES

Abrahams, P.W. 2002. Soils: Their implications to human health. *Science of the Total Environment* 291: 1–32.

Abrahams, P.W. 2005. Geophagy and the involuntary ingestion of soil. *In* O. Selinus, B. Alloway, J.A. Centeno, R.B. Finkelman, R. Fuge, U. Lindh, and P. Smedley (eds), *Essentials of Medical Geology*, pp. 435–458. Elsevier, Amsterdam, The Netherlands.

Abrahams, P.W., M.H. Follansbee, A. Hunt, B. Smith, and J. Wragg. 2006. Iron nutrition and possible lead toxicity: An appraisal of geophagy undertaken by pregnant women of U.K. Asian communities. *Applied Geochemistry* 21: 98–108.

AJN. 1989. Nurses' drug alert. *American Journal of Nursing* 89(8): 1063–1070.

Akabzaa, T.M. and S.M. Yidana. 2011. Evaluation of sources and options for possible clean up of anthropogenic mercury contamination in the Ankobra River Basin in South Western Ghana. *Journal of Environmental Protection* 2: 1295–1302.

Albering, H.J., S.M. van Leusen, E.J.C. Moonen, J.A. Hoogewerff, and J.C.S. Kleinjans. 1999. Human health risk assessment: A case study involving heavy metal soil contamination after the flooding of the River Meuse during the winter of 1993–1994. *Environmental Health Perspectives* 107(1): 37–43.

Allan, J.D., R. Abell, Z. Hogan, C. Revenga, B.W. Taylor, R.L. Welcomme, and K. Winemiller. 2005. Overfishing of inland waters. *BioScience* 55(12): 1041–1051.

Allport, S. 2002. Women who eat dirt. *Gastronomica: The Journal of Food and Culture* 2(2): 28–37.

Appleton, J.D. 2007. Radon: Sources, health risks, and hazard mapping. *Ambio* 36(1): 85–89.

AVMA. 2003. Kaopectate reformulation could be dangerous to cats. *Journal of the American Veterinary Medical Association.* Available at http://www.avma.org/onlnews/javma/nov03/031115b.asp. Accessed 08/31/2011 (verified 3 January 2012).

Baird, C. and M. Cann. 2005. *Environmental Chemistry*, 3rd edn. W.H. Freeman, New York, NY.

Balonov, M.I., E.M. Krisyuk, and C. Ramel. 1999. Environmental radioactivity, population exposure and related health risks in the east Baltic region. *Scandinavian Journal of Work, Environment & Health* 25(3): 17–32.

Bartoš, T., P. Čupr, J. Klánová, and I. Holoubek. 2009. Which compounds contribute most to elevated airborne exposure and corresponding health risks in the Western Balkans? *Environment International* 35: 1066–1071.

Beck, J.N., D.F. Keeley, J.R. Meriwether, and R.H. Thompson. 1986. Environmental radiation exposure rate in Louisiana. *Journal of Environmental Quality* 15: 410–412.

Bourennane, H., F. Douay, T. Sterckeman, E. Villanneau, H. Ciesielski, D. King, and D. Baize. 2010. Mapping of anthropogenic trace elements inputs in agricultural topsoil from northern France using enrichment factors. *Geoderma* 157: 165–174.

Boxall, A.B.A., A. Hardy, S. Beulke, T. Boucard, L. Burgin, P.D. Falloon, P.M. Haygarth, et al. 2009. Impacts of climate change on indirect human exposure to pathogens and chemicals from agriculture. *Environmental Health Perspectives* 117(4): 508–514.

Bradford, S.A., E. Segal, W. Zheng, Q. Wang, and S.R. Hutchins. 2008. Reuse of concentrated animal feeding operation wastewater on agricultural lands. *Journal of Environmental Quality* 37: S97–S115.

Brady, N.C. and R.R. Weil. 2008. *The Nature and Properties of Soils*, 14th edn. Pearson Prentice Hall, Upper Saddle River, NJ.

Brevik, E.C. 2009. Soil, food security, and human health. *In* W. Verheye (ed.), *Soils, Plant Growth and Crop Production*. Encyclopedia of Life Support Systems (EOLSS), Developed under the Auspices of the UNESCO, EOLSS Publishers, Oxford. http://www.eolss.net (verified 3 January 2012).

Briones, M.J.I., P. Ineson, and T.G. Piearce. 1997. Effects of climate change on soil fauna; responses of enchytraeids, Diptera larvae, and tardigards in a transplant experiment. *Applied Soil Ecology* 6: 117–134.

Bultman, M.W., F.S. Fisher, and D. Pappagianis. 2005. The ecology of soil-borne human pathogens. *In* O. Selinus, B. Alloway, J.A. Centeno, R.B. Finkelman, R. Fuge, U. Lindh, and P. Smedley (eds), *Essentials of Medical Geology*, pp. 481–511. Elsevier, Amsterdam, The Netherlands.

Buol, S.W. 2008. *Soils, Land, and Life*. Pearson Prentice Hall, Upper Saddle River, NJ.

Cakmak, I. 2001. Plant nutrition research: Priorities to meet human needs for food in sustainable ways. *In* W. Horst, M.K. Schenk, A. Bürkert, N. Claassen, H. Flessa, W.B. Frommer, H.E. Goldbach, et al. (eds), *Plant Nutrition: Food Security and Sustainability of Agro-Ecosystems Through Basic and Applied Research*. Kluwer Academic Publishers, Dordrecht, The Netherlands.

Cakmak, D., E. Saljnikov, V. Mrvic, M. Jakovljevic, Z. Marjanovic, B. Sikiric, and S. Maksimovic. 2010. Soil properties and trace elements contents following 40 years of phosphate fertilization. *Journal of Environmental Quality* 39: 541–547.

Calabrese, E.J., E.J. Stanek, R.C. James, and S.M. Roberts. 1997. Soil ingestion: A concern for acute toxicity in children. *Environmental Health Perspectives* 105(12): 1354–1358.

Cech, T.V. 2010. *Principles of Water Resources: History, Development, Management, and Policy*, 3rd edn. Wiley, Hoboken, NJ.

Chee-Sanford, J.C., R.I. Mackie, S. Koike, I.G. Krapac, Y.-F. Lin, A.C. Yannarell, S. Maxwell, and R.I. Aminov. 2009. Fate and transport of antibiotic residues and antibiotic resistance genes following land application of manure waste. *Journal of Environmental Quality* 38: 1086–1108.

Cheever, C.L. 2002. Ionizing radiation. *In* B.A. Plog and P.J. Quinlan (eds), *Fundamentals of Industrial Hygiene*, pp. 257–280. National Safety Council Press, Itasca, IL.

Chen, W., N. Krage, L. Wu, G. Pan, M. Khosrivafard, and A.C. Chang. 2008. Arsenic, cadmium, and lead in California cropland soils: Role of phosphate and micronutrient fertilizers. *Journal of Environmental Quality* 37: 689–695.

Chen, W., L. Wu, A.C. Chang, and Z. Hou. 2009. Assessing the effect of long-term crop cultivation on distribution of Cd in the root zone. *Ecological Modelling* 220: 1836–1843.

Clark, D.G. 1971. Inhibition of the absorption of paraquat from the gastrointestinal tract by adsorbents. *British Journal of Industrial Medicine* 28: 186–188.

Combs Jr, G.F. 2005. Geological impacts on nutrition. *In* O. Selinus, B. Alloway, J.A. Centeno, R.B. Finkelman, R. Fuge, U. Lindh, and P. Smedley (eds), *Essentials of Medical Geology*, pp. 161–177. Elsevier, Amsterdam, The Netherlands.

Cupit, M., O. Larsson, C. de Meeûs, G.H. Eduljee, and M. Hutton. 2002. Assessment and management of risks arising from exposure to cadmium in fertilisers—II. *Science of the Total Environment* 291: 189–206.

Cygan, R.T., C.T. Stevens, R.W. Puls, S.B. Yabusaki, R.D. Wauchope, C.J. McGrath, G.P. Curtis, M.D. Siegel, L.A. Veblen, and D.R. Turner. 2007. Research activities at U.S. Government agencies in subsurface reactive transport modeling. *Vadose Zone Journal* 6: 805–822.

Davies, B.E., C. Bowman, T.C. Davies, and O. Selinus. 2005. Medical geology: Perspectives and prospects. *In* O. Selinus, B. Alloway, J.A. Centeno, R.B. Finkelman, R. Fuge, U. Lindh, and P. Smedley (eds), *Essentials of Medical Geology*, pp. 1–14. Elsevier, Amsterdam, The Netherlands.

D'Costa, V.M., K.M. McGrann, D.W. Hughes, and G.D. Wright. 2006. Sampling the antibiotic resistome. *Science* 311: 374–377.

Deckers, J. and E. Steinnes. 2004. State of the art on soil-related geo-medical issues in the world. *In* D.L. Sparks (ed.), *Advances in Agronomy*, Vol. 84, pp. 1–35. Elsevier, Amsterdam, The Netherlands.

Derbyshire, E. 2005. Natural aerosolic mineral dusts and human health. *In* O. Selinus, B. Alloway, J.A. Centeno, R.B. Finkelman, R. Fuge, U. Lindh, and P. Smedley (eds), *Essentials of Medical Geology*, pp. 459–480. Elsevier, Amsterdam, The Netherlands.

Elstner, E.F., R. Fink, W. Höll, E. Lengfelder, and H. Ziegler. Natural and Chernobyl-caused radioactivity in mushrooms, mosses and soil-samples of defined biotops in SW Bavaria. *Oecologia* 73(4): 553–558.

EPA. 1999. *Kaolin* (100104) fact sheet. Environmental Protection Agency, Washington, DC.

FAO. 2003. *Trade Reforms and Food Security: Conceptualizing the Linkages*. Food and Agriculture Organization of the United Nations, Rome, Italy.

Fetter, C.W. 2001. *Applied Hydrogeology*, 4th edn. Prentice-Hall, Upper Saddle River, NJ.

Filippelli, G.M. and M.A.S. Laidlaw. 2010. The elephant in the playground: Confronting lead-contaminated soils as an important source of lead burdens to urban populations. *Perspectives in Biology and Medicine* 53: 31–45.

Finster, M.A., K.A. Gray, and H.J. Binns. 2004. Lead levels of edibles grown in contaminated residential soils: A field survey. *Science of the Total Environment* 320(2–3): 245–257.

Fordyce, F. 2005. Selenium deficiency and toxicity in the environment. *In* O. Selinus, B. Alloway, J.A. Centeno, R.B. Finkelman, R. Fuge, U. Lindh, and P. Smedley (eds), *Essentials of Medical Geology*, pp. 373–415. Elsevier, Amsterdam, The Netherlands.

Fuge, R. 2005a. Soils and iodine deficiency. *In* O. Selinus, B. Alloway, J.A. Centeno, R.B. Finkelman, R. Fuge, U. Lindh, and P. Smedley (eds), *Essentials of Medical Geology*, pp. 417–433. Elsevier, Amsterdam, The Netherlands.

Fuge, R. 2005b. Anthropogenic sources. *In* O. Selinus, B. Alloway, J.A. Centeno, R.B. Finkelman, R. Fuge, U. Lindh, and P. Smedley (eds), *Essentials of Medical Geology*, pp. 43–60. Elsevier, Amsterdam, The Netherlands.

Garrett, R.G. 2005. Natural distribution and abundance of elements. *In* O. Selinus, B. Alloway, J.A. Centeno, R.B. Finkelman, R. Fuge, U. Lindh, and P. Smedley (eds), *Essentials of Medical Geology*, pp. 17–41. Elsevier, Amsterdam, The Netherlands.

Garrison, V.H., E.A. Shinn, W.T. Foreman, D.W. Griffin, C.W. Holmes, C.A. Kellogg, M.S. Majewski, L.L. Richardson, K.B. Ritchie, and G.W. Smith. 2003. African and Asian dust: From desert soils to coral reefs. *BioScience* 53(5): 469–480.

Gilliom, R.J., J.E. Barbash, C.G. Crawford, P.A. Hamilton, J.D. Martin, N. Nakagaki, L.H. Nowell, et al. 2007. *The Quality of Our Nation's Waters: Pesticides in the Nation's Streams and Ground Water, 1992–2001*. Circular 1291, U.S. Geological Survey, Reston, VA. http://pubs.usgs.gov/circ/2005/1291/. Accessed 09/01/2011 (verified 3 January 2012).

Ginsberg, J. 2011. *Selman Waksman and Antibiotics*. American Chemical Society. http://portal.acs.org/portal/acs/corg/content?_nfpb=true&_pageLabel=PP_ARTICLEMAIN&node_id=926&content_id=CTP_004453&use_sec=true&sec_url_var=region1&__uuid=fcc2a69a-83ba-4506-b48f-e9f072e18d54#P8_446. Accessed 08/30/2011 (verified 3 January 2012).

Gorissen, A., A.Tietema, N.N. Joosten, M. Estiarte, J. Peñuelas, A. Sowerby, B.A. Emmett, and C. Beier. 2004. Climate change affects carbon allocation to the soil in shrublands. *Ecosystems* 7: 650–661.

Grad, F.P. 2002. The preamble of the constitution of the World Health Organization. *Bulletin of the World Health Organization* 80(12): 981–984.

Grassman, J.A., S.A. Masten, N.J. Walker, and G.W. Lucier. 1998. Animal models of human response to dioxins. *Environmental Health Perspectives* 106(2): 761–775.

Ha, N.N., T. Agusa, K. Ramu, N.P.C. Tu, S. Murata, K.A. Bulbule, P. Parthasaraty, S. Takahashi, A. Subramanian, and S. Tanabe. 2009. Contamination by trace elements at e-waste recycling sites in Bangalore, India. *Chemosphere* 76: 9–15.

Hamilton, A.J., F. Stagnitti, X. Xiong, S.L. Kreidl, K.K. Benke, and P. Maher. 2007. Wastewater irrigation: The state of play. *Vadose Zone Journal* 6: 823–840.

Handschumacher, P. and D. Schwartz. 2010. Do pedo-epidemiological systems exist? *In* E.R. Landa and C. Feller (eds), *Soil and Culture*, pp. 355–368. Springer, New York, NY.

Harnly, M., R. McLaughlin, A. Bradman, M. Anderson, and R. Gunier. 2005. Correlating agricultural use of organophosphates with outdoor air concentrations: A particular concern for children. *Environmental Health Perspectives* 113(9): 1184–1189.

Hartikainen, H. 2005. Biogeochemistry of selenium and its impact on food chain quality and human health. *Journal of Trace Elements in Medicine and Biology* 18: 309–318.

Hatfield, J.L. 2011. Soil management for increasing water use efficiency in field crops under changing climates. *In* J.L. Hatfield and T.J. Sauer (eds), *Soil Management: Building a Stable Base for Agriculture*, pp. 161–173. Soil Science Society of America, Madison, WI.

Havlin, J.L., J.D. Beaton, S.L. Tisdale, and W.L. Nelson. 2005. *Soil Fertility and Fertilizers: An Introduction to Nutrient Management*, 7th edn. Pearson Prentice Hall, Upper Saddle River, NJ.

Henry, J. and A.M. Kwong. 2003. Why is geophagy treated like dirt? *Deviant Behavior* 24(4): 353–371.

Horrigan, L., R.S. Lawrence, and P. Walker. 2002. How sustainable agriculture can address the environmental and human health harms of industrial agriculture. *Environmental Health Perspectives* 110(5): 445–456.

Hossain, M.K., S.M. Hossain, R. Azim, and A.K.M.M.H. Meaze. 2010. Assessment of radiological contamination of soils due to shipbreaking using HPGe digital gamma-ray spectrometry system. *Journal of Environmental Protection* 1: 10–14.

Hough, R.L. 2007. Soil and human health: An epidemiological review. *European Journal of Soil Science* 58: 1200–1212.

Hu, Q.-H., J.-Q. Weng, and J.-S. Wang. 2010. Sources of anthropogenic radionuclides in the environment: A review. *Journal of Environmental Radioactivity* 101(6): 426–437.

Hungate, B.A., J.S. Dukes, M.R. Shaw, Y. Luo, and C.B. Field. 2003. Nitrogen and climate change. *Science* 302: 1512–1513.

Hurd-Karrer, A.M. 1939. Antagonism of certain elements essential to plants toward chemically related toxic elements. *Plant Physiology* 14(1): 9–29.

IPCC. 2007. Summary for policymakers. *In* S. Solomon, D. Qin, M. Manning, Z. Chen, M. Marquis, K.B. Averyt, M. Tignor, and H.L. Miller (eds), *Climate Change 2007: The Physical Science Basis*, pp. 1–18. Contribution of Working Group I to the Fourth Assessment Report of the Intergovernmental Panel on Climate Change. Cambridge University Press, Cambridge.

Jackson, J.B.C., M.X. Kirby, W.H. Berger, K.A. Bjorndal, L.W. Botsford, B.J. Bourque, R.H. Bradbury, et al. 2001. Historical overfishing and the recent collapse of coastal ecosystems. *Science* 293(5530): 629–638.

Jones, B. 2005. Animals and medical geology. *In* O. Selinus, B. Alloway, J.A. Centeno, R.B. Finkelman, R. Fuge, U. Lindh, and P. Smedley (eds), *Essentials of Medical Geology*, pp. 513–526. Elsevier, Amsterdam, The Netherlands.

Kang, Y., S. Khan, and X. Ma. 2009. Climate change impacts on crop yield, crop water productivity, and food security—A review. *Progress in Natural Science* 19: 1665–1674.

Kardol, P., W.N. Reynolds, R.J. Norby, and A.T. Classen. 2011. Climate change effects on soil microarthropod abundance and community. *Applied Soil Ecology* 47: 37–44.

Keller, A., B. von Steiger, S.E.A.T.M. van der Zee, and R. Schulin. 2001. A stochastic empirical model for regional heavy-metal balances in agroecosystems. *Journal of Environmental Quality* 30: 1976–1989.

Kirkham, M.B. 2011. *Elevated Carbon Dioxide*. CRC Press, Boca Raton, FL.

Klasing, K.C., J.P. Goff, J.L. Greger, J.C. King, S.P. Lall, X.G. Lei, J.G. Linn, F.H. Nielsen, and J.W. Spears. 2005. *Mineral Tolerance of Animals*. The National Academies Press, Washington, DC.

Kresic, N. 2009. *Groundwater Resources: Sustainability, Management, and Restoration*. McGraw Hill, New York, NY.

Lal, R. 2010. Managing soils and ecosystems for mitigating anthropogenic carbon emissions and advancing global food security. *BioScience* 60(9): 708–721.

Lee, W.-Y., W. Iannucci-Berger, B.D. Eitzer, J.C. White, and M.I. Mattina. 2003. Persistent organic pollutants in the environment: Chlordane residues in compost. *Journal of Environmental Quality* 32: 224–231.

Leitzmann, C. 2009. Adequate diet of essential nutrients for healthy people. *In* V. Squires (ed.), *The Role of Food, Agriculture, Forestry and Fisheries in Human Nutrition*. Encyclopedia of Life Support Systems (EOLSS), Developed under the Auspices of the UNESCO, EOLSS Publishers, Oxford. http://www.eolss.net. Accessed 06/09/2010 (verified 3 January 2012).

Lipson, D.A., M. Blair, G. Barron-Gafford, K. Grieve, and R. Murthy. 2006. Relationships between microbial community structure and soil processes under elevated atmospheric carbon dioxide. *Microbial Ecology* 51: 302–314.

Ljung, K., F. Maley, A. Cook, and P. Weinstein. 2009. Acid sulfate soils and human health—A millennium ecosystem assessment. *Environment International* 35: 1234–1242.

Magill, J. and J. Galy. 2005. *Radioactivity, Radionuclides, Radiation.* Springer, Berlin, Germany.

Mannetje, A't., D. McLean, S. Cheng, P. Boffetta, D. Colin, and N. Pearce. 2005. Mortality in New Zealand workers exposed to phenoxy herbicides and dioxins. *Occupational and Environmental Medicine* 62(1): 34–40.

Manuel, J.S. 2000. A toxic house in the country: Building on former farms. *Environmental Health Perspectives* 108(3): A115.

Mbila, M.O. and M.L. Thompson. 2004. Plant-available zinc and lead in mine spoils and soils at the Mines of Spain, Iowa. *Journal of Environmental Quality* 33: 553–558.

Mbila, M.O., M.L. Thompson, J.S.C. Mbagwu, and D.A. Laird. 2001. Distribution and movement of sludge-derived trace metals in selected Nigerian soils. *Journal of Environmental Quality* 30: 1667–1674.

Mehra, R. and M. Singh. 2011. Measurement of radioactivity of ^{238}U, ^{226}Ra, ^{232}Th and ^{40}K in soil of different geological origins in northern India. *Journal of Environmental Protection* 2: 960–966.

Meza-Figueroa, D., R.M. Maier, M. de la O-Villanueva, A. Gómez-Alvarez, A. Moreno-Zazueta, J. Rivera, A. Campillo, C. Grandlic, R. Anaya, and J. Palafox-Reyes. 2009. The impact of unconfined mine tailings in residential areas from a mining town in a semi-arid environment: Nacozari, Sonora, Mexico. *Chemosphere* 77(1): 140–147.

Motarjemi, Y., F. Käferstein, G. Moy, and F. Quevedo. 1994. Contaminated food: A hazard for the very young. *World Health Forum* 15: 69–71.

Nordberg, M. and M.G. Cherian. 2005. Biological responses of elements. *In* O. Selinus, B. Alloway, J.A. Centeno, R.B. Finkelman, R. Fuge, U. Lindh, and P. Smedley (eds), *Essentials of Medical Geology*, pp. 179–200. Elsevier, Amsterdam, The Netherlands.

Norman, M. 2004. Regulation of contaminants in fertilizers in Washington State. *Soil Science Society of America Annual Meeting Abstracts.* Published on CD-ROM.

Olesen, J.E. and M. Bindi. 2002. Consequences of climate change for European agricultural productivity, land use, and policy. *European Journal of Agronomy* 16: 239–262.

Oliver, M.A. 1997. Soil and human health: A review. *European Journal of Soil Science* 48: 573–592.

Osoro, M.K., I.V.S. Rathore, M.J. Mangala, and A.O. Mustapha. 2011. Radioactivity in surface soils around the proposed sites for titanium mining project in Kenya. *Journal of Environmental Protection* 2: 460–464.

Owen, L. 1993. Radon—A new environmental health hazard. *Geography* 78(2): 194–198.

Owens, L.B., J.V. Bonta, M.J. Shipitalo, and S. Rogers. 2011. Effects of winter manure application in Ohio on the quality of surface runoff. *Journal of Environmental Quality* 40(1): 153–165.

Parry, M., C. Rosenzweig, and M. Livermore. 2005. Climate change, global food supply, and risk of hunger. *Philosophical Transactions of the Royal Society B* 360: 2125–2138.

Pepper, I.L., C.P. Gerba, D.T. Newby, and C.W. Rice. 2009. Soil: A public health threat or savior? *Critical Reviews in Environmental Science and Technology* 39: 416–432.

Peterson, R.K.D., P.A. Macedo, and R.S. Davis. 2006. A human-health risk assessment for West Nile virus and insecticides used in mosquito management. *Environmental Health Perspectives* 114(3): 366–372.

Pimentel, D. 2004. Food security and environmental sustainability. *In* R. Lal, P.R. Hobbs, N. Uphoff, and D.O. Hansen (eds), *Sustainable Agriculture and the International Rice–Wheat System*, pp. 37–53. CRC Press, Boca Raton, FL.

Pimentel, D. 2006. Soil erosion: A food and environmental threat. *Environment, Development and Sustainability* 8: 119–137.

Polymenakou, P.N., M. Mandalakis, E.G., Stephanou, and A. Tselepides. 2008. Particle size distribution of airborne microorganisms and pathogens during an intense African dust event in the eastern Mediterranean. *Environmental Health Perspectives* 116(3): 292–296.

Prospero, J.M. and P.J. Lamb. 2003. African droughts and dust transport to the Caribbean: Climate change implications. *Science* 302(5647): 1024–1027.

Ramberg, B. 1986. Learning from Chernobyl. *Foreign Affairs* 65(2): 304–328.

Ratnadass, A., R. Michellon, R. Randriamanantsoa, and L. Séguy. 2006. Effects of soil and plant management on crop pests and diseases. *In* N. Uphoff, A.S. Ball, E. Fernandes, H. Herren, O. Husson, M. Laing, C. Palm, et al. (eds), *Biological Approaches to Sustainable Soil Systems*, pp. 589–602. Taylor & Francis Group, Boca Raton, FL.

Ravi, S., D.D. Breshears, T.E. Huxman, and P. D'Odorico. 2010. Land degradation in drylands: Interactions among hydraulic-aeolian erosion and vegetation dynamics. *Geomorphology* 116: 236–245.

Redd, S.C. 2002. Asthma in the United States: Burden and current theories. *Environmental Health Perspectives* 110(4): 557–560.

Reich, P.B., S.E. Hobbie, T. Lee, D.S. Ellsworth, J.B. West, D. Tilman, J.M. Knops, S. Naeem, and J. Trost. 2006. Nitrogen limitation constrains sustainability of ecosystem response to CO_2. *Nature* 440: 922–925.

Robinson, B.H. 2009. E-waste: An assessment of global production and environmental impacts. *Science of the Total Environment* 408: 183–191.

Rosen, C.J. 2004. Challenges in changing state fertilizer laws: A case study. *Soil Science Society of America Annual Meeting Abstracts*. Published on CD-ROM.

Safe, S.H. 2000. Endocrine disruptors and human health: Is there a problem? An update. *Environmental Health Perspectives* 108(6): 487–493.

Sanchez, P.A. and M.S. Swaminathan. 2005. Hunger in Africa: The link between unhealthy people and unhealthy soils. *The Lancet* 365: 442–444.

Senesi, G.S., G. Baldassarre, N. Senesi, and B. Radina. 1999. Trace element inputs into soils by anthropogenic activities and implications for human health. *Chemosphere* 39(2): 343–377.

Shetty, P. 2009. Fundamentals of human health and nutrition. *In* V. Squires (ed.), *The Role of Food, Agriculture, Forestry and Fisheries in Human Nutrition*. Encyclopedia of Life Support Systems (EOLSS), Developed under the Auspices of the UNESCO, EOLSS Publishers, Oxford. http://www.eolss.net. Accessed 06/09/2010 (verified 3 January 2012).

Sing, D. and C.F. Sing. 2010. Impact of direct soil exposures from airborne dust and geophagy on human health. *International Journal of Environmental Research and Public Health* 7: 1205–1223.

Sparks, D.L. 2003. *Environmental Soil Chemistry*, 2nd edn. Academic Press, Amsterdam, The Netherlands.

Stevens, R.G. and W.L. Pan. 2004. An agronomist perspective on contaminants in fertilizers. *Soil Science Society of America Annual Meeting Abstracts*. Published on CD-ROM.

Taylor, D.A. 2002. Dust in the wind. *Environmental Health Perspectives* 110(2): A80–A87.

Taylor, M.P., A.K. Mackay, K.A. Hudson-Edwards, and E. Holz. 2010. Soil Cd, Cu, Pb, and Zn contaminants around Mount Isa city, Queensland, Australia: Potential sources and risks to human health. *Applied Geochemistry* 25: 841–855.

Tobey, L.E. and T.R. Covington. 1975. Antimicrobial drug interactions. *American Journal of Nursing* 75(9): 1470–1473.

Tracy, B.L., D. Krewski, J. Chen, J.M. Zielinski, K.P. Brand, and D. Meyerhof. 2006. Assessment and management of residential radon health risks: A report from the Health Canada radon workshop. *Journal of Toxicology and Environmental Health, Part A* 69: 735–758.

Troeh, F.R. and L.M. Thompson. 1993. *Soils and Soil Fertility*, 5th edn. Oxford University Press, New York, NY.

Vega, F.A., E.F. Covelo, and M.L. Andrade. 2007. Accidental organochlorine pesticide contamination of soil in Porriño, Spain. *Journal of Environmental Quality* 36: 272–279.

Voisin, A. 1959. *Soil, Grass, and Cancer*. Philosophical Library Inc., New York, NY.

Walsh, L.M., M.E. Sumner, and D.R. Keeney. 1977. Occurrence and distribution of arsenic in soils and plants. *Environmental Health Perspectives* 19: 67–71.

Wan, Y., E. Lin, W. Xiong, Y. Li, and L. Guo. 2011. Modeling the impact of climate change on soil organic carbon stock in upland soils in the 21st century in China. *Agriculture, Ecosystems, and Environment* 141: 23–31.

WHO. 2007a. *Assessing the Iron Status of Populations*, 2nd edn. World Health Organization, Geneva, Switzerland.

WHO. 2007b. *Assessment of Iodine Deficiency Disorders and Monitoring Their Elimination*, 3rd edn. World Health Organization, Geneva, Switzerland.

Wolf, B. and G.H. Snyder. 2003. *Sustainable Soils*. Food Products Press, Binghamton, NY.

Yamada, Y. 2003. Radon exposure and its health effects. *Journal of Health Science* 49(6): 417–422.

Young, S.L., P.W. Sherman, J.B. Lucks, and G.H. Pelto. 2011. Why on Earth? Evaluation hypotheses about the physiological functions of human geophagy. *The Quarterly Review of Biology* 86(2): 97–120.

Zhang, X.C., M.A. Nearing, J.D. Garbrecht, and J.L. Steiner. 2004. Downscaling monthly forecasts to simulate impacts of climate change on soil erosion and wheat production. *Soil Science Society of America Journal* 68: 1376–1385.

Zheljazkov, V.D. and P.R. Warman. 2004. Source-separated municipal solid waste compost application to Swiss chard and basil. *Journal of Environmental Quality* 33: 542–552.

Section II

Human Health as It Relates to Materials Found in Soil

3 Soil, Heavy Metals, and Human Health

Richard Morgan

CONTENTS

3.1 INTRODUCTION

3.1.1 POLLUTION, HEAVY METALS, AND SOIL

Environmental pollution involving heavy metals has attracted a great deal of attention in recent years, driven especially by concerns for human health. Air pollution and water pollution have increased human exposure to heavy metals, and research has revealed important health consequences of such exposure for many of the heavy metals. In many parts of the world, there have also been observable increases in the heavy metal contents of topsoils, but compared with air and water, our understanding of how this translates into increased risk of human exposure and subsequent health effects is still poor (Kibble and Saunders, 2001).

The soil acts as a long-term repository for many heavy metals released into the environment by human activities. In one sense, this provides a degree of protection for the wider environment, by slowing down the transfer of heavy metals into surface water, for example. But it also means that the soil itself can become a significant source of heavy metals for the people using it as a resource, whether actively (e.g., growing crops) or more passively (e.g., erecting a building on a piece of land). Moreover, soil is a complex system that varies physically and chemically from place to place and through time, and this affects our ability to understand the movement of heavy metals through the soil, to assess actual human exposure to specific metals, and therefore to make links to specific health outcomes. Despite the problems with establishing clear cause–effect links between heavy metals in soil and adverse health outcomes in people, there is a growing body of research that suggests that high heavy metal content in soil can contribute to poor human health and that a precautionary approach would be well advised. This is being reflected in the health-related guideline values being developed in many countries in relation to the development of potentially contaminated land.

In this chapter, the nature and sources of heavy metals are briefly outlined, followed by a discussion of soil factors that influence the supply of heavy metals to humans through both direct and indirect means. Broad health issues associated with specific metals are then briefly reviewed, leading into a discussion of the nature and use of soil guideline values using examples from several countries. In the final section, three case studies are presented. The focus of the chapter is on toxicity-related health problems resulting from environmental pollution, but it should be recognized that human health is also affected by natural soil deficiencies of some of the metals (Brevik, 2009; Oliver, 1997).

3.1.2 What Is a Heavy Metal?

Broadly, the term heavy metal is usually applied to those elements with metallic properties and a density of 5 g/cm^3 or higher. However, as van der Perk (2006, p. 125) notes: "... the term heavy metals has no sound terminological or scientific basis." For example, the criteria for defining "heavy" can vary between authors, and not all the heavy metals commonly discussed act as metals in the generally accepted sense; some are classed as semimetals or metalloids (e.g., arsenic and antimony) (Kabata-Pendias and Mukherjee, 2007; Dufuss, 2002). Moreover, some researchers prefer to use the alternative term *trace elements*. However, as Alloway (1995, p. 3) comments, heavy metal is the "most widely recognised and used term" for a large group of trace elements that are "both industrially and biologically important."

Opinions also differ as to which elements make up that large group of trace elements, but for the purpose of this chapter the practical approach advocated by Fergusson (1990) seems apt. He focuses on those elements that are causing the most problems for large numbers of people, either as direct threats to health, or indirectly, by causing environmental changes that have health consequences for people. On this basis, the specific elements considered in this chapter are arsenic* (As), lead (Pb), cadmium (Cd), chromium (Cr), copper (Cu), mercury (Hg), nickel (Ni), and zinc (Zn).

3.1.3 Sources

3.1.3.1 Natural Sources and Biological Functions

Heavy metals are constituents of the rocks that make up Earth's surface. As various forms of erosion and weathering break up the rocks, the elements are incorporated into the soil within inorganic

---

* Arsenic is usually classed as a metalloid, but for ease of expression it will be referred to here as a (heavy) metal.

particles of different sizes, from large rock fragments to sand and silt and finally clay particles. Further weathering, especially forms of chemical weathering, release a proportion of the metals into the soil solution as dissolved species. Once in the soil solution, they can be incorporated into other inorganic or organic forms. The variation in mineral makeup of soil parent material has a profound impact on the natural spatial variation of the heavy metals in soils, which in turn has implications for the supply of certain essential trace elements to living organisms. Hence, soils derived from volcanic parent materials in particular are usually rich in a broad array of heavy metals, while those derived from calcareous materials generally have much lower concentrations of such elements. Moreover, over time, soil formation processes lead to weathered soils that have been depleted of trace elements compared with young soils developing on freshly exposed parent material.

Establishing natural background levels for heavy metals is very difficult, as every location has its own geological and ecological history that will determine which elements are present and in what quantities. However, broad figures have been produced based on studies around the world to provide a sense of the range of natural concentrations of heavy metals in soils of different types, and these are often used to provide a broad context for assessing metal content at a given location when information on local background concentrations is not available (Table 3.1).

As alluded to previously, many of the heavy metals also play important roles in biological systems, albeit in very small quantities. Deficiencies can arise in natural soils due to the nature of the parent material or the age of the soils and the extent of weathering that has taken place. In such cases plant growth can be affected, and this can in turn influence the other organisms in the ecosystem.

It is possible for natural soils to have comparatively high levels of specific heavy metals; one of the best examples is the influence of ultramafic parent material on soils. Ultramafic rocks are comprised of minerals with very high magnesium and iron content, plus high quantities of various other metals (such as Ni, Cu, Co, etc.), and are typically found along tectonic plate boundaries where rocks have been subject to intense hydrothermal processes. When exposed at Earth's surface, the soils developed on ultramafic parent material often have unusual chemical characteristics, one of which is an abundance of one or more heavy metals (Brooks, 1987). Together with certain physical features, the soils pose significant nutrient-supply problems for plant life, including excess levels of one or more heavy metals, and the plant communities developed on these soils are often poorer in species diversity and lower in density than immediately adjacent communities (Keddy, 2007). However, some plant species associated with such areas have evolved mechanisms to cope with soils rich in heavy metals by taking up and locking away large amounts of the metals in their

TABLE 3.1
Average Concentration of Selected Heavy Metals

	Average Crustal Content (mg/kg)	Worldwide Soil (mg/kg)
Arsenic	1.8	4.7
Cadmium	0.1	1.1
Chromium	100	42
Copper	55	14
Lead	14	25
Mercury	0.07	0.1
Nickel	20	18
Zinc	70	62

Source: Based on Kabata-Pendias, A. and Mukherjee, A., *Trace Elements from Soil to Human,* Springer-Verlag, Berlin, Germany, 2007.

biomass. These species, known as hyperaccumulators, can be used by mining companies to identify potential prospecting sites, but they may also provide the means to tackle contaminated sites using phytoremediation methods. Phytoremediation involves growing plants to take up large amounts of available metals from the soil and then cropping and burning the plants followed by either burying the ash in sealed pits or recovering the metals from the ash if concentrations are high enough (Baird and Cann, 2008).

3.1.3.2 Anthropogenic Sources

Human activities over many centuries have resulted in a significant buildup of a wide variety of heavy metals in soils, but with an exponential increase in modern times. There is a temptation to associate heavy metals only with industrial activities and therefore to view them solely as a problem for urban and/or industrial localities. However, various agricultural activities have also contributed to the heavy metal burden in soils across large areas of rural land in many countries.

It is useful to think in terms of three phases of heavy metal release into the environment: the first during extraction and refinement, the second during usage, and the third through disposal. With increasing awareness of the human health risks associated with exposure to heavy metals in recent years, current best practice standards for all three phases are much improved. However, the legacies of past practices are often still present in the landscape (contaminated land around old mines, old industrial areas, old landfills) until cleanup measures are applied. Also, not all countries enforce best practice in extraction, use, and disposal of heavy metals, so some localities may still be experiencing high levels of contamination.

3.1.3.2.1 Extraction

The most important sources of metal contamination of soil are associated with the mining and subsequent smelting of metalliferous ores. Dust from the movement of crushed ore and from mine wastes (tailings) can be dispersed into the surrounding landscape by wind, where it is incorporated into topsoil (Alloway, 1995). Smelting of ores produces particulates that, if not intercepted, will also be dispersed into the surrounding landscape. The Norilsk region of the Russian Arctic has been the location of a major Cu–Ni ore excavation and smelting complex since about 1935; it is reported to be the largest smelting complex in the Arctic, with three smelters currently operating, processing ore from several ore fields (Zhulidov et al., 2011). Environmental damage around the area is extensive, especially to forest and tundra downwind of the complex due to sulfur dioxide emissions, and although specific information is still scarce, preliminary soil studies indicate significant buildup of Cu, Zn, Pb, Hg, Cd, and Ni compared with unpolluted areas of the Taymyr peninsula (Zhulidov et al., 2011). The source of the metals is likely to be a combination of dust from ore processing and particulates from smelter emissions. This example also demonstrates that ore bodies usually contain a range of metals, often in minor quantities compared with the target metal(s), so it is not unusual to see other metals represented in contamination. For example, Cd is present in the ores smelted for Pb–Zn production, and As is often a side product of Cu smelters (Alloway, 1995).

3.1.3.2.2 Use

Table 3.2 summarizes the main uses of the heavy metals being considered in this chapter. Many of the uses do not lead to the direct release of the metal in question into the local environment. Rather, the metals are contained in substances and structures until they are no longer required; then they are disposed of, at which point there may be problems with environmental contamination (see next section). However, some uses do involve environmental release. Examples include the practice (now banned in most countries) of adding tetramethyl lead to petrol/gasoline, the use of As and Cu in animal feedstocks, and pesticides containing As, Pb, and Cu. In addition, some metals can be introduced into the environment inadvertently. A notable example is the distribution of large amounts of Cd onto agricultural land through the use of phosphatic fertilizers manufactured from rock phosphate with high levels of Cd as an impurity (Alloway, 1995). The burning of fossil fuels

TABLE 3.2
Variety of Uses of the Main Heavy Metals

Metal	Uses
Arsenic	Additive to animal feed; wood preservative (copper chrome arsenate); special glasses; ceramics; pesticides; insecticides; herbicides; fungicides; rodenticides; algicides; sheep dip; electronic components (e.g., gallium arsenate semiconductor, integrated circuits, diodes, infrared detectors, and laser technology); nonferrous smelters; metallurgy; coal-fired and geothermal power generation; textiles and tanning; pigments; antifouling paints; fireworks; veterinary medicine.
Cadmium	Ni/Cd batteries; pigments; anticorrosive coatings on metals; plastic stabilizers; alloys; coal combustion; neutron absorber in nuclear reactors.
Chromium	Manufacture of ferroalloys (special steels); plating operations; pigments; textiles; leather tanning; wood preservative; magnetic tapes.
Copper	Electrical cables and wiring; water pipes; roofing; kitchenware; chemical and pharmaceutical equipment; pigments; alloys; fungicides; wood preservative; antifouling agents.
Lead	Antiknock agents (tetramethyl lead); lead–acid batteries; cathode ray tubes; pigments; glassware; ceramics; plastics; paint; alloys; cable sheathing; solder; pipes and tubing.
Mercury	Metal extraction from ores; mobile cathode in chloralkali process (producing $NaOH$ and Cl from brine); electrical and measuring apparatus; fungicides; catalysts; pharmaceuticals; dental materials; scientific instruments; rectifiers; oscillators; electrodes; mercury vapor lamps; long-life bulbs; X-ray tubes; solders.
Nickel	Alloy in steel industry; electroplating; Ni/Cd batteries; arc-welding rods; paint and ceramic pigments; surgical and dental prostheses; molds for glass and ceramic containers; computer components; catalysts.
Zinc	Zinc alloys (bronze, brass); anticorrosion coating; batteries; cans; PVC stabilizers; precipitate Au from cyanide solution; medicines; chemicals; rubber; paints; soldering and welding fluxes.

Source: Based on Siegel, F.R., *Environmental Geochemistry of Potentially Toxic Metals*, Springer, Berlin, 2002.

also falls into this category, with a wide range of metals being released by the combustion of coal and petroleum products (Alloway, 1995).

3.1.3.2.3 Disposal

A number of forms of disposal can result in the release of heavy metals into the environment. One very problematic form is *"in situ"* disposal and involves the removal of old paint from houses and other structures. Although many countries now ban the use of Pb-based paints for general use, old paint can still be Pb-based and sanding to remove that paint disperses Pb into the local area very effectively (Pierzynski et al., 2005). More generally, metallic products are now routinely recycled in most countries and are smelted to recover the constituent metals. Where best practice methods are used this should not result in heavy metal contamination. However, this does not occur everywhere and in some countries "cottage industry" recycling of computer components (e-waste) has resulted in elevated heavy metal contamination (Leung et al., 2008). Disposal of industrial solid waste onto land can result in heavy metal release into local areas from dust and leaching; the metals involved will reflect the industrial source of the waste (Alloway, 1995). Better environmental management methods, including improved disposal techniques and greater use of land restoration and management, can mitigate this problem.

Many products containing heavy metals still go to landfills and this can result in a number of metals, including Cd, Cu, Pb, and Zn, leaching into local water bodies or into groundwater (Alloway, 1995). In some situations this may lead to deposition and accumulation on local floodplains. The other major form of solid waste disposal is sewage sludge, also known as biosolids. Produced by wastewater treatment plants, the solids are subject to various treatments to make them suitable for disposal onto land or into landfills. The heavy metal content of sewage sludge varies according to the source of the original waste, and different industry types can have a significant impact on the metals present in the final sludge and their concentration (Alloway, 1995; Antoniadis et al., 2006). For example, municipal sludge can have comparatively high levels of Zn, Cr, and Cu, while sludge from the textile industry can have much higher Cr levels but lower amounts of the other main heavy metals (Tan, 2009). Many countries set guidelines for the application of sewage sludge to manage the potential for heavy metal accumulation from repeated applications (Pierzynski et al., 2005). A related problem is the use of composted agricultural manures as fertilizers. If the animal feed contained certain metal supplements, such as As or Cu, these will be present in the organic fertilizer produced from the manure (Pierzynski et al., 2005).

Finally, incineration of refuse is an increasingly popular option in some countries, partly to reduce pressure from expanding landfills on scarce land resources. It also provides an alternative and comparatively low-cost energy source, so in some countries biosolids are burnt, not only providing energy, but also solving the sensitive problem of how to dispose of human waste. Incineration can result in the emission of heavy metals into the atmosphere, depending on the characteristics of the waste being burnt and the pollution control methods employed by the incinerator operator. Most of the heavy metals of concern are associated with particulates, so good incinerator practice can control these. However, Hg control requires the treatment of flue gases (European Commission, 2006).

3.2   SOIL PROCESSES AND HEAVY METALS

In broad terms the impacts of heavy metals on human health, and indeed on wider ecosystem health, are a function of the supply of heavy metals to the soil (a combination of natural sources and inputs from human activities) and the rate of release into the soil from the source material (essentially breakdown processes governed by temperature, moisture, and, in the case of organics, microbial activity). The latter is the domain of soil processes that affect long- and short-term storage of the metals, rates of release back into the soil solution, mobility through the soil profile, and incorporation of a proportion of the metals into living organisms.

Figure 3.1 provides a simple overview of the key compartments in the soil system where heavy metals can be held. Central to the figure is the soil solution, which occupies the pore spaces in the soil and bathes the surfaces of the organic and inorganic particles and the root system. Biological uptake of the metals requires them to be in ionic form in the solution. For most metals this is a simple, positively charged cation (e.g., Cu^{2+}, Zn^{2+}), but some of the metalloid elements, such as arsenic, form complex oxyanions (e.g., AsO_3^{2-}), which have a negative charge.

Some of the available metal ions are taken up by soil biota and higher plants to meet the normal trace element needs of these organisms, hence only a very small quantity of these elements need to be available in the soil solution at any given time. If there is a comparative abundance of the metal ions in the soil solution, some plants may be able to take up larger amounts of a given element, which can then lead to toxicity problems for the plant unless the metal can be sequestered in the plant and prevented from damaging key growth processes. Consumption of plants with elevated metal content can then pose toxicity problems for higher organisms.

Metal ions that are not used almost immediately for biological purposes are subject to other physical and chemical processes that can take the elements out of solution for shorter or longer periods of time and therefore affect their bioavailability:

1. Some elements are taken into complex organic compounds in the soil. The major functional groups involved in formation of metal complexes are the carboxyl (—COOH) and phenolic (—OH) groups (Sparks, 2003); Cu has a particular propensity to form such complexes with soil organic matter (SOM) (Pierzynski et al., 2005). Release back into solution can come from the action, for example, of hydrogen ions (protons) disrupting the bonds holding the metal, a process that would be encouraged by increased acidification of the soil.

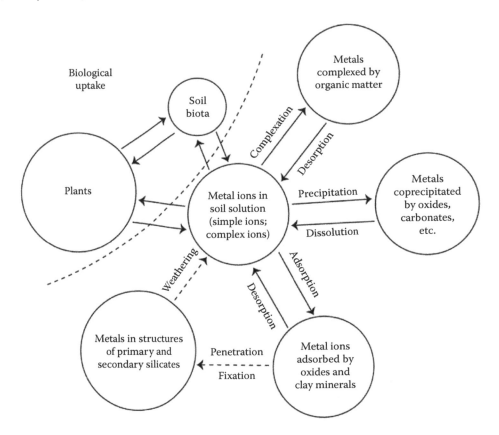

FIGURE 3.1 Forms and dynamics of heavy metals in soil. (Adapted from McLaren, R.G. and Cameron, K.C., *Soil Science,* Oxford University Press, Melbourne, Australia, 1996; Pierzynski, G.M., Sims, J.T., and Vance, G.F., *Soils and Environmental Quality,* CRC Press, Boca Raton, FL, 2005.)

2. Clay and humus particles generally carry significant amounts of surface charge, mostly negative, which attracts positively charged ions from solution and creates dense layers of adsorbed ions close to the surfaces of these particles. Most of the charge sites are satisfied by nonspecific adsorption, with the dominant cations in solution (e.g., Ca, Mg, K, H, etc.) competing to satisfy the surface charges on the clay and humus particles, broadly in proportion to their concentration in solution. Heavy metals also exploit specific adsorption sites, especially on the surfaces of iron (Fe) and manganese (Mn) hydrous oxides, but also on the edges of the clays and on some organic surfaces (Alloway, 1995). The metal ions form covalent bonds (i.e., sharing electrons) with particular structures on these surfaces, especially OH groups on the hydrous oxides. The process is strongly pH dependent, occurring more readily in less acidic soils, while acidic conditions promote desorption and release of the metals back into solution. Arsenic, in particular, has a very strong affinity for Fe oxides (McLaren et al., 2006).

3. Most of the metal ions can precipitate out of solution as sulfides, carbonates, phosphates, hydroxides, etc., depending on solution conditions, pH levels, and the presence of relevant anions (Baird and Cann, 2008). Continuing growth of the precipitates effectively locks a proportion of the metals away more permanently. In addition, those ions initially attracted to specific surface charges on hydrous oxides or clays can be incorporated into those developing structures following further precipitation of the mineral from solution. For a period the occluded metals can be released back into solution by chemical breakdown of the

precipitate; however, the longer the process goes on, the more permanent the occlusion, especially if the metal ions migrate further into the crystalline structure of the developing mineral (Wild, 1993).

4. Finally, fixation processes can see metal ions moving from the surface of crystalline minerals, notably clays such as illite and vermiculite, and penetrating and being held within the internal structure of the mineral. Release into soil solution requires significant breakdown of the mineral structure by chemical weathering (McLaren and Cameron, 1996).

The chemical characteristics of the various heavy metals ensure that, at any given time, only a very small proportion of any particular metal is in the soil solution (typically at levels varying from 0.1 to 0.01 μg/L) (McLaren and Cameron, 1996). The bulk of the metals in the soil will be complexed, precipitated, occluded, or fixed by other soil constituents. However, as anthropogenic activities increase the input of heavy metals to soils, even that very small proportion can become a significant quantity for biological systems that only require very small amounts of these elements. In terms of exposure pathways that have implications for human health, this means most of the heavy metals are associated with soil particles and will be transported along with soil particles by wind (as dust) or water (mainly as suspended sediments in the short term). However, heavy metals dissolved in the soil solution are more mobile within the soil and can be transported down the profile by leaching processes, possibly into groundwater and/or surface water, or they might be taken up by plants. Plant uptake is particularly important in the movement of Cd from soils into the human food chain, while leaching processes can be important in the movement of As through the soil system.

Two soil factors, pH and the redox potential of the soil system, have important effects on the processes outlined earlier, and especially on heavy metal mobility within, and in some cases out of, the soil system:

1. Most heavy metals in an acid soil environment increase their concentration in soil solution. This is partly because there are fewer hydroxyl (OH) groups to coordinate with, thus reducing precipitation from solution, but it also reflects the direct action of hydrogen ions on recently precipitated compounds and organic complexes, releasing metals back into solution. The more acidic the soil, the higher the proportion of many heavy metals in solution, usually as simple cations. The chemistry of arsenic is more complex; it forms complex oxyanions (negatively charged, complexed with several oxygens and one or more hydrogens), with the specific form varying according to the pH of the soil. In acid soils, the arsenic oxyanions tend to be adsorbed by Fe oxides and in alkaline soils, by Ca compounds (McLaren et al., 2006).

2. As biological systems, soils generally require a good supply of oxygen. Consequently, the organic and inorganic chemistry of the soil is adjusted to a largely aerated environment, with oxygen playing a central role in many biochemical processes as the most effective electron acceptor available. When oxygen supply is restricted, by flooding or compaction for example, the soil system starts to respond. Oxygen-rich compounds in the soil (such as oxides) are used by microorganisms as a replacement for atmospheric oxygen, using the oxygen in the compounds as electron acceptors, and thereby changing the chemical nature of the compounds (Pierzynski et al., 2005). Metals with more than one oxidation state are vulnerable to this process in reducing (low oxygen) conditions, especially Fe and Mn oxides. But it is important to note that several heavy metals, including Cr, Cu, and As, also have more than one oxidation state, which can affect their behavior in the soil. For example, the stable form of Cr in the soil is Cr(III), which is nontoxic and acts as an essential trace element. The oxidized hexavalent form, Cr(VI), is much more mobile and more toxic to humans, but fortunately is not the common form in soils, although it can be a significant water contaminant (Naja and Volesky, 2009).

Reducing conditions affect the stability of some important soil constituents, especially those containing oxygen such as nitrates and sulfates. From the perspective of heavy metals, several changes are important. First, Fe and Mn hydrous oxides break down and the reduced forms of these elements disperse in the soil solution. Any heavy metals that had been coprecipitated or occluded in the hydrous oxides as they developed will be returned to the soil solution, under conditions that will favor a larger proportion remaining in solution. Second, some heavy metals may be reduced to their lower oxidation states (e.g., Cu(II) to Cu(I)), promoting their release from compounds into the soil solution. Third, in strongly reducing soils, the reduction of sulfates noted earlier also promotes the precipitation of sulfides of many of the heavy metal elements, including Cd, Pb, and Zn, thus removing the metals from solution. Finally, arsenic cannot only be reduced to its lower oxidation state, from As(V) to As(III), which is more soluble, but can also be subject to the process of methylation, in which microbial action leads to one or more methyl groups being attached to As(III) oxyanions (of which there are also several main forms) to produce complex methylated arsenic compounds, which are soluble and can be leached down the profile into groundwater (Naja and Volesky, 2009; Huang, 2008; Kabata-Pendias and Mukherjee, 2007; Alloway, 2001; Kabata-Pendias and Pendias, 2001; O'Neill, 1995).

Clearly the interaction of pH and redox potential can produce complex outcomes for the various metal elements, depending on the chemical nature of the element and the presence or absence of various soil constituents, including a thriving microbial system. This complexity has major implications for managing soils with significant concentrations of one or more of the heavy metals.

The following is a very brief overview of each of the eight heavy metals of interest in this chapter, their natural and anthropogenic sources, and their broad behavior in soil. All have demonstrated biological functions except cadmium, mercury, and lead.

3.2.1 Arsenic (As)

Arsenic is present in many metalliferous ores, especially Cu and Pb ore bodies and commonly sulfide ores, so that mining and smelting of many ores can release As into the surrounding environment. It can also be released from burning fossil fuels, especially coal. Arsenic has been, and to a lesser extent still is, used in pesticides (animal dips, orchard sprays, etc.) and is a major component of one common form of timber preservative (copper chromium arsenate, CCA) (Table 3.2). It is also widely used in the electronics industry, especially in semiconductor production (O'Neill, 1995). In soil, it occurs predominantly in the arsenate (As(V)) form, mainly as complex oxyanions that are adsorbed especially to Fe and Mn hydrous oxides, but also to clays and humus. Available As is usually very low, but the complexity of transfers between different forms, including the reduced As(III) form, makes it difficult to predict the behavior of arsenic in different soils and under different conditions (McLaren et al., 2006). It seems to be mobilized by the breakdown of Fe and Mn oxides under reducing conditions; mobility has also been linked to the role of complexation with dissolved organic matter (Garcia-Sanchez et al., 2010). Arsenic has a similar chemistry to phosphorus and in the soil there is evidence they compete for sorption sites, especially on hydrous oxides (McLaren et al., 2006). It has also been suggested that application of phosphate fertilizers might reduce As bioavailability, but there is other evidence that indicates the opposite might occur (Kabata-Pendias and Pendias, 2001). Finally, As is subject to the methylation process (described earlier), which converts a proportion of the soil As into a soluble, mobile, and plant-available form.

3.2.2 Cadmium (Cd)

Cadmium occurs in low quantities in a wide range of rock types and is mined for use in batteries, pigments, etc. (Table 3.2). The main source of environmental Cd is probably from smelting ores of metals such as Zn (Baird and Cann, 2008). It is also a contaminant of some phosphatic fertilizers due to high natural concentrations of Cd in certain rock phosphates. The chemistry of Cd is closely related to that of Zn, but it is more mobile in acid conditions and shows stronger affinity for sulfur (Kabata-Pendias and Pendias, 2001). It can form a variety of complex inorganic cations in solution depending on the concentrations of other soluble constituents, but usually the soil solution Cd tends to be in the simple Cd^{2+} form. It also forms complexes with various organic acids. Broadly, Cd becomes more available in acid soils as desorption takes place, but in reducing conditions the metal precipitates as a sulfide and so becomes less available. Maximum bioavailability of Cd is between about pH 4.5 and 5.5 (Kabata-Pendias and Pendias, 2001). Because it is very similar to Zn in its characteristics, Cd is easily taken up by plants even though it has no identified role in plant metabolism. This has important implications for human health as Cd is recognized as one of the most toxic of the heavy metals (Kabata-Pendias and Mukherjee, 2007).

3.2.3 CHROMIUM (Cr)

The soil concentration of Cr generally reflects the local parent material, with high to extremely high levels in soils derived from ultramafic and mafic rock types (i.e., rocks with high magnesium and iron content), less in soils with sedimentary and acid igneous rock parent materials, and low in soils from calcareous parent material (Alloway, 1995). Atmospheric input to soil can come from metallurgical industries, coal burning, brick production, and iron and steel production. Input can also come from liquid and solid wastes from tanneries and use of sewage sludge as a soil conditioner; Cr is also an ingredient in CCA-treated timber (Kabata-Pendias and Pendias, 2001). With two principal oxidation states (Cr(III) and Cr(VI)), the metal forms many inorganic complexes, both cationic and anionic, and organic complexes in the soil. The Cr(III) state is more stable in the soil than the Cr(VI) oxidation state and consequently dominates in most soils. Manganese compounds can oxidize Cr(III) to Cr(VI), but this is normally more than countered by the ability of soil organic matter to reduce Cr(VI) to Cr(III) (Kabata-Pendias and Pendias, 2001). Chromium has low availability under normal conditions, with precipitation, adsorption, and fixation processes taking most Cr(III) out of solution in soils with pH 5.5 or above. Even below that pH level, only a small proportion of Cr is present in the soil solution (Kabata-Pendias and Pendias, 2001). The Cr(VI) form is more soluble and mobile as well as being more toxic to plants, animals, and humans. However, when added to soils through human actions, Cr(VI) is usually rapidly reduced to the Cr(III) form by the action of soil organic matter (Kožuh et al., 2000).

3.2.4 COPPER (Cu)

Copper is widely used in industry (Table 3.2), so mining and smelting are important sources of environmental copper buildup, as are metal-using industries. However, copper is also widely used in agriculture as a pesticide (especially fungicide sprays in orchards and vineyards) and as a feed additive, and in timber treatment (CCA). It is present in sewage sludge and animal manures (if Cu feed additives are used), which may be used as soil conditioners (Kabata-Pendias and Pendias, 2001). The metal occurs in solution as Cu(II) in aerated soils and shows a strong affinity for organic matter and hydrous oxides of Fe and Mn. It is also adsorbed onto clays and subject to some degree of fixation by silicate minerals. Copper complexes with certain organic structures are called chelates. Chelates have a ring structure with the metal ion held by two or more separate bonds. They are soluble, promoting mobility in the soil, but can also release the metals for plant uptake, so influence bioavailability (Kabata-Pendias and Pendias, 2001). Copper is one of the least mobile of the heavy metals in soil (Kabata-Pendias and Pendias, 2001) and therefore shows marked accumulation in the topsoil where anthropogenic input is important. On the other hand, Cu is well represented in the soil solution in a variety of soils, especially as soluble chelated Cu (Kabata-Pendias and Pendias, 2001). In high concentrations, such as in contaminated soils, Cu can seriously affect plant growth and the soil biota and damage the long-term health of the soil system (Komárek et al., 2008).

3.2.5 LEAD (Pb)

Natural background levels of lead in soil reflect the particular parent material of the soil, with acid igneous and sedimentary rocks usually higher in natural Pb than either the ultramafic or calcareous rock types. However, such has been the extent of Pb use by humans that most topsoils in inhabited parts of the globe will have been enriched to some extent by Pb, most often from the use of leaded petrol/gasoline. Mining and smelting of Pb ore can result in contamination of the surrounding environment by Pb but also various other heavy metals; the length of mining activities in some countries has left a significant legacy of Pb-contaminated land for local communities to deal with (e.g., see the cases reported by Taylor et al., 2010; Pruvot et al., 2006). Apart from its previously widespread use as a fuel additive, the metal has also been used in paints, pesticides, and glazes and is still widely employed in Pb–acid batteries (Table 3.2). Once in the soil, Pb persists as it is readily adsorbed by organic matter, Fe and Mn oxides, and clays and is incorporated into carbonate and phosphate compounds (Kabata-Pendias and Mukherjee, 2007). Consequently there is usually a low concentration of Pb in the soil solution, although this can be increased in acidic conditions. Pb is considered to be the least mobile of all the heavy metals and the particular affinity of Pb for organic matter in the soil is one reason cited for the accumulation of the metal in topsoils in polluted environments (Kabata-Pendias and Pendias, 2001).

3.2.6 MERCURY (Hg)

Mercury has a strong affinity for organic matter; hence parent materials with high organic content, such as shales and similar sedimentary deposits and coal, often have high Hg content compared to most other rock types. However, Hg also accumulates naturally in topsoils through atmospheric deposition due to volatilization of elemental Hg by soils and plants and its emission by volcanic activity. Natural sources have been supplemented by various anthropogenic sources, especially mining and smelting of Zn and Cu, coal burning, and industrial processes such as the production of chlorine and caustic soda (chlor-alkali plants that use Hg as a key part of the production process) (Table 3.2). Locally, in developing countries, it is still used in gold mining. In recent years the rise in popularity of long-life fluorescent bulbs has caused an increase in the disposal of Hg in landfills (Baird and Cann, 2008). As with Pb, the extent of Hg pollution around the world is such that it is difficult to establish original background concentrations in many soils (Kabata-Pendias and Mukherjee, 2007). The form of Hg in the soil varies according to pH, redox potential, organic matter content and the presence of other ions such as chlorine and sulfur. The simple cation form seems to dominate in typical soils, forming complexes with a variety of organic and inorganic ligands or being adsorbed by hydrous oxides (Kabata-Pendias and Pendias, 2001). In strongly reducing soils Hg precipitates out as sulfides. Soil solution concentration of Hg is normally low, but a small proportion of soil Hg may be methylated by soil microorganisms to produce soluble and mobile organic forms, which plants are able to take up and can also be leached from soils. In Hg-polluted marine systems, methyl Hg has been a major form absorbed by fish, with subsequent mercury poisoning of local people consuming the fish (e.g., Minimata disease in Japan) (Naja and Volesky, 2009). Volatilization of a proportion of Hg from soil can also occur, encouraged by high microbial activity in moist environments (Kabata-Pendias and Mukherjee, 2007). Overall, Hg accumulates in topsoils due to the affinity with organic matter, with comparatively low mobility within the soil system unless methylated (Resi et al., 2009).

3.2.7 NICKEL (Ni)

Nickel is similar to chromium in its parent material characteristics. It is strongly associated with ultramafic rocks and much less so with acidic igneous or calcareous materials. Therefore natural soil levels usually reflect local geology and soil-forming processes (Kabata-Pendias and Pendias, 2001).

Human activities have greatly increased Ni input to soils through atmospheric deposition associated with mining and smelting, oil and coal combustion, and iron and steel production (Table 3.2). Nickel may also be present in sewage sludge applied to agricultural land and in various forms of solid waste (including Ni–Cd batteries). In the soil nickel is usually found in its simple Ni(II) form under the normal range of soil conditions, although it can form complex ions and has several oxidation states. It is adsorbed onto, and coprecipitates with, Fe and Mn hydrous oxides and can be incorporated into certain types of clays; however, in topsoils, Ni is mainly found in association with organic matter (Kabata-Pendias and Mukherjee, 2007). Although most of the organically bound Ni is immobile, a proportion is incorporated into soluble organic chelates, which promote mobility and can enhance plant uptake. There is also some evidence that soil organic matter can remobilize Ni from inorganic compounds, particularly in more acidic environments (Kabata-Pendias and Mukherjee, 2007). Overall, Ni seems to be more mobile than many heavy metals in moving between the various components of the soil system (hydrous oxides, organic matter, clays, etc.). Its concentration in solution is usually low, but as with many of the metals, increases with increasing soil acidity.

3.2.8   ZINC (Zn)

Zinc is found in most rock types, with rather higher levels in igneous rocks and clay-rich sedimentary rocks compared with limestones and sandstones (Kabata-Pendias and Pendias, 2001). Natural sources have been supplemented by human activities, with atmospheric inputs from mining and smelting (nonferrous ores), coal and oil burning, metal-processing industries, fertilizers, and pesticides, and other inputs from the use of sewage sludge on agricultural land (Table 3.2). The factors affecting Zn in the soil are similar to Cu, but Zn is one of the more soluble and mobile of the heavy metals, especially in acid soils (Kabata-Pendias and Pendias, 2001). It can be found in simple or complex cationic and anionic forms depending on soil pH, the presence of other ions, and the redox characteristics of the soil. Zinc forms precipitates with a wide variety of ions and this increases as the pH of the soil solution rises. Clays are particularly important as adsorption sites for Zn, as is organic matter, and to a lesser extent Fe and Mn hydrous oxides (Kabata-Pendias and Pendias, 2001).

3.3   HUMAN HEALTH AND HEAVY METALS

3.3.1   PATHWAYS AND FORMS OF EXPOSURE

Unlike water and air, there are several potential exposure pathways by which people can be exposed to soil-based heavy metals, which complicates investigations of the links between soil contamination and specific health outcomes. Moreover, the relative importance of the various exposure pathways varies according to factors such as the environmental setting (e.g., the physical, biological, and chemical characteristics of the soil, local climate, topographical position, etc.), the particular metals in the soil, the current land use, and the characteristics of the local population. There are three main pathways: ingestion, inhalation, and dermal contact, and each can be broken down into more specific pathways (based on Kibble and Saunders, 2001):

   Ingestion of
      Contaminated soil
      Dust containing contaminated soil particles
      Food grown in contaminated soil
      Water affected by contaminated soil
   Inhalation of
      Contaminated soil particles
      Volatile chemicals from contaminated soil

Dermal contact with
    Contaminated soil
    Dust containing contaminated soil particles
    Water affected by contaminated soil.

The main pathways are briefly described in the following subsections.

3.3.1.1 Direct Ingestion of Soil

It is commonly believed that this is the most important pathway for human exposure, probably outweighing all other forms of exposure combined (Kibble and Saunders, 2001), although other specific pathways may assume a degree of importance in certain situations.

Deliberate ingestion of soil (geophagia or soil pica) tends to be associated with very young children playing outside their homesand experimenting with tastes and textures of different objects and substances. With adults, soil ingestion may be due to accidental intake, such as eating vegetables that have not had all the soil removed. However, in some parts of the world there may be cultural reasons for soil consumption, while it has also been observed in some communities subject to extreme hunger (Henry and Cring, Chapter 8, this volume; Kibble and Saunders, 2001; Oliver, 1997).

Rates of ingestion can vary according to age and propensity for soil pica. Young children may ingest up to 200 mg soil/day; adults, less than 100 mg/day. However, for young children with strong soil pica tendencies, rates between 1 and 10 g soil/day have been reported, and in extreme cases as high as 60 g/day (Kibble and Saunders, 2001). As young children are usually considered to be the most sensitive to heavy metal contaminants, the potential for such high rates of ingestion is of concern. On the other hand, as Kibble and Saunders (2001) note, intake rates do not equate with uptake rates by the body, and in fact it is very difficult to assess the actual uptake of metals through mechanisms such as soil ingestion. Not only is it difficult to estimate the amount of soil ingested by an individual, it is also difficult to estimate the degree and speed of uptake of soil-bound metals compared with metals freely available in solution. In addition, age again seems to play a part. For example, children absorb Pb through the gastrointestinal tract up to five times more efficiently than adults (Kibble and Saunders, 2001).

3.3.1.2 Inhaling Resuspended Particles

Working with soil during cultivation, earthworks, etc., releases particles into the air that may be inhaled by workers and any other people close by. Through this process heavy metals that may have been in the soil for many years can be reintroduced into the atmosphere and become a risk for humans, emphasizing the ongoing legacy of heavy metal pollution. In many countries Pb is no longer used as an additive in petrol/gasoline because of its health effects on people. Yet the Pb that was released into the environment over many years of use and accumulated in topsoils is still there. Resuspension allows some of that Pb to enter the atmosphere again to be redistributed and to pose a renewed health risk to people.

Inhaled particulate matter enters the respiratory system, with very small particles lodging in the bronchial membranes in the lung. Here there is a chance that metals at or near the surface of the particles can be absorbed into the blood system. Compared with ingestion, inhalation of resuspended particles is unlikely to be a significant source of heavy metal uptake (Kibble and Saunders, 2001), but it may be an important form of exposure over long periods of time. Inhalation may be an important pathway for As and Pb uptake in particular (Oliver, 1997).

3.3.1.3 Drinking Water

Under certain conditions many of the heavy metals can move from the soil and into either groundwater or surface water. In some cases this can lead to the contamination of drinking water supplies. Changes in soil conditions such as waterlogging, which produces reducing conditions, and increased acidity can promote mobility in many of the heavy metals, as will the action of certain microorganisms able to methylate metals such as As and Hg. The presence of organic chelates that take up metal ions and hold them in more mobile and available forms in the soil solution will be important in soils with higher organic matter content under moderately acid conditions. Processes of these kinds can lead to the leaching of a variety of heavy metals from contaminated soils and locations such as landfills into surface-water bodies or groundwater.

Arsenic-enriched drinking water is a major problem around the world, especially in Bangladesh, but also parts of India, Chile, Argentina, and a number of countries in south and east Asia (Baird and Cann, 2008). In most situations the problem is a natural one, with high As levels in the local geology leading to high concentrations in groundwater. The provision of wells in countries such as Bangladesh has allowed millions of people greater access to underground water supplies in an attempt to reduce waterborne diseases caused by microbial contamination of surface-water supplies. Unfortunately, about half the wells on the Bengal Delta were supplying water with very high As levels, leading to health problems for local people; many have developed skin lesions and the rates of lung cancer have risen markedly (Baird and Cann, 2008). There have been suggestions that the use of groundwater for irrigation may have contributed to the problem. According to this view, following depletion of the groundwater for irrigation the subsequent recharge of the aquifer during the monsoon season may increase the supply of dissolved carbon from the soil surface to the groundwater zone. This might then have promoted the reduction of Fe(III) compounds, a major soil storage location for As, thereby releasing As into the soil solution and subsequently into the groundwater (Baird and Cann, 2008).

There is also growing evidence that anthropogenic As in soil, from activities such as sheep dips or timber treatment using CCA, can also move through the soil profile and enter local groundwater (McLaren et al., 2006).

3.3.1.4   Indirect Ingestion

Indirect ingestion refers to metals taken into the body through the consumption of food or water containing elevated levels of the metals. For example, the use of As-enriched water for irrigation in Bangladesh and West Bengal seems to have increased As intake in local communities through the consumption of rice and vegetables grown in the irrigated fields (Correll et al., 2006).

All plants need a range of trace elements and of the eight metals being discussed here, Zn, Ni, and Cu are essential to healthy plant growth. These elements will be taken up by plants to meet their physiological needs, but plants can regulate their uptake so that high soil concentrations do not usually lead to high concentrations in the plants. The other metals, Cd, Cr, Pb, Hg, and As, do not appear to have a metabolic role in plants, although Cr is required in human metabolism (Kabata-Pendias and Mukherjee, 2007). Some of these elements can be taken up relatively easily by plants, especially Cd, and to a lesser extent As. The uptake of Cd in food crops is of particular concern as it accumulates in the human body and has been linked to a number of health problems, including renal complaints. The other metals are usually only taken up in small quantities and tend to be sequestered in the roots, usually by the formation of complex organic or inorganic compounds. However, in soils with high concentrations of metals such as Pb, some crop plants can show a significant rise in metal content in both leaves and roots (Kabata-Pendias and Mukherjee, 2007).

A recent study of urban horticulture in northern Nigeria examined heavy metal content of lettuce and amaranthus crops grown on land irrigated by domestic and industrial wastewaters and found significant accumulation of a number of metals, including Pb, Cr, Cu, Ni, and Zn in those plants (Agbenin et al., 2009). A study of heavy metal accumulation in leafy vegetables grown near two Australian smelters (one Pb–Zn, the second Cu) similarly found elevated levels of Cd and Pb near the first smelter and elevated Cu levels near the second (Kachenko and Singh, 2006), to levels above Australian Food Standard guidelines.

Indirect ingestion of food plants is generally seen as secondary to direct ingestion of soil, especially for metals such as Pb. However, it can be a significant source for Cd, and in the case of As it can be an important additional source when combined with drinking water intake (Kabata-Pendias and Mukherjee, 2007).

3.3.1.5 Dermal Contact with Soil

Absorption through the skin (dermal absorption) tends to favor the more volatile organic compounds, so it is thought to be very low for metals in their inorganic forms. However, Cr(VI), the oxidized and more toxic form of chromium, does produce skin problems on contact; fortunately it is not the usual form in normal soil conditions. Mercury (inorganic) and Ni can also cause skin contact dermatitis, but absorption *per se* does not seem to be an issue (Kabata-Pendias and Mukherjee, 2007).

3.3.2 Main Health Issues

A number of the heavy metal elements, for example, Zn and Cu, are crucial to human health and are necessary components of a healthy diet. Most seem to work by facilitating biochemical processes, usually as part of or in association with specific enzymes, where they act as electron acceptors or donors within specific biochemical reactions. As such, they are only required in trace amounts and the lack of one or more trace elements in the diet can cause health problems of various kinds. Once diagnosed, the deficiencies can be dealt with using dietary supplements.

Of concern in this chapter are the health consequences of soils contaminated by high concentrations of heavy metals due to human activities. In general, those metals that perform a metabolic function in the human body (such as Cu, Zn, and Ni) are regulated to avoid excess supply, so toxicity is not usually a problem, even in highly polluted environments. Toxicity problems usually involve those elements that have no apparent biological function, with Cd, Pb, Hg, and As being the most important in this group. There is still much to learn about the specific processes by which the various metals affect human health, especially for some of the less commonly studied metals, but Table 3.3 summarizes the broad health concerns associated with the main heavy metals covered in this chapter.

TABLE 3.3
Main Health Symptoms Associated with Excess Amounts of Specific Elements in Humans

Element	Main Symptoms	Essential/Nonessential for Human Health/ADSTR Ranking[a]
Arsenic	Variable disorders including nervous system disorders, liver and kidney failure, intestine tract distress, anemia, skin cancer	Nonessential/ADSTR: No. 3
Cadmium	Cardiomyopathy, liver and kidney damage, gastroenteritis, pneumonitis, osteomalacia, cancer	Nonessential/ADSTR: No. 6
Chromium	Lesions in skin, intestinal mucosa, pulmonary edema, lung cancer	Essential
Copper	Effects rare, but liver and kidney damage in infants possible if prolonged exposure	Essential
Lead	Nervous system disorders, brain damage, hematologic effects, kidney disease, intestine tract distress, hypertension	Nonessential/ADSTR: No. 1
Mercury	Nervous and gastric system disorders, kidney and pulmonary damage, a potent teratogen	Nonessential/ADSTR: No. 2
Nickel	Gastric, liver, and kidney defects; neurological effects; emphysema; lung cancer	Nonessential
Zinc	Anemia, tissue lesions	Essential

Source: Adapted from Kabata-Pendias, A. and Mukherjee, A., *Trace Elements from Soil to Human*, Springer-Verlag, Berlin, Germany, 2007.

Note: These symptoms relate to excess levels due to any form of exposure; they are not intended to indicate the potential outcome of soil-based exposure.

[a] ADSTR (U.S. Agency for Toxic Substances and Disease Registry): Ranking of most toxic substances in environmental waste.

Toxicologists categorize chemicals in a number of ways with respect to their potential for affecting human health based on dose–response curves. One common distinction is between threshold and nonthreshold models. Threshold models, which would apply to the trace elements, assume there is a lower threshold of concentration of the element that poses no health risk to humans; in fact, in the case of the trace elements, low doses confer health benefits. However, above the threshold, adverse effects can be observed and related to the level of exposure or dose of the element in question. Nonthreshold models are based on the assumption that even the lowest doses carry some degree of health risk. Using this basic distinction, arsenic can be viewed as a nonthreshold contaminant, while the other seven metals in our list would be threshold contaminants (note this includes Pb, Cd, and Hg, which are not considered to have specific biological functions).

In terms of specific effects on human health, Hu (2002, p. 66) makes the following points:

The toxicity of metals most commonly involves the brain and the kidney, but other manifestations occur, and some metals, such as arsenic, are clearly capable of causing cancer. An individual with metals toxicity, even if high dose and acute, typically has very general symptoms, such as weakness or headache. This makes the diagnosis of metals toxicity in a clinical setting very difficult unless a clinician has the knowledge and training to suspect the diagnosis and is able to order the correct diagnostic test. Chronic exposure to metals at a high enough level to cause chronic toxicity effects (such as hypertension in individuals exposed to lead and renal toxicity in individuals exposed to cadmium) can also occur in individuals who have no symptoms. Much about metals toxicity, such as the genetic factors that may render some individuals especially vulnerable to metals toxicity, remains a subject of intense investigation. It is possible that low-level metals exposure contributes much more towards the causation of chronic disease and impaired functioning than previously thought.

3.3.3 SOIL GUIDELINE VALUES FOR HEAVY METALS

Environmental standards are familiar management tools. They indicate the level of a particular substance within a given medium that is considered safe in terms of human or ecological health, based on current understanding of the processes involved. Drinking water standards are a good example. With soil, setting standards is much more difficult. Unlike water or air, soil is a highly heterogeneous medium that also varies markedly from place to place. Within it, contaminants such as heavy metals can be more available, less available, or totally unavailable, depending on their form in the soil fabric. Therefore, it can be very difficult to equate an amount of a metal in the soil with a particular level of risk to human health or to wider ecological health. In addition, the potential exposure pathways are numerous, and will vary according to factors such as the nature of the local land use and the size and characteristics of the local population. Also, as noted in the previous section, the actual link with specific health consequences is not well understood for many of the metals. For these reasons, most countries refer not to soil standards but to soil guidelines values, intervention values, screening values, investigation levels, or a similar form of wording (Nathanail and Earl, 2001). Their purpose is to suggest when a possible health risk may exist at a specific locality that might warrant further investigation rather than to be used as the basis for specific action. Soil guideline values (or equivalent) for the eight heavy metals addressed in this chapter for a selection of countries are shown in Table 3.4.

Many approaches to establishing soil guideline values based on human health utilize exposure scenarios. Toxicologists estimate a daily intake for each exposure pathway, expressed as milligrams per kilogram body weight, as a level (for threshold substances) at or below which there should be no appreciable rise in risk of adverse health outcomes over the lifetime of a person, or (for nonthreshold substances) for which there is a minimal, acceptable level of risk. These estimates are based on complex calculations of the likely amount of a substance transferred from the soil into the human subject via the various exposure pathways (ingestion, inhalation, skin contact, etc.). These are then used to set levels for soil content of the contaminants for each exposure scenario. Hence, as soil ingestion by adults and children is likely to be more important in residential areas where some

TABLE 3.4
Soil Guideline Values (mg/kg dry weight soil) for Selected Countries

	United States[a]		Australia[b]			New Zealand[c]		The Netherlands[d]	Canada[e]		United Kingdom[f]	
	EPA RSL Res	EPA RSL Ind	HIL (A)	HIL (D)	EIL	HD Res	10% Prod Res	SIV	SQG-HH	SQG-EH	SGV resid	SGV alltm
As	0.39	1.6	100	400	20	45	20	55	12	17	32	43
Cd	70	800	20	80	3	230	3	12	14	10	10	1.8
Cr(VI)	0.29	5.6	100	400	1	1,500	460	380	0.4			
Cr(III)	120,000	1,500,000	120,000	480,000	400	nl	nl					
Cr total									220	64		
Cu	3,100	41,000	1,000	4,000	100	nl	nl	190	1,100	63		
Hg inorg	10	43	15	60	1	1,000	310	10	6.6	12	170	80
Hg methyl	7.8	100	10	40							11	8
Pb	400	800	300	1,200	600	500	210	530	140	300		
Ni	3,700–3,800	44,000–47,000	600	2,400	60			210		50	130	230
Zn	23,000	310,000	7,000	28,000	200			720		200		

Sources: United States: EPA, Regional Screening Levels, Pacific Southwest, Region 9, 2011. http://www.epa.gov/region9/superfund/prg/ (verified 22 February 2012).

Australia: Imray, P. and Langley, A., *Health Based Soil-Investigation Levels*, enHealth Council, Canberra, Australia, 2001. http://www.health.gov.au/internet/main/publishing.nsf/content/66E7D805C1C1AD69CA2573CC0013EA68/$File/env_soil.pdf (verified 22 February 2012).

New Zealand: Ministry for the Environment, *Draft Users' Guide: National Environmental Standard for Assessing and Managing Contaminants in Soil to Protect Human Health*, Ministry for the Environment, Wellington, New Zealand, 2011. http://www.mfe.govt.nz/publications/rma/draft-users-guide-nes-for-assessing-managing-contaminants-in-soil/ (verified 22 February 2012).

Canada: Canadian Council of Ministers for the Environment, *Canadian Soil Quality Guidelines for the Protection of Environmental and Human Health*, 1999 (separate fact sheet for each metal). http://ceqg-rcqe.ccme.ca/ (verified 22 February 2012).

The Netherlands: Dutch Environment Ministry, *Circular on Target Values and Intervention Values for Soil Remediation*, Dutch National Institute of Public Health and the Environment (RIVM), The Hague, The Netherlands, 2000, values taken from Appendix 3, Carlon (2007).

United Kingdom: Environment Agency, *Soil Guideline Values*, 2009 (separate technical notes for each metal or compound). http://www.environment-agency.gov.uk/research/planning/64015.aspx (verified 22 February 2012).

Note: nl, No limit.

[a] EPA Regional Screening Levels (RSL) 2009. RSL Res, Residential soil; RSL Ind, industrial soil.

[b] National Environmental Protection Measures (NEPM) 1999. HIL, health investigation levels; (A), high-density residential; (D), recreational; EIL, ecological investigation level.

[c] National Environmental Standards (soil contaminants and human health) (draft) 2011. HD Res, high-density residential; 10% Prod Res, 10% produce residential.

[d] Ministry of Housing, Spatial Planning and Environment (VROM) 2000. SIV, Soil intervention values for soil remediation.

[e] Canadian Soil Quality Guidelines (SQG) for the Protection of Environmental and Human Health 2001. SQG-HH, human health; SQG-EH, environmental health.

[f] Environment Agency, land contamination 2009. SGV, Soils guideline value. SGV resid, residential; SGV allmt, allotments (urban gardens).

home-grown produce is consumed by the residents, and since ingested soil is the most effective way to transfer comparatively large quantities of heavy metals into the human body, the soil guideline values for heavy metals in those soils are usually much lower than for other land uses. Table 3.4 shows the Australian and New Zealand guidelines use exposure scenarios of this kind (not all are included in the table due to space constraints), while the U.K. approach uses a simpler dichotomy between residential areas and allotments (urban or periurban gardens used by some families for growing vegetables for domestic consumption). The US EPA Regional Screening Levels are the most complex, with a number of scenarios including residential, industrial, and recreational soils (residential and industrial levels are shown in this table). Within each category specific screening levels for ingestion, inhalation, and dermal contact are given in relation to carcinogenic and non-carcinogenic health risks respectively. Levels for other land-use scenarios can be calculated from formulae provided by the US EPA.

Levels recommended to protect human health are not intended to be used as wider environmental standards. For example, soil guideline values for Cu in Table 3.4 can be very high (3000–4000 mg/kg) and in some cases there is no recommended upper limit. This reflects the assessment of the specific risk to human health from Cu through ingestion, inhalation, and skin contact; from a human health perspective Cu is not toxic. Yet soil concentrations of Cu above about 2000 mg/kg would usually indicate serious industrial pollution (Alloway, 1995) and the soil would almost certainly show signs of significant damage to biological processes from the elevated Cu levels. As a consequence, ecological guidelines usually set much lower soil Cu levels. In Table 3.4, the Australian and Canadian guidelines include ecological or environmental health values as well as human health values. For Canada, the environmental Soil Quality Value (SQV-EH) for Cu is 63 mg/kg, contrasting with 1100 mg/kg for the human health value (SQV-HH). This provides a useful reminder that human health is also affected by wider environmental and social determinants, and loss of soil productivity through contamination with heavy metals may be as important in the long term as direct toxicity effects.

The values for As levels in residential soils in Table 3.4 vary from 0.39 mg/kg (US EPA) to 100 mg/kg (Australia), and similar levels of variability can be seen in some of the other metals in the table. This must cause concern, and no little confusion, among potential users of the information. Teaf et al. (2010) examined guidelines for soil As levels at state and federal levels in the United States (as well as a number of other countries) and found that levels within the United States ranged from 0.039 mg/kg in Wisconsin to 40 mg/kg in Montana. The variations reflect the methods used to determine appropriate levels. For example, in Wisconsin the suggested level is based on a risk of cancer using a 10^{-7} risk level, compared with the more usual 10^{-6} level used by the US EPA. Many other states simply take the natural background level of As (which varies between about 7 and 40 mg/kg) and in effect require soil As to stay at that background level (Teaf et al., 2010). A recent review of EU national practices in setting soil screening values found similar variability in approaches and suggested levels among member states (Carlon, 2007). Setting standards or guidelines for soil heavy metals is fraught with uncertainty through lack of knowledge about the behavior of metals in the soil, the factors that influence their uptake by humans, and the effects the metals actually have on human health. Moreover, different risk assessment approaches may well suggest different guideline values, so users need to be aware of how the values were derived and take note of any "health warnings" attached to the guideline values.

3.4 CASE STUDIES

3.4.1 Shipham, United Kingdom

In the late 1970s a national geochemical survey in the United Kingdom drew attention to the very high levels of cadmium, lead, and zinc in soils in and around the small mining town of Shipham,

Somerset, in the west of England. Postwar expansion of the town had seen housing developed on land previously mined for zinc. Concentrations of cadmium in garden soils reached as high as 360 mg/kg (median 91 mg/kg) compared with a median of <1 mg/kg for agricultural soils in England and Wales (Morgan and Simms, 1988). Maximum levels for Zn and Pb were also very high at 37,200 and 6,540 mg/kg, respectively (Alloway, 2001). In the Jinzu Valley in Japan, soil Cd levels of about 3 mg/kg were associated with problems with skeletal development in newborn babies and osteoporosis in adult women. Hence the soil Cd levels in Shipham caused immediate concern and triggered a series of studies into the sources, levels, and availability of Cd in the soil, its uptake by vegetables grown and consumed by local people, and any evidence of Cd impacts on the health of the local community.

Despite the very high Cd levels in the soil and evidence of Cd uptake by various vegetable crops that contributed to the local diet, studies at the time and subsequently have produced no conclusive evidence of health effects on the population of Shipham (Elliott et al., 2000; Strehlow and Barltrop, 1988). A number of reasons have been suggested for this rather surprising, counterintuitive finding.

First, despite the very high levels of Cd in the soil, the bioavailability of the metal was comparatively low due to the soil pH ranging between 7.0 and 7.8 (Thornton, 2010). Morgan and Simms (1988) quote a soil solution level of 0.158 mg/L for a Shipham soil, compared with the figure for Jinzu Valley of 0.119 mg/L. Second, the uptake of Cd by plants, while elevated, was probably lower than it might have been due to interference by other elements such as Ca, which is abundant in the weakly alkaline soils.

Third, the main route for Cd ingestion by people was through the locally grown vegetables, and the average concentration of Cd in those vegetables (0.25 mg/kg dry matter) was almost 17 times the national average (Alloway, 2001). However, local people had a varied diet, so their weekly intake of Cd (200 µg) was only moderately elevated above the national average of 140 µg/week, and still well below the WHO upper threshold of tolerable intake of 450–500 µg/week (Thornton, 2010). In contrast, the weekly dietary intake of Cd in contaminated areas in Japan was estimated to be in the region of 1400 µg.

Fourth, the lack of evidence of health problems does not mean there have been no problems or that problems will not arise in future. For example, Elliott et al. (2000) looked at mortality figures and incidence of cancers; other, less extreme health outcomes were not investigated. The original health studies reported by Strehlow and Barltrop (1988) asked volunteers about their health and took blood and hair samples. Longer-term residents were underrepresented in the survey and the samples taken could only provide certain indicators for health effects. Moreover, there are a number of factors that can affect Cd levels in the body, including smoking, gender, age, and nutrient interactions, and the relatively small sample size (around 500 residents) means it is difficult to reach definitive conclusions.

One key message from this case study is the complexity of the links between soil heavy metal content and human health outcomes. Despite very high soil levels of a potentially toxic element, the local population seem to show very few, if any, signs of health problems due largely to the wider soil environment limiting the movement of the element into the food chain. It is a cautionary tale; we cannot always assume high levels of heavy metals in the soils will inevitably cause adverse outcomes for human health. On the other hand, they should certainly alert us to investigate the possibility and we would be wise to maintain a watching brief, even if health problems are not immediately evident.

3.4.2 OMAHA LEAD SITE

For 127 years, until its closure in 1997, the American Smelting and Refining Company operated a lead refinery in the city of Omaha, Nebraska, United States, on the west bank of the Missouri River. The refinery itself occupied 23 acres (9.3 ha) (Dearwent et al., 2006), but was the main source of soil lead contamination over a significantly larger area. In 1999 the city council contacted the US EPA because a significant proportion of children under the age of 6 years living in parts of the city close to the smelter site had elevated blood lead levels; soil lead concentrations seemed to be the major source of the lead ingested by the children. Following initial assessment of environmental and health data, an initial area of about 9000 acres (3642 ha), which included residential properties, child-care centers, schools, and recreational and amenity areas, was given National Priority List status in 2003 by the US EPA under the Superfund law to allow further investigations and environmental remediation to be carried out under federal funding. The Agency for Toxic Substances and Disease Registry (ATSDR) conducted a public health assessment for the EPA on the Omaha Lead Site, confirming the problem of elevated blood lead levels in a comparatively large proportion of young children.

The final Omaha Lead Site has a proposed area of 17,291 acres (7,000 ha) and contains about 125,000 residents, of whom just over 14,000 are children under the age of 7 years. Initial work in 1999 targeted child-care centers with soil Pb levels above 400 mg/kg (the US EPA-established screening level), removing topsoil, replacing it with clean soil, and reestablishing a grass sward. They then extended the program to highly contaminated sites, initially defined as >2500 mg/kg soil Pb, then redefined in 2003 as >1200 mg/kg, and over the subsequent years lowered the threshold for remediation to 800 mg/kg, then 400 mg/kg (this last in 2009). By the end of 2009, 5,000 properties had had soil replaced and another 10,000 with soil Pb levels above the 400 mg/kg threshold were eligible for soil remediation. In parallel, the EPA and the city council were addressing the issue of lead-based paint to ensure no recontamination of soil can occur.

Clearly this has been and continues to be a massive environmental cleanup, indicating the extent of soil contamination across the city due to past industrial practices. Lead is classed by the ATSDR as the most important environmental contaminant and, as this example demonstrates, the major concern is for the health and well-being of young children, the group most vulnerable to the effects of lead poisoning. The Public Health Assessment for the Omaha Lead Site summarizes the health consequences:

> The literature suggests that children with blood lead levels of 10–20 μg/dL are at risk of having decreases in IQ of up to 11 points, and slightly impaired hearing and growth. Those children with levels from 20 μg/dL to 40 μg/dL could experience problems in metabolizing vitamin D, which is important in bone development. Children with levels greater than 40 μg/dL could experience anemia and other blood-related problems. Colic, kidney disease, and diseases of the brain have been observed in children with blood lead levels greater than 60 μg/dL. (ATSDR, 2005, p. 6)

Tests on about 9600 children up to the age of 6 years, residing in the initial Omaha Lead Site area between 2000 and 2003, showed 6.2% had blood lead levels (BLL) at or above 10 μg/dL, which is double the national average (and three times the Nebraska average). Of the 484 children with elevated levels, 320 had levels of 10–14 μg/dL, 159 had levels ranging between 15 and 44 μg/dL, 4 had levels between 45 and 69 μg/dL, and 1 was higher than 69 μg/dL (ATSDR, 2005). As the initial assessment covered about a third of the children in the 1–6 years age group, there would have been approximately 1600 children at risk of significant health effects due to soil lead contamination in that period (ATSDR, 2005). Since then the area of the Superfund site has almost doubled and includes 14,000 children in the vulnerable age group (EPA, 2010). An elevated BLL rate of 6.2% (which may be conservative, as an earlier assessment suggested a rate of over 9%) would suggest in any given year approximately 900 children are at risk of health effects, though this will decline as an increasing proportion of contaminated locations are treated. As of 2009, about one-third of affected properties had been treated. In the meantime, the cohort of young children continues to renew itself, with more young children being exposed to the risk of lead poisoning until all sites are treated, which may take at least another 10 years. With this in mind, a major public education program was launched in 2005 to raise community awareness of the problem and to indicate strategies for reducing the chances of children ingesting soil or contaminated house dust.

3.4.3 ELECTRONIC WASTE RECYCLING IN CHINA

Electronic technology is integral to modern economic growth and globalization has seen the rapid rise of a new industry, recycling electronic waste.

> ... e-waste refers to end-of-life electronic products, including televisions, monitors, computers, audio and stereo equipment, video cameras, telephones, fax and photocopy machines and printers, mobile phones, wireless devices, integrated circuits (chips), motherboards, cathode ray tubes (CRTs), and other peripheral items. (Frazzoli et al., 2011, p. 269)

In some countries recycling e-waste, which comprises up to 3% of global waste (Robinson, 2009), is well organized and managed responsibly. In others, an informal and ad hoc approach has resulted in poorly managed recycling that has had severe impacts on local environments, including soil resources, and most probably on local people. China is a major center for this industry, but significant amounts of recycling also takes place in several other countries including India, Pakistan, Vietnam, the Philippines, Malaysia, Nigeria, and Ghana (Robinson, 2009).

Around the small town of Guiyu, in Guangdong Province, about 250 km northeast of Hong Kong, is a series of small villages, each with e-waste recycling activities based on waste brought in from other parts of China and overseas, especially South Korea (Li et al., 2011). Local people process e-waste in numerous mid- and small-sized workshops, carrying out solder recovery from printed circuit boards, acid extraction of precious metals from various types of e-waste, larger-scale metal recovery by pyrometallurgy methods (incineration and smelting), and shredding and processing of scrap plastic (Li et al., 2011; Robinson, 2009). The industry has developed over the last two decades with little government supervision until recent years, with the result that land, water, and air have been affected by a range of contaminants including polychlorinated biphenyls (PCB) and related compounds, polycyclic aromatic hydrocarbons (PAH), and heavy metals. Although practices are slowly improving, there is still a significant legacy of abandoned sites, waste tips, and contaminated soils and water (Li et al., 2011).

Research into the environmental and human health effects of this industry in China has started to reveal the extent and severity of the problems. Li et al. (2011) measured heavy metal content in various locations around the Guiyu area and found clear evidence of soil contamination in an area previously used for open burning of waste, most probably including e-waste. A former acid extraction facility also had elevated levels, though to a lesser extent than the burn site (Table 3.5).

In another part of Guangdong Province, Luo et al. (2011) also found the soils of former e-waste incineration sites to have the highest heavy metal concentrations in the area. In addition, Cd levels in the soils of local paddy fields and vegetable gardens were also elevated, and this was reflected in the Cd content of edible vegetables from those sites, which exceeded Chinese standards for maximum allowable levels of Cd in food. Samples of polished rice from another e-waste processing area, Taizhou in Zeijiang Province, had a significantly higher concentration of Cd than rice from a nonpolluted area, reflecting the elevated Cd content of the paddy soils (Fu et al., 2008).

TABLE 3.5
Mean Heavy Metal Content (mg/kg) in Soils Contaminated by E-Waste Processing, Guiyu, China

	As	Cd	Cr	Cu	Hg	Ni	Pb	Zn
Burn site	52.1	10.02	320	12,700	0.19	1,100	480	3,500
Acid site	26.03	1.21	2,600	4,800	0.21	480	150	330
Local background	5.82	0.154	10.3	38.7	0.163	33.2	41.6	68

Source: Figures extracted from Li, J., Duan, H., and Shi, P., *Waste Management & Research*, 29, 727–738, 2011.

In terms of links to human health, Wang et al. (2009) observed very high levels of Cu and Pb, and to a lesser extent Cd, in the hair of Taizhou residents compared with a control group, and attributed this to contamination from the e-waste recycling activities. More worrying, Guo et al. (2010) found evidence of elevated Pb levels in human placentas in Guiyu, which they attributed to pollution from the e-waste processing activities. In neither study was the actual form of human exposure determined; there were probably several pathways, including direct inhalation of particulates from recycling processes. However, Guo et al. (2010) consider contaminated soils to be a likely contributor to Pb exposure through inhalation of resuspended particles, dietary uptake, and dermal contact.

The consequences of two decades of largely uncontrolled e-waste recycling, and the legacy it has left behind in parts of southeastern China, will take some time to unfold. The research studies described here show the complexities of trying to evaluate exposure pathways and levels of risk to human health. The example also emphasizes that such concerns take time to manifest themselves: the real health costs of e-waste recycling may not be evident for some time, and it may never be possible to link particular health problems to specific metals or to particular exposure pathways. However, the precautionary principle tells us not to allow the lack of definitive scientific evidence to delay taking sensible precautions when the consequences of not doing so could be severe. This would seem to be a sensible approach with regard to the general concerns over heavy metal contamination of soils and the possible implications for human health.

REFERENCES

Agbenin, J.O., M. Danko, and G. Weld. 2009. Soil and vegetable compositional relationships of eight potentially toxic metals in urban garden fields from northern Nigeria. *Journal of the Science of Food and Agriculture* 89: 49–54.

Alloway, B.J. 1995. Soil processes and the behaviour of heavy metals. *In* B.J. Alloway (ed.), *Heavy Metals in Soils*, 2nd edn, pp. 11–37. Blackie Academic & Professional, London.

Alloway, B.J. 2001. Soil pollution and land contamination. *In* R.M. Harrison (ed.), *Pollution: Causes, Effects and Control*, 4th edn, pp. 352–377. The Royal Society of Chemistry, Cambridge.

Antoniadis, V., C.D. Tsadilas, V. Samaras, and J. Sgouras. 2006. Availability of heavy metals applied to soil through sewage sludge. *In* M.N.V. Prasad, K.S. Sajwan, and R. Naidu (eds), *Trace Elements in the Environment: Biogeochemistry, Biotechnology, and Bioremediation*, pp. 39–61. CRC Press, Boca Raton, FL.

ATSDR. 2005. Public health assessment for Omaha lead, Omaha, Douglas County, Nebraska. EPA Facility ID: NESFN0703481 April 28 2005. U.S. Department of Health and Human Services, Agency for Toxic Substances and Disease Registry, Division of Health Assessment and Consultation, Atlanta, GA.

Baird, C. and M. Cann. 2008. *Environmental Chemistry*, 4th edn. W.H. Freeman, New York, NY.

Brevik, E.C. 2009. Soil, food security, and human health. *In* W. Verheye (ed.), *Soils, Plant Growth and Crop Production*. Encyclopedia of Life Support Systems (EOLSS), Developed under the Auspices of the UNESCO, EOLSS Publishers, Oxford. http://www.eolss.net (verified 22 February 2012).

Brooks, R.R. 1987. *Serpentine and Its Vegetation*. Croom Helm, London.

Carlon, C. (ed.). 2007. *Derivation Methods of Soil Screening Values in Europe. A Review and Evaluation of National Procedures Towards Harmonization*. European Commission, Joint Research Centre EUR 22805-EN, Ispra, Italy. http://ies.jrc.ec.europa.eu/uploads/fileadmin/Documentation/Reports/RWER/EUR_2006-2007/EUR22805-EN.pdf (verified 22 February 2012).

Correll, R., S.M. Imamul Huq, E. Smith, G. Owens, and R. Naidu. 2006. Dietary intake of arsenic from crops. *In* R. Naidu, E. Smith, G. Owens, P. Bhattacharya, and P. Nadebaum (eds), *Managing Arsenic in the Environment: From Soil to Human Health*, pp. 255–271. CSIRO, Collingwood, Victoria, Australia.

Dearwent, S., M.M. Mumtaz, G. Godfrey, T. Sinks, and H. Falk. 2006. Health effects of hazardous waste. *Annals of the New York Academy of Sciences* 1076: 439–448.

Dufuss, J.H. 2002. "Heavy metals"—A meaningless term? (IUPAC Report). *Pure and Applied Chemistry* 74: 793–807.

Elliott, P., R. Arnold, S. Cockings, N. Eaton, L. Jarup, J. Jones, M. Quinn, et al. 2000. Risk of mortality, cancer incidence, and stroke in a population potentially exposed to cadmium. *Occupational and Environmental Medicine* 57: 94–97.

EPA. 2010. Omaha lead. EPA ID# NESFN0703481. http://www.epa.gov/Region7/cleanup/npl_files/nesfn0703481.pdf (verified 22 February 2012).

European Commission. 2006. *Integrated Pollution Prevention and Control Reference Document on the Best Available Techniques for Waste Incineration.* http://www.ineris.fr/ippc/sites/default/files/files/wi_bref_0806.pdf (verified 22 February 2012).

Fergusson, J.E. 1990. *The Heavy Elements: Chemistry, Environmental Impact and Health Effects.* Pergamon Press, Oxford.

Frazzoli, C., A. Mantovani, and O.E. Orisakwe. 2011. Electronic waste and human health. *In* J.O. Nriagu (ed.), *Encyclopedia of Environmental Health*, pp. 269–281. Elsevier, Amsterdam, The Netherlands.

Fu, J., Q. Zhou, J. Liu, W. Liu, T. Wang, Q. Zhang, and G. Jiang. 2008. High levels of heavy metals in rice (*Oryza sativa* L.) from a typical e-waste recycling area in southeast China and its potential risk to human health. *Chemosphere* 71: 1269–1275.

Garcia-Sanchez, A., P. Alonso-Rojo, and F. Santos-Francés. 2010. Distribution and mobility of arsenic in soils of a mining area (Western Spain). *Science of the Total Environment* 408: 4194–4201.

Guo, Y., X. Huo, Y. Li, K. Wu, J. Liu, J. Huang, G. Zheng, et al. 2010. Monitoring of lead, cadmium, chromium and nickel in placenta from an e-waste recycling town in China. *Science of the Total Environment* 408: 3113–3117.

Hu, H. 2002. Human health and heavy metals exposure. *In* M. McCally (ed.), *Life Support: The Environment and Human Health*, pp. 65–81. MIT Press, Boston, MA.

Huang, P.M. 2008. Impacts of physicochemical–biological interactions on metal and metalloid transformations in soils: An overview. *In* A. Violante, P.M. Huang, and G.M. Gadd (eds), *Biophysico-Chemical Processes of Heavy Metals and Metalloids in Soil Environments*, pp. 3–52. Wiley, Hoboken, NJ.

Kabata-Pendias, A. and A. Mukherjee. 2007. *Trace Elements from Soil to Human.* Springer-Verlag, Berlin, Germany.

Kabata-Pendias, A. and H. Pendias. 2001. *Trace Elements in Soils and Plants*, 3rd edn. CRC Press, Boca Raton, FL.

Kachenko, A.G. and B. Singh. 2006. Heavy metals contamination in vegetables grown in urban and metal smelter contaminated sites in Australia. *Water, Air, and Soil Pollution* 169: 101–123.

Keddy, P.A. 2007. *Plants and Vegetation. Origins, Processes, Consequences.* Cambridge University Press, Cambridge.

Kibble, A.J. and P.J. Saunders. 2001. Contaminated land and the link to health. *In* R.E. Hester and R.H. Harrison (eds), *Assessment and Reclamation of Contaminated Land. Issues in Environmental Science and Technology*, pp. 65–84. Royal Society of Chemistry, Cambridge.

Komárek, M., J. Száková, M. Rohoškova, H. Javorská, V. Chrastný, and J. Balík. 2008. Copper contamination of vineyard soils from small wine producers: A case study from the Czech Republic. *Geoderma* 147: 16–22.

Kožuh, N., J. Štupar, and B. Gorenc. 2000. Reduction and oxidation processes of chromium in soils. *Environmental Science & Technology* 34: 112–119.

Leung, A.O.W., N. Duzgoren-Aydin, K.C. Cheung, and M.H. Wong. 2008. Heavy metals concentrations of surface dust from e-waste recycling and its human health implications in Southeast China. *Environmental Science and Technology* 42: 2674–2680.

Li, J., H. Duan, and P. Shi. 2011. Heavy metals contamination of surface soil in electronic waste dismantling area: Site investigation and source-apportionment analysis. *Waste Management & Research* 29: 727–738.

Luo, C., C. Liu, Y. Wang, X. Liu, F. Li, G. Zhang, and X. Li. 2011. Heavy metal contamination in soils and vegetables near an e-waste processing site, south China. *Journal of Hazardous Materials* 186: 481–490.

McLaren, R.G. and K.C. Cameron. 1996. *Soil Science*, 2nd edn. Oxford University Press, Melbourne, Australia.

McLaren, R.G., M. Megharaj, and R. Naidu. 2006. Fate of arsenic in the soil environment. *In* R. Naidu, E. Smith, G. Owens, P. Bhattacharya, and P. Nadebaum (eds), *Managing Arsenic in the Environment: From Soil to Human Health*, pp. 157–182. CSIRO, Collingwood, Victoria, Australia.

Morgan, H. and D.L. Simms. 1988. [The Shipham Report] Discussion and conclusions. *Science of the Total Environment* 75: 135–143 (Special issue).

Naja, G. and B. Volesky. 2009. Toxicity and sources of Pb, Cd, Hg, As and radionuclides in the environment. *In* L.K. Wang, J.P. Chen, Y.T. Hung, and N.K. Shammas (eds), *Heavy Metals in the Environment*, pp. 13–62. CRC Press, Boca Raton, FL.

Nathanail, C.P. and N. Earl. 2001. Human health risk assessment: Guideline values and magic numbers. *In* R.E. Hester and R.H. Harrison (eds), *Assessment and Reclamation of Contaminated Land. Issues in Environmental Science and Technology*, pp. 85–102. Royal Society of Chemistry, Cambridge.

Oliver, M.A. 1997. Soil and human health: A review. *European Journal of Soil Science* 48: 573–592.

O'Neill, P. 1995. Arsenic. *In* B.J. Alloway (ed.), *Heavy Metals in Soils*, 2nd edn, pp. 105–121. Blackie Academic & Professional, London.

Pierzynski, G.M., J.T. Sims, and G.F. Vance. 2005. *Soils and Environmental Quality*, 3rd edn. CRC Press, Boca Raton, FL.

Pruvot, C., F. Douay, F. Hervé, and C. Waterlot. 2006. Heavy metals in soil, crops and grass as a source of human exposure in the former mining areas. *Journal of Soils and Sediments* 6: 215–220.

Resi, A.T., S.M. Rodrigues, C. Araúlo, J.P. Coelho, E. Pereira, and A.C. Duarte. 2009. Mercury contamination in the vicinity of a chlor-alkali plant and potential risks to local population. *Science of the Total Environment* 407: 2689–2700.

Robinson, B.H. 2009. E-waste: An assessment of global production and environmental impacts. *Science of the Total Environment* 408: 183–191.

Siegel, F.R. 2002. *Environmental Geochemistry of Potentially Toxic Metals*. Springer, Berlin, Germany.

Sparks, D.L. 2003. *Environmental Soil Chemistry*, 2nd edn. Academic Press, San Diego, CA.

Strehlow, C.D. and D. Barltrop. 1988. [The Shipham Report] Health studies. *Science of the Total Environment* 75: 101–133 (Special issue).

Tan, K.H. 2009. *Environmental Soil Science*, 3rd edn. CRC Press, Boca Raton, FL.

Taylor, M.P., A.K. Mackay, K.A. Hudson-Edwards, and E. Holz. 2010. Soil Cd, Cu, Pb, and Zn contaminants around Mount Isa City, Queensland, Australia: Potential sources and risks to human health. *Applied Geochemistry* 25: 841–855.

Teaf, C.M., D.J. Covert, P.A. Teaf, E. Page, and M.J. Starks. 2010. Arsenic cleanup criteria for soils in the US and abroad: Comparing guidelines and understanding inconsistencies. *Proceedings of the Annual International Conference on Soils, Sediments, Water and Energy* 15(1): 94–102. http://scholarworks.umass.edu/soilsproceedings/vol15/iss1/10 (verified 22 February 2012).

Thornton, I. 2010. Research in applied environmental geochemistry, with particular reference to geochemistry and health. *Geochemistry, Exploration, Environment, Analysis* 10: 317–329.

van der Perk, M. 2006. *Soil and Water Contamination: From Molecular to Catchment Scale*. Taylor & Francis, Hoboken, NJ.

Wang, T., J. Fu, Y. Wang, C. Liao, Y. Tao, and G. Jiang. 2009. Use of scalp hair as indicator of human exposure to heavy metals in an electronic waste recycling area. *Environmental Pollution* 157: 2445–2451.

Wild, A. 1993. *Soils and the Environment: An Introduction*. Cambridge University Press, Cambridge.

Zhulidov, A.V., R.D. Robarts, D.F. Pavlov, J. Kämäri, T.Y. Gurtovaya, J.J. Meriläinen, and I.N. Pospelov. 2011. Long-term changes of heavy metal and sulphur concentrations in ecosystems of the Taymyr Peninsula (Russian Federation) north of the Norilsk Industrial Complex. *Environmental Monitoring & Assessment* 181: 539–553.

4 Organic Pollutants in Soil

Lynn C. Burgess

CONTENTS

4.1 INTRODUCTION

Soil has always been an important depository of the various organic chemicals produced naturally or anthropogenically. Organic pollutants that were once released into the air or water will someday end up in the soil. The exception might be those chemicals deposited at the bottom of the oceans. Soil contamination is a serious human and environmental problem for both industrialized and non-industrialized nations (Aelion, 2004). Soil has been recognized as a potentially important medium of exposure. A large body of evidence has shown the risks of adverse health effects with the exposure of humans to contaminated soil. Children appear to have a much higher level of exposure to toxic substances in the soil than adults. This is thought to be due to pica and mouthing behavior of younger children (Sedman, 1989). One would think, with the large quantities of organic chemicals used in agriculture for the production of crops, in particular pesticides (Niu and Yu, 2004), that urban areas would have less pollution-related health problems. However, urban areas have a legacy of environmental pollution linked to industrial activities, coal burning, motor vehicle emissions, waste incineration, and waste dumping (Leake et al., 2009). In agricultural areas, because of the effort to provide people with adequate quantities of agricultural products, farmers have been using an increasing amount of organic chemicals, but the resulting pollution has enormous potential for environmental damage (Niu and Yu, 2004).

The types of organic pollutants commonly found in soils are polychlorinated biphenyls (PCBs), polybrominated biphenyls, polychlorinated dibenzofurans, polycyclic aromatic hydrocarbons

(PAHs), organophosphorus and carbamate insecticides, herbicides, and organic fuels, especially gasoline and diesel. Another source of soil pollution is the complex mixture of organic chemicals, metals, and microorganisms in the effluent from septic systems (Pettry et al., 1973); animal wastes (Figure 4.1); and other sources of biowaste. Biowaste is an attractive source of fertilizer for agricultural use that provides a method of recycling this waste; however, there are potential serious adverse effects on the soil due to the heavy metals and organic contaminants in the biowaste. Recently, the problem of pharmaceutical residues and their metabolites in biowaste has been revealed. Pharmaceuticals and their metabolites in the feces and urine of humans and animals are a source of pollution, with the main impact being from antibiotics, hormones, and antiparasitic drugs (Albihn, 2001).

These environmental contaminants come from a huge variety of sources; contaminants from one country are often transported to another country via air, water, foodstuffs, manufactured products, and travelers (Carpenter et al., 1998). Soil is continuously being made airborne and then transported through the atmosphere by global mechanisms of weather and climate. This deflated soil, or dust, can be deposited great distances from the point of origin, and this soil may carry multiple contaminants from the point of origin (Sing and Sing, 2010). The soils of the world are a vast mixture of chemicals, and although conditions are such that an individual is rarely exposed to a single compound, the great majority of people are exposed to a vast chemical mixture of organics, their metabolites, and other compounds at low concentrations (Carpenter et al., 2002).

The toxicological study of chemical mixtures is inherently difficult, and the science of mixtures is not refined or highly codified (Feron et al., 1998). Toxicology is primarily a science of individual poisons, with the understanding of the risks from multiple agents, particularly at low levels of exposure, remaining as the greatest challenge to the toxicologist today (Mauderly, 1993; Feron et al., 2002). Human exposure to organic pollutants in the soil is an area of toxicology that is very difficult to study due to the previously stated reasons and the low concentration of the pollutants. Most people are not directly exposed to toxicants in the soil; most of their exposure is indirect (Sing and Sing, 2010). The toxicological studies of single organic pollutants found in soils are limited and the research conducted on the metabolites of these pollutants and of any chemical mixtures is very limited. The single compound exposure of these compounds and their metabolites is well researched, with exposure levels determined, but the effects of the soil and soil organisms

FIGURE 4.1   A manure slurry being applied to an agricultural field in Arkansas, United States, as a means of disposing of waste products and adding nutrients to the soil. (Photo by Tim McCabe, USDA NRCS.)

on organic pollutants have not yet been adequately studied and remain unstudied for most organic pollutants. The majority of toxicological studies are conducted at relatively high doses and for short periods of exposure. This makes the application of this data to exposure from something like soil very difficult, with the exposure from soil usually being chronic and at very low concentrations (Hough, 2007). There is scattered evidence that human exposure to very low concentrations of some toxicants can be beneficial and not produce any detectable adverse effects (Calabrese and Baldwin, 1998). This phenomenon is called hormesis.

This chapter presents a basic overview of the various organic pollutants that are commonly found in the soil, their known toxicity to humans, and health concerns related to them. The methods of contamination of the soil by organics, their movement into and out of the soil, and the potential effects of human exposure to them are discussed. Where research is available, a discussion of removal of the organics from the soil has been included.

4.2   TOXICITY AND HUMAN HEALTH CONCERNS OF ORGANIC POLLUTANTS

The organic pollutants of the soil are carbon-based compounds that are usually synthesized and therefore unnatural. These compounds are referred to as xenobiotics, from the Greek term "xeno" for "strange." The differences between these synthesized organic compounds and their natural parent compounds are the common insertion of halogen atoms (chlorine, fluorine, bromine) or multivalent nonmetal atoms (such as sulfur or nitrogen) into the structures (Calabrese and Baldwin, 1998). Since these xenobiotics are unnatural and contain types of atoms in locations not seen in any natural compounds, organisms have not evolved adequate biotransformation pathways to deal with these compounds. Therefore, the synthetic organics are very resistant to biological decay and usually highly toxic to organisms, even at extremely low doses.

Enormous quantities of these compounds are manufactured (or were previously manufactured) every year. They include plastics and plasticizers, lubricants, refrigerants, fuels, solvents, pesticides, and preservatives (Calabrese and Baldwin, 1998). Many of the most toxic organic pollutants were never the intended commercial product. For example, neither polychlorinated dibenzo-p-dioxins (PCDDs or dioxin) nor polychlorinated dibenzofurans (PCDFs) were produced commercially in the United States but are unintended by-products of certain chemical processes involving chlorine, as well as combustion and incineration processes (Franzblau et al., 2008; Petrosyan, 2004).

Many organic pollutants have been applied directly to plants growing in the soil, indirectly to the soil, or directly to the soil (Figure 4.2). Other pollutants are deposited by atmospheric fallout or precipitation on the soil (Figure 4.3). These organic pollutants can either be mobile or remain in the soil, with the hydrophobic organic contaminants being strongly associated with the soil and the more water-soluble contaminants being transported long distances while they are partitioned primarily in the aqueous phase (Aelion, 2004). Generally, metabolites of these organics tend to be more hydrophobic than the parent compounds and partition more to the aqueous phases. This may make them more biologically available and therefore more toxic than their parent compounds (Klaassen, 2008).

Many organics were not intentionally applied to the soil but reach it unintentionally. For example, when pesticides were applied aerially to forests, about 25% reached the tree foliage, about 1% reached the target insects, about 30% reached the soil, and the rest were likely to be lost to the air or in runoff water, which may in turn deposit them somewhere on the soil (Calabrese and Baldwin, 1998).

The term *persistent organic pollutant* (POP) is used extensively with agrochemicals and some industrial products. These compounds generally biodegrade very poorly and most will bioaccumulate in tissues of organisms. The group includes 12 compounds, known as the "dirty dozen." These are aldrin, chlorodane, DDT (1,1,1-trichloro-2,2-bis(p-chlorophenyl)ethane), dieldrin, endrin, mirex, heptachlor, hexachlorobenzene, toxaphene, PCBs, dioxins, and furans (Jorgenson, 2001). Young children are the most susceptible to POPs because their bodies are immature and developing rapidly and because their behavior, such as mouthing objects, may increase their exposure. This exposure can come from air, dust, soil, and food (Wilson et al., 2003).

FIGURE 4.2 Application of pesticides to a lettuce crop in Arizona, United States. In this case, the pesticide is applied directly to both the crop and exposed soil in between the crop rows. (Photo by Jeff Vanuga, USDA NRCS.)

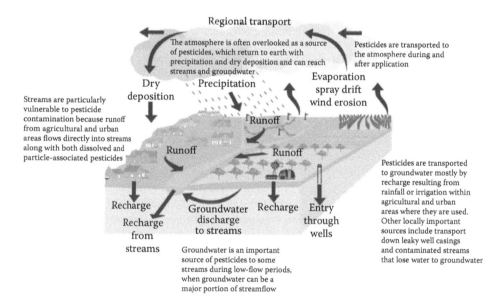

FIGURE 4.3 Routes of pesticide transport. (From Gilliom, R.J., Barbash, J.E., Crawford, C.G., Hamilton, P.A., Martin, J.D., Nakagaki, N., Nowell, L.H., et al., *The Quality of Our Nation's Waters: Pesticides in the Nation's Streams and Ground Water, 1992–2001.* Circular 1291, U.S. Geological Survey, Reston, VA, 2007. http://pubs.usgs.gov/circ/2005/1291/, verified 6 January 2012.)

4.2.1 Pesticides

The organic soil pollutants that are most often used intentionally are the pesticides. Pesticides are natural or synthetic substances that kill or otherwise control some targeted organism. All pesticides share the ability to block a vital metabolic process in these organisms (Niu and Yu, 2004). Pesticides include about 100,000 formulations of about 1,000 chemicals. Between 1960 and 2000 about 3.63 metric tons of active compounds were produced (Petrosyan, 2004).

Pesticides are usually organic, although a few are inorganic. Inorganic pesticides generally use a heavy metal or a halogen as their toxic element. Organic pesticides were developed mostly during and after World War II. They are usually much less toxic to humans than inorganic pesticides. The organic pesticides were originally thought to be biodegradable and therefore an improvement over the nonbiodegradable inorganic pesticides, but most organic pesticides were found to be also nonbiodegradable. Organic pesticides generally act to either overactivate or inhibit some normal metabolic process in the target organism, in contrast to the inorganic pesticides, which generally damage cells (Niu and Yu, 2004).

Pesticides are grouped by their targeted organisms and then their common parent compound. There are insecticides, herbicides, rodenticides, nematodicides, and fungicides. The closer in evolution and function the targeted organism is to humans, the more toxic the pesticide will be to humans; therefore, rodenticides are the most toxic to humans, followed by insecticides, with most herbicides posing little risk to humans (Klaassen, 2008).

Insecticides and their impurities pose the largest health risk to humans when associated with soil. Rodenticides are usually not intentionally applied to soil in significant quantities like insecticides are, so rodenticides would rarely be considered soil pollutants.

4.2.2 Insecticides

Insecticides are grouped in four categories based on their chemical composition: organochlorine, organophosphorus, carbamate, and natural pyrethrins and pyrethroids (Niu and Yu, 2004). The organochlorines are hydrocarbon compounds with chlorine atoms substituted for various hydrogen atoms. These groups include DDT, methoxychlor, toxaphene, benzene hexachloride (lindane), chlorodecane (Kepone), and chlorinated cyclopentadienes (aldrin and dieldrin) (Niu and Yu, 2004).

The organophosphorus insecticides contain one or two phosphorus atoms, with additional oxygen, sulfur, or nitrogen atoms. These compounds include chlorphos (trichlorophosphone), carbophos (malathion), thiophos (parathion), and metaphos (parathion-methyl) (Petrosyan, 2004). The next group is the carbamate insecticides, which are derivatives of carbamic acid, H_2NCOOH. This group includes carbofuran, carbaryl (Sevin), aldicarb, and Primicarb (Niu and Yu, 2004).

The last group is the naturally plant-produced pyrethrins and pyrethroids. These compounds are obtained from chrysanthemum or pyrethrum flowers. Synthetic analogs are also produced. The first of these was adethrin and then fenvalerate (Niu and Yu, 2004). Many plants produce compounds to act as pesticides; other examples of these are nicotine found in tobacco and rotenone from certain legume roots.

4.2.2.1 Organochlorine Insecticides

In an attempt to develop insecticides less toxic to humans and more effective at lower doses than the inorganic and organometallic compounds, the chemical industries started to produce organochlorines in the 1940s (Niu and Yu, 2004). These new insecticides were initially thought to be biodegradable, but most were not. They were less toxic than organophosphates but were found to bioaccumulate in living organisms. They have reached dangerous levels in many species, especially those high up the food chain (Vega et al., 2007).

The first organochlorine to be produced in large quantities was DDT. DDT was first synthesized by O. Zerdlin in 1874 and was discovered in 1939 by Paul Mueller to be an insecticide (Klaassen, 2008). The first mass use was in 1943 in Naples, Italy, to combat a typhoid epidemic (Niu and Yu, 2004). DDT was used extensively against mosquitos to control malaria for the next 30 years. By the 1960s, however, evidence indicated that these chemicals were accumulating in the ecosystem with undesirable side effects. Due to its liposoluble properties, it was being readily absorbed and building up to toxic levels in fish, mammals, birds, and humans as it passed up the food chain. DDT's toxic action was first seen as a disruption of hormone systems (Vega et al., 2007). DDT was banned in the United States in 1972 but is still highly persistent and globally distributed; it is also still being used in many parts of the world (Klaassen, 2008). Like all organochlorines, not only does it accumulate in animal tissues, but it is extremely stable and persists in soils and in plants. DDT has reached most groundwater and surface-water sources (Vega et al., 2007). DDT has a reported half-life of >60 years outside organisms (Turner and Sharpe, 1997).

The principle long-lived metabolite of DDT in the human body is p,p'-dichlorodiphenyldichloro-ethylene (DDE) (Turner and Sharpe, 1997), which is created by the removal of HCl from DDT. DDE is nondegradable biologically; it is the most important metabolite of DDT toxicologically since this form acts as a potent antiandrogen rather than having the estrogenic effects of o,p'-DDT (Turner and Sharpe, 1997; Toppari et al., 1996). DDT and its metabolites bind in $vitro$ to rat androgen receptors, significantly inhibiting the specific binding of [^{3}H] 5α-dihydrotestosterone (DHT) (Cocco, 2002).

DDE was the metabolite that led to the ban of DDT in most of the developed world. DDE interferes with estrogen, which regulates the distribution of calcium, so it affects the thickness of eggshells, and this was a major problem for birds of prey, being at the top of the food chain (Niu and Yu, 2004).

The antiandrogenic effects of p,p'-DDE are more difficult to research and validate, and they have more variation. The toxic effect seen here, most notably in fish, amphibians, and reptiles, is the reduced maleness or sexual potency of the males. There is reason to believe this is also affecting humans by reducing the function of male testosterone. There are other weakly estrogenic synthetic compounds, including organochlorine pesticides, PCBs, phenolic compounds, and phthalate esters (Turner and Sharpe, 1997).

Acute exposure to DDT is a rare event, and if the soil is the source of exposure, then only chronic, low-dose exposure to DDT needs to be considered. The nonestrogenic toxicity of DDT and DDE increases liver weight, causes hepatic cell hypertrophy and necrosis, and induces some P450s. DDT has been shown to be carcinogenic to the liver and lungs of lab animals, and is classified as a possible human carcinogen by the International Agency for Research on Cancer (IARC). The evidence is still inconclusive but there is increasing evidence for the carcinogenicity of DDT and DDE in hormone-sensitive cancers, such as breast, endometrium, and prostate cancers (Safe, 2000).

4.2.2.2 Chlorinated Cyclopentadienes (Aldrin and Dieldrin)

The first cyclopentadiene pesticides were aldrin and dieldrin, in 1948, produced by Shell Chemical (Jorgenson, 2001). Along with DDT, these two insecticides were the most widely used agricultural pesticides in the United States in the 1960s (Jorgenson, 2001; Niu and Yu, 2004). Agricultural use of aldrin and dieldrin was ended in the 1970s in the United States and in most countries by the mid-1980s because of environmental and human health considerations (Niu and Yu, 2004).

Aldrin and dieldrin were formulated from a waste product of synthetic rubber, cyclopentadiene. They were made by a chemical process called the Diels–Alder reaction, which gave them their names (Jorgenson, 2001).

Aldrin and dieldrin have been used as soil poisons for rootworms, cutworms, and wireworms. Dieldrin is more expensive to produce than aldrin, so the production of dieldrin was about 10% that of aldrin. Both were used on about 90 different crops. Both bind to the soil, but aldrin is more volatile than dieldrin and readily degrades into dieldrin in the soil (Jorgenson, 2001).

Aldrin and dieldrin have a moderate to high acute oral toxicity. They differ from DDT in that they are easily absorbed through the skin. Their toxicity is seen in the central nervous system (CNS), with convulsions as the main symptom of poisoning (Klaassen, 2008). These compounds are slow to metabolize and they bioaccumulate in the human liver, body fat, breast milk, and semen (Klaassen, 2008; Jorgenson, 2001). Dieldrin can cause liver enlargement and liver cancer, can be a tumor promoter, and have endocrine-disrupting properties (Klaassen, 2008). It has been predicted that dieldrin will be present in the environment until 2030 (Jorgenson, 2001), due to its half-life in soil of about 25 years (Matsumoto et al., 2009).

Aldrin and dieldrin have reportedly been detected in agricultural crops and therefore could be candidates for phytoremediation. Phytoremediation is the biological removal, degradation, or immobilization of organic compounds by plants. The family *Cucubitacease*, which includes zucchini and cucumbers, took up more dieldrin than other plants. The mechanism for this is unknown (Matsumoto et al., 2009). Phytoremediation is a relatively safe method for removal of these insecticides from contaminated soils.

4.2.2.3 Organophosphorus and Carbamate Insecticides

The insecticide groups most commonly used presently are the organophosphorus compounds and carbamates. Organophosphorus compounds are usually esters, amides, or thiol derivatives of phosphoric acid. There are about 50,000 compounds in this group (Kamanyire and Karalliedde, 2004). Carbamates are derivatives of carbamic acid. They have the same mechanisms of action as organophosphorus compounds and are clinically indistinguishable. The difference in their function is that the chemical bonding is completely reversible, giving these compounds a significantly shorter toxic half-life when not at a lethal dose (Leibson and Lifshitz, 2008).

Common organophosphorus compounds used today are parathion, malathion, chlorpyrifos, and dichloros. The common carbamates in use are carbaryl (Sevin), aldicarb, mancozeb, and sodium methyl diothiocarbamate (Klaassen, 2008; Figure 4.4). Both organophosphorus and carbamate insecticides have acute toxic effects on human parasympathetic, sympathetic, and central nervous systems. These insecticides interfere with the metabolism of acetylcholine by inhibiting the enzyme that hydrolyzes it, acetylcholinesterase (Eskenazi et al., 1999). This inhibition is irreversible in the organophosphorus compounds; therefore, the enzyme is unable to degrade the neurotransmitter acetylcholine to choline and acetic acid. This leads to an accumulation of acetylcholine throughout the nervous system, resulting in overstimulation of the muscarinic and nicotinic receptors (Leibson and Lifshitz, 2008). The carbamates differ only in that their toxicity is considerably shorter and more transient in duration due to the hydrolysis of the carbamate molecule (Risher et al., 1987).

The signs and symptoms of acute organophosphorus and carbamate poisonings are the same and include miosis (constricted pupils), increased urination, diarrhea, salivation, muscle fasciculation, and CNS effects (dizziness, lethargy, fatigue, headache, confusion, convulsions, and coma) (Klaassen, 2008). Acute organophosphorus compound exposure, but not carbamate exposure, also manifests two late developing syndromes. The first is the intermediate syndrome, which develops one to several days after poisoning. This syndrome is marked by weakness of the respiratory, neck, and proximal limb muscles. The resulting respiratory paralysis may be fatal in 15%–40% of patients and recovery of the surviving patients usually takes up to 15 days. The mechanism of this syndrome is unknown and there is no specific treatment other than supportive care. The other syndrome is called organophosphate-induced delayed polyneuropathy (OPIDP). Signs and symptoms are tingling in the hands and feet, followed by sensory loss, progressive muscle weakness and flaccidity of the distal skeletal muscles of the extremities, and ataxia. These may be seen 2–3 weeks after poisoning after both the acute cholinergic syndrome and the intermediate syndrome have passed. The mechanism is a breakdown of the axons and their terminals, leading to the breakdown of the neuritic segments and myelin sheaths. The target of the organophosphorus compounds in the syndrome is not the acetylcholinesterase, but another enzyme called neuropathy target esterase. The research is still not clear at this time (Klaassen, 2008). The effects of OPIDP are generally not reversible and have been observed more than 10 years after poisoning (Karmel and Hoppin, 2004).

Organophosphorus compounds and carbamates do not exhibit the chronic effects and bioaccumulation abilities of organochlorines. There is limited research into the long-term effects of these compounds. Most of these studies have shown that either there were no measurable effects when compared to untreated experimental animals or the effects were similar to low-dose acute poisoning features (Eskenazi et al., 1999).

Children have been shown to be much more affected by these insecticides (Cohen, 2007). A California study showed that farmworkers' children are highly exposed to pesticides from

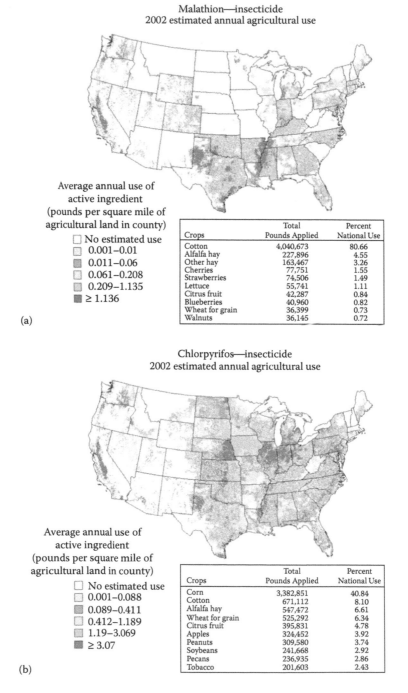

Malathion—insecticide
2002 estimated annual agricultural use

Average annual use of
active ingredient
(pounds per square mile of
agricultural land in county)

☐ No estimated use
☐ 0.001–0.01
☐ 0.011–0.06
☐ 0.061–0.208
☐ 0.209–1.135
■ ≥ 1.136

Crops	Total Pounds Applied	Percent National Use
Cotton	4,040,673	80.66
Alfalfa hay	227,896	4.55
Other hay	163,467	3.26
Cherries	77,751	1.55
Strawberries	74,506	1.49
Lettuce	55,741	1.11
Citrus fruit	42,287	0.84
Blueberries	40,960	0.82
Wheat for grain	36,399	0.73
Walnuts	36,145	0.72

(a)

Chlorpyrifos—insecticide
2002 estimated annual agricultural use

Average annual use of
active ingredient
(pounds per square mile of
agricultural land in county)

☐ No estimated use
☐ 0.001–0.088
☐ 0.089–0.411
☐ 0.412–1.189
☐ 1.19–3.069
■ ≥ 3.07

Crops	Total Pounds Applied	Percent National Use
Corn	3,382,851	40.84
Cotton	671,112	8.10
Alfalfa hay	547,472	6.61
Wheat for grain	525,292	6.34
Citrus fruit	395,831	4.78
Apples	324,452	3.92
Peanuts	309,580	3.74
Soybeans	241,668	2.92
Pecans	236,935	2.86
Tobacco	201,603	2.43

(b)

FIGURE 4.4   (a–d) The use of selected organophosphorus and carbamate insecticides in the United States in 2002. (Courtesy of U.S. Geological Survey.)

contaminated food, household pesticides, drift of pesticides from nearby agricultural applications, contaminated breast milk from farmworker mothers, the fields that they play in, and pesticides tracked into the house by household members working in the fields (Eskenazi et al., 1999). These insecticides, along with several other types, should be considered to have neurodevelopmental toxicity in children (Bjorling-Poulsen et al., 2008).

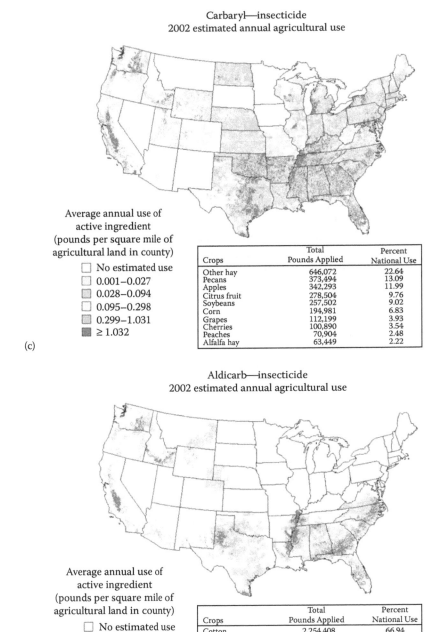

Carbaryl—insecticide
2002 estimated annual agricultural use

Average annual use of
active ingredient
(pounds per square mile of
agricultural land in county)

☐ No estimated use
☐ 0.001–0.027
☐ 0.028–0.094
☐ 0.095–0.298
☐ 0.299–1.031
■ ≥ 1.032

(c)

Crops	Total Pounds Applied	Percent National Use
Other hay	646,072	22.64
Pecans	373,494	13.09
Apples	342,293	11.99
Citrus fruit	278,504	9.76
Soybeans	257,502	9.02
Corn	194,981	6.83
Grapes	112,199	3.93
Cherries	100,890	3.54
Peaches	70,904	2.48
Alfalfa hay	63,449	2.22

Aldicarb—insecticide
2002 estimated annual agricultural use

Average annual use of
active ingredient
(pounds per square mile of
agricultural land in county)

☐ No estimated use
☐ 0.001–0.034
☐ 0.035–0.183
☐ 0.184–0.867
☐ 0.868–4.092
■ ≥ 4.093

(d)

Crops	Total Pounds Applied	Percent National Use
Cotton	2,254,408	66.94
Peanuts	376,522	11.18
Citrus fruit	256,409	7.61
Potatoes	226,503	6.73
Sugarbeets	179,724	5.34
Pecans	37,936	1.13
Tobacco	26,187	0.78
Sweet potatoes	6,851	0.20
Dry beans	3,430	0.10

FIGURE 4.4   (Continued)

4.2.3 HERBICIDES

An *herbicide* is a compound that is used to kill or regulate, typically adversely, the growth of any unwanted plant (Kramer and Baker, 2004). Herbicides represent an extremely broad array of chemical classes and in turn act at a large number of sites of metabolic function and energy transfer in plant cells. These sites include amino acid biosynthesis, cell division, carotenoid biosynthesis, and photo bleachers (Duke, 1990). Most of these targeted sites are exclusive to plants; therefore, herbicides are generally less toxic than other pesticides and are thought to pose little risk to humans. Most exposure to herbicides occurs due to intentional ingestion (Roberts et al., 2010), an accident during handling, misidentification, or inappropriate use. Within these circumstances, some classes of herbicides are associated with significant toxicity to humans (Kramer and Baker, 2004). The exceptions to the generally low toxicity to humans are the herbicides paraquat and diquat. The major toxicological concern for paraquat is its acute systemic effects, particularly in the lungs and secondarily in the kidneys (Klaassen, 2008).

A lethal dose of paraquat causes damage to the alveolar epithelial cells within 24 h after exposure. After 2–4 days, large areas of alveolar epithelium are completely lost. This is followed by alveolar edema, extensive infiltration of inflammatory cells into the alveolar spaces, and finally death due to severe anoxia (Klaassen, 2008).

Diquat is chemically similar to paraquat, but has a different toxicological profile. Acute toxicity is somewhat lower. The chronic exposure target organs for toxicity are the gastrointestinal tract, kidneys, and eyes. Clinical symptoms include nausea, vomiting, diarrhea, ulceration of the mouth and esophagus, decline of renal function, and neurologic effects, but no pulmonary fibrosis (Klaassen, 2008).

Herbicides can be classified by their application. Preplanting herbicides are applied to the soil before the appearance of any weeds, and postemergent herbicides are applied to the soil or foliage after germination of the crop or weeds. Herbicides also can be classified by the manner in which they are applied to the plants. Contact herbicides affect the plants that they are applied to, while translocated herbicides are applied to the soil or plants and are absorbed and distributed to distant tissues. There are also nonselective herbicides, which may kill all forms of vegetation, and selective herbicides, which are used to kill weeds without harming crops (Figure 4.5) (Klaassen, 2008).

Soils are generally exposed to herbicides due to their methods of application. What has not been studied is the toxicity of the herbicides' metabolites that are produced in the soils; this has not been considered as worthy of study. Since most herbicides are considered to have a low toxicity to humans and other animals, and the concentrations of these metabolites in the soils are low, the metabolites are considered to be even less of a toxicity problem than the parent compounds. This is an assumption without any scientific evidence and we may find it to be incorrect. There is some justification for concern. Consider the situation of the toxicity of the class of chlorophenoxy herbicides, which was due to an impurity produced during synthesis of the herbicide. The toxicity was due to 2,3,7,8-tetrachlorodibenzo-p-dioxin (TCDD or dioxin) (Kramer and Baker, 2004). Dioxin is extremely toxic (it will be discussed in the next section) and this is due to the addition of chlorine atoms, a process that is assumed never to occur to a metabolite naturally. This shows that toxifying processes may be occurring without our knowledge.

4.3 POLYCHLORINATED DIBENZODIOXINS AND POLYCHLORINATED DIBENZOFURANS

PCDDs and PCDFs are one of the groups of POPs that are ubiquitous in Earth's environment. These compounds are referred to as dioxins and dioxin-like compounds. In total, there are 7 PCDDs and 10 PCDFs, which are considered dioxin or dioxin-like compounds by the WHO (van den Berg et al., 2006). Compounds that have dioxin-like properties have lateral halogenation at more than three sites. These may be chlorinated, brominated, or a combination of the two (White and Birnbaum, 2009). While very small quantities of dioxins are generated naturally, by way of forest fires and volcanic eruptions, major sources are waste incineration, the reprocessing metal industry, and the paper and pulp industry. The main source, however, was from contaminated herbicides (White and Birnbaum, 2009; Kogevinas, 2001).

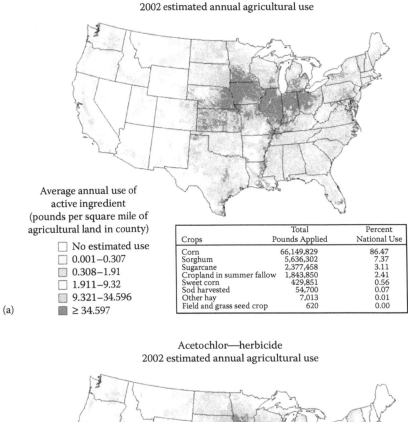

Atrazine—herbicide
2002 estimated annual agricultural use

Average annual use of
active ingredient
(pounds per square mile of
agricultural land in county)

☐ No estimated use
☐ 0.001–0.307
▨ 0.308–1.91
☐ 1.911–9.32
▨ 9.321–34.596
■ ≥ 34.597

(a)

Crops	Total Pounds Applied	Percent National Use
Corn	66,149,829	86.47
Sorghum	5,636,302	7.37
Sugarcane	2,377,458	3.11
Cropland in summer fallow	1,843,850	2.41
Sweet corn	429,851	0.56
Sod harvested	54,700	0.07
Other hay	7,013	0.01
Field and grass seed crop	620	0.00

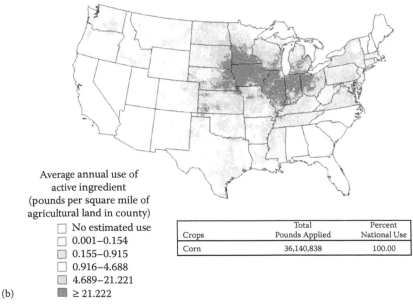

Acetochlor—herbicide
2002 estimated annual agricultural use

Average annual use of
active ingredient
(pounds per square mile of
agricultural land in county)

☐ No estimated use
☐ 0.001–0.154
▨ 0.155–0.915
☐ 0.916–4.688
▨ 4.689–21.221
■ ≥ 21.222

(b)

Crops	Total Pounds Applied	Percent National Use
Corn	36,140,838	100.00

FIGURE 4.5 (a, b) The use of the herbicides atrazine and acetochlor (Roundup) in the United States in 2002. Atrazine is a selective herbicide that is most commonly used to kill weeds in maize (corn); it is the most heavily used herbicide in the United States. Acetochlor is a nonselective herbicide that is commonly used on maize that has been genetically modified to tolerate it. (Courtesy of U.S. Geological Survey.)

Most of these hydrophobic and lipophilic compounds are very resistant to being metabolized in vertebrates, including humans. Due to these properties, these compounds are biomagnified through the food chain and the highest tissue concentrations are found in the species at the top of the food chain. Most of the toxic effects of these compounds are mediated through the aryl hydrocarbon receptor (AhR). The AhR is highly conserved in most vertebrate species (van den Berg et al., 2006). The most toxic member of the dioxin group is 2,3,7,8-tetrachlorodibenzo-p-dioxin; this compound has the greatest affinity for the Ah receptor and therefore is considered the most toxic chemical produced by humans (Tuyet-Hanh et al., 2010).

Dioxins binding to the AhR may disrupt its endogenous functions. This is a physiological activator of the AhR that likely induces rapid on/off signaling through the receptor. This prevents the AhR from functioning to maintain homeostasis (White and Birnbaum, 2009). TCDD is therefore considered by the IARC to be a human carcinogen (Kogevinas, 2001), a teratogen (White and Birnbaum, 2009), and a severe disruptor of the immune system (Klaassen, 2008; Sulentic and Kaminski, 2011). Animal studies have shown that dioxin results in damage to the liver, reproductive system, nervous system, hormonal system, cardiovascular system, and lungs (Tuyet-Hanh et al., 2010). Dioxin was reported to delay mammary development due to exposure during early development, again during peripubertal period of the mammary gland development, and during the start of lactation (Fenton, 2006). The effect of TCDD on the immune system has been reported to be due to the impairment of the B-cells, brought about by impairing B-cell maturation and reducing IgM and IgG secretion (Sulentic and Kaminski, 2011). TCDD may cause increases in food allergies by preventing stable induction of oral tolerance and affecting dendritic cells and CD4[+] immune cells in the mesenteric lymph nodes (Chmill et al., 2010). Chloroacne is the only consistent sign of toxicity in humans due to any of the polyhalogenated aromatic hydrocarbons under discussion in this chapter. Chloroacne takes weeks to develop after an acute exposure (Tuyet-Hanh et al., 2010).

Human exposure to TCDD is nearly entirely through the diet, particularly milk and other dairy products, fish, and meat (Kogevinas, 2001; Roeder et al., 1998; Rideout and Teschke, 2004; Pohl et al., 1995). Dioxin gets into food through a complex and involved process. There are many sources for dioxins, with the largest contributor being the past application of contaminated herbicides on agricultural soils (Roeder et al., 1998). Other significant sources are incineration of waste, industrial processes, and deposition on the soil by atmospheric fallout, water (White and Birnbaum, 2009), or the application of sewage sludge, the solid by-product of municipal sewage or wastewater treatment processes. Research with sewage sludge used on agricultural lands showed that it had a negligible impact on most aboveground plants, as it was only detected in the peel of root crops. However, application on grazing or forage land could significantly increase human dietary exposure to PCDDs (Rideout and Teschke, 2004).

Once dioxins are present in the soil, within the top 0.1 cm of the surface they have a half-life of 9–15 years (Tuyet-Hanh et al., 2010) and below 0.1 cm in the subsurface soil the half-life is 25–100 years (Tuyet-Hanh et al., 2010; Dai and Oyana, 2008). Dioxins have a half-life in the human body of 5.8–11.3 years (Dai and Oyana, 2008) and this makes the correlation of human tissue concentration to local soils very difficult, especially considering the contribution of outside sources of dioxin.

There is an unfortunate example of environmental dioxin soil contamination in Vietnam that has provided much information on the nature of this pollutant and its effects on a human population. During the Vietnam War, the U.S. military carried out an operation code-named "Ranch Hand," which sprayed over 73.8 million L of herbicides over Vietnam, Laos, and Cambodia to defoliate large areas of crops and forest. Agent Orange, a 50:50 mixture of the herbicides 2,4,5-trichlorophenoxyacetic acid (2,4,5-T) and 2,4-dichlorophenoxyacetic acid (2,4-D), accounted for two-thirds of the defoliant used. The 2,4,5-T, however, was contaminated with TCDD (Tuyet-Hanh et al., 2010; Nhu et al., 2009). Large amounts of these herbicides were also spilled onto the local soils, causing considerable soil, water, sediment, and food contamination by dioxin at the Bien Hoa Airbase, where operation Ranch Hand took place (Tuyet-Hanh et al., 2010). Dioxin contaminated the local fish

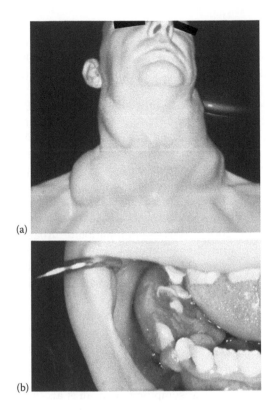

(a)

(b)

FIGURE 4.6   (a) A man suffering from Hodgkin's disease; (b) a child with non-Hodgkin's lymphoma in the mouth. (Courtesy of Centers for Disease Control and Prevention, images #12630 and #12629.)

and some still show dioxin levels 100,000 times higher than the surrounding environment (Tuyet-Hanh et al., 2010; Nhu et al., 2009). Dioxin is found on soil particles and dust attached to grass, vegetables, and crops. It is also concentrated in the meat of the local animals that fed on the area's grasses (Tuyet-Hanh et al., 2010). People living in areas sprayed with Agent Orange have dioxin levels that are 135 times that of people in adjacent uncontaminated areas (Nhu et al., 2009). Persons in these dioxin hot spots are showing adverse health effects. There is sufficient evidence to link dioxin exposure to chronic lymphocytic leukemia, soft-tissue sarcoma, non-Hodgkin's lymphoma, Hodgkin's disease (Figure 4.6), and chloroacne. There is also "limited or suggestive" evidence of an association between dioxin exposure and laryngeal cancer, cancers of the lung, bronchus, trachea, and prostate, as well as multiple myeloma, amyloid light-chain amyloidosis, early-onset transient peripheral neuropathy, porphyria cutanea tarda, hypertension, type 2 diabetes (mellitus), and spina bifida (Tuyet-Hanh et al., 2010).

4.4   POLYCHLORINATED BIPHENYLS

PCBs are toxic synthetic aromatic compounds that make up a group of 209 different congeners or related chemicals, some containing up to 10 chlorine atoms (McGuinness and Dowling, 2009). PCBs are either oily liquids or solids and are colorless to light yellow. They have no smell or taste (Koplan, 2000). PCBs are nearly water-insoluble, nonpolar, lipid-soluble, and inert (Chaudhry and Chapalamadugu, 1991). Due to their noninflammability, chemical stability, and high electrical constants, technical PCB mixtures were used widely for industrial and commercial applications (Aken et al., 2010). These applications include uses as heat transfer fluids, hydraulic lubricants, dielectric fluids for transformers and capacitors, plasticizers, wax extenders, adhesives, organic diluents, dedusting agents, pesticide extenders, cutting oil, carbonless reproducing paper, and flame retardants (Safe, 1992). PCBs were manufactured from 1929 to 1979 and it is estimated that 1.4 million metric tons were produced worldwide (McGuinness and Dowling, 2009; Aken et al., 2010; Rocca and Mantovani, 2006). PCBs were banned in most countries in 1979 (McGuinness and Dowling, 2009; Aken et al., 2010).

PCBs are relatively stable and the more highly chlorinated the PCBs, the more lipophilic they are. PCBs are widely distributed and transported throughout the environment, and their residues have been identified in air, water, aquatic and marine sediments, fish and wildlife, and human tissue (Safe, 1992). PCBs have entered the environment from accidental spills and leaks during transportation or from electrical transformers, capacitors, or other products containing PCBs. Presently, PCBs are still being released from poorly maintained hazardous waste sites, illegal or improper dumping of PCB wastes, such as old transformers or their liquid, disposal of PCB products in municipal landfills not designed to handle hazardous wastes, and by burning of PCB wastes in incinerators (Koplan, 2000). PCB contamination of urban soils outside of industrial areas has been shown to be from buildings where PCBs were used in caulking around windows and in expansion joints (Herrick et al., 2007). In farmlands, PCBs usually entered the soil compartment through atmospheric deposition or unintentionally through contaminated biosolids or irrigation sources (Wang et al., 2010).

PCBs are highly hydrophobic, leading to their bioaccumulation in living organisms. In humans, PCBs are commonly detected in breast milk and blood, with concentrations increasing with age (Aken et al., 2010). The persistent higher chlorinated PCBs (Cl_8–Cl_{10}) are major contaminants in human tissues (Safe, 1992).

Acute PCB exposure can cause chloroacne, darkened skin and nail pigmentation, hearing loss, eye disorders, jaundice, and neurological symptoms (Curran et al., 2011). Chronic PCB exposure shows much more variability in symptoms than acute exposure. PCBs usually cause increased cancer in lymphatic and hemopoietic tissues, and the gastrointestinal tract, but in many cases the liver is the primary target site (Safe, 1992). Most of the toxic effects of PCBs are due to a disruption of the AhR, with the same effects as seen with dioxin (Curran et al., 2011).

Studies have shown that PCBs form covalent adducts with proteins, DNA, and RNA and induce DNA strand breaks and DNA repair, with these effects dependent on the metabolism of PCBs into intermediates such as arene oxides, which alkylate cellular macromolecules (Safe, 1992). PCBs, like dioxins, are suspected endocrine disruptors and have been associated with the feminization of males (McGuinness and Dowling, 2009).

PCBs are divided into two groups based on their toxicological mechanisms. The first group is similar in action to dioxins; these PCBs can bind to the AhR and their toxicity is via AhR activation. The other group is the non-dioxin-like PCBs; these have been shown to be associated with reprotoxic, neurotoxic, and carcinogenic effects. These PCBs also antagonize androgen receptors and inhibit gap junctional intercellular communication (Hamers et al., 2011).

The toxicity of PCBs has been known since the 1930s (Aken et al., 2010), and in the 1970s in-depth rodent studies on PCBs set a "no observable adverse effect level" (NOAEL) at 10–26 mg/kg/day. Studies with monkeys then showed adverse learning and behavioral effects with exposures over a hundred times lower than in rats. Human epidemiological studies of children exposed to PCBs prenatally have revealed effects such as decreased IQ, poorer performance on neurobehavioral tests, and delayed reading ability at levels about a thousand times lower than those that caused detectable impairment in monkeys. The NOAEL in humans is now estimated to be about 0.0001 mg/kg/day or about a million times lower than levels based on rat studies (Solomon and Huddle, 2002).

Plants constitute the major route of entry of PCBs into the food chain (Aken et al., 2010). Feeds are the major way that PCBs transfer to milk and eggs. The molecular form with six chlorine atoms shows the highest carryover into milk and eggs (Rocca and Mantovani, 2006). The use of animal feeds that contain rendered animal fats, tissues, and oils is a significant source of exposure to both PCBs and dioxins among food-production animals (Sapkota et al., 2007).

4.5 POLYCYCLIC AROMATIC HYDROCARBONS

PAHs are compounds that are found naturally and can also be man-made (EPA, 2008). They are produced through the incomplete combustion and pyrolysis of organic matter. PAHs can be released from sources such as forest fires, volcanic emissions, residential wood burning, petroleum catalytic cracking, industrial combustion of fossil fuels (Lau et al., 2010), asphalt, tobacco smoke, agricultural burning, eating foods grown on contaminated soil, or eating grilled meats and other grilled foods (EPA, 2008).

All PAHs contain at least two fused aromatic rings in linear, angular, or cluster arrangements. Chemical reactivity, aqueous solubility, and volatility of PAHs decrease with increasing molecular weight. As a result, PAHs differ in their transport, distribution, and fate in the environment and in their effect on biological systems (Seo et al., 2009).

The presence of PAHs in soils is an issue of concern due to their carcinogenic, mutagenic, and teratogenic properties (Lau et al., 2010). The USEPA has identified six PAHs as priority pollutants. High-molecular-weight PAHs are important as they are recalcitrant. They are difficult to remove from contaminated soil (Seo et al., 2009).

PAHs have been found to be a suspected carcinogen in humans (Godschalk et al., 2003). They have been shown to be associated with p53 mutations and increases in cancer in humans (Mordukhovich et al., 2010; Mattsson et al., 2009). PAHs form PAH-DNA adducts that serve a role in the transformation of normal cells into cancerous cells (Godschalk et al., 2003). PAHs are also shown to be immunosuppressive and to suppress both humoral (B-cell-mediated) and cellular (T-cell-mediated) immune responses. These are mediated by the AhR (Mattsson et al., 2009). The mechanism of PAH immunosuppression can also be associated with the metabolites of the PAHs that are produced through the action of cytochrome p450 isozymes, CYPIAI, CYPIAZ, and CYPIBI, which are transcriptionally induced by AhR (Klaassen, 2008). PAH studies have shown that these compounds alter normal organismal development and can induce a teratogenic phenotype similar to that caused by dioxin (Billiard et al., 2008).

The highest contributors to human PAH exposure are not linked to agricultural soils. Urban and forest soils generally have higher levels of PAHs than agricultural soils (Placha et al., 2009). Much of the human exposure, outside of industrial sources where workers are handling PAHs, is due to cooking (grilling and smoking) of food and cigarette smoke (Mordukhovich et al., 2010).

4.6 PETROLEUM-BASED FUELS

There are two petroleum fuels that have commonly contaminated soil. These are diesel fuels and gasoline. Because of their massive production and use, they are among the most common sources of organic pollutants for the surface soil (Zanaroli et al., 2010).

4.6.1 GASOLINE

Gasoline is a petroleum fuel that is a complex mixture of over 500 saturated or unsaturated hydrocarbons, having 3–12 carbons (Caprino and Togna, 1998). Diesel fuels are also a complex mixture, primarily of parafilms, including N-, iso-, and cycloparaffins, and aromatic hydrocarbons, including naphthalene and alkylbenzenes. These are obtained from the middle distillate gas–oil fraction during petroleum separation (Zanaroli et al., 2010). Gasoline usually contains additives and other organics depending on the origin of the crude oil, processing techniques, season-to-season changes, and country of origin (Caprino and Togna, 1998). Diesel usually contains fewer additives but can be more complex and variable. The additives in gasoline can be organic lead, oxygenates (such as methanol, ethanol, ethyl tertiary butyl ether, tertiary butyl alcohol, and tertiary amyl methyl ether), ethylene dibromide, ethyl dichloride, methylcyclopentadienyl manganese tricaronyl, benzene, and alkylbenzene (toluene and xylene) (Caprino and Togna, 1998). All of these additives have their own toxicity.

Additionally, gasoline and diesel both contain PAHs and other natural organics. Soil contamination with these hydrocarbons causes extensive damage of local systems since accumulation of pollutants in animal and plant tissue may cause death or mutation (Das and Chandran, 2011).

It is difficult to generalize about the toxicity of gasoline and diesel due to their variable composition. Aromatic hydrocarbons are more toxic than naphthalenes, which are more toxic than paraffins. Gasoline has been studied much more intensely in the noncombustible form than diesel, which has minimal toxicological studies on its oil form.

Human exposure to diesel and gasoline, especially from a soil source, will usually be by inhalation and skin contact. The principal target organ for gasoline toxicity is the CNS. The effect is CNS depression, which has similar symptoms to ethanol inebriation. Acute gasoline exposure will result in flushing of the face, ataxia, staggering, vertigo, mental confusion, headaches, blurred vision, slurred speech, and difficulty in swallowing. At high concentrations coma and death can occur in a few minutes without any signs of respiratory struggle or post mortem signs of anoxia (Reese and Kimbrough, 1993).

Chronic gasoline exposure can lead to renal and liver cancer, acute myeloid leukemia, myeloma heart disease, changes to the CNS, skin alterations, including melanoma (Figure 4.7), and modifications of the mucous membranes (Caprino and Togna, 1998).

4.6.2 Diesel

Diesel may cause many of the same acute and chronic toxicities as gasoline, since diesel contains many of the same hydrocarbons such as PAHs. Diesel is not classified as a human carcinogen but there is evidence that diesel is a carcinogen (IARC, 1989). Inhalation of diesel can cause headaches and slight giddiness, and ingestion causes symptoms very similar to acute gasoline exposure (U.S. Coast Guard, 1984). Skin contact can cause dermatitis (Figure 4.8; International Labour Office, 1983).

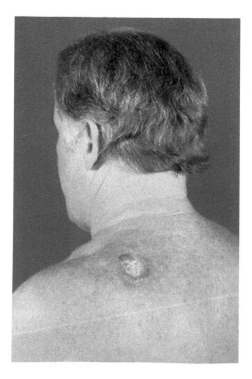

FIGURE 4.7 A melanoma on a man's shoulder. (Courtesy of Centers for Disease Control and Prevention, image #13416.)

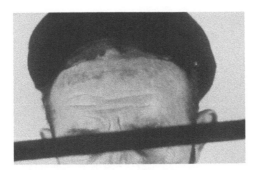

FIGURE 4.8 Contact dermatitis along the hairline. (Courtesy of Centers for Disease Control and Prevention, image #4486.)

Other toxicities that are similar to gasoline exposure are possible but toxicological studies of diesel oil are very limited. The study of diesel exhaust has shown this form to be very toxic to humans (Barath et al., 2010).

4.7 REMOVAL OF ORGANIC POLLUTANTS FROM THE SOIL

Human activities have led to land disposal of an increasing number of the organic chemicals produced. These chemicals are dispersed into soils, groundwater, and surface water (Calvet, 1989). Transport, transformation, and biological effects of organic chemicals in soils depend strongly on their retention by the solid organomineral phase. This retention is accomplished by several interaction mechanisms with the soil surface. These interactions include ion exchanges, interaction with metallic cations, polar interactions, charge transfers, and London–van der Waals dispersion forces/hydrogen effect (Calvet, 1989). These interactions with the soil will determine the fate of the organic chemicals. The strength of these interactions will control not only the potential of the pollutant to reach groundwater but also its bioavailability to plants and microorganisms. If any covalent bonds are created, the pollutant forms bound residues with soil that are no longer extractable or bioavailable. If the pollutant interacts with weak bonds, then it is likely to be more bioavailable for microbes or plants or be transported to groundwater (Chamignon et al., 2008).

4.7.1 REMEDIATION OF SOIL POLLUTANTS

Remediation is the degradation or removal of environmental pollutants from the soil (Das and Chandran, 2011). The traditional technologies used for this process included excavation, transport to specialized landfills, incineration, stabilization, and vitrification. Recently, there has been much interest in bioremediation technologies that use plants and microorganisms to degrade the pollutants in the soil to less-toxic and/or nontoxic substances (McGuinness and Dowling, 2009; Seo et al., 2009).

Bioremediation technologies have many advantages over traditional remediation technologies as they can be applied *in situ* without the need for removal and transport of the polluted soil (Figure 4.9), are usually less expensive and less labor-intensive, and have a higher level of public acceptance (Seo et al., 2009). There are conditions where *in situ* bioremediation is not adequate and the soil must be handled *ex situ* by placing it into a slurry bioreactor. These conditions are when (1) oily sludges, soils, or sediments are polluted with high concentrations of recalcitrant contaminants, for example, polynuclear aromatic hydrocarbons, diesel, explosives, pesticides, and chlorinated organic pollutants; (2) clayey and stratified soils with low hydraulic conductivity and low permeability are accompanied by high contents of organic matter; (3) the soils are in regions where

FIGURE 4.9 A worker monitoring the performance of an *in situ* bioremediation project in Maryland, United States. Note the relative lack of soil disturbance. (Courtesy of U.S. Geological Survey.)

environmental conditions are adverse to biological processes, for example, low temperatures; or (4) the contaminated sites require a short remediation time. Slurry bioreactor technology provides mixing, bioaugmentation, addition of surfactants and solvents, and enhanced microorganisms to biodegrade the soil pollutants (Robles-Gonzalez et al., 2008).

Unfortunately, bioremediation is a very complex technology that is not yet fully understood. The number of natural or man-made organic compounds present in the biosphere is between 8 and 16 million molecular species, of which as many as 40,000 are predominant in our daily lives. However, after 50 years of research and biodegradation, detailed knowledge about the degrading pathways is available for only about 900 chemical species (Gomez et al., 2007). The process of biodegradation of these organics is complicated by the ever-increasing numbers of anthropogenic chemicals in the environment. Adverse effects from these anthropogenic chemicals are seen in the environment because they are not readily degraded by microbes (Copley, 2009). This microbial metabolism not only cannot eliminate many of these chemicals but also can generate chemical species that are as toxic or as persistent as the original ones (Gomez et al., 2007). These physio-chemical and biological processes to remove pollutants from the environment are, in general, a slow and unpredictable way of counteracting anthropogenic pollution and irreversible damage to the biosphere. Bacteria, which have been evolving for more than 3 billion years, have developed strategies for obtaining energy from virtually every group of compounds. On the other hand, anthropogenic compounds have been in contact with these bacteria for only about 100 years; therefore some of them are still poorly degraded (Diaz, 2004).

The catabolism of environmental pollutants requires the deployment of a series of enzymes (Wackett, 2004). These microbial enzymes that act upon the anthropogenic compounds arise from promiscuous activities of previously existing enzymes that served different roles in other pathways. These enzymes need to be available when their promiscuous activities are needed and at the same time they must continue to serve their original functions. The affinity for the normal substrate for the active site is likely to be higher than the promiscuous substrate, thereby reducing any activity on any compound except the natural substrate (Copley, 2009).

Bacteria are the most active agents in bioremediation. Fungal genera are the next most active organisms in degradation of organic pollutants (Tigini et al., 2009). Though algae and protozoa are important members of the microbial community in the aquatic and terrestrial ecosystems, they are scanty in their involvement in hydrocarbon biodegradation. Some reports have not shown protozoa have any ability to utilize hydrocarbons (Das and Chandran, 2011).

There are several factors that influence the degradation of organic pollutants. Temperature is the most important physical factor, with the highest rates of degradation at 30°C–40°C in the soil environment. Nutrients are very important, especially nitrogen, phosphorous, and, in some cases, iron. When the supply of carbon is significantly increased, the availabilities of nitrogen and phosphorous generally become the limiting factors in degradation (Das and Chandran, 2011). Other factors affecting degradation of organic chemicals in soil include pH, time, adsorption, moisture, and soil types (Laveglia and Dahm, 1977). Other toxic pollutants in the soil, especially metals, can strongly inhibit the biodegradation by microorganisms when the toxicants interact with enzymes directly involved in biodegradation or with enzymes involved in general metabolism (Sandrin and Maier, 2003).

4.7.2 Phytoremediation of Contaminated Soils

Phytoremediation uses vegetation for *in situ* treatment of contaminated soils, sediments, and water. It is used at sites where the pollutants can be accessed by the roots of plants and can be sequestered, degraded, immobilized, or metabolized in place (Dietz and Schnoor, 2001). There has been some use of plants alone for bioremediation but the use of plants in conjunction with plant-associated bacteria offers much potential for bioremediation (Das and Chandran, 2011).

Phytoremediation is popular because of its cost-effectiveness, aesthetic advantages, and long-term applicability (Dietz and Schnoor, 2001). Phytoremediation can be cost-effective (a) for large sites with shallow residual levels of contamination by organic, nutrient, or metal pollutants, where contamination does not pose an imminent danger and only "polishing treatment" is required; and (b) where vegetation is used as a final cap and closure of the site (Das and Chandran, 2011).

Plants have shown the capacity to tolerate high concentrations of organic pollutants without adverse effects, and they can take up and convert chemicals to less toxic metabolites. Additionally, they degrade chemicals in the rhizosphere by the release of root exudates and enzymes and cause a buildup of carbon in the soil (Dietz and Schnoor, 2001). This involvement of the rhizosphere and the microbial contribution to the degradation of organic contaminants have been shown to be the most significant mechanism for removal of petroleum products in vegetated, contaminated soils (Das and Chandran, 2011). As with bacteria, only certain plants are capable of or efficient at phytoremediation of organics from the soil. The selection of capable plants for particular pollutants, the association of mycorrhizal fungi and bacteria with the plants' roots, and the development of transgenic plants that express bacterial enzymes will improve plant tolerance and metabolism of toxic organic compounds in the soil (McGuinness and Dowling, 2009).

4.8 CONCLUSIONS

The planet's soils are the final depositary for most anthropogenic and natural organic pollutants. Many of these were applied directly to the soil or the plants growing in the soil. Others have found their way there through air or water. The vastness of the soil has led to the dilution of these pollutants, and most of the pollutants remain on or near the surface of the soil unless they have been moved by the action of water, organisms, or mechanical mixing such as plowing. This dilution has reduced the toxicity of these pollutants but the unknown factor is the action of the soil, its chemistry, and the combined action of all the microorganisms, plants, and invertebrates that live in the soil. This biological action, combined with the influences of the soil components, has the potential to create new metabolites and chemicals. The science of toxicology is just beginning to recognize the need to study chemical mixtures and to expand these studies to include the POPs and the organic pollutants that can bioaccumulate in organisms (Feron et al., 1998). These types of studies have not fit well into the standard toxicological methodology, which deals with one compound at a relatively high concentration in one organism. The ability to study a single compound at a very low dose and for a very long period of time is extremely difficult, and this certainly does not fit into our present grant system. The combination of multiple compounds acting in association with hundreds of species in a complex chemical medium like the soil is very difficult to study.

If we look at the example of dioxin, we get a possible scenario for what might be occurring in the soil. The chemistry that produced dioxin would never occur in soil, since dioxin is produced during the production of organic compounds, like herbicides, at high temperatures and with available halogens. But dioxin is a by-product and is produced at very low concentrations, and still it is so extremely toxic that after much further dilution, it is still a problem. We do not know if the addition of other organic chemicals to the soil is creating other very toxic xenobiotics and at very low concentrations but with important health effects on humans and other organisms. These unknown compounds could be accumulating in plants that we use for food or as forage for our livestock and then bioaccumulating in the livestock and then in us. This process has been demonstrated with dioxin (Tuyet-Hanh et al., 2010). Unless someone looks for these unknown xenobiotics, we will never know.

There have been vast increases in several diseases since the mass introduction of organic pesticides after World War II. One example is autism: the number of diagnosed cases has risen at an unheard of rate and the best guess for the etiology of autism is something in the environment (Szpir, 2006; Daniels, 2006). No one knows if the environmental cause of autism is being produced in our soils.

The soil and all the xenobiotics that we allow to accumulate there need to be considered as a serious and potential threat to our health. We cannot dismiss the problem until we have scientifically investigated what is occurring in the soil.

REFERENCES

Aelion, C.M. 2004. Soil contamination monitoring. *In* H.I. Inyang and J.L. Daniels (eds), *Environmental Monitoring*. Encyclopedia of Life Support Systems (EOLSS), Developed under the Auspices of the UNESCO, EOLSS Publishers, Oxford. http://www.eolss.net (verified 6 January 2012).

Aken, B.V., P.A. Correa, and J.L. Schnoor. 2010. Phytoremediation of polychlorinated biphenyls: New trends and promises. *Environmental Science & Technology* 44(8): 2767–2776.

Albihn, A. 2001. Recycling biowaste—Human and animal health problems. *Acta Veterinaria Scandinavica* (Suppl. 95): 69–75.

Barath, S., N.L. Mills, M. Lundback, H. Tornqvist, A.J. Lucking, J.P. Langrish, S. Soderberg, et al. 2010. Impaired vascular function after exposure to diesel exhaust generated at urban transient running conditions. *Particle and Fibre Toxicology*. http://www.particleandfibretoxicology.com/content/7/1/19. Accessed 4 January 2012 (verified 6 January 2012).

Billiard, S.M., J.N. Meyer, D.M. Wassenberg, P.V. Hodson, and R.T. Di Giulio. 2008. Nonadditive effects of PAHs on early vertebrate development: Mechanisms and implications for risk assessment. *Toxicological Sciences* 105(1): 5–23.

Bjorling-Poulsen, M., H.R. Anderson, and P. Grandjean. 2008. Potential developmental neurotoxicity of pesticides used in Europe. *Enviromental Health* 7(50). http://www.ehjournal.net/content/7/1/50. Accessed 13 July 2011 (verified 6 January 2012).

Calabrese, E.J. and L.A. Baldwin. 1998. Hormesis as a biological hypothesis. *Environmental Health Perspectives* 106(Suppl. 1): 357–362.

Calvet, R. 1989. Adsorption of organic chemicals in soils. *Environmental Health Perspectives* 88: 145–177.

Caprino, L. and G.I. Togna. 1998. Potential health effects of gasoline and its constituents: A review of current literature (1990–1997) on toxicological data. *Environmental Health Perspectives* 106(3): 115–125.

Carpenter, D.O., K.F. Arcaro, B. Bush, W.D. Niemi, S. Pang, and D.D. Vakharia. 1998. Human health and chemical mixtures: An overview. *Environmental Health Perspectives* 106(Suppl. 6): 1263–1270.

Carpenter, D.O., K.F. Arcaro, and D.C. Spink. 2002. Understanding the human effects of chemical mixtures. *Environmental Health Perspectives* 110(Suppl. 1): 25–42.

Chamignon, C., N. Haroune, C. Forano, A.-M. Delort, P. Besse-Hoggan, and B. Combourieu. 2008. Mobility of organic pollutants in soil components. What role can magic angle spinning NMR play? *European Journal of Soil Science* 59: 572–583.

Chaudhry, G.R. and S. Chapalamadugu. 1991. Biodegradation of halogenated organic compounds. *Microbiological Reviews* 55(1): 59–79.

Chmill, S., S. Kadow, M. Winter, H. Weighardt, and C. Esser. 2010. 2,3,7,8-Tetrachlorodibenzo-*p*-dioxin impairs stable establishment of oral tolerance in mice. *Toxicological Sciences* 118(1): 98–107.

Cocco, P. 2002. On the rumors about the silent spring. Review of the scientific evidence linking occupational and environmental pesticide exposure to endocrine disruption health effects. *Cadernos de Saúde Pública* 18(2): 379–402.

Cohen, M. 2007. Environmental toxins and health. *Australian Family Physician* 36(12): 1002–1004.

Copley, S.D. 2009. Evolution of efficient pathways for degradation of anthropogenic chemicals. *Nature Chemistry Biology* 5(8): 559–566.

Curran, C.P., C.V. Vorhees, M.T. Williams, M.B. Genter, M.L. Miller, and D.W. Nebert. 2011. *In utero* and lactational exposure to a complex mixture of polychlorinated biphenyls: Toxicity in pups dependent on the Cyp1a2 and Ahr genotypes. *Toxicological Sciences* 119(1): 189–208.

Dai, D. and T.J. Oyana. 2008. Spatial variations in the incidence of breast concer and potential risks associated with soil dioxin contamination in Midland, Saginaw, and Bay Counties, Michigan, USA. *Environmental Health* 7(49): 7–49. http://www.ehjournal.net/content/7/1/49. Accessed 13 July 2011 (verified 6 January 2012).

Daniels, J.L. 2006. Autism and the environment. *Enviromental Health Perspectives* 114(7): A396.

Das, N. and P. Chandran. 2011. Microbial degradation of petroleum hydrocarbon contaminants: An overview. *Biotechnology Research International* 2011: 1–13.

Diaz, E. 2004. Bacterial degradation of aromatic pollutants: A paradigm of metabolic versatility. *International Microbiology* 7: 173–180.

Dietz, A.C. and J.L. Schnoor. 2001. Advances in phytoremediation. *Environmental Health Perspectives* 109(Suppl. 1): 163–168.

Duke, S.O. 1990. Overview of herbicide mechanisms of action. *Environmental Health Perspectives* 87: 263–271.

EPA. 2008. Polycyclic Aromatic Hydrocarbons (PAHs). United States Environmental Protection Agency, Office of Solid Waste. http://www.epa.gov/wastes/hazard/wastemin/minimize/factshts/pahs.pdf (verified 6 January 2012).

Eskenazi, B., A. Bradman, and R. Castorina. 1999. Exposures of children to organophosphate pesticides and their potential adverse health effects. *Environmental Health Perspectives* 107(Suppl. 3): 409–419.

Fenton, S.E. 2006. Endocrine-disruption compounds and mammary gland development: Early exposure and later life consequences. *Endocrinology* 147(Suppl. 6): S18–S24.

Feron, V.J., F.R. Cassee, and J.P. Groten. 1998. Toxicology of chemical mixtures: International perspective. *Enviromental Health Perspectives* 106(Suppl. 6): 1281–1289.

Feron, V.J., F.R. Cassec, J.P. Groten, P.W. van Vilet, and J.A. van Zorge. 2002. International issues on human health effects of exposure to chemical mixtures. *Enviromental Health Perspectives* 110(Suppl. 6): 893–899.

Franzblau, A., E. Hedgeman, Q. Chen, S.-Y. Lee, P. Adriaens, A. Demond, D. Garabrant, et al. 2008. Case report: Human exposure to dioxins from clay. *Environmental Health Perspectives* 116(2): 238–242.

Gilliom, R.J., J.E. Barbash, C.G. Crawford, P.A. Hamilton, J.D. Martin, N. Nakagaki, L.H. Nowell, et al. 2007. *The Quality of Our Nation's Waters: Pesticides in the Nation's Streams and Ground Water, 1992–2001*. Circular 1291, U.S. Geological Survey, Reston, VA. http://pubs.usgs.gov/circ/2005/1291/ (verified 6 January 2012).

Godschalk, R.W.L., F.-J. Van Schooten, and H. Bartsch. 2003. A critical evaluation of DNA adducts as biological markers for human exposure to polycyclic aromatic compounds. *Journal of Biochemistry and Molecular Biology* 36(1): 1–11.

Gomez, M.J., F. Pazos, F.J. Guijarro, V. de Lorenzo, and A. Valencia. 2007. The environmental fate of organic pollutants through the global microbial metabolism. *Molecular Systems Biology* 3(114): 1–11.

Hamers, T., J.H. Kamstra, P.H. Cenijn, K. Pencikova, L. Palkova, P. Simeckova, J. Vondracek, P.L. Andersson, M. Stenberg, and M. Machala. 2011. *In vitro* toxicity profiling of ultrapure non-dioxin-like polychlorinated biphenyl congeners and their relative toxic contribution to PCB mixtures in humans. *Toxicological Sciences* 121(1): 88–100.

Herrick, R.F., D.J. Lefkowitz, and G.A. Weymouth. 2007. Soil contamination from PCB-containing buildings. *Enviromental Health Perspectives* 115(2): 173–175.

Hough, R.L. 2007. Soil and human health: An epidemiological review. *European Journal of Soil Science* 58: 1200–1212.

IARC. 1989. Occupational exposures in petroleum refining; crude oil and major petroleum fuels. *Monographs on the Evaluation of the Carcinogenic Risk of Chemicals to Man*, Vol. 45. World Health Organization, International Agency for Research on Cancer, Geneva, Switzerland.

International Labour Office. 1983. *Encyclopedia of Occupational Health and Safety*, Vols I and II, 3rd edn. International Labour Office, Geneva, Switzerland.

Jorgenson, J.L. 2001. Aldrin and dieldrin: A review of research on their production, environmental deposition and fate, bioaccumulation, toxicology, and epidemiology in the United States. *Environmental Health Perspectives* 109(Suppl. 1): 113–139.

Kamanyire, R. and L. Karalliedde. 2004. Organophosphate toxicity and occupational exosure. *Occupational Medicine* 54(2): 69–75.

Karmel, F. and J.A. Hoppin. 2004. Association of pesticide exposure with neurologic dysfunction and disease. *Environmental Health Perspectives* 112(9): 950–958.

Klaassen, C. (ed.). 2008. *Casarett and Doull's Toxicology: The Basic Science of Poisons*, 7th edn. McGraw-Hill, New York, NY.

Kogevinas, M. 2001. Human health effects of dioxins: Cancer, reproductive and endocrine system effects. *European Society of Human Reproduction and Embryology* 7(3): 331–339.

Koplan, J.P. 2000. Toxicological profiles for polychlorinated biphenyls (PCBs). U.S. Department of Health and Human Services, Public Health Service, Agency for Toxic Substances and Disease Registry, Atlanta, GA.

Kramer, R.E. and R.C. Baker. 2004. Herbicides. *In* T. Satoh (ed.), *Environmental Toxicology and Human Health*. EOLSS Publishers, Oxford.

Lau, E.V., S. Gan and H.K. Ng. 2010. Extraction Techniques for Polycyclic aromatic hydrocarbons in soils. *International Journal of Analytical Chemistry* 2010: 1–9.

Laveglia, J. and P.A. Dahm. 1977. Degradation of organphosphorus and carbamate insecticides in the soil and by soil microorganisms. *Annual Review of Entomology* 22: 483–513.

Leake, J.R., A. Adam-Bradford, and J. Rigby. 2009. Health benefits of "grow your own food" in urban areas: Implications for contaminated land risk assessment and risk management. *Environmental Health* 8(Suppl. 1): S6. http://www.ehjournal.net/content/8S1/S6. Accessed 12 July 2011 (verified 6 January 2012).

Leibson, T. and M. Lifshitz. 2008. Organophosphate and carbamate poisoning: Review of the current literature and summary of clinical and laboratory experience in southern Israel. *Israel Medical Association Journal* 10: 767–770.

Matsumoto, E., Y. Kawanaka, S.-J. Yun, and H. Oyaizu. 2009. Bioremediation of the organochlorine pesticides, dieldrin and endrin, and their occurence in the environment. *Applied Microbiology and Biotechnology* 84: 205–216.

Mattsson, A., S. Lundstedt, and U. Stenius. 2009. Exposure of HepG2 cells to low levels of PAH-containing extracts from contaminated soils results in unpredictable genotoxic stress responses. *Environmental and Molecular Mutagenesis* 50: 337–348.

Mauderly, J.L. 1993. Toxicological approaches to complex mixtures. *Environmental Health Perspectives Supplements* 101(Suppl. 4): 155–165.

McGuinness, M. and D. Dowling. 2009. Plant-associated bacterial degradation of toxic organic compounds in soil. *International Journal of Environmental Research and Public Health* 6: 2226–2247.

Mordukhovich, I., P. Rossner Jr, M.B. Terry, R. Santella, Y.-J. Zhang, H. Hibshoosh, L. Memeo, et al. 2010. Associations between polycyclic aromatic hydrocarbon-related exposures and p53 mutations in breast tumors. *Environmental Health Perspectives* 118(4): 511–518.

Nhu, D.D., T. Kido, R. Naganuma, N. Swano, K. Tawara, M. Nishijo, H. Nakagawa, N.N. Hung, and L.T.H. Thom. 2009. A GIS study of dioxin contamination in a Vietnamese region sprayed with herbicide. *Environmental Health and Preventive Medicine* 14: 353–360.

Niu, J. and G. Yu. 2004. Agricultural chemicals. *In* Q. Yi (ed.), *Environmental and Ecological Sciences, Engineering and Technology Resources*. Encyclopedia of Life Support Systems (EOLSS), Developed under the Auspices of the UNESCO, EOLSS Publishers, Oxford. http://www.eolss.net (verified 6 January 2012).

Petrosyan, V.S. 2004. Organic pollution from agrochemicals. *In* H.I. Inyang and J.L. Daniels (eds), *Environmental Monitoring*. Encyclopedia of Life Support Systems (EOLSS), Developed under the Auspices of the UNESCO, EOLSS Publishers, Oxford. http://www.eolss.net (verified 6 January 2012).

Pettry, D.E., R.B. Reneau Jr, M.I. Shanholtz, S.A. Graham Jr, and C.W. Weston. 1973. Soil pollution and environmental health. *Health Services Reports* 88(4): 323–327.

Placha, D., H. Raclavska, D. Matysek, and M.H. Rummeli. 2009. The polycyclic aromatic hydrocarbon concentration in soils in the Region of Valasske Merzirici, the Czech Republic. *Geochemical Transactions* 10(12): 1–21.

Pohl, H., C. DeRosa, and J. Holler. 1995. Public health assessment for dioxin exposure from soil. *Chemosphere* 31(1): 2437–2454.

Reese, E. and R.D. Kimbrough. 1993. Acute toxicity of gasoline and some additives. *Environmental Health Perspectives* 101(Suppl. 6): 115–131.

Rideout, K. and K. Teschke. 2004. Potential for increased human foodborne exposure to PCDD/F when recycling sewage sludge on agricultural land. *Environmental Health Perspectives* 112(9): 959–969.

Risher, J.F., F.L. Mink, and J.F. Stara. 1987. The toxicologic effects of the carbamate insecticide aldicarb in mammals: A review. *Environmental Health Perspectives* 72: 267–281.

Roberts, D.M., N.A. Buckley, F. Mohamed, M. Eddleston, D.A. Goldstein, A. Mehrsheikh, M.S. Bleeke, and A.H. Dawson. 2010. A prospective observational study of the clinical toxicology of glyphosate-containing herbicides in adults with acute self-poisoning. *Clinical Toxicology (Phila)* 48(2): 129–136.

Robles-Gonzalez, I.V., F. Fava, and H.M. Poggi-Varaldo. 2008. A review on slurry bioreactors for bioremediation of soils and sediments. *Microbial Cell Factories* 7(5). http://www.microbialcellfactories.com/content/7/1/5 (verified 6 January 2012).

Rocca, C.L. and A. Mantovani. 2006. From environment to food: The case of PCB. *Annals of the Italian National Institute of Health* 42(4): 410–416.

Roeder, R.A., M.J. Garber, and G.T. Schelling. 1998. Assessment of dioxins in foods from animal origins. *Journal of Animal Science* 76: 142–151.

Safe, S.H. 1992. Toxicology, structure–function relationship, and human and environmental health impacts of polychlorinated biphenyls: Progress and problems. *Environmental Health Perspectives* 100: 259–268.

Safe, S.H. 2000. Endocrine disruptors and human health—Is there a problem? An update. *Environmental Health Perspectives* 108(6): 487–493.

Sandrin, T.R. and R.M. Maier. 2003. Impacts of metals on the biodegradation of organic pollutants. *Environmental Health Perspectives* 111(8): 1093–1101.

Sapkota, A.R., L.Y. Lefferts, S. McKenzie, and P. Walker. 2007. What do we feed to food-production animals? A review of animal feed ingredients and their potential impact on human health. *Environmental Health Perspectives* 115(5): 663–670.

Sedman, R.M. 1989. The development of applied action levels for soil contact: A scenario for the exposure of humans to soil in a residential setting. *Enviromental Health Perspectives* 79: 291–313.

Seo, J.-S., Y.-S. Keum, and Q.X. Li. 2009. Bacterial degradation of aromatic compounds. *International Journal of Analytical Chemistry* 6: 278–309.

Sing, D. and C.F. Sing. 2010. Impact of direct soil exposures from airborne dust and geophagy on human health. *International Journal of Environmental Research and Public Health* 7: 1205–1223.

Solomon, G.M. and A.M. Huddle. 2002. Low levels of persistent organic pollutants raise concerns for future generations. *Journal of Epidemiological Community Health* 56: 826–827.

Sulentic, C.E.W. and N.E. Kaminski. 2011. The long winding road toward undestanding the molecular mechanisms for B-cell suppression by 2,3,7,7-tetrachlorodibenzo-*p*-dioxin. *Toxicological Sciences* 120(S1): S171–S191.

Szpir, M. 2006. Tracing the origins of autism: A spectrum of new studies. *Enviromental Health Perspectives* 114(7): A412–A418.

Tigini, V., V. Prigione, S. Di Toro, F. Fava, and G.C. Varese. 2009. Isolation and characterisation of polychlorinated biphenyl (PCB) degrading fungi from a historically contaminated soil. *Microbial Cell Factories* 8(5). http://www.microbialcellfactories.com/content/8/1/5 (verified 6 January 2012).

Toppari, J., J.C. Larsen, P. Christiansen, A. Giwercman, P. Grandjean, L.J. Guillette Jr, B. Jegou, et al. 1996. Male reproductive health and environmental xenoestrogens. *Environmental Health Perspectives* 104(Suppl. 4): 741–803.

Turner, K.J. and R.M. Sharpe. 1997. Environmental oestrogens-present understanding. *Journal of Repouction and Fertility* 2: 69–73.

Tuyet-Hanh, T.T., L. Vu-Anh, N. Ngoc-Bich, and T. Tenkate. 2010. Environmental health risk assessment of dioxin exposure through foods in a dioxin hot spot—Bien Hoa City, Vietnam. *International Journal of Environmental Research and Public Health* 7: 2395–2406.

U.S. Coast Guard. 1984. *CHRIS—Hazardous Chemical Data*. U.S. Government Printing Office, Washington, DC.

van den Berg, M., L.S. Birnbaum, M. Denison, M. De Vito, W. Farland, M. Feeley, H. Fiedler, et al. 2006. The 2005 World Health Organization reevaluation of human and mammalian toxic equivalency factors for dioxins and dioxin-like compounds. *Toxicological Sciences* 93(2): 223–241.

Vega, F.A., E.F. Covelo, and M.L. Andrade. 2007. Accidental organochlorine pesticide contamination of soil in Porrino, Spain. *Journal of Environmental Quality* 36: 272–279.

Wackett, L.P. 2004. Evolution of enzymes for the metabolism of new chemical inputs into the environment. *The Journal of Biological Chemistry* 279(40): 41259–41262.

Wang, T., Y. Want, J. Fu, P. Wang, Y. Li, Q. Zhang, and G. Jiang. 2010. Characteristic accumulation and soil penetration of polychlorinated biphenyls and polybrominated diphenyl ethers in wastewater irrigated farmlands. *Chemosphere* 81: 1045–1051.

White, S.S. and L.S. Birnbaum. 2009. An overview of the effects of dioxins and dioxin-like compounds on vertebrates, as documented in human and ecological epidemiology. *Journal of Environmental Science and Health Part C: Environmental Carcinogenesis & Ecotoxicology Reviews* 27(4): 197–211.

Wilson, N.K., J.C. Chuang, C. Lyu, R. Menton, and M.K. Morgan. 2003. Aggregate exposures of nine preschool children to persistent organic pollutants at day care and at home. *Journal of Exposure Analysis and Environmental Epidemiology* 13: 187–202.

Zanaroli, G., S. Di Toro, D. Todaro, G.C. Varese, A. Bertolotto, and F. Fava. 2010. Characterization of two diesel fuel degrading microbial consortia enriched from a non acclimated, complex source of microorganisms. *Microbial Cell Factories* 9(10). http://www.microbialcellfactories.com/content/9/1/10 (verified 6 January 2012).

5 Human Disease from Introduced and Resident Soilborne Pathogens

Thomas E. Loynachan

CONTENTS

5.1 INTRODUCTION

When we were young, most of us were told to wash our hands and face. Dirt (soil out of place) was something to be avoided. This was especially true when dirty objects were near the mouth and nose. In relatively recent times (beginning about 150 years ago), science showed that microorganisms living or surviving in the soil can cause human health problems. We now know that these organisms can be quite diverse and include viruses, bacteria, fungi, protozoa, and some nasty parasitic nematodes. Soil can be a major source for transmission of disease. It is not necessary for humans to physically eat soil to come into contact with infectious organisms. We accidentally ingest soil when we put our dirty fingers in our mouths or eat fruits or vegetables that have not been thoroughly washed and cooked. Preharvest contamination may occur in agricultural products subjected to irrigation with reclaimed wastewater, crop fertilization with sewage sludge, or fecal pollution in the soils in which they are grown (one needs to be especially careful where human feces are directly applied to soils). We may even swallow parasites in unfiltered water that has washed off or passed through contaminated soils. Soil particles may be picked up by the wind and inhaled. For some diseases (especially caused by worms), we need to avoid physical contact with soil to avoid the disease. Avoiding contact with soil, however, for many people is not feasible. We live upon soil, dig in soil, breath air that may be filled with soil, use soil to dispose of human and animal wastes, use soil to purify drinking water, grow foodstuffs in soil, and consume animals that live upon soil. Thus humans have developed a close relationship with soil and, in fact, we depend upon soil for our very existence.

Some pathogens naturally occur in the soils of the world and others are transitory, often following contamination by wastes from diseased humans or animals. In the latter case, the soil serves as a means of transmission to foster new infections. These organisms may not grow in soil but can survive until contact is made with a new host. Advances in food safety, personal hygiene, and water treatment have greatly reduced the threat of pathogens being transmitted by soil, especially in developed countries of the world.

Disease-causing organisms may infect humans by (1) passing through the skin (e.g., several nematodes such as hookworms), (2) contaminating a surface skin abrasion (e.g., *Clostridium tetani*), (3) being ingested orally (e.g., *Clostridium botulinum*), or (4) being inhaled into the respiratory system (e.g., *Aspergillus fumigatus*). Enteric organisms (found in the intestines of humans and animals) must survive in the soil until contacting a new host. There are many bacteria (*Salmonella* spp., *Shigella* spp., *Campylobacter* spp., *Yersinia* spp.) and viruses (hepatitis A, hepatitis E, adenoviruses) in this category. Protozoa (e.g., *Entamoeba histolytica*) and worms (e.g., *Ascaris lumbricoides*) are also enteric and can survive for periods of time in soil. Soil conditions are important in determining the length of survival for many of the enterics.

5.2 PATHOGEN IMPACT ON HUMANS

5.2.1 FACULTATIVE PARASITES AND PATHOGENS

Organisms may be native soil inhabitants (geophilic) and when conditions are favorable cause disease within a host. These may be normal flora or fauna of the soil. When coming into contact with a host, they may or may not cause disease. The condition of the host and the dose level received are important. The former are sometimes called opportunistic pathogens and people with compromised immune systems are especially susceptible.

The infective dose is the number of virulent organisms entering the host required to cause disease. Human feeding studies suggest, with several diseases, that about 400–500 bacteria may cause illness in some individuals. Other organisms, often those that produce enterotoxins, such as *Shigella* spp. and *Escherichia coli* O157:H7, may require as few as 10 cells to cause infection. Enteric viruses have very low infectious doses in the order of tens to hundreds of virions. In contrast, ingested *Salmonella* spp. may require between 1 million and 1 billion bacteria to cause infection. Thus, just because an infective organism comes into contact with a host does not necessarily mean infection will result, and the threshold varies by disease organism and host.

5.2.2 PATHWAY TO HUMANS

5.2.2.1 Ingestion (Contaminated Food, Water, or Soil)

Ingestion is the most common means of infection of humans. This is particularly true when the source has recently been contaminated by fecal material from an infected individual.

5.2.2.2 Inhalation

Humans breathe on average approximately 25 pounds of air daily (Griffin, 2007). Dust may contain high concentrations of microorganisms in the form of fungal spores, bacteria, and viruses. Also, microbes may adhere to fine aquatic sprays. Griffin (2007) states that studies have identified a wide range of dustborne pathogens that can move great distances through the atmosphere, and "it is clear that we have only begun to grasp the true numbers of microorganisms capable of using the atmosphere as an infectious route or as a means of extending the limit of their dispersion."

Tanner et al. (2008) evaluated the atmosphere downwind from biosolid land-application sites. Risks were primarily related to aerosols, with viruses posing the greatest threat. The study suggested that aerosols from land application of biosolids pose a detectable but manageable risk to biosolid workers.

5.2.2.3 Surface Contact and Trauma

For many pathogens the skin serves as an effective barrier. Some organisms (especially nematodes) can penetrate intact skin, while other organisms penetrate the skin only when it has been torn or cut. Deep puncture wounds are particularly troublesome because they often bleed little and thus pathogens are not washed away from the wound. Deep puncture wounds also may scab-over quickly and contain the pathogens within the body where damage occurs.

5.3 TYPES OF SOILBORNE PATHOGENS

The principal soilborne microbial pathogens that infect humans can be divided into five types (from smallest to largest): viruses, bacteria, fungi, protozoa, and helminths. Specific traits of the more common soilborne pathogens are discussed in each category.

5.3.1 Viruses

Viruses typically range from 20 to 300 nm in length. They are simple organisms that contain either RNA or DNA surrounded by a protein coat (capsid). They may or may not have an attachment tail. A unique feature of viruses is they cannot reproduce independently of a host; that is, their life cycle must go through a host for them to multiply. Most living organisms (from bacteria to plants and animals, including humans) can serve as a host.

Viruses are a major cause of gastroenteritis worldwide, especially in infants and young children. Many human viruses infect the gastrointestinal tract and are excreted in feces into the environment. Numbers of viral particles may be 10^{10}–10^{11}/g of feces of infected individuals. These are often called enteric viruses because they are associated with human intestines.

The longer a virus survives in the environment outside its host, the greater the risk of transmission to a new host. Survival varies with conditions of the environment such as heat, moisture, and pH. Survival in the environment often leads to incidences of human infection through contaminated food or water (both drinking and recreational waters). Because of their small size, viruses, of all human pathogens, are most likely to contaminate groundwater. Viruses can move considerable distances (hundreds of meters) in subsurface environments (Borchardt et al., 2003). In a U.S. study, 72% of groundwater sites were positive for human enteric viruses (Fout et al., 2003). The enteric viruses most frequently associated with disease outbreaks are noroviruses and hepatitis A (Borchardt et al., 2003).

Current safety standards for food and water quality typically do not specify levels of viruses considered acceptable (Okoh et al., 2010). Viruses are thought to be more stable than common bacterial indicators in the environment and human adenoviruses have been proposed as a suitable index for environmental contamination (Okoh et al., 2010). They are prevalent and stable; they are considered human specific and are not detected in animal wastewaters or slaughterhouse sewage; and PCR-based procedures show adequate sensitivity to detect specific serotypes associated only with human contamination (Okoh et al., 2010).

The following subsections contain a brief summary of the major viral pathogens that survive in soil.

5.3.1.1 Hepatitis A

Viral hepatitis results in liver inflammation and is often identified as one of five hepatitis viruses (A, B, C, D, or E). Hepatitis A is caused by the hepatitis A virus (HAV). It is most commonly transmitted by the fecal–oral route through food or water contamination. Hepatitis A is different from other types of hepatitis because it typically is less serious and does not develop into chronic hepatitis. The average incubation period is approximately 1 month, so the long time between contamination and occurrence makes diagnosis difficult. In many less-developed countries, occurrence of infection is high and the native population develops resistance against the disease. Symptoms include fatigue, fever, nausea, diarrhea, weight loss, and light-colored stools.

In column studies where water was leached through soil, HAVs were retained in clay loam soils but passed relatively easily through coarse sand and muck (organic-rich) soils (Sobsey et al., 1995). The clay loam columns removed greater than 99.98% of the viruses in 10 cm of soil.

5.3.1.2 Hepatitis E

Hepatitis B and C are often transmitted through blood or sexual contact, so they do not normally contact soil. Hepatitis D seemingly can only propagate in the presence of the hepatitis B virus. Hepatitis E, like hepatitis A, has a fecal–oral transmission route and is caused by the hepatitis E virus (HEV). Symptoms are similar to those of hepatitis A but may be more severe, especially in women who are pregnant. Animals appear to serve as a reservoir of the virus. It is common in developing countries, especially those with hot climates. As stated in a World Health Organization (WHO, 2005) article, "As no specific therapy is capable of altering the course of acute hepatitis E infection, prevention is the most effective approach against the disease." Swine manure may serve as an important source of HEV when applied to soils (Fernández-Barredo et al., 2006).

5.3.1.3 Noroviruses

These are single-stranded RNA viruses shown in recent years to be a major cause of acute gastroenteritis in humans. The illness is usually short and symptoms include nausea, vomiting, diarrhea, and stomach cramping. Viruses are transmitted by direct contact or the fecal–oral route and affect people of all ages. The viruses are highly infective and as few as 10 viral particles can cause disease. Noroviruses have been recovered from soil and water environments.

5.3.1.4 Enteric Adenoviruses

Adenoviruses are approximately 70–90 nm long and contain double-stranded DNA (Figure 5.1). These viruses commonly cause respiratory illness but may also cause other symptoms such as gastroenteritis and diarrhea. Common transmission is by direct contact, fecal–oral pathways, or contaminated water. Adenoviruses are common and persistent in wastewaters. They have high resident times and may be good indicators of fecal contamination of water because of their persistence

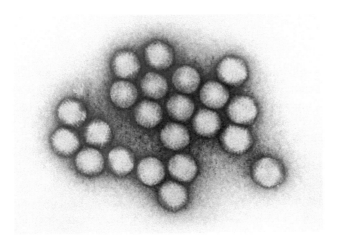

FIGURE 5.1 Viral particles are extremely small (70–90 nm) and consist of a protein coat (capsid) surrounding either RNA or DNA nucleic acid. Shown here is an electron micrograph of adenoviruses that contain double-stranded DNA. (Courtesy of Centers for Disease Control and Prevention, Public Health Image Library, image #10010.)

(Hundesa et al., 2006). Adenovirus DNA was recently found by real-time PCR in groundwater where fecal bacteria were not detected (Ogorzaly et al., 2010). The authors mention that this cool, dark environment may be conducive for DNA survival but caution that the presence of adenovirus DNA may not indicate a direct human health risk.

5.3.1.5 Poliovirus Types 1 and 2

These crippling viruses are often spread from person to person by the fecal–oral route. Polio was one of the most dreaded diseases of the first half of the twentieth century and crippled thousands of people; once contracted, there is no cure for polio. In some infections the viral particles enter the central nervous system and lead to poor development and coordination of skeletal muscles. The different types of polio viruses have slightly different compositions of their capsids (protein coat surrounding the RNA). Vaccines developed in the 1950s have been effective in preventing the spread of the disease in developed countries.

5.3.1.6 Echoviruses

These enteric RNA viruses cause symptoms ranging from rashes and diarrhea to bronchitis and inflammation of soft tissue (including the membranes surrounding the heart and brain). Echoviruses are highly infectious and can cause death in children. Wang et al. (1981) spiked secondary sewage effluents with echoviruses and continuously applied the effluent to the top of a soil column. Virus removal appeared to be related mainly to hydraulic flow rates: the faster the flow rate, the lower the rate of removal. Most viral particles were removed in the upper few centimeters of the column.

5.3.1.7 Coxsackievirus

These viruses contain single-stranded RNA and live in the human digestive tract. Like many viruses, the common mode of infection is by fecal–oral contamination. Most infections occur in children and often cause flu-like symptoms but can lead to more serious infections. The virus was named for Coxsackie, New York, where it was isolated from human feces in 1948. Jorgensen (1985) showed soil to be a good filter of coxsackievirus under unsaturated-flow conditions but the viruses penetrated much deeper into the soil column under saturated flow. He surmised that, under unsaturated flow, water moved mainly through micropores, which allowed good soil–virus contact. Under saturated flow, the viral particles moved mainly through macropores, with little soil–virus contact, thus not allowing adsorption.

5.3.2 BACTERIA

Bacteria are prokaryotic cells that are typically 1–3 μm in length. Soil is the home for many bacteria that are active in carbon and nutrient cycling. However, other bacteria that routinely live in soil can cause disease when they gain entry into a human host (Table 5.1). Still other bacteria originate from an infected host and survive in soil long enough to enter a new host by food, water, or direct soil contact.

5.3.2.1 *E. coli* O157:H7

E. coli O157:H7 is a Gram-negative slender rod-shaped bacterium. *E. coli* are normally harmless flora of the intestinal tract of humans and other mammals, often making up 0.1% of the total bacteria in the adult intestinal tract (Science Encyclopedia, 2005). The vast majority of *E. coli* are not only harmless to humans but are beneficial in aiding food digestion. These organisms usually survive for a short time outside the gut; thus, *E. coli* are often used as an indicator of recent human fecal contamination. One particular *E. coli* strain (O157:H7), however, contains a unique somatic antigen (O157) and flagella antigen (H7) that results in toxin production causing severe illness and possible death in humans. Those who recover, especially children, may have permanent kidney damage.

TABLE 5.1

Some Disease-Causing Bacteria and Actinomycetes Associated with Soil

Organism	Gateway	Disease Name	Associated Problems	Soil Residency[a]	Incidence	Geographic Distribution
Actinomadura spp.	Skin trauma	Maduromycosis, actinomycetoma	Disfigurement	Permanent	300–400/year	Tropical Africa, India, South and Central America
Bacillus anthracis	Respiration, skin trauma, ingestion	Anthrax	Ulcer w/black necrotic center, severe breathing problems, shock, nausea, fever, vomiting, diarrhea, abdominal pain	Periodic	Rare	South and Central America, southern and eastern Europe, Asia, Africa, the Caribbean, the Middle East
Burkholderia (Pseudomonas) pseudomallei	Respiration, ingestion, skin trauma	Melioidosis	Fever, aches, cough, diarrhea	Permanent	Common in Thailand	Worldwide, most common in tropics and subtropics
Clostridium tetani	Skin trauma	Tetanus (lockjaw)	Painful muscle tightening	Permanent	500,000 annual deaths	Worldwide
Clostridium botulinum	Ingestion	Botulism	Vision problems, slurred speech, dry mouth, difficulty swallowing, muscle weakness	Permanent	Rare	Worldwide
Clostridium spp. (other than the aforementioned two)	Skin trauma	Gas gangrene	Air under skin, blisters, fever, drainage, pain	Permanent	Common before antibiotics	Worldwide
Escherichia coli	Ingestion	Diarrhea	Bloody diarrhea, abdominal pain, vomiting, fever	Incidental	5 million/year	Worldwide
Rickettsia spp.	Tick bite	Rocky Mountain spotted fever, other tick fevers, and spotted fevers	Fever, nausea, vomiting, severe headache, pain, rash	Periodic	250–1200 cases/year in the United States	Worldwide
Salmonella spp.	Ingestion	Salmonellosis, typhoid fever, paratyphoid fever	Diarrhea, fever, abdominal pain, and cramps	Incidental	2–4 million/year in the United States	Worldwide
Streptomyces spp.	Skin trauma	Skin infection	Redness, pain, skin becomes hot and swollen	Permanent	Extremely rare	Worldwide

Source: Brevik, E.C., *Soils, Plant Growth and Crop Production*, Encyclopedia of Life Support Systems (EOLSS), Developed under the Auspices of the UNESCO, EOLSS Publishers, Oxford, 2009, http://www.eolss.net, verified 8 January 2012; based on *Essentials of Medical Geology*, Bultman, M.W., Fisher, F.S., and Pappagianis, D., The ecology of soil-borne human pathogens, pp. 481–511, Copyright 2005, from Elsevier; http://www.cdc.gov/.

[a] *Permanent:* the organism is a permanent inhabitant of the soil that is capable of completing its entire life cycle in the soil; *periodic:* the organism requires part of its life cycle to be completed in the soil; *transient:* the organism may be found in the soil naturally, but the soil is not required to complete its life cycle; *incidental:* organisms that are introduced into the soil through anthropogenic activities.

E. coli O157:H7 is often found naturally in cow intestines and contamination of human food with cow fecal materials can potentially cause the disease. As few as 10 cells of *E. coli* O157:H7 can cause infection (U.S. Food and Drug Administration, 2007a).

5.3.2.2 *Campylobacter* spp.

Campylobacteriosis is a disease caused by *Campylobacter* spp., which are Gram-negative, curved or spiral rods. The disease is commonly caused by *C. jejuni*, which is normally present in swine, chicken, turkey, and cattle manures. This common bacterial infection of humans produces diarrhea and dysentery, including cramps, fever, and pain. Surveys show that *C. jejuni* is the leading cause of bacterial diarrhea in the United States and causes more disease than *Shigella* spp. and *Salmonella* spp. combined (U.S. Food and Drug Administration, 2007b). Contaminated food and water are the main sources of the bacteria. *Campylobacter* spp. appear to survive in soil but not reproduce. In four New Zealand soils, 99% of applied *C. jejuni* were retained in the top 5 cm, where they survived for at least 25 days at 10°C (Ross and Donnison, 2006). Analysis of *Campylobacter* spp. on three broiler farms showed that soil in the area around the poultry house was a potential source of *Campylobacter* contamination (Rivoal et al., 2005).

5.3.2.3 *Salmonella* spp.

Members of the genus *Salmonella* are non-spore-forming, Gram-negative rods. The genus is large but only a few species cause human disease. These are often zoonotic and can transfer between humans and animals (especially poultry). Reptiles such as turtles, lizards, and snakes often harbor *Salmonella* spp. The U.S. Food and Drug Administration banned the sale of turtles smaller than 4 in. in 1975 to prevent the spread of salmonella. Organisms normally enter humans by ingestion and often large numbers are required to cause disease in healthy adults. The main symptom of this disease is diarrhea (sometimes severe). One species (*Salmonella typhi*) is mainly restricted to humans in developing countries and causes typhoid fever, which can lead to death. Once infected, patients may continue to excrete the bacteria with their feces for a year or longer. It is commonly accepted that between 1 million and 1 billion bacteria are required to cause infection, although researchers suggest some people may be infected by far fewer bacteria (MedicineNet, 2010).

When manure containing *Salmonella* spp. was added to soil, there were large population decreases in the first week but detectable levels remained after 180 days (Holley et al., 2006). Higher soil moisture, higher manure additions, and higher clay contents increased *Salmonella* survival.

5.3.2.4 *Shigella* spp.

These are Gram-negative, nonmotile, rod-shaped bacteria that primarily reside in humans and other primates such as chimpanzees and monkeys. Human feces are the common mode of transmission. The organisms cause diarrhea, fever, cramps, and blood/mucus in stools. Some strains produce an enterotoxin similar to that of *E. coli* O157:H7. The infective dose is as few as 10 cells, depending on the age and condition of the host (U.S. Food and Drug Administration, 2007c).

5.3.2.5 *Bacillus anthracis* (Anthrax)

Anthrax is a deadly disease caused by spores of *Bacillus anthracis*, a rod-shaped, Gram-positive bacterium (Figure 5.2). The endospores can survive in soil for many, many years and the resulting disease is often lethal in victims. Spores may enter humans by being inhaled or ingested, or by skin contact, where the spores germinate and multiply rapidly, causing the disease. Animals may ingest spores while grazing and become infected, and humans who eat the meat of these animals can then become infected.

This pathogen is present in soils worldwide. *B. anthracis* sporulates with greater frequency in low-lying marshy areas. Animal disease is often associated with neutral or alkaline soils having

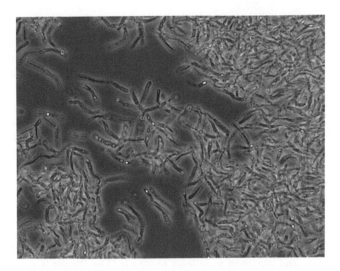

FIGURE 5.2 *Bacillus anthracis* spores (ca. 1 µm in diameter) appear as lighter spheres in this phase contrast micrograph. (Courtesy of Centers for Disease Control and Prevention, Public Health Image Library, image #1892.)

free calcium carbonate, where the spores appear to germinate more readily and multiply to infectious levels when moisture, temperature, and nutrition are optimal (Animal Disease Diagnostic Laboratory, 2005). In 2001, anthrax spores were used for bioterrorism in letters delivered by the U.S. Postal Service.

5.3.2.6 *Clostridium tetani* (Tetanus)

The next three bacterial pathogens belong to the *Clostridium* genus, which are Gram-positive rods that have one end larger than the other end due to the presence of terminal spores. Because of their importance (and toxicity), each will be discussed separately. *Clostridium tetani* is an obligate anaerobe that causes tetanus by producing an extremely potent toxin when it colonizes the intestinal tract of humans. This disease results in muscular spasms and treatment must be sought early or the disease is often fatal. The spores produced by the organism can survive in soil for long periods and enter the host through an abrasion or break in the skin. Effective vaccines are available to prevent tetanus and it is important that everyone stay current on vaccination to avoid this deadly disease.

Not all soils contain *C. tetani*. In a 2006 study in Japan, *C. tetani* occurred in 8 of 35 soils tested (Haneda et al., 2006), and in an earlier study consisting of 290 soil samples, 18.6% were reported positive for *C. tetani* (Kobayashi et al., 1992). The contamination level of selected soils in South Africa was 28% (Wilkins et al., 1988). Thus, from these three studies, it can be generalized that approximately 25% of world soils contain this bacterium.

5.3.2.7 *Clostridium botulinum* (Botulism)

Similar to *C. tetani*, *C. botulinum* is an anaerobic spore-former commonly found in soil. The organism produces a potent neurotoxin during growth that affects the central nervous system of humans. The spores most commonly originate from improperly stored food. The disease incidence is low but mortality of those infected can be high if treatment is delayed.

Wobeser et al. (1987) reported that 38% of 326 soil samples from 28 Saskatchewan wetlands contained *C. botulinum*. Smith (1975) analyzed 21 widely dispersed soils from across the United States and reported the presence of *C. botulinum* in 8 of those samples.

5.3.2.8   *Clostridium perfringens* (Gas Gangrene)

C. perfringens is naturally present in many environments including soil. It can cause food poisoning and gas gangrene. It is an anaerobic, spore-forming rod that can persist for long periods in soil. *C. perfringens* is a source of commonly reported foodborne illness and is often associated with buffets where food is left warm for long periods (Rohrs, 1998).

Gas gangrene is mainly caused by *C. perfringens* and results from the bacteria colonizing and producing toxins in healthy muscle tissue that is adjacent to traumatized muscle or soft tissue. It results in inadequate blood supply causing death of the tissue. One symptom may be gas under the skin. The onset of gas gangrene is sudden and the disease spreads quickly, perhaps requiring amputation of the infected body part to save the victim's life.

Li et al. (2007) sampled soils for the presence of *C. perfringens*. Most of the 502 soils collected in the survey contained the bacterium.

5.3.2.9   *Vibrio cholerae* (Cholera)

Cholera is usually thought of as a waterborne disease in developing countries (Figure 5.3). Severe fluid loss due to vomiting and diarrhea may lead to death. The illness is caused by ingestion of *V. cholerae*, which travels to the small intestine, attaches to the intestinal wall, and produces the cholera toxin. Approximately 10^6 cells are required to cause the disease in a healthy adult (U.S. Food and Drug Administration, 2007d).

Gerichter et al. (1975) showed that *V. cholerae* survived longer in moist than in dry soil but died off within days. More concern likely occurs when fresh fruits and vegetables are irrigated with sewage water. Additionally, fish and shellfish living in water contaminated with human wastes are a source of the disease.

5.3.2.10   *Legionella* spp.

In July 1976, many attending the American Legion convention in Philadelphia became ill, with 34 eventually dying. Investigators later isolated the causative organism, a fastidious Gram-negative bacillus. It is mainly a waterborne organism found where water is stored at higher than ambient temperatures such as cooling towers, fountains, and whirlpools. It can, however, be isolated from streams, lakes, moist soil, and mud. Infection commonly occurs by aerosols or mist droplets containing the bacteria. Early symptoms are flu-like and can include nausea and diarrhea. In some cases this may develop into pneumonia, especially in older adults or those with compromised immune systems.

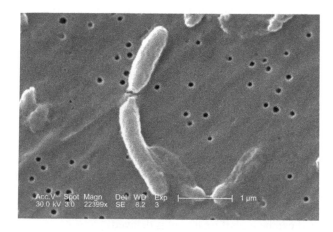

FIGURE 5.3   Cholera is caused by the bacterium *Vibrio cholerae*, which is seen here in an electron micrograph (magnification approximately ×22,000). (Courtesy of Centers for Disease Control and Prevention, Public Health Image Library, image #7816.)

Legionella spp. was isolated from 6.3% (86 of 1362) of soil samples widely collected in Japan (Katsunori et al., 2002). The organism was found in higher numbers in the fall but persisted throughout the year. Interestingly, Rowbotham (1980) reported that *Legionella* spp. can be isolated from soil amoebae.

5.3.2.11 *Mycobacterium leprae* (Leprosy)

Leprosy has been one of the most dreaded diseases over the centuries and, although now largely controlled, new cases continue to develop. India has over 50% of the world's leprosy cases. Infection may be by human-to-human contact (but the disease is not highly contagious) and some researchers suggest that the causative organism might be a soil chemoautotroph that can live both in soil and in the human body (Chakrabarty and Dastidar, 2002). The disease causes weakened muscles, skin lesions and sores, nerve damage, and progressive debilitation with loss of control of hands, arms, and legs.

Lavania et al. (2008) used PCR amplification techniques to evaluate Indian soils for the presence of *M. leprae*. They found soils from areas of high occurrence of leprosy were 55% positive for the bacterium but only 15% positive in areas of low occurrence.

5.3.2.12 *Mycobacterium tuberculosis* (Tuberculosis)

Tuberculosis was once the leading causes of death in the United States and continues to be a major world disease. In 2006, 9.2 million new cases were reported resulting in 3 million deaths (Merck Manuals, 2009). The causative organism is a small, slow-growing, aerobic bacillus. Lung infection often accompanies the disease and is most often transmitted when the infected person coughs, expelling infectious droplets into the air. Organisms may remain suspended in air for several hours. Latent tuberculosis is common and occurs when individuals harbor the disease organisms in their bodies but fail to become sick. Disease may later develop when the individual becomes susceptible.

Mycobacterium tuberculosis appears not to survive long periods in soil, but a related organism (*M. bovis*) that infects cattle (and may also infect humans who drink unpasteurized milk) has been reported to survive for up to 21 months in soil (Young et al., 2005). Soil, however, has had a profound effect on the disease caused by *M. tuberculosis*. Treatment of tuberculosis with the antibiotic streptomycin became common in the 1950s; the antibiotic was isolated by Selman A. Waksman from the soil actinomycetes *Streptomyces griseus*. Waksman won the Nobel Prize in 1952 for this discovery.

5.3.3 FUNGI

Fungi are better known as causative agents of crop and plant diseases but some also can cause human diseases. Relatively few fungi of this large kingdom cause human disease (Table 5.2) and most are superficial, with few resulting in death. Fungal diseases are usually not epidemic as are many bacterial and viral diseases. Furthermore, most antibiotics that control bacterial pathogens are not effective in treating fungal infections because fungi are eukaryotic rather than prokaryotic cells. Being eukaryotic organisms, finding a treatment that will kill the fungus and not harm human cells is more challenging. The typical fungal spore is 1–40 µm in length. The diseases of warm-blooded animals caused by fungi are known as mycoses (singular = mycosis).

Many human fungal pathogens appear to be common soil fungi that can infect humans when conditions are appropriate such as through an open sore or wound. These are called opportunistic fungi or facultative parasites. An example is corn smut (*Ustilago maydis*), which is eaten as a delicacy in some parts of the world but can cause skin lesions under appropriate conditions.

5.3.3.1 Dermatophytes

Dermatophytes are fungi that cause skin, nail, or hair diseases. These are superficial infections that typically are easy to recognize and often cause noted discomfort in the individual. Common names

TABLE 5.2

Some Disease-Causing Fungi Associated with Soil

Organism	Disease Name	Associated Problems	Soil Residency[a]	Incidence[b]	Geographic Distribution	Gateway
Coccidioides	Coccidioidomycosis (valley fever)	Flu-like symptoms, fever, cough, rash, headache, muscle aches	Permanent	15/100,000	Southwestern United States, northern Mexico, microfoci in Central and South America	Respiratory, sometimes trauma to skin
Histoplasma capsulatum	Histoplasmosis	Respiratory symptoms, fever, chest pains, dry or nonproductive cough	Permanent	80% living in endemic areas have positive skin test	Eastern and central United States, microfoci in Central and South America, Africa, India, Southeast Asia	Respiratory
Blastomyces dermatitidis	Blastomycosis	Fever, chills, productive cough, myalgia, arthralgia, pleuritic chest pain	Permanent	1–2/100,000	South-central, southeastern, and midwestern United States, microfoci in Central and South America and Africa	Respiratory
Aspergillus spp.	Aspergillosis	Wheezing, fever, cough, allergic sinusitis, chest pain, shortness of breath	Permanent	1–2/100,000	Worldwide	Respiratory, possible through contaminated biomedical devices
Sporothrix schenckii	Sporotrichosis	Bumps or nodules that open and may resemble boils; eventually lesions look like open sores	Permanent	Uncommon and sporadic	North, Central, and South America, Africa, Europe	Trauma to skin
Trichophyton spp., *Microsporum* spp., *Epidermophyton* spp.	Tinea corporis (ringworm)	Itchy, red circular rash with healthy-looking skin in the middle		Common	Worldwide	Skin contact

Source: Brevik, E.C., *Soils, Plant Growth and Crop Production*, Encyclopedia of Life Support Systems (EOLSS), Developed under the Auspices of the UNESCO, EOLSS Publishers, Oxford, 2009, http://www.eolss.net, verified 8 January 2012; based on *Essentials of Medical Geology*, Bultman, M.W., Fisher, F.S., and Pappagianis, D., The ecology of soil-borne human pathogens, pp. 481–511, Copyright 2005, from Elsevier; http://www.cdc.gov/.

[a] See footnote for Table 5.1.

[b] Some of the incidence rates are given as the number of confirmed cases per 100,000 population.

for these infections include ringworm and athlete's foot. "Jock itch" also may be caused by the ringworm fungus, with infections in the groin area. Despite the name "ringworm," it is caused by fungi. The name was given by the Romans, who tended to attribute many early diseases to worms.

5.3.3.1.1 Trichophyton spp. (Ringworm)

Ringworm is caused by two main fungal genera: *Trichophyton* and *Microsporum* (Figure 5.4). Both are Ascomycota. The symptoms of ringworm include itchy, red, raised, scaly patches that may blister and ooze, which are often redder around the outside with normal skin tone in the center (thus forming a ring). The fungi that cause this parasitic infection feed on keratin, the material found in the outer layer of skin, hair, and nails.

5.3.3.1.2 Microsporum spp. (Ringworm)

Microsporum are also a causative agent of ringworm. One interesting feature of *Trichophyton* and *Microsporum* is their production of both small and large spores (macroconidia) that may grow up to 160 μm long. Members of both genera appear closely related.

5.3.3.2 Systematic Infections

Internal infections occur deeper within the body in vital organs or the central nervous system. Entry is usually through a surface wound or inhalation and then the organism is circulated in the body by blood or the respiratory system. These fungi often grow in soil as saprophytes and derive their nutrition from nonliving organic material.

5.3.3.2.1 Histoplasma capsulatum

This organism causes the disease histoplasmosis, which primarily affects the lungs but can affect other body parts. The organism grows in soil, especially neutral-pH soil covered with bat or bird feces such as in caves or bird roosts. The excrement enriches the soil for growth of the fungal mycelium. The spores or mycelial fragments, which can travel several hundred feet, are transmitted by breathing into the lungs. The organism is dimorphic, having a mycelium form in soil and a yeast form in humans at 37°C. Disease susceptibility increases as the quantity of inoculum increases. Many people harbor the organisms but they cause no damage or disease symptoms. Persons who have compromised immune systems are most at risk and death can occur. The disease is particularly common in the Ohio River Valley in the United States.

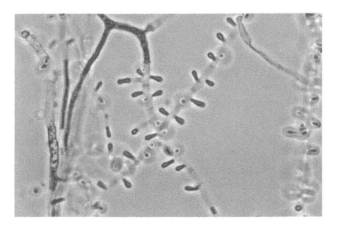

FIGURE 5.4 Fungal conidia (spores) of *Trichophyton soudanense* seen in this view cause "ringworm" in humans (×10,000 magnification). (Courtesy of Centers for Disease Control and Prevention, Public Health Image Library, image #11008.)

5.3.3.2.2 *Coccidioides immitis*

This is a filamentous fungus that inhabits soil (usually in drier regions) and can cause the human respiratory disease coccidioidomycosis (sometimes called "valley fever"). The fungus produces asexual spores, which can be breathed into the lungs, by mycelium fragmentation. Valley fever causes flu-like symptoms, which, in mild cases, are treated only by bed rest. Approximately 60% of those infected show no symptoms. In other individuals, the symptoms are more severe and individuals with compromised immune systems seem most at risk. This disease-causing organism is dimorphic, where the form in soil (mycelium) is different than the form in the host (round spherules). Understandably, these different forms caused early confusion in identifying the organism. *C. immitis* is unevenly distributed in soil, and its presence appears unrelated to vegetation but may be related to soil texture: the fungus is most often found in very fine sand and silt soils and can survive in soil for up to 40 years (Fisher et al., 2007).

5.3.3.2.3 *Aspergillus fumigatus*

This fungus is common in soil and in compost heaps. It produces prolific spores that are ubiquitous in the air we breathe, and humans inhale several hundred spores per day. In healthy individuals these are quickly eliminated by the immune system. Problems are encountered in individuals with weak immune systems that are overtaxed by the fungus causing aspergillosis. Mortality can be as high as 50%. This is a disease of the lungs or occasionally deeper tissue and is relatively rare in humans (it is more common in birds and other animals). A related member of the same genus (*A. flavus*) can cause similar symptoms.

5.3.4 PROTOZOA

Protozoa are single-celled, eukaryotic organisms. They are common in soil and are thought to be important in maintaining ecosystem balance by feeding on smaller organisms such as bacteria. Most soil protozoa are not harmful to humans but some can be human parasites causing disease (Table 5.3). They vary in size from approximately 10–50 µm but some (especially aquatic forms) can be much larger. Several protozoa have an active phase (trophozoite stage) and a dormant (cystic) phase. The cystic phase is capable of surviving in unfavorable environments for extended periods. Protozoa are normally classified based on their means of locomotion. Most human pathogens multiply in the human gut and are excreted into the environment where they survive in water or soil until coming into contact with a new host. The mode of transmission is mainly fecal–oral.

5.3.4.1 *Entamoeba histolytic*

Entamoeba histolytic is a cause of amoebic dysentery in humans. Diarrhea, fatigue, and severe weight loss often are symptoms. The organisms may pass through the intestinal wall and reach the bloodstream, where they infect other vital organs, especially the liver, causing death if not treated. Infected humans excrete both active and cystic forms but it is the cystic forms that survive outside the host in water, soil, and sewage.

5.3.4.2 *Giardia intestinalis*

Giardia intestinalis causes a common waterborne disease that results in diarrhea in its host (Figure 5.5). The organism is a flagellated protozoan that is protected by an outer shell and as such can survive outside the body for long periods. The parasite lives in the gut of humans and numerous other animals and passes, sometimes in the millions, in the feces. *Giardia* is found worldwide. The cysts, when swallowed, pass to the small intestine, where they adhere to the surface and reproduce. As they move to the drier environment of the colon, they encyst and can be passed out in the cystic form.

TABLE 5.3

Common Disease-Causing Protozoa Associated with Soil

Organism	Point of Infection	Disease Name	Associated Problems	Soil Residency[a]	Incidence	Geographic Distribution
Cryptosporidium parvum	Epithelial cells of gastrointestinal tract	Cryptosporidiosis	Diarrhea	Transient–incidental	Not known	Worldwide
Cyclospora cayetanensis	Epithelial cells of gastrointestinal tract	Cyclosporiasis	Diarrhea	Incidental	15,000/year in the United States	Worldwide, most common in tropics and subtropics
Entamoeba histolytica	Intestine	Amebiasis	Diarrhea, dysentery, liver abscesses	Incidental	40 million	Worldwide
Balantidium coli	Intestine	Balantidiasis	Acute hemorrhagic diarrhea, ulceration of the colon	Transient–incidental	Rare	Worldwide
Giardia lamblia	Large intestine	Giardiasis	Diarrhea, abdominal cramps, bloating, fatigue, weight loss	Transient–incidental	500,000	Worldwide
Isospora belli	Small intestine	Isosporiasis	Chronic diarrhea, abdominal pain, weight loss	Transient–incidental	Rare	Worldwide, most common in tropics and subtropics
Toxoplasma gondii	Eye, brain, heart, skeletal muscles	Toxoplasmosis	Flu-like symptoms, birth defects, hepatitis, pneumonia, blindness neurological disorders, myocarditis	Transient	60 million in the United States are probably carriers	Worldwide, most common in warm climate, low altitudes
Dientamoeba fragilis	Gastrointestinal tract	*D. fragilis* infection	Mild abdominal pain, gas, diarrhea	Unknown	2%–4% of the population in developed countries	Worldwide

Source: Brevik, E.C., *Soils, Plant Growth and Crop Production*, Encyclopedia of Life Support Systems (EOLSS), Developed under the Auspices of the UNESCO, EOLSS Publishers, Oxford, 2009, http://www.eolss.net, verified 8 January 2012; based on *Essentials of Medical Geology*, Bultman, M.W., Fisher, F.S., and Pappagianis, D., The ecology of soil-borne human pathogens, pp. 481–511, Copyright 2005, from Elsevier; http://www.cdc.gov/.

[a]   See footnote for Table 5.1.

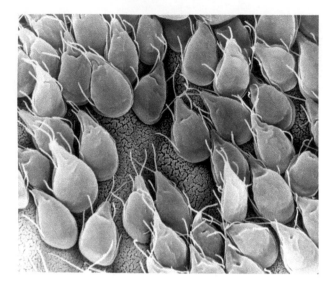

FIGURE 5.5 Numerous *Giardia* sp. (5–15 μm in diameter) attached to the surface of the small intestine as seen in this electron micrograph. (Courtesy of Centers for Disease Control and Prevention, Public Health Image Library, image #11632.)

Olson et al. (1999) showed that *Giardia* cysts survived in soil at 4°C for up to 8 weeks but the cysts were inactivated at 1 week at 25°C. Olson et al. (1999) showed in their study that *Cryptosporidium* spp. survived better in soil than *Giardia* spp. at all temperatures.

5.3.4.3 *Cryptosporidium parvum*

Cryptosporidiosis is a parasitic disease caused by a protozoan that infects the intestines of mammals including humans. It is transmitted as hardy oocysts (a thick-walled cyst) by the fecal–oral pathway and often causes watery diarrhea in its host. The parasite is found throughout the world and is a common cause of diarrhea in humans in the United States.

Jenkins et al. (2002) studied survival of the oocysts in three soils that were maintained at three temperatures and three water potentials. Temperature affected survival the most (greater survival at lower temperature). Soil texture had some effect but moisture tensions between 0.033 and 1.5 MPa had little effect on survival. The authors concluded that animal wastes can be a long-term reservoir of oocysts in watersheds.

5.3.4.4 *Acantamoeba* spp.

Acantamoeba spp. are amoebic protozoa that live in many soils of the world and can cause human disease, especially in individuals with weakened immune systems. The protozoa often enter the body through cuts or the nasal passage and spread by the circulatory system to the central nervous system or eye. Hosts with the disease display an altered mental status, headaches, and other neurological signs. Symptoms can lead to death. Under laboratory conditions, cysts have been shown to remain viable for 24 years.

5.3.4.5 *Naegleria fowleri*

Naegleria fowleri is an amoebic protozoan that lives in soil and warm water, and on vegetation worldwide. Although the organism is common in the environment, it rarely infects humans. Infection usually occurs when diving or jumping into warm water; the organism enters the body through the nose and then travels to the brain. The resulting disease (amoebic meningoencephalitis) has a 95% mortality rate.

5.3.5 PARASITIC WORMS (HELMINTHS)

Most parasitic worms reproduce by eggs. The eggs pass through the stools of infected humans and persist in soil (Table 5.4). A new host then comes into contact with the soil containing eggs, becomes infected, and the cycle repeats. Societies that have higher sanitation standards often have lower levels of all intestinal parasites including worms.

> If you are feeling low and discouraged, check out the life cycle of the parasitic nematode "guinea worm" (*Dracunculus medinensis*). Diagnosis is usually made by seeing the adult worm protruding from your skin. You likely will conclude life is good if you don't have this parasite!

A parasite that kills its host before the parasite is ready to reproduce is likely to quickly become extinct. Therefore, most parasites must complete all or a part of their life cycle within the host. Some of the life cycles of the parasitic worms are fascinating (and maybe a little revolting).

5.3.5.1 Trematoda (Fluke)

Trematodes are flatworms of the class Trematoda, which are often called flukes. There are several different flukes and their life cycles are complex, often involving freshwater mollusks or vertebrates (including humans). They have two suckers, one near the mouth and one on the underside of the worm. Eggs are usually contained in the feces of infected individuals and can persist in soil anywhere human wastes are used as fertilizer. The health threat to humans is mostly in developing countries, and the disease causes weakness, diarrhea, and other symptoms. The liver fluke (*Fasciola hepatica*) has an alternate snail host and, along with the blood fluke, causes serious diseases in humans. Several flukes appear to survive in moist soil.

5.3.5.2 Cestoda (Tapeworm)

Humans contract tapeworms mainly by eating raw meat but also by eating foods contaminated with infected feces (Figure 5.6). These are parasitic flatworms that live in the digestive tracts of hosts. Most need at least two hosts to complete their life cycles. Tapeworms normally attach to the intestinal wall and absorb nutrients through their skin. As they grow, eggs are produced (each worm contains both male and female sex organs) that pass in the feces into the environment, where they survive until consumed by a new host.

5.3.5.3 Nematoda (Roundworm)

5.3.5.3.1 Ascaris spp.

Ascaris spp. are large intestinal roundworms (nematodes) most common in warm humid regions of the world. Some eggs appear to need a period in soil to mature before they cause human disease. These eggs have thick shells, are chemically and physically resistant, and must lie in soil at around 28°C for 20 days until they become infective (WormLearn Nematodes, 2006). There are five major species of intestinal nematodes found in humans. *Ascaris lumbricoides* is probably the most common and infects people when they swallow soil containing the eggs. The eggs pass through the stomach unharmed. Inside the small intestine, the eggs hatch and the larvae emerge. They burrow through the lining of the intestine and enter the bloodstream, where they are carried to the lungs. Here they pass through several stages of development and eventually are coughed up. Some are swallowed and the mature worms can pass the acid stomach unharmed. Once in the small intestine again, they mature and produce eggs to be defecated outside the body. This whole process takes 2–3 months. The worms can live up to a year and an adult female can produce up to 6 million eggs in her lifetime. An adult female is longer than the average earthworm (up to 45 cm long).

TABLE 5.4
Common Disease-Causing Helminths Associated with Soil

Organism	Common Name	Point of Infection	Disease Name	Associated Problems	Soil Residency[a]	Incidence	Geographic Distribution
Nematodes							
Ancylostoma duodenale	Hookworm	Intestine	Ancylostomiasis	Chronic anemia, diarrhea, cramps	Periodic	1.2 billion/year (includes *N. americanus*)	North Africa, southern Europe, northern Asia, western South America
Necator americanus	Hookworm	Intestine	Ancylostomiasis	Chronic anemia, diarrhea, cramps	Periodic	1.2 billion/year (includes *A. duodenale*)	Central and South America, southern Asia, Australia, Pacific Islands
Ascaris lumbricoides	Roundworm	Small intestine	Ascariasis	Stunted growth, abdominal pain, intestinal obstruction, cough	Periodic	1.5 billion/year	Worldwide
Strongyloides stercoralis	Roundworm	Small intestine	Strongyloidiasis	Abdominal pain, diarrhea, rash	Permanent	100 million/year	Tropical, subtropical, and temperate regions
Toxocara canis, Toxocara cati	Roundworm	Organs	Toxocariasis	Fever, abdominal pain, hives, cough, wheezing, loss of vision, crossed eyes	Transient	10,000/year	Worldwide
Enterobius vermicularis	Pinworm	Colon	Enterobiasis	Anal itching	Incidental	200 million total	Temperate regions
Trichuris trichiura	Whipworm	Large intestine	Trichuriasis	Bloody diarrhea, iron-deficiency anemia, rectal prolapse	Incidental	800 million total	Worldwide
Trematodes							
Schistosoma spp.	Fluke	Mesenteric veins outside the liver	Schistosomiasis	Damage to the liver, intestines, lungs, or bladder	Periodic	200 million total	Tropical regions
Cestodes							
Taenia saginata	Beef tapeworm	Small intestine		Vitamin deficiency, abdominal pain, weakness, change in appetite, weight loss	Transient	50 million/year	Worldwide
Taenia solium	Pork tapeworm	Small intestine	Taeniasis and cysticercosis	Pain, paralysis, optical and psychic disturbances, convulsions	Transient	50 million/year	Worldwide

Source: Brevik, E.C., *Soils, Plant Growth and Crop Production,* Encyclopedia of Life Support Systems (EOLSS), Developed under the Auspices of the UNESCO, EOLSS Publishers, Oxford, 2009, http://www.eolss.net, verified 8 January 2012; based on *Essentials of Medical Geology,* Bultman, M.W., Fisher, F.S., and Pappagianis, D., The ecology of soil-borne human pathogens, pp. 481–511, Copyright 2005, from Elsevier; http://www.cdc.gov/.

[a] See footnote for Table 5.1.

FIGURE 5.6 This view shows the head of a tapeworm (*Taenia solium*) with the hooks for attachment to the intestine clearly visible. Also visible are two suckers near the head where the worm feeds. The worms rarely are larger than 5 mm in diameter but can grow up to 10 m long. (Courtesy of Centers for Disease Control and Prevention, Public Health Image Library, image #1515.)

5.3.5.3.2 *Hookworm*

These are particularly nasty worms that gain access to the human body without being ingested. Larvae of these parasites living in contaminated soil can penetrate bare skin. They are called hookworms because their heads are bent forward forming a slight hook (Figure 5.7). This allows the worm to lie flat along the intestine while being attached. After the eggs are passed out of the body in the feces, they can hatch within 48 h. The larval hookworms live for a time in the soil where they wait for a host. If the larval worms contact the skin of a host, they immediately burrow in, entering the bloodstream. They travel with the blood to the lungs where they develop, erupt out of the lungs, and climb up the trachea. Here, they are swallowed and pass to the intestine to mature. Hookworms need specific soil conditions for the larvae form to survive: neither too dry nor too wet. Hookworms are limited to the tropics or subtropical regions. Humans with chronic infections often become anemic. Humans are infected by two species, *Necator americanus* and *Ancylostoma duodenale*, which have similar life cycles but differ in geographic ranges and effectiveness of treatments.

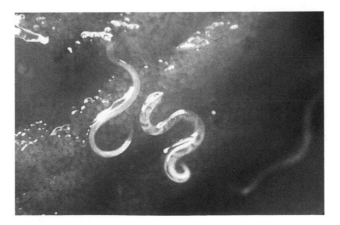

FIGURE 5.7 This photograph shows hookworms attached to the intestine. They attach by means of sharp teeth. Adult females are approximately 0.4 mm in diameter and 8–13 mm long. (Courtesy of Centers for Disease Control and Prevention, Public Health Image Library, image #5205.)

5.3.5.3.3   Enterobius vermicularis *(Pinworm)*

Members of this species infect humans but similar worms can infect other animals. Pinworms are widespread throughout the world, from the tropics to the Arctic. The worms often infect children as they are careless about putting their fingers into their mouths. The female pinworm is 0.8–1.3 cm long and can be seen with the unaided eye. Eggs, after being swallowed, pass to the small intestine, where they hatch, releasing the larvae (Figure 5.8). The larvae travel along the small intestine into the large intestine and eventually exit the anus. Eggs may be laid on the skin or passed in the stool. They become distributed in bedding and spread by hands to other inanimate objects in the home. They also can become airborne and settle with dust.

5.3.5.3.4   Strongyloides stercoralis

This organism is a normal resident of soil but appears to be an opportunistic pathogen that can infect humans and causes "threadworm." The nematode penetrates skin that comes into contact with the soil (Figure 5.9). The female can live and feed in the small intestine and produce eggs (without a male) that soon hatch. Mild gastrointestinal symptoms are most common. The larvae are passed with fecal material. Infection is uncommon in developed countries but is endemic in tropical and subtropical countries, with incidences as high as 40% in some regions.

5.3.5.3.5   Trichinella spiralis *(Trichinosis)*

Trichinosis is caused in humans mainly by eating raw or undercooked meat and becoming infected with the nematode *Trichinella spiralis*. The cysts of the organism hatch in the intestines and the small worms move to the blood system, where they are carried to the muscles and can remain for years. There is no known cure once the larvae invade the muscles, but the symptoms can be treated to lessen the associated pain. Jovic et al. (2001) buried infected pig muscle in soil at three depths and periodically dug up the muscle to check infectivity of the larvae. Larvae were found to be infective at all depths for up to 90 days when the study was terminated.

5.3.5.3.6   Trichuris trichiura *(Whipworm)*

These worms are 3–4 cm long and enter the body by ingestion of the eggs through contaminated soil or fecal-contaminated food. The eggs hatch and worms attach to the lining of the intestine where they produce eggs that are passed with stools. They are called whipworms because about two-thirds of their length is thin and one-third is thick, and thus the worm resembles a whip. It is the narrow end that burrows into the intestinal wall. The burrowing of the worms can cause severe

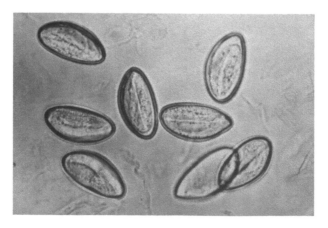

FIGURE 5.8   Many parasitic worms reproduce by eggs. Shown here are eggs of *Enterobius vermicularis*, which cause pinworms in humans (eggs are approximately 50–60 μm long). (Courtesy of Centers for Disease Control and Prevention, Public Health Image Library, image #4818.)

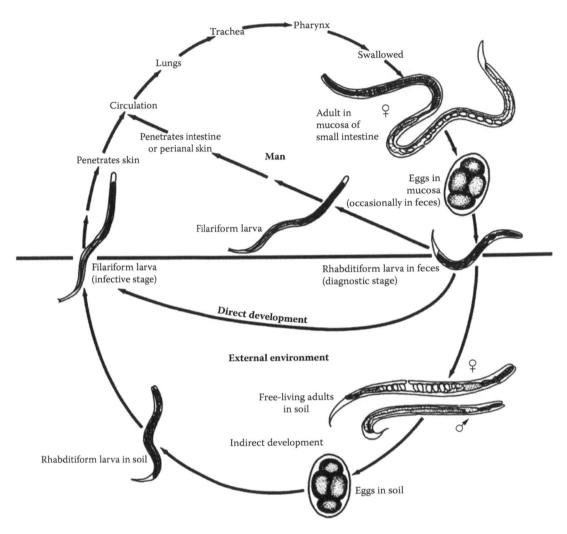

FIGURE 5.9 The life cycle of *Strongyloides stercoralis* involves the external soil environment, eggs in the soil, germination of the eggs, and development of the infective nematodes. (Courtesy of Centers for Disease Control and Prevention, Public Health Image Library, image #5223.)

irritation of the intestine, diarrhea, nausea, and abdominal pain. Similar to *Ascaris*, these worms appear to need a period (2 or 3 weeks) in a warm moist soil to mature to the point of becoming dangerous to humans.

5.4 PATHOGEN INTRODUCTION, SURVIVAL, AND MOVEMENT

Disease organisms can be introduced into the soil by many pathways but most commonly enter soil by contaminated water, animal or human feces, or sewage sludge. Relatively few pathogens are permanent residents of soil, where they can serve as opportunistic pathogens. If a pathogen is not passed directly (hand shake, cough) to a new host, the pathogen must survive for a period away from the host. This may take place in air, water, or soil. Many pathogens exit the host by oral or respiratory pathways or by feces. Fecal materials may be applied to the soil surface or be incorporated to varying extents into the soil. Once in the soil the pathogen must survive until contact with a potential host or may move with water into surface water or groundwater. Thus, the fresher the contamination, the greater is the potential for human infection. One needs to be particularly cautious when eating food grown in soil fertilized with human wastes or irrigated with water that contains human pathogens. When water is not properly treated, the potential exists for transmission of human pathogens. The pathogens may remain on the surface of the vegetation or in the soil that contaminates the vegetation when picked and processed.

Land application is a common, environmentally friendly practice to recycle nutrients contained in wastes. If the wastes contain human pathogens, understanding factors contributing to pathogen survival in soil is important. Most pathogens do not grow in soil but just survive, often with an exponential rate of die-off.

Understanding the mobility of pathogens through soil may also be important for human health. Movement is often related to the size of organisms. Helminths are approximately the size of sand particles, protozoa the size of silt particles, bacteria the size of fine silt and coarse clay, and viruses the size of very fine clay. The larger helminths and protozoa are physically caught by the soil matrix and their movement is often limited unless there are large continuous pores with rapid water movement, such as with preferential water flow. Bacteria and especially viruses would have a much greater potential to move through soil but they appear often to be sorbed by physical and chemical properties on mineral surfaces.

5.4.1 SOURCES

5.4.1.1 Night Soil

Use of night soil is commonly practiced in many parts of the world, especially in developing countries. Human wastes are spread on land as a means of disposal and recycling of nutrients. If the night soil comes from a household with sick family members, the night soil likely contains the causative pathogen.

5.4.1.2 Sewage Sludge

The wastes of many municipalities in developed countries are treated by primary, secondary, and tertiary treatments. Raw wastes arriving at the treatment plants often contain high levels of pathogens. Based on the type of treatment involved, pathogen loads may be reduced but seldom are eliminated. Sewage sludges often contain significant numbers of viral, bacterial, protozoan, fungal, and helminthic pathogens (Straub et al., 1993). When sewage sludges are applied to soil, the pathogens may survive and move into surface water or groundwater, creating the potential for disease outbreak.

5.4.1.3 Municipal Wastewater

Municipal wastewater often is a mixture of human excreta (sewage), solid materials that are suspended, and a variety of organic and inorganic chemicals that originate from residential, commercial, and industrial activities. The wastewater may be a major carrier of enteric pathogens.

5.4.1.4 Septic Tanks

Wastewater from many rural households is treated in septic tanks with the liquid portion moving to a drainfield for disposal. Pathogens are expected to die either in the septic tank or drainfield soil. Problems can develop in the drainfield if the liquid stays in contact with soil for insufficient time. This happens when the effluent moves too quickly to surface water or groundwater. An adequate column of aerobic soil, with moderate clay for sorption, minimizes pathogen movement. In aerobic soils, survival of septic bacteria and viruses is usually poor because they do not compete well with natural soil microbes. Brown et al. (1979) found that 120 cm of soil was adequate to minimize groundwater pollution by fecal coliform or coliphages (enteric viruses) from septic tank effluents.

5.4.1.5 Animal Wastes (Zoonoses)

A wide variety of pathogenic viruses, bacteria, and parasites that may infect humans can be found in the feces of wild and domestic animals. Zoonoses are infectious agents that can be transmitted from animals to humans or from humans to animals. Animal vectors include most warm-blooded animals (birds, cats, dogs, cattle, hogs, mice, monkeys, etc.) and some unusual vectors such as bats, fleas, lice, mosquitos, snails, and ticks. For example, the Black Death (bubonic plague) killed about 25 million Europeans in the fourteenth century. This disease is caused by the Gram-negative bacterium *Yersinia pestis*, which often was transmitted by fleas carried on the backs of rodents. Close proximity between humans and animals can cause transformations of the disease-causing organism. In Malaysia in 1999, fruits bats carrying the Nipah virus caused the infection of pigs. The pig population acted as an amplifier host, which eventually resulted in infection of pig farmers, resulting in 105 human deaths (Field et al., 2001).

Although more than 130 microbial pathogens have been identified from all animal species that may be transmitted to humans by various routes, the most significant manure-borne zoonotic pathogens are the protozoan parasites *Cryptosporidium parvum* and *Giardia duodenalis* and the bacterial pathogens *Salmonella* spp., *Campylobacter* spp., *E. coli* O157:H7, and *Listeria monocytogenes* (Gerba and Smith, 2005). Viruses of potential concern include poliovirus, coxsackievirus, echovirus, hepatitis A, rotavirus, and norovirus. Domestic and wild animals contaminate soil and water by their excrement, through runoff from soil, or through seepage of water through soil that contains large amounts of animal feces. These pathogens are a great concern to the public, who are usually exposed through consumption of fecal-contaminated food or water.

5.4.1.6 Bioterrorism Possibilities

Following the Battle of Carthage in 149 BC, the victorious Romans supposedly "salted the earth" to prevent food production and limit the possible comeback of the city. History clearly has examples of biological agents causing devastating effects on whole populations of people. Two noted examples are the bubonic plague, as noted earlier, that occurred in Europe in the fourteenth century and the introduction of smallpox by Europeans to Native Americans in North America. Although well beyond the scope of this review, bioterrorism microorganisms are a threat to human health and well-being. The most likely transmission mode is probably direct contact through the respiratory system or food and water where the organisms (or their spores) are seeded into the environment. Also possible is the transmission of plant or animal pathogens (much like the Romans when salting the earth) to destroy or incapacitate a country's ability to produce food. The Centers for Disease Control and Prevention has compiled a list of bioterrorism agents in categories based on their risk factor, several of which may persist in soil. Diseases in Category A include anthrax, botulism, plague, smallpox, tularemia, and viral hemorrhagic fevers (Centers for Disease Control and Prevention, 2010).

5.4.2 Survival (Ecological Competence)

Ecological competence is the ability of an organism to survive and possibly grow in a new environment. Pathogens in soil include (a) those organisms that are part of the natural soil flora and fauna and that opportunistically cause disease in humans and (b) those organisms that are natural pathogens of humans and animals but must survive the environment when being transmitted to a new host. Soil conditions affect survival rates (Table 5.5). As a group, helminths and the spores of bacteria and fungi persist longer in soil than other organisms.

5.4.2.1 Plant Contact Vs. Soil Contact

Although survival time varies considerably among organisms and depends upon the environment, soil is a better medium for survival than plant surfaces (Table 5.6). The soil is more protected from

TABLE 5.5

Soil Factors that Control or Influence Transport of Pathogenic Organisms through Soil

Factor	Effects
Soil	Fine-textured soils higher in clays retain more pathogens than coarse-textured soils higher in sands. Iron oxides increase sorption of soils. Muck soils retain viruses poorly.
Filtration	Bacteria are strained at the soil surface, limiting their movement.
pH	Sorption generally increases as pH decreases.
Cations	Sorption increases in the presence of cations (helps reduce the negative charges on soil particles and microbes at neutral pHs). Rainwater may desorb viruses from soil because of its low conductivity.
Soluble organics	Organics may complete with pathogens for sorption sites. Humic and fulvic acids reduce virus sorption to soils.
Microbial type	Sorption varies with microbial type and strain. For example, some viruses move easily through soil, whereas others are retained.
Flow rate	Higher flow rates of water through soil more easily move pathogens through soil.
Saturated vs. unsaturated flow	Movement is generally less under unsaturated flow because of greater soil-to-pathogen contact.
Preferential flow	Rapid water movement through large channels bypasses soil contact and even large pathogens can move quickly through soil.

Source: Adapted from Britten, G., *Wastewater Microbiology*, Wiley-Liss, a John Wiley & Sons, Inc., New York, NY, 1999.

UV radiation and is not subject to rapid desiccation. A pathogen that does survive on a plant surface, however, may be particularly troublesome because humans directly consume plants but do not typically directly consume soil.

5.4.2.2 Time

Recent contamination poses greater risks than older contamination. The death phase of organisms is often exponential with time. Finer-textured soils appear to rapidly sorb pathogenic organisms, especially under conditions of unsaturated flow. The smallest organisms are viruses, which might be the most hazardous. Sobsey et al. (1980), however, showed the majority of enteric viral particles were sorbed within 15 min in soil. Clayey materials more efficiently sorbed viruses than sands or organic soil components. Sorption was effective over a range of pH values (Sobsey et al., 1980). Clearly if organisms are not growing in soil but are only surviving, death of microorganisms can occur during long retention periods.

TABLE 5.6

Approximate Survival Time of Pathogens in Soils and on Plants

	Soil		Plants	
Pathogen	Absolute Maximum	Common Maximum	Absolute Maximum	Common Maximum
Bacteria	1 year	2 month	6 month	1 month
Viruses	6 month	3 month	2 month	1 month
Protozoa	10 day	2 day	5 day	2 day
Helminths	7 year	2 year	5 month	1 month

Source: Gerba, C.P. and Smith Jr, J.E., *Journal of Environmental Quality*, 34, 42–48, 2005.

5.4.2.3 Soil Conditions

5.4.2.3.1 Sunlight

Sunlight contains UV light that inactivates pathogens. Also sunlight tends to dry the environment and organisms normally survive better under moist than dry conditions. Thus, pathogens would be expected to survive for longer periods when added and incorporated into the soil versus left on the soil surface.

5.4.2.3.2 Moisture

Pathogens survive better in moist soils under conditions of high relative humidity. Viruses were not recovered after 25 days in an air-dried sandy soil (Bagdasaryan, 1964). Hurst et al. (1980b) found greater die-off in dry soils than in soils kept flooded. They concluded that allowing the soil to dry between effluent application cycles should help prevent buildup of viruses in the soil and lessen the potential of viral contamination of groundwater.

5.4.2.3.3 Temperature

Pathogens generally survive better at cooler temperatures but some pathogens do not tolerate freezing. In a loamy soil at pH 7.5, poliovirus and echovirus could be recovered after 110–130 days at 3–10°C and 40–90 days at 18–23°C (Bagdasaryan, 1964). Tierney et al. (1977) showed that polioviruses survived in field studies for 123 days in the winter but only 11 days during summer months. Sobsey et al. (1980) reported that viruses in soil survived much better at 5°C than at 25°C. The high temperatures generated in composting are effective in destroying many pathogens in sewage sludges (Epstein et al., 1976).

5.4.2.3.4 pH

In general, microorganisms prefer a neutral soil pH although exceptions occur. Viruses survived 110–170 days in soil at pH 7.5 but only 25–60 days at pH 5.0 (Bagdasaryan, 1964). Farrell et al. (1974) showed that liming sewage sludge to pH 11.5 effectively reduced the number of bacterial pathogens, although nematode ova (eggs) survived.

5.4.2.3.5 Organic Matter

In addition to clays, organic matter is a source of cation exchange capacity in soils. Organic matter appears not to be related, however, to a soil's ability to sorb pathogens. Muck (high organic matter) soils have low affinity for viral sorption (Sobsey et al., 1980). Soils with higher organic matter contents, however, may allow greater survival of pathogens due to moisture or nutrient relationships.

5.4.2.3.6 Texture

Pathogens generally are thought to move through soil in the order sand > silt > clay. Sand grains are the coarsest particles in soil and clay the finest. Hagedorn et al. (1981) concluded after reviewing many studies that physical sorption is the main factor in the soil retention of bacteria, which becomes more effective in soils with increasing clay contents. Hurst et al. (1980a) suggested that sorption of viral particles to clay surfaces may lengthen the survival time in soil. But, with higher clay contents and related filtering capacity, viral transport to groundwaters should be minimal in clay soils.

5.4.2.3.7 Predatory Fauna

Many soil nematodes and protozoa consume protozoa and bacteria but there is little evidence the feeding can be directed to reduce pathogen survival in soil.

5.4.2.4 Gene Transfer from Pathogen to Nonpathogen

Horizontal gene transfer is the transfer of genetic material between organisms independent of reproduction (i.e., not necessarily passed from mother to daughter cells). Horizontal gene transfer in soil is generally accepted; however, the movement of pathogenic genes from a pathogen to a nonpathogen is mainly speculative at this time.

5.4.2.5 Invertebrates as Possible Vectors

Most studies have considered animal pathogens as the prime vectors for human disease but much less is known about the role of invertebrates, which constitute >95% of the animal kingdom. This appears to be an emerging field. For example, Gerrard et al. (2006) showed that *Photorhabdus asybiotica*, a bacterium that causes bacterial infections in humans and whose source was previously unknown, was found in a symbiotic association with a soil nematode. They state that with the dominance of invertebrates in the biosphere, more invertebrate pathogens will likely emerge as vectors for human infection in the future.

5.4.3 MOVEMENT

Among pathogenic microorganisms, viruses are of major concern because they are smaller than bacteria and protozoa and far more mobile in the subsurface environment (Schijven and Hassanizadeh, 2000). Numerous studies have shown that microbes in surface-applied effluent travel through soils in different ways depending on soil properties. Chu et al. (2003) indicated that there is much variability in virus retention and transport through different soils and aquifer materials, depending on their properties (metal oxides, organic matter, pH, etc.) and the degree of water saturation. They found that the presence of *in situ* metal oxides was a significant factor responsible for virus sorption and inactivation. Differences among viruses, allowing some to move more readily through soil than others, could be related to surface charge properties of the viral coat (capsid).

5.4.3.1 Water

Water may pass quickly through soil when all of the pores are saturated (saturated flow) or pass more slowly when many pores are filled with air (unsaturated flow). In unsaturated flow, the water moves mainly by differences in soil moisture tensions. One additional type of water flow through soil is preferential flow where water bypasses the soil matrix and passes quickly by the pull of gravity through large cracks or fauna channels.

5.4.3.1.1 Saturated Flow

Microbes are often attached to soil colloids and do not necessarily move rapidly through soil. For larger microbes, their size often restricts passage. Sobsey et.al. (1980) showed that sandy and organic soil materials were poor in virus removal when suspended in wastewater and many viruses could pass through soil columns during simulated rainfall and saturated flow. Under the same conditions, clayey soil material removed ≥99.9995% of the viruses from applied wastewater. For sorption to be effective, close contact between the soil and viral particles is required. When water flows through soil too rapidly, especially in coarse-textured soils, this close contact does not happen.

5.4.3.1.2 Unsaturated Flow

Hijnen et al. (2005) studied movement of pathogens in gravel and a sandy soil and showed that movement was governed by the size of the organisms (viruses > bacteria > protozoa) and the hydraulic conductivity of the soil (the faster the flow, the more pathogen movement). Also, the surface properties of both organisms and soils controlled retention and release of organisms into flowing water. Several workers have shown that there is a clear increase in the retention of viruses as the saturation of soil decreases (Torkzabana et al., 2006).

5.4.3.1.3 Preferential Flow

When water passes by gravity rapidly through soil by large cracks, old root channels, or fauna channels, there is little opportunity for the soil to serve as a filter to remove pathogens. It is under these conditions that pathogens can move great distances. Also, soils with good structure have large macropores that facilitate pathogen movement, whereas soils with less-developed soil structure have a low potential for microbial bypass flow. McLeod et al. (2008) evaluated microbial movement through 12 New Zealand soils and showed that soils with well-developed soil structure had a high potential for microbial bypass flow.

5.4.3.2 pH and Isoelectric Point

The isoelectric point is the pH at which a particle does not move in a charged field (no charge or equal positive and negative charges). Most bacteria and viruses have an isoelectric point between pH 2 and 3. Thus, bacterial and viral particles are mostly negatively charged at the neutral pHs of many soils. Soil pH may affect the surface chemistry of the microbes, the soil mineral components, and the soil organic fractions, all of which can impact rates of microbe retention and release.

5.4.3.3 Cation Exchange Capacity

Soil clays and organic matter are negatively charged in temperate soils of the world at neutral pHs and can attract and hold positively charged pathogens as effluent moves through soil. When both the soil particles and microorganisms are negatively charged, divalent cations such as Ca^{2+} and Mg^{2+} may serve to bind the two together. Once held, the pathogens are available to the enzymes of other soil organisms and to predation where die-off can occur.

5.4.3.4 Cation Composition and Strength

As noted, the surfaces of most viruses and bacteria are negatively charged at the pH of most soils. Therefore, a proper electrolyte concentration is needed so that the cations of the salts can neutralize the negative charge on the organisms and soil particles. Duboise et al. (1976) showed that sewage effluent allowed sorption of viral particles in a sandy soil column whereas distilled water caused release of viral particles and transport down the column. Thus, viruses sorb poorly to soils in low-ionic-strength solutions. Duboise et al. (1976) concluded that the retention of viruses in a soil column increases as water saturation decreases when the chemical conditions are favorable for sorption (i.e., neutral pH and relatively high ionic strength).

5.5 CONCLUSIONS

A wide variety of pathogens can live or survive in soil, so your mother's advice to keep soil away from your mouth was sound. Human pathogens from the soil include viruses, bacteria, fungi, protozoa, and a variety of worms. Death can result from infection by any of these organisms. Some produce deadly toxins that have a high mortality rate. Other organisms stress the human body and those individuals with suppressed immune systems are most at risk. Some of these organisms are natural residents of soil and are opportunistic pathogens when conditions favor their growth and development. Other pathogens originate from infected hosts and survive in soil long enough to come into contact with another host to cause infection again. If the time between release of an organism from an infected host and contact with a new suitable host is short, the organism may only need to survive briefly in the soil. The most common mode of transmission is by the fecal–oral route.

For enteric organisms that normally grow in the warm, moist environment of the human body, soil can be inhospitable. Soils can vary in temperature, pH, moisture, and nutrition. Survival in soil usually decreases the longer the organism remains in the soil environment.

Soil also is a great purifier. Water that passes slowly through only a meter or so of soil is often free of pathogens. Soil at neutral pH, with some clay content, which has close contact with the water passing through it, is most effective at pathogen removal. Viruses, because of their small size, are more likely to escape the filtering process.

REFERENCES

Animal Disease Diagnostic Laboratory. 2005. *Bacillus anthracis.* http://www.addl.purdue.edu/newsletters/2005/Spring/bacillisanthracis.htm. Accessed 2 July 2010 (verified 8 January 2012).

Bagdasaryan, G.A. 1964. Survival of viruses of the enterovirus group (poliomyelitis, echo, coxsackie) in soil and on vegetables. *Journal of Hygiene, Epidemiology, Microbiology & Immunology* 8: 497–505.

Borchardt, M.A., P.D. Bertz, S.K. Spencer, and D.A. Battigelli. 2003. Incidence of enteric viruses in groundwater from household wells in Wisconsin. *Applied and Environmental Microbiology* 69: 1172–1180.

Brevik, E.C. 2009. Soil, food security, and human health. *In* W. Verheye (ed.), *Soils, Plant Growth and Crop Production.* Encyclopedia of Life Support Systems (EOLSS), Developed under the Auspices of the UNESCO, EOLSS Publishers, Oxford. http://www.eolss.net (verified 8 January 2012).

Britten, G. 1999. *Wastewater Microbiology,* 2nd edn. Wiley-Liss, a John Wiley & Sons, Inc., New York, NY.

Brown, K.W., H.W. Wolf, K.C. Donnelly, and J.F. Slowey. 1979. The movement of fecal coliforms and coliphages below septic lines. *Journal of Environmental Quality* 9: 121–125.

Bultman, M.W., F.S. Fisher, and D. Pappagianis. 2005. The ecology of soil-borne human pathogens. *In* O. Selinus, B. Alloway, J.A. Centeno, R.B. Finkelman, R. Fuge, U. Lindh, and P. Smedley (eds), *Essentials of Medical Geology,* pp. 481–511. Elsevier, Amsterdam, The Netherlands.

Centers for Disease Control and Prevention. 2010. *Bioterrorism Agents/Disease.* http://www.bt.cdc.gov/agent/agentlist-category.asp#. Accessed 23 July 2010 (verified 8 January 2012).

Chakrabarty, A.N. and S.G. Dastidar. 2002. Is soil an alternative source of leprosy infection? *Acta Leprologica* 12: 79–84.

Chu, Y., Y. Jin, T. Baumann, and M.V. Yates. 2003. Effect of soil properties on saturated and unsaturated virus transport through columns. *Journal of Environmental Quality* 32: 2017–2025.

Duboise, S.M., B.E. Moore, and B.P. Sagik. 1976. Poliovirus survival and movement in a sandy forest soil. *Applied and Environmental Microbiology* 31: 536–543.

Epstein, E., G.B. Willson, W.B. Burge, D.C. Mullen, and N.K. Enkiri. 1976. A forced aeration system for composting wastewater sludge. *Journal of the Water Pollution Control Federation* 48: 688–694.

Farrell, J.B., J.E. Smith Jr, S.W. Hathaway, and R.B. Dean. 1974. Lime stabilization of primary sludges. *Journal of the Water Pollution Control Federation* 46: 113–121.

Fernández-Barredo, S., C. Galiana, A. García, S. Vega, M.T. Gómez, and M.T. Pérez-Gracia. 2006. Detection of hepatitis E virus shedding in feces of pigs at different stages of production using reverse transcription-polymerase chain reaction. *Journal of Veterinary Diagnostic Investigation* 18: 462–465.

Field, H., P. Young, J.M. Yob, J. Mills, L. Hall, and J. Mackenzie. 2001. The natural history of Hendra and Nipah viruses. *Microbes and Infection* 3: 307–314.

Fisher, F.S., M.W. Bultman, S.M. Johnson, D. Pappagianis, and E. Zaborsky. 2007. Coccidioides niches and habitat parameters in the southwestern United States: A matter of scale. *Annals of the New York Academy of Sciences* 1111: 47–72.

Fout, G.S., B.C. Martinson, M.W.N. Moyer, and D.R. Dahling. 2003. A multiplex reverse transcription-PCR method for detection of human enteric viruses in groundwater. *Applied and Environmental Microbiology* 69: 3158–3164.

Gerba, C.P. and J.E. Smith Jr. 2005. Sources of pathogenic microorganisms and their fate during land application of wastes. *Journal of Environmental Quality* 34: 42–48.

Gerichter, C.B., I. Sechter, A. Gavish, and D. Cahan. 1975. Viability of *Vibrio cholerae* biotype El Tor and of cholera phage on vegetables. *Israel Journal of Medical Sciences* 11: 889–895.

Griffin, D.W. 2007. Atmospheric movement of microorganisms in clouds of desert dust and implications for human health. *Clinical Microbiology Reviews* 20: 459–477.

Gerrard, J.G., S.A. Joyce, D.J. Clarke, R.H. Ffrench-Constant, G.R. Nimmo, D.F.M. Looke, E.J. Feil, L. Pearce, and N.R. Waterfield. 2006. Nematode symbiont for *Photorhabdus asymbiotica. Emerging Infectious Diseases* 12: 1562–1564.

Hagedorn, C., E.L. McCoy, and T.M. Rahe. 1981. The potential for ground water contamination from septic effluents. *Journal of Environmental Quality* 10: 1–8.

aneda, J., Y. Shiobara, M. Inui, T. Sekiguchi, Y. Sato, Y. Takayama, R. Kikuno, S. Okuda, M. Inoue, and T. Sasahara. 2006. Distribution of *Clostridium tetani* in topsoil from Sagamihara, central Japan. *Kansenshogaku Zasshi* 80: 690–693 (abstract in English).

Hijnen, W.A.M., A.J. Brouwer-Hanzens, K.J. Charles, and G.J. Medema. 2005. Transport of MS2 phage, *Escherichia coli, Clostridium perfringens, Cryptosporidium parvum,* and *Giardia intestinalis* in a gravel and a sandy soil. *Environmental Science & Technology* 39: 7860–7868.

Holley, R.A., K.M. Arrus, K.H. Ominski, M. Tenuta, and G. Blank. 2006. Salmonella survival in manure-treated soils during simulated seasonal temperature exposure. *Journal of Environmental Quality* 35: 1170–1180.

Hundesa, A., C. Maluquer de Motes, S. Bofill-Mas, N. Albinana-Gimenez, and R. Girones. 2006. Identification of human and animal adenoviruses and polyomaviruses for determination of sources of fecal contamination in the environment. *Applied and Environmental Microbiology* 72: 7886–7893.

Hurst, C.J., C.P. Gerba, and I. Cech. 1980a. Effects of environmental variables and soil characteristics on virus survival in soil. *Applied and Environmental Microbiology* 40: 1067–1079.

Hurst, C.J., C.P. Gerba, J.C. Lance, and R.C. Rice. 1980b. Survival of enteroviruses in rapid-infiltration basins during the land application of wastewater. *Applied and Environmental Microbiology* 40: 192–200.

Jenkins, M.B., D.D. Bowman, E.A. Fogarty, and W.C. Ghiorse. 2002. *Cryptosporidium parvum* oocyst inactivation in three soil types at various temperatures and water potentials. *Soil Biology & Biochemistry* 34: 1101–1109.

Jorgensen, P.H. 1985. Examination of the penetration of enteric viruses in soil under simulated conditions in the laboratory. *Water Science and Technology* 17: 197–199.

Jovic, S., M. Djordjevic, Z. Kulisic, S. Pavlovic, and B. Radenkovic. 2001. Infectivity of *Trichinella spiralis* larvae in pork buried in the ground. *Parasite* (2 Suppl.): S213–S215.

Katsunori, F., O. Yasuho, D. Chikaku, H. Motonobu, and F. Masafumi. 2002. Isolation of *Legionella* spp. from soils in Japan. *Journal of Antibacterial and Antifungal Agents* 30: 555–556.

Kobayashi, T., K. Watanabe, and K. Ueno. 1992. Distribution of *Clostridium botulinum* and *Clostridium tetani* in Okinawa Prefecture. *Kansenshogaku Zasshi* 66: 1639–1644 (abstract in English).

Lavania, M., K. Katoch, V.M. Katoch, et al. 2008. Detection of viable *Mycobacterium leprae* in soil samples: Insights into possible sources of transmission of leprosy. *Infection, Genetics and Evolution* 8: 627–631.

Li, J., S. Sayeed, and B.A. McClane. 2007. Prevalence of enterotoxigenic *Clostridium perfringens* isolates in Pittsburgh (Pennsylvania) area soils and home kitchens. *Applied and Environmental Microbiology* 73: 7218–7224.

McLeod, M., J. Aislabie, J. Ryburn, and A. McGill. 2008. Regionalizing potential for microbial bypass flow through New Zealand soils. *Journal of Environmental Quality* 37: 1959–1967.

MedicineNet. 2010. *Salmonella.* http://www.medicinenet.com/salmonella/page2.htm. Accessed 2 July 2010 (verified 8 January 2012).

Merck Manuals. 2009. *Tuberculosis.* http://www.merck.com/mmhe/sec17/ch193/ch193a.html. Accessed 19 July 2010 (verified 8 January 2012).

Ogorzaly, L., I. Bertrand, M. Paris, A. Maul, and C. Gantzer. 2010. Occurrence, survival, and persistence of human adenoviruses and F-specific RNA phages in raw groundwater. *Applied and Environmental Microbiology* 76: 8019–8025.

Okoh, A.I., T. Sibanda, and S.S. Gusha. 2010. Inadequately treated wastewater as a source of human enteric viruses in the environment. *International Journal of Environmental Research and Public Health* 7: 2620–2637.

Olson, M.E., J. Goh, M. Philips, N. Guselle, and T.A. McAllister. 1999. *Giardia* cyst and *Cryptosporidium* oocyst survival in water, soil, and cattle feces. *Journal of Environmental Quality* 28: 1991–1996.

Rivoal, K., C. Ragimbeau, G. Salvat, P. Colin, and G. Ermel. 2005. Genomic diversity of *Campylobacter coli* and *Campylobacter jejuni* with isolates recovered from free-range broiler farms and comparison with isolates of various origins. *Applied and Environmental Microbiology* 71: 6216–6227.

Rohrs, B. 1998. *Clostridium perfringens,* not the 24 hour flu. http://ohioline.osu.edu/hyg-fact/5000/5568.html. Accessed 15 July 2010 (verified 8 January 2012).

Ross, C.M. and A.M. Donnison. 2006. *Campylobacter jejuni* inactivation in New Zealand soils. *Journal of Applied Microbiology* 101: 1188–1197.

Rowbotham, T.J. 1980. Preliminary report on the pathogenicity of *Legionella pneumophila* for freshwater and soil amoebae. *Journal of Clinical Pathology* 33: 1179–1183.

Schijven, J.F. and S.M. Hassanizadeh. 2000. Removal of viruses by soil passage: Overview of modeling, processes, and parameters. *Critical Reviews in Environmental Science and Technology* 30: 49–127.

Science Encyclopedia. 2005. *Escherichia coli.* http://science.jrank.org/pages/2571/Escherichia-coli.html. Accessed 2 July 2010 (verified 8 January 2012).

Sobsey, M.D., C.H. Dean, M.E. Knuckles, and R.A. Wagner. 1980. Interactions and survival of enteric viruses in soil materials. *Applied and Environmental Microbiology* 40: 92–101.

Sobsey, M.D., R.M. Hall, and R.L. Hazard. 1995. Comparative reductions of hepatitis A virus, enteroviruses and coliphage MS2 in miniature soil columns. *Water Science and Technology* 31: 203–209.

Smith, L.D. 1975. Common mesophilic anaerobes, including *Clostridium botulinum* and *Clostridium tetani*, in 21 soil specimens. *Applied Microbiology* 29: 590–594.

Straub T.M., I.L. Pepper, and C.P. Gerba. 1993. Hazards from pathogenic microorganisms in land-disposed sewage sludge. *Reviews of Environmental Contamination & Toxicology* 132: 55–91.

Tanner, B.D., J.P. Brooks, C.P. Gerba, C.N. Haas, K.L. Josephson, and I.L. Pepper. 2008. Estimated occupational risk from bioaerosols generated during land application of class B biosolids. *Journal of Environmental Quality* 37: 2311–2321.

Tierney, J.T., R. Sullivan, and E.P. Larkin. 1977. Persistence of poliovirus 1 in soil and on vegetables grown in soil previously flooded with inoculated sewage sludge or effluent. *Applied and Environmental Microbiology* 33: 109–113.

Torkzabana, S., S.M. Hassanizadeh, J.F. Schijven, H.A.M. de Bruin, and A.M. de Roda Husman. 2006. Virus transport in saturated and unsaturated sand columns. *Vadose Zone Journal* 5: 877–885.

U.S. Food and Drug Administration. 2007a. BBB—*Escherichia coli* O157:H7. http://www.fda.gov/food/foodsafety/foodborneillness/foodborneillnessfoodbornepathogensnaturaltoxins/badbugbook/ucm071284.htm. Accessed 2 July 2010 (verified 8 January 2012).

U.S. Food and Drug Administration. 2007b. BBB—*Campylobacter jejuni*. http://www.fda.gov/Food/FoodSafety/FoodborneIllness/FoodborneIllnessFoodbornePathogensNaturalToxins/BadBugBook/ucm070024.htm. Accessed 2 July 2010 (verified 8 January 2012).

U.S. Food and Drug Administration. 2007c. BBB—*Shigella* spp. http://www.fda.gov/Food/FoodSafety/FoodborneIllness/FoodborneIllnessFoodbornePathogensNaturalToxins/BadBugBook/ucm070563.htm. Accessed 2 July 2010 (verified 8 January 2012).

U.S. Food and Drug Administration. 2007d. BBB—*Vibrio cholerae* serogroup O1. http://www.fda.gov/Food/FoodSafety/FoodborneIllness/FoodborneIllnessFoodbornePathogensNaturalToxins/BadBugBook/ucm070071.htm. Accessed 15 July 2010 (verified 8 January 2012).

Wang, D.S., C.P. Gerba, and J.C. Lance. 1981. Effect of soil permeability on virus removal through soil columns. *Applied and Environmental Microbiology* 42: 83–88.

Wilkins, C.A., M.B. Richter, W.B. Hobbs, M. Whitcomb, N. Bergh, and J. Carstens. 1988. Occurrence of *Clostridium tetani* in soil and horses. *South African Medical Journal* 73: 718–720.

Wobeser, G., S. Marsden, and R.J. MacFarlane. 1987. Occurrence of toxigenic *Clostridium botulinum* type C in the soil of wetlands in Saskatchewan. *Journal of Wildlife Diseases* 23: 67–76.

World Health Organization (WHO). 2005. *Hepatitis E*, fact sheet 280. http://www.who.int/mediacentre/factsheets/fs280/en/. Accessed 2 July 2010 (verified 8 January 2012).

WormLearn Nematodes. 2006. Nematodes: The roundworms. http://home.austarnet.com.au/wormman/wlnema.htm. Accessed 2 July 2010 (verified 8 January 2012).

Young, J.S., E. Gormley, and E.M.H. Wellington. 2005. Molecular detection of *Mycobacterium bovis* and *Mycobacterium bovis* BCG (Pasteur) in soil. *Applied and Environmental Microbiology* 71: 1946–1952.

6 Radioactive Elements in Soil
Interactions, Health Risks, Remediation, and Monitoring

Charles E. Turick, Anna S. Knox, and Wendy W. Kuhne

CONTENTS

6.1 INTRODUCTION

Ionizing radiation is a form of electromagnetic radiation that can produce ions or charged particles. Types of ionizing radiation include α-, β-, γ-, and x-rays. The metric unit of measure for radioactive decay is Becquerel (Bq), which is equal to one disintegration per second. One Curie (Ci), an older, nonmetric unit of measure, is equal to 3.7×10^{10} Bq.

Depending on the type and intensity of ionizing radiation, damage to living tissues may result, thus making it necessary to limit exposure. Radiation is measured in several ways in relation to exposure. The amount of energy from ionizing radiation that is adsorbed by tissue is measured using a unit of measurement known as a rad or a gray (Gy) in metric units (1 Gy = 100 rad). The biological risk associated with radiation exposure is measured using the unit rem or the metric unit Sievert (Sv, 1 Sv = 100 rem).

Radiation is ubiquitous and can vary based on such factors as geological conditions and altitude. Very generally speaking, typical background levels for the United States are about 360 mrem/year. Background radiation is comprised largely (~81%) of naturally occurring radiation emanating from soils, groundwater, or cosmic sources. While about 19% of background radiation is man-made, a large portion of this is from medical use. Less than 1% of background radiation (ca. 2 mrem/year) is derived from nuclear energy production, nuclear weapons construction, and fallout combined (Washington State Department of Health, 2010; Tang and Saling, 1990).

6.2 SOURCES OF RADIONUCLIDES IN SOIL

Radionuclides in soils can originate from both naturally occurring and man-made sources. Naturally occurring radionuclides primarily have two distinct sources: (1) ^{238}U and ^{232}Th decay series and (2) cosmic-ray interactions in the atmosphere.

Generally, the background radiation level is fairly constant around the world, being 8–15 μrad/h. However, there are some areas with sizable populations that have high background radiation levels. The highest levels are found primary in Brazil, India, and China. These high radiation levels are due to high concentrations of radioactive materials in soils. One such mineral, monazite, is a highly insoluble rare earth mineral that occurs in beach sand together with the mineral ilmenite. The principal radionuclides in monazite are from the ^{232}Th series, but there is also some uranium progeny, ^{226}Ra. In Brazil, monazite sand deposits are found along certain beaches. The external radiation levels on these black sands range up to 5 mrad/h, which is almost 400 times the normal background level in the United States (Kathren, 1991). On the southwest coast of India, the monazite deposits are larger than those in Brazil. However, the dose from external radiation is, on average, similar to the doses reported in Brazil, 500–600 mrad/year. Daughter products of ^{238}U and ^{232}Th decay include radium, which decays to radon, an inert gas often associated with soils formed in granitic parent materials. As a noble gas, radon diffuses from the soil into the air. Large variations of radon can be found in the atmosphere. Inhalation of radon-containing aerosols contributes to the majority of radiation exposure in humans. Since radon exposure is related to the local geology as well as building materials, exposure is higher indoors than outdoors. Efforts to minimize radon exposure include sealing floors and improving ventilation (Hewitt, 2001).

Uranium and Th were initially concentrated in plant biomass millions of years ago, prior to conversion to coal, and hence are released into the environment during coal burning and smelting of nonferrous metals. ^{210}Pb and ^{214}Pb are also part of the ^{238}U decay chain. ^{210}Pb is of more concern due to its much longer half-life of 22 years. Mining activities as well as nuclear weapons fabrication activities included U; hence, U isotopes have made their way into the environment. Pu and Tc are man-made radionuclides that have also entered the environment through nuclear weapons testing and accidental spills during fabrication.

As mentioned earlier, humans have used radioactivity for more than 100 years and, through its use, have added to the natural inventories. The amounts are small compared to the natural amounts discussed earlier and, due to the shorter half-life of many of the nuclides, have a marked decrease since the halting of aboveground testing of nuclear weapons. Some examples of human-produced nuclides and their half-lives are presented in Table 6.1. The major sources of human-produced nuclides are those produced as a result of nuclear weapons production or testing (e.g., ^{137}Cs, ^{90}Sr, and ^{131}I) and accidental airborne and effluent releases from nuclear facilities such as those from Chernobyl and Fukushima.

immobilization is strongly pH dependent. Andersson et al. (1982) showed that U was not sorbed strongly by apatite at low pH; for example, the Kd value for U at pH 3.3 was 207 but increased to 24,660 at pH 7.7 (Table 6.2).

6.3.6 PLUTONIUM

As a man-made element, Pu is an integral part of the nuclear weapons program and hence has become an anthropogenic element that has also been integrated into part of the environment, ranging in concentration from parts per billion to parts per hundred (EPA, 2011; Riley and Zachara, 1992). This is largely due to aboveground nuclear testing in the mid-twentieth century. While aboveground tests have stopped, the resulting nuclear fallout remains a legacy of that time. In this regard, the most noteworthy isotope is ^{239}Pu, which has a half-life of 2.41×10^4 years and, as a result, could pose a long-term threat to the environment if it exists in elevated concentrations.

Pu chemistry is complex such that in the environment Pu can exist in the +3, +4, +5, and +6 oxidation states, including multiple oxidation states at one time (Choppin, 2003; Choppin et al., 1997). Pu also binds easily to surfaces in the soil matrix and thereby becomes immobilized. However, strong tendencies for Pu to complex with organic ligands in groundwater can increase its solubility and thus its mobility in the environment (Rusin et al., 1994). Several studies have been conducted on the sorption of Pu on minerals, soils, and other geological materials. A review of data from the literature indicated that Kd values for Pu typically range over 4 orders of magnitude (Thibault et al., 1990; Table 6.2). Plutonium adsorption to soil should depend on the plutonium redox state due to the multiple oxidation states of Pu. However, a review of the literature does not indicate that soil redox potential plays a significant role in Pu sorption. Bell and Bates (1988) studied Pu sorption by soils with different pH and concluded that the adsorption of Pu varied as a function of pH, with maximum adsorption occurring at a pH value of about 6 (Table 6.2). Similarly, Glover et al. (1976) showed that Pu adsorption by different soils from different states is pH dependent.

6.4 EFFECTS OF REDOX AND pH ON RADIONUCLIDES IN SOILS

In the natural environment, retention or mobility of radionuclides depends largely on the redox conditions and pH of the system and can be rather complex. Figure 6.1 provides a general overview of relevant soil–chemical reactions associated with redox conditions. For instance, typical agricultural and horticultural soils exhibit Eh values that range from 100 to 600 mV and pH values of 4 to 8 (Koch-Steindl and Pröhl, 2001). Generally speaking, in soil, NO_3^- reduction begins between 550 and 450 mV, while Mn_2^+ formation is initiated between 450 and 350 mV. In wet soils, O_2 and NO_3^- are no longer detectable at 330 and 220 mV, respectively. Iron(II) begins to form around 150 mV. For waterlogged soil, SO_4^{2-} and sulfide formation commences at −50 mV, while methane begins to form at −120 mV and by −180 mV sulfate is no longer detectable. So, using Tc as an example, at neutral pH values Tc(VII) is reduced between 200 and 100 mV (Lieser and Bauscher, 1987), which is approximately the point at which nitrate disappears and Fe(II) begins to form.

This behavior is likely because of control by biogeochemical processes or reduction caused by the presence of organic matter. However, contaminants like technetium are easily oxidized even though the surrounding redox conditions may be more conducive to reduction (Icenhower et al., 2008). The primary remediation treatment of Tc has been to reduce it from mobile Tc(VII) to relatively immobile Tc(IV). However, if it is immobilized to TcO_2, its solubility may exceed the EPA's maximum contaminant level (MCL) of 900 pCi/L (5×10^{-10} M Tc). Therefore, simply reducing Tc to an oxide, while it may greatly reduce Tc(VII) concentrations, is not sufficient from a regulatory point of view. Tc reduction to form a sulfide phase does maintain Tc well below the Tc MCL to a level of 5×10^{-17} M (8×10^{-5} pCi/L, assuming groundwater conditions and a solid phase of Tc_2S_7). Another important issue with regards to long-term stewardship of remediated Tc is slowing down the rate of reoxidation of these reduced phases.

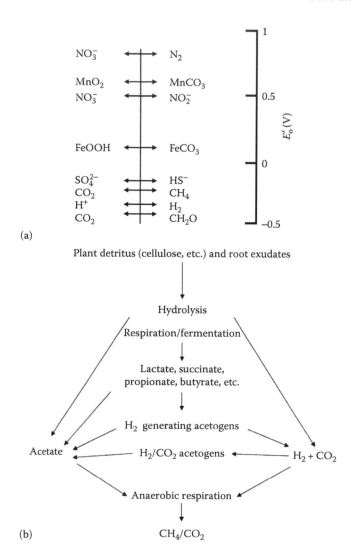

(a)

(b)

FIGURE 6.1 (a) Redox potentials (pH 7) for common microbial terminal electron acceptors used during anaerobic respiration, their redox couples, and free energy changes involving two redox couples in respiratory processes; (b) generalized organization and carbon flow of subsurface microbial activity.

6.5 EFFECTS OF MICROBIAL ACTIVITY ON SOIL RADIONUCLIDE FATE AND MOBILITY

Geochemical changes, including the chemical transformation and fate of radionuclides, are often catalyzed by microbial activity in the subsurface. Although ionizing radiation is damaging to living cells, including microorganisms, soil microorganisms demonstrate a considerable resistance to radioactive environments. Numerous mechanisms have been identified that increase resistance (Daly et al., 2004) as well as enhance growth in the presence of ionizing radiation (Turick et al., 2011; Bagwell et al., 2008; Dadachova and Casadevall, 2008). Consequently, the radioactive environment associated with nuclear contamination does not present a significant roadblock to microbial activity.

Microbial activity is a function of available organic carbon, which fuels microbial growth and metabolic performance. In the soil environment, the main sources of organic carbon are detritus (such as leaf litter and other decaying biomass) as well as root exudates. Nutrient breakdown and subsequent flow down to the saturated zones (aquifers) provides subsurface microorganisms with food for growth (Figure 6.1b). Through their metabolic activity, naturally occurring microbes have the potential to contribute significantly to the transformation of radionuclides in the subsurface. Unlike organic pollutants, which can often be completely decomposed to CO_2, most inorganic pollutants, including radionuclides, cannot be removed from the soil. Instead, microbes play a role in chemically transforming radionuclides and thereby altering their mobility and toxicity. Hence, chemical transformation, not degradation, is the outcome of radionuclide/microbe interactions.

6.5.1 INDIRECT MICROBIAL TRANSFORMATIONS

There are several indirect ways that microbes influence the biogeochemical behavior of radionuclides in soils by initiating changes in the aquifer such as pH, redox conditions, and increased CO_2 concentrations. Decreases in pH and redox contribute to reducing conditions and may also enhance abiotic sorption of contaminants to the surrounding minerals. Increased CO_2 concentrations as a result of microbial activity may be sufficient to precipitate some radionuclides due to increased pH. Of course, the depletion of O_2 by microbial activity can lower the redox conditions. Other examples of microbial activities that influence biogeochemical dynamics include SO_4 reduction to H_2S and Fe(III) reduction to Fe(II). In both cases the end products can react with inorganic contaminants and result in insoluble minerals. For instance, Fe(II) from bacterial Fe(III) reduction reacts with Sr(II) and enhances solid phase partitioning of Sr (Parmar et al., 2000). In some environments, especially where SO_4 concentrations are relatively high, sulfate-reducing bacteria will produce H_2S. As a result, indirect geochemical transformations by H_2S may be the predominant mechanism of inorganic immobilization resulting in sulfide mineral formation.

6.5.2 DIRECT MICROBIAL TRANSFORMATIONS

Microbes also directly influence radionuclide behavior if the radionuclide is intimately involved in microbial energy metabolism. This energy dependence occurs with redox-active radionuclides. Under anaerobic conditions, some radionuclides can accept electrons and hence serve as electron sinks, linking this chemical reducing activity to microbial growth. Alternatively, some radionuclides may serve as electron donors and link this oxidizing activity to microbial growth.

When redox-active radionuclides are soluble, they are usually highly bioavailable and demonstrate relatively rapid transformation rates. Bioavailability decreases, however, as a result of inorganic sorption onto surface minerals in an aquifer. Microbes can utilize their own genetic strategies to produce soluble or cell-surface-associated electron shuttles to transform contaminants associated with mineral surfaces (Turick et al., 2002, 2003, 2009; Marsili et al., 2008; Gorby et al., 2006; Reguera et al., 2005). While contaminant sorption to the solid phase may substantially limit transformation, these strategies enhance transformation rates.

6.5.3 ENERGY INPUT FOR MICROBIAL TRANSFORMATION OF RADIONUCLIDES

The breakdown of organic carbon originating as plant detritus is a stepwise process that contributes to decreased O_2 concentrations in the vadose and upper aquifer zones as well as supplying carbon and energy sources to the aquifer during recharge events (Figure 6.1b). This process contributes to the energy input into the system that is responsible for driving biogeochemical changes as a result of microbial activity. So, while the respiratory and fermentative microbes responsible for this breakdown may not contribute to inorganic transformations directly, their activity is vital to the process as a whole. Conversion of complex organics to nonfermentable carbon sources (i.e., acetate) provides a carbon source targeted at anaerobic respiring bacteria and thus favors the transformation of inorganics.

Rates of microbial transformation of inorganics depend on (1) density, physiological status, and adaptive capacity of subsurface microbes; (2) O_2 concentrations and other potential terminal electron acceptors that would compete with oxidized inorganics; (3) concentrations of available organic carbon and/or H_2 that serve as energy sources; (4) the bioavailability and concentration of the redox-active inorganics; (5) naturally occurring electron shuttles; and (6) microbial inhibitors. Perhaps the most significant rate-limiting step is decreased contaminant bioavailability.

6.6 HEALTH RISKS

Soils have always been important to humans because they provide valuable trace elements necessary for maintaining good health and serve as a source of nutrients and building materials. Polluted soils can have adverse effects on human health if they become contaminated with constituents including metals, chemicals, biological agents, and radionuclides (Brevik, 2009; Abrahams, 2002; Kibble and Saunders, 2001).

The relationship between soil and human health is primarily dependent on the characteristics of the radionuclide and the pathway of exposure including ingestion (deliberate and incidental) (Ferguson and Marsh, 1993), inhalation, and dermal contact (Kibble and Saunders, 2001). Some human health effects of radionuclide contamination are forms of cancer and genetic mutations (Entry et al., 1997). Understanding the routes of exposure; movement of contaminants through environmental media including air, water, soils, plants, and animals; and properties of the radionuclides present is critically important for estimating the ultimate risk to human health.

The degree of risk associated with a particular contaminant involves not just the toxicity of the chemical. In order to estimate the risk an environmental contaminant poses to human health, its transport, fate, and potential biomagnification must be considered (Alexander, 1994). This chapter will not address specifics of risk management, but the factors that contribute to risk assessment of radioactive contaminants will be discussed.

6.6.1 Pathway of Exposure

Releases of radionuclides to the environment from activities associated with the nuclear industry and processing facilities or from accidents like Chernobyl and Fukushima have typically been to water and air. These releases have been in the form of suspended or dissolved materials released in liquid effluents or as gases or particles to air. An important depositional process that removes radionuclides from the atmosphere and transports them to the terrestrial ecosystem is precipitation events. Events such as rainfall remove particles that are submicron in size where washout removes larger particles, generally greater than 1 μm, by adhering to the surface of ice crystals or water droplets that fall to Earth's surface. Once the radionuclides are deposited in terrestrial ecosystems they are very likely to become incorporated into the near-surface soil layer. This is particularly true for radionuclides with physical half-lives greater than a few weeks (Englemann and Schmel, 1974).

6.6.2 Exposure from the Soil Surface Layer

6.6.2.1 Inhalation

As radionuclides reside in the near-surface layer, an initial concern is with inhalation of radionuclides that are bound to the soil particles. The smallest particles (<50 μm) can easily be resuspended by wind and made available for direct inhalation by humans and animals (Whicker and Schultz, 1982). The inhalation pathway may be more important than ingestion in some cases, such as comparatively insoluble radionuclides in arid, dusty environments.

6.6.2.2 Ingestion

Ingestion of radionuclides can occur either as a deliberate act or incidentally when eating contaminated food (Entry et al., 1997). Deliberate ingestion of soil particles is of particular concern and more common in young children because of their activities playing outdoors, lack of proper hygiene (hand washing), and mouthing behaviors including placing objects in their mouths (Calabrese et al., 1997). These activities are less likely to occur in older children and adults. Incidental ingestion of radionuclides through contaminated food can be through simple soil-to-plant transfers or more complex pathways where radionuclides move through the levels of the food chain. Since radionuclides cannot be degraded, they have the potential to accumulate in plant species and increase in concentration as they make their way up the food chain (Entry et al., 1997). An example of movement through the food chain can be from soils to plants to herbivores to humans. The potential for the uptake and bioaccumulation of radionuclides into plants, animals, and aquatic species used as food sources is driven by the bioavailability.

6.6.2.3 Pathway from Soil to Plants

Bioavailability of radionuclides from soils to plants is driven by the following factors: (1) the chemical nature and reactivity of the isotope as these may affect the availability of the isotope within soil pore water within the rhizosphere, (2) the route of exposure (e.g., root vs. foliar exposure), (3) the plant species itself (physical stature, root–shoot ratio, growing or dormant), and (4) the nutrient requirements of the plant (chemical similarity of the isotope to a nutrient). Radionuclides in air or water can be deposited on or otherwise attached to plant surfaces directly, without passage through the soil. Frequently radioactive materials in soil adhere to the root or shoot surfaces of plants, but are not actually incorporated into plant tissues. Radionuclides bound to soil particles can adhere directly to plant surfaces through soil splash, deposition from air, or through harvest practices (Sheppard and Evenden, 1992). Such material may be held only temporarily on the surface and later washed or blown off or attach quite strongly if the plant has "hairy" leaves that act to collect and trap contamination. Some surface-deposited material may even be assimilated into plant tissues through stomata or epidermal tissues. Uptake from soil by plants refers to the passage of atoms or molecules from the soil solution through the root membrane and into plant tissues, where translocation can take place.

6.6.2.4 Pathway from Plants to Herbivores

Passage of radioactive material from plants to herbivores is accomplished mainly by ingestion. If entire leaves are consumed, then both internally incorporated and externally attached material will enter the consumer. In some cases the radioactive material incorporated within the plant may be in a more soluble form than surface material and thus a higher fraction may be assimilated by the consumer. Herbivores also ingest radionuclides associated with soils or sediments. In some cases, soil ingestion by animals is deliberate, but measurable quantities can also be inadvertently ingested in the process of foraging on low-growing plants and licking and preening of fur, feathers, or young (Whicker and Schultz, 1982). This transfer of radionuclides from plants to herbivores is particularly important if the herbivore is then used as a food source for humans. An example of this process can be seen with a plant that has accumulated ^{137}Cs that is eaten by a deer which humans then hunt for food.

6.7 *IN SITU* REMEDIATION OF SOILS CONTAMINATED WITH RADIONUCLIDES

Radioactive contamination exists in a number of subsurface locations including contamination of government as well as private lands. A number of sites with radioactive waste in the United States have been mapped and characterized (EPA, 2012; Riley and Zachara, 1992). Mitigation of the environmental and public health threats from radioactive contaminants can be approached through either physical removal of contaminants or by controlling environmental conditions to decrease contaminant mobility in the environment leading to contaminants being sequestered in place. Physical removal is labor-intensive and is often only cost-effective for higher contaminant concentrations. It includes obvious approaches such as trucking contaminants away or pump-and-treat approaches for groundwater as well as the use of absorbent media (Gu et al., 1998), soil-washing techniques, filtration, and a variety of other approaches (IAEA, 2006). More sophisticated approaches incorporate technologies like electrokinetic techniques for the efficient removal of contaminants from soils (Agnew et al., 2011; Virkutyte et al., 2002).

Physicochemical removal of radioactive contaminants involves chemical conversion of contaminants followed by sequestration in the environment. A number of techniques have been developed for this purpose and include reactive precipitation that involves chemical conversion of a contaminant to a less soluble form followed by its precipitation from the groundwater. The resulting precipitate is then considered immobilized and hence no longer a risk, as long as it remains immobile. An example of this approach is the use of zerovalent iron for the precipitation of radioactive contaminants *in situ* (Gu et al., 1998).

The primary treatment of ^{99}Tc has been to reduce it from mobile Tc(VII) to relatively immobile Tc(IV). There are synthetic resins, for example, Bio-Rad AG-OH, that show very high Kd values of 1360 mL/g (Table 6.2) and have high potential to remove ^{99}Tc effectively. But synthetic resins are considered by many as expensive and not robust enough for *in situ* environmental applications. Therefore, zerovalent iron or activated carbon offer more effective and economical approaches than resins in removing Tc (Table 6.2).

Once the highest concentrations of contaminants are removed from the ground, the diffusion of contaminants into the subsurface will be minimized. These lower concentrations of contaminants are often not cost-effective to remove but could be sequestered *in situ* and thereby prevented from entering the food chain. Plant uptake of radionuclides can also be used as a remediation tool in the form of phytoremediation.

6.7.1 Phytoremediation

Phytoremediation incorporates a radioecology approach by leveraging the ability of plants to incorporate metals and radionuclides into their tissues as part of their metabolic activity (Caldwell et al., 2011; Willey and Collins, 2007; Negri and Hinchman, 2000; Entry et al., 1999). Phytoremediation offers a low-cost approach to soil remediation and appears best suited for contaminants at or near soil surfaces. Radionuclide-accumulating plants could be grown and harvested to remove contaminants. The dried plants could then be reduced in mass through combustion or composting and disposed of as radioactive waste. This approach offers a reduced waste burden compared to other technologies such as heap leaching and soil washing (Zhu and Chen, 2009). An interesting niche for this technology may be at seep zones where contaminated groundwater emerges at the terrestrial surface.

While many radionuclides are taken up by plants as part of their normal metabolic activity, interestingly uranium uptake has been documented in some plants (Huang et al., 1998) even though it is not similar to any essential ion in the plants' metabolism (Ebbs et al., 1998). Another interesting example of phytoremediation involves the bioavailability and plant uptake of the anthropogenic radionuclide technetium. The pertechnetate anion can be transported from soil to components of the plant, whereas the reduced form of technetium, Tc(IV), is immobile (Yanagisawa et al., 1999; Yanagisawa and Muramatsu, 1995). Plant growth is optimal in soils with pH between 7.5 and 5.5 and Eh values between 600 and 100 mV (Koch-Steindl and Pröhl, 2001). These conditions are also favorable for mobility of many radionuclides, including technetium as the pertechnetate anion.

6.7.2 Immobilization

Radionuclide sorption onto soil components like clays, minerals, organic colloids, microorganisms, or any combination of these has been demonstrated to potentially immobilize or at least lower the rate of contaminant transport in the subsurface. However, where the conditions are not optimal, radionuclide immobilization can be enhanced by physical or chemical control of soil conditions. Numerous techniques have been developed for manipulating soil conditions to improve radionuclide contaminant immobilization (Hamby, 1996) and include capping, *in situ* grouting, addition of chemicals to enhance chelation, and immobilization by introducing chemicals that restrict radionuclide migration.

6.7.3   BIOREMEDIATION

Through their physiological activity, microorganisms can be used to sequester radionuclides in a number of ways. Through the addition of carbon sources (biostimulation), the activity of indigenous microorganisms can sometimes be enough to sequester radionuclides. If indigenous populations are insufficient to carry out remedial activities, populations of useful microorganisms can be introduced (bioaugmentation) in order to achieve the desired results. In many cases the amount of nutrient input from the natural environment through plant litter decomposition may provide enough nutrients to transform radioactive contaminants in the subsurface to an immobile state (natural attenuation).

There are numerous ways microorganisms can alter the subsurface environment to control the mobility of radioactive contaminants. Through stimulation of microbial growth via the addition of organic carbon sources, pH and Eh can easily be altered as a result of the activity of the bulk microbial population. In addition, microbial activity also influences mineral formation, especially in relation to radionuclides. For instance, ^{90}Sr has been shown to be incorporated into calcite during ureolysis by *Bacillus pasteurii* (Fujita et al., 2004). In another example, tyrosine supplementation to soil resulted in microbial production of pyomelanin, which ultimately led to U immobilization in the soil matrix (Turick et al., 2008).

The ability of some microorganisms to respire anaerobically by using redox-active radionuclides as terminal electron acceptors in lieu of oxygen provides a useful tool for biotransformation of many oxidized radioactive contaminants to a reduced and immobile form. The degree and complexity of microbial/radionuclide transformations have been reviewed extensively (Ehrlich and Newman, 2009; Lloyd and Macaskie, 2000), and they offer considerable potential for environmental stabilization.

In the subsurface a significant degree of microbial activity occurs on surfaces in large mixed-species communities of bacteria known as biofilms. Conditions in the biofilms may not be representative of the bulk pore water due to the complex ecology of the biofilm communities. For instance, although sulfate-reducing bacteria require anaerobic conditions above pH 5.5, sulfate reduction occurs at lower pH values in aerobic waters. The reason is that environmental alterations within biofilm communities provide a conducive environment for sulfate reducers in an otherwise hostile environment.

6.7.4   MONITORING APPROACHES

Once radioactive contaminants are successfully sequestered in the subsurface, environmental conditions must be monitored in order to ensure that the contaminants do not remobilize to the level of public concern. The amount of time required to monitor the stability of contaminants is largely dependent on the half-life of the contaminant. The chemical, physical, and biological factors that contribute to contaminant immobilization are the same factors that, when reversed (i.e., reducing conditions change to oxidizing conditions or significant shifts in pH), threaten to cause remobilization of contaminants. As a result, organized long-term monitoring schemes are required in order to maintain public health. Monitoring environmental activities over the long term provides several challenges that must be met in order ensure public safety. One aspect is the potential loss of societal memory as to the location and/or existence of contamination (Okrent and Xing, 1993).

Computerized models of subsurface activities may remedy this concern and are currently useful in predicting the behavior of sequestered contaminants. Validation of the predicted models will be prudent over time to ensure that public health risks are minimized. This can be done by monitoring specific aspects of the subsurface that are likely to control solubility/insolubility dynamics at any given site.

The choice of environmental parameters and methods for analysis of the stability of subsurface conditions relative to contaminant sequestration will vary from site to site. In addition, the cost must be considered and, hence, only a minimum of samples and parameters are likely to be chosen. Consequently, data from the site should be able to provide meaningful insight into the biogeochemical character of as much of the subsurface as possible. Potentially important biogeochemical parameters to consider are Eh, pH, and dissolved gasses including oxygen, carbon dioxide, hydrogen sulfide, nitrous oxide, and methane. Other chemicals could include dissolved organic matter, as a means to determine the amount and type of available carbon sources that could support microbial growth, and dissolved inorganics, such as ferrous iron, that indicate ongoing biogenic metal reduction. Molecular and biochemical approaches can also be used to determine specific characteristics about the bacterial population in relation to the geochemistry and stability of radionuclides.

Each of the parameters listed provides a valuable "piece of the puzzle" and together all the data should produce a clear picture of the biogeochemical activity in the subsurface that will be insightful to forecast contaminant fate. The cost of an extensive analytical approach can be prohibitive, but minimizing the amount of data acquired blurs the big picture and reduces the confidence of any forecast of biogeochemical activity. Comprehensive approaches are needed that incorporate numerous biogeochemical parameters over an extended area of the subsurface. Such approaches are being developed through the field of "biogeophysics," which is concerned with the "geophysical signatures of microbial interactions with geologic media" (Atekwana and Slater, 2009). Biogeophysical techniques incorporate geophysical methods to evaluate microbial processes like biofilm formation, microbe-mediated redox processes, biogeochemistry, and microbe/mineral transformations at spatial scales not achievable with conventional microbiological or geochemical techniques (Atekwana and Slater, 2009).

6.8   CONCLUDING REMARKS

Soils are a tremendously important resource and influence our health and well-being. Radioactive soil constituents, both those that are naturally occurring and those that exist through anthropogenic activity, play an important role in our health in the long term. By improving our understanding of their geochemical and biogeochemical behavior in various soils, we are better positioned to understand and control the impact of radionuclides on human health. The maturation of the environmental remediation industry as well as risk assessment of contaminated sites (Ferguson, 1996) has contributed significantly toward our understanding, management, and cleanup of radionuclides in soils. As far as future directions, a multidisciplinary approach to understanding and mitigating risks of radionuclides in soils will be needed (Abrahams, 2002). This will include scientists, engineers, and educators to develop and implement new technologies for understanding, monitoring, and minimizing potential risks from environmental radioactivity.

REFERENCES

Abrahams, P.W. 2002. Soils: Their implications to human health. *Science of the Total Environment* 291: 1–32.
Agnew, K., A.B. Cundy, L. Hopkinson, I.W. Croudace, P.E. Warwick, and P. Purdie. 2011. Electrokinetic remediation of plutonium-contaminated site wastes: Results from a pilot-scale on-site trial. *Journal of Hazardous Materials* 186: 1405–1414.
Alexander, M. 1994. Introduction. *In* M. Alexander (ed.), *Biodegradation and Bioremediation*, pp. 1–3. Academic Press, San Diego, CA.

Ames, L.L., J.E. McGarrah, B.A. Walker, and P.F. Salter. 1983. Uranium and radium sorption on amorphous ferric oxyhydrooxide. *Chemical Geology* 40: 135–148.

Andersson, K., B. Torstenfelt, and B. Allard. 1982. Sorption behavior of long-lived radionuclides in igneous rock. *In Proceeding of an International Symposium on Migration in the Terrestrial Environment of Long-Lived Radionuclides from the Nuclear Fuel*, pp. 111–131, 27–31 July 1981, Knoxville, TN, IAEA-SM-257/20. International Atomic Energy, Vienna, Austria.

Atekwana, E. and L.D. Slater. 2009. Biogeophysics: A new frontier in earth science research. *Reviews of Geophysics* 47: RG4004.

Baes, C.F. and R.D Sharp. 1983. A proposal for estimation of soil leaching and leaching constants for use in assessment models. *Journal of Environmental Quality* 12: 17–28.

Bagwell, C.E., C.E. Milliken, S. Ghoshroy, and D.A. Blom. 2008. Intracellular copper accumulation enhances the growth of *Kineococcus radiotolerans* during chronic irradiation. *Applied and Environmental Microbiology* 74: 1376–1384.

Balsley, S., P. Brady, J. Krumhansl, and H. Anderson. 1996. Iodide retention by metal sulfide surfaces: Cinnabar and chalcocite. *Environmental Science & Technology* 30: 3025–3027.

Bell, J. and T.H. Bates. 1988. Distribution coefficients of radionuclides between soils and groundwaters and their dependence on various test parameters. *Science of the Total Environment* 69: 297–317.

Bostick, D.T., W.D. Arnold, M.W. Burgess, D.J. Davidson, J.H. Wilson, W.D. Bostick, T.A. Dillow, et al. 1995. FY 1995 separation studies for liquid low-level waste treatment at Oak Ridge National Laboratory. DOE Tech. Report No. ORNL/TM-13101. OSTI ID: 206373; Legacy ID: DE96006025.

Brevik, E.C. 2009. Soil, food security, and human health. *In* W. Verheye (ed.), *Soils, Plant Growth and Crop Production*. Encyclopedia of Life Support Systems (EOLSS), Developed under the Auspices of the UNESCO, EOLSS Publishers, Oxford. http://www.eolss.net (verified 15 January 2012).

Bugai, D.A., R.D. Waters, S.P. Dzhepo, and A.S. Skalskij. 1996. Risks from radionuclide migration to groundwater in the Chernobyl 30-km zone. *Health Physics* 71: 9–18.

Calabrese, E.J., E.J. Stanek, R.C. James, and S.M. Roberts. 1997. Soil ingestion: A concern for acute toxicity in children. *Environmental Health Perspectives* 105: 1354–1358.

Caldwell, E.F., M.C. Duff, C.E. Ferguson, and D.P. Coughlin. 2011. Plants as bio-monitors for Cs-137, Pu-238, Pu-239, 240 and K-40 at the Savannah River Site. *Journal of Environmental Monitoring* 13: 1410–1421.

Choppin, G. 2003. Actinide speciation in the environment. *Radiochimica Acta* 91: 645–649.

Choppin G.R., A.H. Bond, and P.M. Hromadka. 1997. Redox speciation of Pu. *Journal of Radioanalytical and Nuclear Chemistry* 219: 203–210.

Dadachova, E. and A. Casadevall. 2008. Ionizing radiation: How fungi cope, adapt, and exploit with the help of melanin. *Current Opinion in Microbiology* 11: 525–531.

Daly, M.J., E.K. Gaidamakova, V.Y. Matrosova, A.Vasilenko, M. Zhai, A. Venkateswaran, M. Hess, et al. 2004. Accumulation of Mn(II) in *Deinococcus radiodurans* facilitates gamma-radiation resistance. *Science* 306: 1025–1028.

Ebbs, S., D.J. Brady, and L.V. Kochain. 1998. Role of uranium speciation in the uptake and translocation of uranium by plants. *Journal of Experimental Botany* 49: 1183–1190.

Ehrlich, H.L. and D.K. Newman. 2009. Geomicrobial interactions with chromium, molybdenum, vanadium, uranium, polonium and plutonium. *In* H.L. Ehrlich and D.K. Newman (eds), *Geomicrobiology*, pp. 429–438. CRC Press, Boca Raton, FL.

Englemann, R.J. and G.A. Schmel (eds). 1974. Particulate and gaseous pollutants. *In Proceedings of symposium at Richland*, 4–6 September, Washington, DC. Technical Information Center, Office of Public Affairs ERDA.

Entry, J.A., L.S. Watrud, R.S. Manasse, and N.C. Vance. 1997. Phytoremediation and reclamation of soils contaminated with radionuclides. *In* E.L. Kruger, T.A. Anderson, and J.R. Coats (eds), *Phytoremediation of Soil and Water Contaminants*, pp. 299–306. American Chemical Society, New York, NY.

Entry, J.A., L.S. Watrud, and M. Reeves. 1999. Accumulation of ^{137}Cs and ^{90}Sr from contaminated soil by three grass species inoculated with mycorrhizal fungi. *Environmental Pollution* 104: 449–457.

EPA. 2011. Radiation protection: Commonly encountered radionuclides. http://www.epa.gov/radiation/radionuclides/ (verified 15 January 2012).

EPA. 2012. Radioactively contaminated national priorities list (NPL) sites. http://www.epa.gov/radiation/cleanup/npl_sites.html (verified 15 January 2012).

Ferguson, C.C. 1996. Assessing human health risks from exposure to contaminated land. *Land Contamination & Reclamation* 4: 159–170.

Ferguson, C.C. and J.A. Marsh. 1993. Assessing human health risks from ingestion of contaminated soil. *Land Contamination & Reclamation* 1: 177–185.

Fioravanti, M. and A. Makhijani. 1997. *Containing the Cold War Mess: Restructuring the Environmental Management of the U.S. Nuclear Weapons Complex.* Institute for Energy and Environmental Research, Takoma Park, MD.

Fredrickson, J.K., J.M. Zachara, D.L. Balkwill, D. Kennedy, S.M. Li, H.M. Kostandarithes, M.J. Daly, M.F. Romine, and F.J. Brockman. 2004. Geomicrobiology of high-level nuclear waste-contaminated vadose sediments at the Hanford site, Washington State. *Applied and Environmental Microbiology* 70: 4230–4241.

Fujita, Y., G.D. Redden, J.C. Ingram, M.M. Cortez, F.G. Ferris, and R.W. Smith. 2004. Strontium incorporation into calcite generated by bacterial ureolysis. *Geochimica et Cosmochimica Acta* 68(15): 3261–3270.

Garnier-Laplace, J., K. Beaugelin-Seiller, and T.G. Hinton. 2011. Fukushima wildlife dose reconstruction signals ecological consequences. *Environmental Science & Technology* 45: 5077–5078.

Gaschak, S.P., Y.A. Makliuk, A.M. Maksimenko, M.D. Bondarkov, I. Chizhevsky, E.F. Caldwell, G.T. Jannik, and E.B. Farfán. 2011. Frequency distributions of ^{90}Sr and ^{137}Cs concentrations in an ecosystem of the "Red Forest" area in the Chernobyl exclusion zone. *Health Physics* 101: 409–415.

Glover, P.A., F.J. Miner, and W.O. Polzer. 1976. Plutonium and americium behavior in the soil/water environment. I: Sorption of plutonium and americium by soils. *In Proceedings of Actinide-Sediment Reactions Working Meeting*, pp. 225–254, Seattle, Washington, DC, BNWL-2117. Battelle Pacific Northwest Laboratories, Richland, WA.

Gorby, Y.A., S. Yanina, J.S. McLean, K.M. Rosso, D. Moyles, A. Dohnalkova, T.J. Beveridge, et al. 2006. Electrically conductive bacterial nanowires produced by *Shewanella oneidensis* strain MR-1 and other microorganisms. *Proceedings of the National Academy of Sciences of the United States of America* 103(30): 11358–11363.

Gu, G., L. Liang, M.J. Dickey, X. Yin, and S. Dai. 1998. Reductive precipitation of uranium(VI) by zero-valent iron. *Environmental Science & Technology* 32: 3366–3373.

Hamby, D.M. 1996. Site remediation techniques supporting environmental restoration activities—A review. *Science of the Total Environment* 191: 203–224.

Hewitt, C.N. 2001. Radioactivity in the environment. *In* R.M. Harrison (ed.), *Pollution: Causes, Effects and Control*, pp. 474–499. The Royal Society of Chemistry, Cambridge.

Hinton, T.G., D.I. Kaplan, A.S. Knox, D.P. Coughlin, R.V. Nascimento, S.I. Watson, D.E. Fletcher, and B.-J. Koo. 2006. Use of illite clay for *in situ* remediation of ^{137}Cs-contaminated water bodies: Field demonstration of reduced biological uptake. *Environmental Science & Technology* 40: 4500–4505.

Hu, Q.-H., J.-Q. Weng, and J.-S. Wang. 2011. Sources of anthropogenic radionuclides in the environment: A review. *Journal of Environmental Radioactivity* 101: 426–437.

Huang, J.W., M.J. Blaylock, Y. Kapulnik, and B.D. Ensley. 1998. Phytoremediation of uranium-contaminated soils: Role of organic acids in triggering uranium hyperaccumulation in plants. *Environmental Science & Technology* 32: 2004–2008.

IAEA. 2006. *Remediation of Sites with Mixed Contamination of Radioactive and Other Hazardous Substances.* International Atomic Energy Agency, Vienna, Austria.

Icenhower, J.P., W.J. Martin, N.P. Qafoku, and J.M. Zachara. 2008. The geochemistry of technetium: A summary of the behavior of an artificial element in the natural environment. Pacific Northwest National Laboratory, Richland, WA. http://www.pnl.gov/main/publications/external/technical_reports/PNNL-18139.pdf (verified 15 January 2012).

Icenhower, J.P., N.P. Qafoku, J.M. Zachara, and W.J. Martin. 2010. The biogeochemistry of technetium: A review of the behavior of an artificial element in the natural environment. *American Journal of Science* 310: 721–752.

Kaplan, D., G. Iverson, S. Mattigod, and K. Parker. 2000. I-129 test and research to support disposal decisions. DOE Tech. Report No. WSRC-TR-2000-00283.

Kathren, R. 1991. *Radioactivity in the Environment.* Taylor and Francis, Boca Raton, FL.

Kibble, A.J. and P.J. Saunders. 2001. Contaminated land and the links with health. *In* R.E. Hester and R.M. Harrison (eds). *Assessment and Reclamation of Contaminated Land, Issues in Environmental Science and Technology*, Issue 16, pp. 65–84. The Royal Society of Chemistry, Cambridge.

Kim, B.T., H.K. Lee, H. Moon, and K.J. Lee. 1995. Adsorption of radionuclides from aqueous solutions by inorganic adsorbents. *Separation Science and Technology* 30: 3165–3182.

Koch-Steindl, H. and G. Pröhl. 2001. Considerations on the behaviour of long-lived radionuclides in the soil. *Radiation and Environmental Biophysics* 40: 93–104.

Liang, L., B. Gu, and X. Yin. 1996. Removal of technetium-99 from contaminated groundwater with sorbents and reductive materials. *Separation Technology* 6: 111–122.

Lieser, K.H. and C. Bauscher. 1987. Technetium in the hydrosphere and in the geosphere. I: Chemistry of technetium and iron in natural waters and influence of the redox potential on the sorption of technetium. *Radiochimica Acta* 42: 205–213.

Lieser, K.H., B. Gleitsmann, and T. Steinkopff. 1986. Sorption of trace elements or radionuclides in natural systems containing groundwater and sediments. *Radiochimica Acta* 40: 33–37.

Lloyd, J.R. and L.E. Macaskie. 2000. Bioremediation of radionuclide-containing wastewaters. *In* D.R. Lovley (ed.), *Environmental Microbe–Metal Interactions*, pp. 277–327. ASM Press, Washington, DC.

Marsili, E., D.B. Baron, I.D. Shikhare, D. Coursolle, J.A. Gralnick, and D.R. Bond. 2008. *Shewanella* secretes flavins that mediate extracellular electron transfer. *Proceedings of the National Academy of Sciences of the United States of America* 105: 3968–3973.

Negri, C.M. and R.R. Hinchman. 2000. The use of plants for the treatment of radionuclides. *In* I. Raskin and B.D. Ensley (eds), *Phytoremediation of Toxic Metals: Using Plants to Clean Up the Environment*, pp. 107–132. Wiley-Interscience, New York, NY.

Okrent, D. and L. Xing. 1993. Future risk from a hypothesized RCRA site disposing of carcinogenic metals should a loss of societal memory occur. *Journal of Hazardous Materials* 33: 363–384.

Otosaka, S., K.A. Schwehr, D.I. Kaplan, K.A. Roberts, S. Zhang, C. Xu, H.-P. Li, et al. 2011. Factors controlling mobility of ^{127}I and ^{129}I species in an acidic groundwater plume at the Savannah River Site. *Science of the Total Environment* 409: 3857–3865.

Parmar, N., L.A. Warren, E.E. Roden, and F.G. Ferris. 2000. Solid phase capture of strontium by the iron reducing bacteria *Shewanella alga* strain BrY. *Chemical Geology* 169: 281–288.

Reguera, G., K.D. McCarthy, T. Mehta, J.S. Nicoll, M.T. Tuominen, and D.R. Lovley. 2005. Extracellular electron transfer via microbial nanowires. *Nature* 435: 1098–1101.

Riley, R.G. and J.M. Zachara. 1992. Chemical contaminants on DOE lands and selection of contaminant mixtures for subsurface science research. Office of Energy Research, Subsurface Science Program, U.S. Department of Energy, Washington, DC.

Rusin, P.A., L. Quintana, J.R. Brainard, B.A. Striemeier, C.D. Talt, S.A. Ekberg, P.D. Palmer, T.W. Newton, and D.L. Clark. 1994. Solubilization of plutonium hydrous oxide by iron-reducing bacteria. *Environmental Science & Technology* 28: 1686–1690.

Sheppard, S.C. and W.G. Evenden. 1992. Concentration enrichment of sparingly soluble contaminants (U, Th and Pb) by erosion and by soil adhesion to plants and skin. *Environmental Geochemistry and Health* 14: 121–131.

Song, K.C., H.D. Kim, H.K. Lee, H.S. Park, and K.J. Lee. 1997. Adsorption characteristics of radiotoxic cesium and iodine from low-level liquid wastes. *Journal of Radioanalytical and Nuclear Chemistry* 223: 199–205.

Tang, Y.S. and J.H. Saling. 1990. *Radioactive Waste Management*. Hemisphere Publishing Corporation, New York, NY.

Thibault, D.H., M.I. Sheppard, and P.A. Smith. 1990. *A Critical Compilation and Review of Default Soil Solid/Liquid Partition Coefficients, Kd, for Use in Environmental Assessments*, AECL-10125. Whiteshell Nuclear Research Establishment, Pinawa, MB.

Turick, C.E., L.S. Tisa, and F. Caccavo Jr. 2002. Melanin production and use as a soluble electron shuttle for Fe(III) oxide reduction and as a terminal electron acceptor by *Shewanella algae* BrY. *Applied and Environmental Microbiology* 68: 2436–2444.

Turick, C.E., F. Caccavo Jr, and L.S. Tisa. 2003. Electron transfer to *Shewanella algae* BrY to HFO is mediated by cell-associated melanin. *FEMS Microbiology Letters* 220: 99–104.

Turick, C.E., A.S. Knox, C.L. Leverette, and Y.G. Kritzas. 2008. *In-situ* uranium immobilization by microbial metabolites. *Journal of Environmental Radioactivity* 99: 890–899.

Turick, C.E., A.S. Beliaev, B.A. Zakrajsek, C.L. Reardon, T. Poppy, A. Maloney, D.A. Lowy, and A.A. Ekechukwu. 2009. The role of 4-hydroxyphenylpyruvate dioxygenase in enhancement of solid-phase electron transfer by *Shewanella oneidensis* MR-1. *FEMS Microbiology Ecology* 68: 223–235.

Turick, C.E., A.A. Ekechukwu, C.E. Milliken, A. Casadevall, and E. Dadachova. 2011. Gamma radiation interacts with melanin to alter its oxidation–reduction potential. *Bioelectrochemistry* 82: 69–73.

Vine, E.N., R.D. Aquilar, B.P. Bayhurst, W.R. Daniels, S.J. DeVilliers, B.R. Erdal, F.O. Lawrence, et al. 1980. Sorption–desorption studies on tuff. II: A continuation of studies with samples from Jackass Flats, Nevada and initial studies with samples from Yucca Mountain, Nevada, Informal Report LA-8110-MS. Los Alamos Scientific Laboratory, Los Alamos, NM.

Virkutyte, J., M. Sillanpää, and P. Latostenmaa. 2002. Electrokinetic soil remediation—Critical review. *Science of the Total Environment* 289: 97–121.

Washington State Department of Health. 2010. *Background Radiation*. Washington State Department of Health Office of Radiation Protection Fact Sheet Number 10. http://www.doh.wa.gov/ehp/rp/factsheets/factsheets-htm/fs10bkvsman.htm (verified 15 January 2012).

Whicker, F.W. and V.E. Schultz. 1982. *Radioecology: Nuclear Energy and the Environment*, Vol. II. CRC Press, Boca Raton, FL.

Willey, N. and C. Collins. 2007. Phytoremediation of contaminated soils. *Radioactivity in the Environment* 10: 43–69.

Yanagisawa, K. and Y. Muramatsu. 1995. Transfer of technetium from soil to paddy and upland rice. *Journal of Radiation Research* 36: 171–178.

Yanagisawa, K., H. Takeda, K. Miyamoto, and S. Fuma. 1999. Transfer of technetium from paddy soil to rice seedling. *Journal of Radioanalytical and Nuclear Chemistry* 243: 403–408.

Zhu, Y.-G. and B.-D. Chen. 2009. Principles and technologies for remediation of uranium-contaminated environments. *Radioactivity in the Environment* 14: 358–374.

7 Soil's Influence on Water Quality and Human Health

Martin F. Helmke and Russell L. Losco

CONTENTS

7.1 INTRODUCTION

Humans require access to abundant, clean water to sustain health. Water quality directly affects human health, and few earth systems have a more controlling influence on water quality than soil. Most water available to humans, namely groundwater accessed by wells or surface water in streams or lakes, passes through soil. Soil is capable of removing pathogens and toxic compounds from water. Soil stores water, which reduces flooding frequency and serves as a water resource reservoir. These processes are not unlimited, however. Society must understand the benefits and limitations of soil to protect water resources for the benefit of human health. Creative and strategic utilization of soil will continue to play a key role in sustainable maintenance of global water quality.

The three critical components for understanding environmental risk to human health are contaminant source, pathway, and exposure. Mitigating any of these will reduce health risk. Soil plays an important role as a pathway because it controls migration of contaminants to groundwater and surface water (Figure 7.1). Soil can be bypassed, however, by runoff and anthropogenic activities such as agricultural drainage tiles, injection wells, and storm sewers.

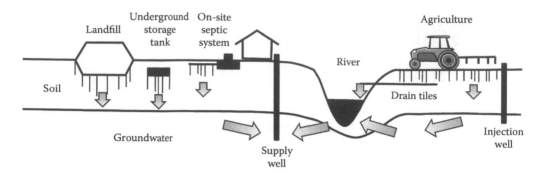

FIGURE 7.1 Pathways water may follow through soil to groundwater and surface-water sources.

Conversely, soil may be a source of contaminants or facilitate contaminant transport, and humans may extract water contaminated by soil for distribution.

The most widespread contaminants in water supplies are enteric pathogens, nitrate, pesticides, heavy metals, and industrial carcinogens. Of these, enteric pathogens pose the greatest threat to human health. Dysentery, an acute gastrointestinal disorder caused by ingestion of water rendered unsanitary by human and animal waste, is the third-leading cause of death in low-income countries (WHO, 2008a). Worldwide, 1.7 million people die each year from diarrheal diseases associated with drinking contaminated water. In 2006, an estimated 884 million people lacked access to an "improved" drinking water source (i.e., groundwater source or municipal water supply system) and 2.5 billion lived without adequate sanitation infrastructure (UNICEF and WHO, 2008). In 2008, 1.3 million children between the ages of 1 and 59 months died from diarrheal diseases, which accounted for 24% of all deaths in this age group worldwide (Black et al., 2010). Child mortality from dysentery has been dramatically reduced in the United States by (a) requiring wastewater to be passed through soil via on-site septic systems, (b) treating municipal wastewater in sewage treatment plants, (c) chlorinating public water, and (d) improving access to health care. Nonetheless, the annual number of reported illnesses in the United States resulting from contact with waterborne pathogens is estimated to be between 1 and 7 million (Craun, 1985). Fortunately, these are generally nonfatal due, in part, to access to advanced health care. Using soil to isolate the contaminant source (sewage) from receptors provides a comparatively inexpensive means of solving one of the world's greatest health problems.

The global health effect of ingesting nitrates, pesticides, heavy metals, and anthropogenic carcinogens in drinking water has not been fully quantified. However, in an attempt to minimize this risk, countries such as the United States and the World Health Organization (WHO, 2008b) have established maximum safe concentrations of these compounds in drinking water based on expected lifetime consumption rates. Careful soil management may be used to facilitate meeting health-risk benchmarks for drinking water quality.

7.2 SOIL PROCESSES THAT CONTROL WATER QUALITY

In a natural system, precipitation is the source of most groundwater. This precipitation is absorbed into the soil and percolates downward through the unsaturated zone. In an environment impacted by humans, this process is modified by agricultural, industrial, residential, and transportation-oriented activities that disrupt the pathway to groundwater recharge and introduce pollutants. The soil mantle serves as a filter that reduces direct introduction of polluted water to the groundwater flow system and, by extension, surface-water bodies that are recharged by groundwater. Groundwater may be further remediated by geologic materials below soil; however, soil is an excellent filter medium and can protect groundwater resources if used wisely. Soil filters and purifies water by three basic mechanisms: physical filtration, chemical sorption, and biodegradation. All of these must be considered if we are to understand the relationship between soil, water quality, and human health. Collectively, these three mechanisms are referred to as the renovative ability of soil. A complete review of solute fate and transport in soil is beyond the scope of this chapter. However, the processes of water movement through soil, fate and transport of pathogens, sorption, and biodegradation largely govern soil's influence on water quality.

7.2.1 MOVEMENT OF WATER THROUGH SOIL

The degree to which soil mitigates contaminant transport is strongly dependent on pore water storage and residence time. Most surficial soils contain 25%–60% porosity (Brady and Weil, 2004), which allows soils to store large volumes of water during precipitation events, followed by subsequent slow release to groundwater or surface-water systems. Under saturated conditions, water may move rapidly downward and laterally through macropores, resulting in short soil residence times in the order of minutes. Short residence time allows dissolved contaminants to pass rapidly through soil unabated. Under unsaturated conditions, large pores may drain, resulting in a reduction in hydraulic conductivity of several orders of magnitude, an increase in water residence time (days), and a reversal of hydraulic gradient.

Physical filtration is the physical capture and retention of pollutants within pore spaces in the soil. The level of physical filtration is governed by a number of parameters, including porosity, pore size, pore tortuosity, pore interconnectivity, soil thickness, and soil permeability (Lal and Shukla, 2004). Porosity refers to the volume of total pore space as a percentage of the volume of the soil. Porosity can be related to permeability but the two are separate aspects of a soil. Generally speaking, clays have a higher porosity than sands, yet the pores in clays are much smaller. Smaller pores transmit fluids more slowly than larger pores, resulting in a slow, yet effective, filtration process. Interconnected pores lead to greater permeability. Large pores will not contribute significantly to permeability if they "dead-end" and are therefore poorly connected. The curvature of pores, known as tortuosity, also impacts filtration. A straight pore will transmit water faster than one that provides a convoluted path; therefore a soil with high tortuosity will be a more effective filter. A thicker soil profile with a given porosity will increase the level of filtration by the sheer volume of the soil as a filter medium.

Permeability influences filtration in a number of ways. Rapid velocities can allow pollutants to remain in suspension as water passes through the filter medium. Slower velocities can enhance entrapment of pollutants; however, a permeability that is too slow allows water to pond and run off on the surface before infiltrating into the soil or to move horizontally out to the surface. Permeabilities that are too slow can also allow the development of an anoxic environment, which impacts chemical and biological filtration. Filtration quality often depends as much on the pollutant as it does on the permeability; however, in general a moderate permeability is the most ideal for effective physical filtration.

7.2.2 FATE AND TRANSPORT OF PATHOGENS

Human pathogens such as protozoa, bacteria, and enteroviruses all threaten water quality and pose great risks to human health. Fortunately, soil significantly reduces long-distance, viable transport of these organisms by size exclusion, adsorption, and die-off.

Pathogens too large to pass physically through water-filled pores are rendered immobile by the process of size exclusion. Most pathogens are smaller than sand-sized grains, but many are larger than clay particles (Figure 7.2). The cysts of pathogenic protozoa such as *Giardia lamblia* and *Cryptosporidium parvum* are relatively large (8–13 and 3–5 μm diameter, respectively). Bacteria, such as fecal coliform, are smaller, in the order of 0.2–1.0 μm. Viruses are some of the smallest biocolloids, ranging in diameter between 20 and 300 nm (Azadpour-Keeley et al., 2003). A clay-textured soil with clay-sized (2 μm or smaller) pores would, theoretically, prevent large pathogens

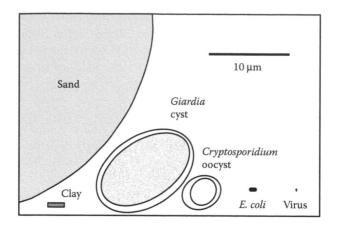

FIGURE 7.2 Relative sizes of sand- and clay-sized particles compared with enteric pathogens *Giardia*, *Cryptosporidium*, *E. coli*, and viruses.

from passing through by size exclusion. Most natural soils and aquifer materials, however, contain macropores that allow rapid transport of pathogenic biocolloids when large pores fill under saturated conditions (Clothier et al., 2008). Recent studies have documented rapid transport of *Cryptosporidium* (Harter et al., 2000; Nasser et al., 2003), fecal coliform bacteria (Jamieson et al., 2002), and viruses (Bales et al., 1995) through soil under saturated conditions. When soil is unsaturated, advection occurs along thin films of water occupying partially saturated pores. Under these conditions, biocolloids may be too large to travel within the water films or may accumulate at the air/water interface. Sobsey et al. (1995) reported a 79%–100% removal of viruses from water passing through a clay loam under unsaturated conditions. One must be cautious, however, as pathogens may travel quickly through unsaturated soil if the water content is sufficient or may be remobilized as the water content is increased. Transient recharge events, such as rainfall, snowmelt, or anthropogenic actions, can mobilize biocolloids collected at the air/water or air/water/soil interface and can flush them through soil at great velocity (El-Farhan et al., 2000).

Die-off, or inactivation, of pathogens within soil water serves to effectively remediate the water for later human consumption. Studies have reported that *Giardia* and *Cryptosporidium* oocysts, fecal coliform bacteria, and enteroviruses remain active in soil for up to 160 (Kato et al., 2003), 100 (Mubiru et al., 2000), and 100 days (Stramer, 1984), respectively, under cool, saturated conditions. Most studies of pathogen survival report a first-order decrease in population with time, with a first-order die-off rate ranging from 0.001 to 17.3/day (Azadpour-Keeley et al., 2003). Pathogen survival in soil is significantly reduced under unsaturated conditions. Unsaturated soil allows aerobic microbes to rapidly deplete soluble organic matter and nutrients, thereby outcompeting pathogens adapted to an anaerobic environment (Reddy et al., 1981).

7.2.3 Sorption of Contaminants

Chemical constituents in water may be attracted to particle surfaces in soil by the process of adsorption. Clay minerals present in most soils present a net negative surface charge that causes dissolved cations to adsorb to the soil matrix. Toxic cations, most notably heavy metals, sorb strongly to soils by this mechanism. Organic, nonpolar compounds preferentially adsorb to soil organic carbon. The properties of the contaminant, cation exchange capacity (CEC), and fraction of organic carbon present in the soil all affect the magnitude of this sorption process (Lal and Shukla, 2004).

Chemical filtration or sorption is accomplished by two related mechanisms, cation exchange and anion exchange. The ability of a given soil to exchange cations and anions is known as the cation or anion exchange capacity. While every soil particle has a certain level of either cation or anion exchange capacity, it is clay minerals that provide the vast majority of this capacity. The exchange of cations and anions is accomplished due to the isomorphic substitutions and functional groups present in clay minerals, which provide exchange sites (Gardiner and Miller, 2004). In general, CEC greatly exceeds anion exchange capacity. This is why, for example, nitrate contamination of groundwater is a problem in agricultural areas. Nitrate is an anion and is not readily sorbed to clay minerals. When it is applied in amounts that exceed the uptake by plants, the excess nitrate is rapidly transported through the soil column and into the groundwater, with very little of it sorbed to the soil.

Adsorption/desorption may be characterized by the partitioning of a chemical between the sorbed and dissolved phases. A traditional approach for describing this relationship is to use the Freundlich sorption isotherm:

$$C_s = K_d C_l^{1/n},\tag{7.1}$$

where
C_s is the sorbed concentration
K_d is the sorption coefficient
C_l is the dissolved concentration
n is a slope correction factor (Freundlich, 1909)

The parameter K_d varies over many orders of magnitude, from 0 for a nonsorbing solute to well over 1×10^6 L/kg for strongly sorbing contaminants in an organic-rich soil (U.S. EPA, 1996). For cases when $n = 1$, the equation is equivalent to the linear sorption isotherm.

A common misconception is that adsorption permanently removes contaminants from water. Equation 7.1 implies that a dynamic equilibrium exists between the dissolved and sorbed phases. If the aqueous concentration is sufficient, contaminants will preferentially sorb to soil particles, thereby reducing the concentration in soil water. If the system is later rinsed with clean water, however, the soil will recontaminate the water by the reverse process. Moreover, a contaminated soil can introduce contaminants to groundwater and surface water by leaching.

Most solute transport models treat sorption as a reversible, instantaneous, and constant process. There are notable limitations to these assumptions, however. Some contaminant/soil combinations display sorption/desorption hysteresis, resulting in a difference between adsorption and desorption coefficients. Burgos et al. (1999) determined experimentally that desorption K_d of phenanthrene (67–493 L/kg OC) was two to three orders of magnitude greater than adsorption (3.2–3.7 L/kg OC). This would promote effective remediation of soil water and would reduce leaching rebound. Sorption is not an instantaneous process, as time is required for contaminants to migrate and sorb to soil-binding sites. Zhang and Selim (2005) reported that sorption of arsenate was twice as great after 21 days compared to the traditional 24-h tests. In natural soil systems water may pass through soil in a matter of minutes, thereby reducing the opportunity for adsorption/desorption to take place. Finally, there is a limit to how much chemical can be adsorbed to a soil. Bolton and Evans (1996), for example, conducted batch equilibrium experiments to measure the maximum capacity for soils in Ontario, Canada, to adsorb cadmium. They applied the Langmuir isotherm (Langmuir, 1916) and demonstrated that the maximum cadmium adsorption capacity (q_{max}) ranged from 560 to 11,200 mg/kg, depending on clay content, organic content, and pH.

7.2.4 BIODEGRADATION OF CONTAMINANTS IN SOIL WATER

Many contaminants are readily utilized as sources of energy for microbial populations in soil. This presents a tremendous resource for remediation of toxic compounds in soil water. Under aerobic conditions, soil bacteria utilize dissolved organic compounds such as organic matter from sewage, petroleum compounds, and some pesticides as electron donors and rapidly break them down as part of their metabolic process (Wise et al., 2000). Other compounds, such as nitrate, are used as electron acceptors under reducing conditions if a natural electron donor is present. Chlorinated compounds, such as chlorinated ethenes, organochlorine pesticides, and dioxin are not easily metabolized by bacteria but may be cometabolized by facultative anaerobes (Hemond and Fechner, 1994). Biodegradation allows soil to sustainably remediate water by rendering contaminants harmless so long as the rate of contaminant introduction does not exceed the soil's remedial capacity.

7.2.5 COLLOID-FACILITATED TRANSPORT

An intriguing transport mechanism occurs when strongly sorbing contaminants, which would otherwise be relatively immobile, attach to soil colloids that are carried by mobile soil water. Colloids are small particles between 1 nm and 1 μm diameter. Common colloidal-sized particles include clay minerals, metal oxides, humic acids, bacteria, and viruses (Kretzschmar et al., 1999). Soils generate concentrations of mobile colloids that exceed 1 g/L in soil water, presenting an effective vehicle for transporting contaminants (DeNovio et al., 2004). Colloids move rapidly by advection through mobile macropores. However, numerous processes slow the migration of colloids, including physical deposition and chemical attachment to soil surfaces. Under unsaturated conditions, colloid migration is further retarded when these particles are trapped at the gas/water interface. Deposition, attachment, and trapping greatly reduce the mass of contaminant likely to be transmitted through soil. Therefore, colloid-facilitated transport is of greatest concern for contaminants that sorb strongly and are highly toxic, such as radionuclides, heavy metals, and hydrophobic organic compounds.

7.3 SOIL AND WATER QUALITY

7.3.1 GROUNDWATER

The importance of groundwater as a natural resource cannot be overstated. Over 98% of Earth's available fresh water is groundwater (Fetter, 2001). The projected population increase by 2050 will expand the demand for water and will heighten the use of groundwater. In the United States, 51.7% of the population relies on groundwater as the primary source of water (Solley et al., 1993) through 140,000 municipal wells and 15 million private wells (U.S. Census Bureau, 2008). Soil is the material that stands between surficial contaminant sources and groundwater. More than 43 million Americans obtain untreated water from residential wells (Hutson et al., 2004), trusting that soil will function effectively to remove contaminants.

Although the United States has one of the most advanced source-water systems in the world, recent findings by the U.S. Geological Survey NAWQA Program (Toccalino and Hopple, 2010) reveal that many public groundwater supplies in the United States are contaminated. The USGS reports that 22% of samples collected from public wells exceed public health benchmarks. Interestingly, 74% of these contaminant exceedances were a result of natural sources, such as arsenic, manganese, strontium, boron, and radionuclides. Contaminants associated with anthropogenic sources, such as nitrates, pesticides, and volatile organic compounds (VOCs), accounted for 26% of samples with concentrations elevated above health-risk benchmarks.

The vast majority of water-supply wells in the United States are private and are not subject to federal monitoring requirements under the Safe Drinking Water Act. Between 1991 and 2004, the U.S. Geological Survey collected groundwater samples from 2100 domestic wells across the United States to determine drinking water quality (DeSimone et al., 2009). Groundwater contaminants exceeded drinking water health benchmarks in 23% of the wells. Most of these contaminants were inorganic and of natural origin (radon, arsenic, uranium, and fluoride). Nitrate exceeded the U.S. EPA drinking water standard (10 mg/L NO_3–N) in 4.4% of the wells. Anthropogenic organic compounds (herbicides, insecticides, solvents, disinfection by-products, hydrocarbons and oxygenates, refrigerants, and fumigants) were detected in 60% of sampled wells, but concentrations rarely exceeded health-risk benchmarks. Microbial analysis revealed that 34% of the wells contained coliform bacteria and 7.9% were contaminated by *Escherichia coli*. Elevated nitrate and microbial concentrations were associated with shallow wells located in areas of extensive agriculture.

7.3.2 SURFACE WATER

Surface-water systems, including rivers, streams, and lakes, account for 0.26% of fresh water on Earth and approximately 0.83% of available fresh water (Shiklomanov, 2000). The majority of municipal and private water systems utilize surface water as their primary source of water due to ease of access. Unfortunately, surface water is also the most easily contaminated. There are over 5.6 million km of rivers and streams in the United States. In a recent survey conducted by the U.S. EPA, 44% of waterways were classified as impaired (U.S. EPA, 2009). The most common contaminant, pathogens, was identified in 28% of impaired waterways. Nitrate and pesticides were also detected in a large number of samples, and were attributed to agriculture and urbanization.

7.3.3 SOIL AS A FILTER

Most soils have a tremendous capacity to remove contaminants from water. This is most notably the case for metals and organic compounds that adsorb strongly to soils rich in organic matter and clay. For this reason, organochlorine pesticides, PCBs, heavy metals, and dioxins are often found in soil where they are discarded and are rarely detected in groundwater or surface water leaving the soil system.

An emerging case study that underscores the importance of soil as a filter medium is the use of chlorinated herbicides during the Vietnam War (Figure 7.3). Between 1961 and 1971, the United States and South Vietnamese militaries sprayed over 76.5 million L of herbicides over Vietnam to destroy enemy food crops and increase visibility near U.S. and allied military bases during Operation "Ranch Hand" (Stellman et al., 2004). Sixty-five percent of the deployed agents included 2,4,5-trichlorophenoxyacetic acid (2,4,5-T), which was contaminated by 2,3,7,8-tetrachlorodibenzo-*p*-dioxin (TCDD). The notorious "Agent Orange" was the most widely used herbicide, containing 50% 2,4,5-T and approximately 3 mg/L TCDD. Herbicides were sprayed over 1.7 million ha at a concentration 10 times the application rate for domestic weed control.

Remarkably, TCDD is still detectable in Vietnam after 40 years, at concentrations that are near or slightly above the health-risk standard. TCDD is a well-documented teratogen and carcinogen. The U.S. EPA has established TCDD risk-based exposure standards of 4.5 ng/kg for residential soil and 0.5 pg/L for tap water (U.S. EPA, 2010). Recent investigations in Vietnam report soil TCDD concentrations ranging between 1 and 41 ng/kg, with an arithmetic mean of 8.8 ng/kg in heavily sprayed areas along the Ho Chi Minh Trail (Hofmann and Wendelborn, 2007). The greatest concentrations of TCDD in soil are found on former U.S. military bases where herbicides were stored and loaded onto aircraft. Soil samples collected in 2009 from a former aircraft loading site at Da Nang Airfield revealed a TCDD concentration of 13,400 ng/kg in surface soil (0–10 cm depth) (Hatfield Consultants, 2009). Concentration of TCDD diminished to 4.15 ng/kg at a depth of 115 cm, demonstrating how strongly this compound adsorbs to soil after 40 years. Groundwater collected from shallow wells at this site revealed no detectable TCDD, reinforcing the assertion that dioxin is strongly sorbed to soil and is generally immobile (Hofmann and Wendelborn, 2007).

TCDD continues to present a health risk in Vietnam. Human exposure studies in the 1990s identified TCDD concentrations in duck and fish fat ranging from 34 to 92 ng/kg. TCDD was not detected in agricultural products such as rice, manioc, or vegetable oil. Concentrations of TCDD in human blood serum was between nondetectable and 44 ng/kg, and blood TCDD concentrations were twice as great in individuals who lived in sprayed villages during the 1960s and 1970s (Hatfield Consultants, 2009). Given that TCDD was not detected in groundwater, the most likely exposure route to the Vietnamese was dermal contact with contaminated soil on former

FIGURE 7.3 (a) Deployment of Agent Orange by UC-123 aircraft during Operation "Ranch Hand" of the Vietnam War. (b) Agent Orange and other herbicides were shipped in 208 L (55 gal) drums in Vietnam. (From Buckingham, W.A., *Operation Ranch Hand: The USAF and Herbicides in Southeast Asia, 1961–1971.* Air Force Historical Studies Office, U.S. Government Printing Office, Washington, DC, 1989.)

military bases or ingestion of TCDD that bioaccumulated in fatty tissue of aquatic organisms (Young et al., 2008). We may conclude, therefore, that soil played a critical role in mitigating the spread of TCDD. If TCDD had been mobilized in groundwater the health toll would have been far more severe. Vietnamese researchers have demonstrated that bioremediation can be used to effectively remove TCDD from soil and are actively remediating soil at the former Da Nang Airbase (Dang et al., 2008).

7.3.4 SOIL AS A SOURCE OF CONTAMINANTS

Although soil is usually associated with the natural remediation of contaminated water, it may also be a source of contaminants. A classic and unfortunate case of soil-derived contamination is arsenic in Bangladesh groundwater. Bangladesh has a population of approximately 175 million and is one of the poorest countries in Asia. In the 1960s, most residents accessed water from surface-water sources that were heavily contaminated by enteric pathogens, resulting in an unusually high infant mortality rate of 163/1000 live births (UNICEF, 2010). The Bangladesh government, with financial assistance from WHO, UNICEF, and numerous human relief organizations, initiated a program to provide groundwater as a safe alternative source of water. By the mid-1990s, 6–11 million "tube" wells were installed, serving 90% of the population (Kinley and Hossain, 2003). Tube wells are constructed from 5 cm diameter steel tubes that are driven up to 50 m deep and attached to a small hand pump. The health benefit of switching the population from surface water to groundwater was immediate, resulting in a lowering of infant mortality to 41/1000.

By the 1990s, Bangladeshis reported an increase in chronic skin lesions, including melanosis and keratosis. A groundwater investigation conducted in 1993 reported high arsenic concentrations in tube wells. A more exhaustive study completed in 2000 revealed that 57% of wells exceeded the health-risk standard of 10 µg/L arsenic in drinking water adopted by the United States and WHO (Figure 7.4), leaving 75 million "at risk" for exposure to arsenic (Yu et al., 2003). Many wells (10%) contained over 200 µg/L arsenic, and some as high as 14,000 µg/L (Figure 7.4). The international scientific community was perplexed by the rapid and widespread contamination of groundwater by arsenic in Bangladesh. Soil arsenic concentrations in Bangladesh soils are typically less than 10 mg/kg, and this arsenic is oxidized and bound with iron as a precipitate. Recent studies suggest that groundwater recharge containing high concentrations of dissolved organic matter creates a local reducing environment that mobilizes arsenic in soil, thereby contaminating shallow groundwater. Ironically, the likely source of this organic matter is agricultural ponds that are typically filled with human waste, the very source of water the Bangladeshis worked so hard to avoid (Harvey et al., 2006).

Natural and anthropogenic arsenic contamination of groundwater is a global phenomenon. A recent study of 18,850 wells conducted by the U.S. Geological Survey revealed that 8% of public

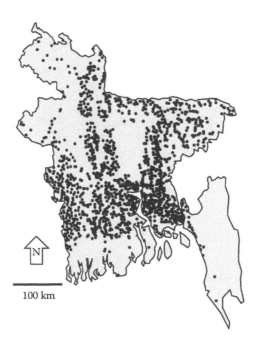

FIGURE 7.4 Arsenic concentration in shallow groundwater of Bangladesh. The international drinking water standard for arsenic is 10 µg/L. (From Kinniburgh, D.G. and Smedley, P.L., *Arsenic Contamination of Groundwater in Bangladesh, Vol. 2: Final Report*. British Geological Survey, Nottingham, 2001.)

water-supply systems in the United States are contaminated by arsenic above the 10 μg/L MCL established by the U.S. EPA (2000; Focazio et al., 2000).

7.3.5 Agriculture, Soil, and Water Quality

Agriculture is the foundation of the global economy and will play a critical role in supporting the projected world population of 9 billion by 2050. Efforts to improve agricultural productivity in the second half of the twentieth century led to the use of large quantities of nutrients and pesticides. Overapplication and a poor understanding of environmental and health impacts resulted in pollution of groundwater and surface-water resources. Fertilizer use in the United States has increased 10-fold since World War II (Figure 7.5). Typical nitrogen application rates in the Midwest exceed 160 kg/ha/year (USDA, 2008), and the total mass of nitrogen fertilizer applied to agricultural land exceeds 12 Mt/year in the United States (Dubrovsky and Hamilton, 2010). Fertilizer has resulted in an increase in nutrient concentrations in surface water and groundwater. A recent study documented that 30% of agricultural streams in the United States exceeded the U.S. EPA MCL of 10 mg/L NO_3–N and 90% of streams draining agricultural areas tested positive for nutrient concentrations greater than background concentrations (Dubrovsky and Hamilton, 2010). Eighty–three percent of samples collected from shallow groundwater wells in agricultural regions of the United States exceeded the MCL for nitrate. Fortunately, only 3% of public supply wells exceeded the MCL.

The correlation between elevated concentrations of nutrients in streams and groundwater in agricultural areas suggests that application of fertilizer and animal waste to fields is responsible for much of the contamination observed. Soil appears to be ineffective at removing nutrients in these areas. Responsible nutrient management, such as applying the appropriate quantity of fertilizer during maximum plant uptake, can reduce nutrient emissions and reduce costs (Sawyer et al., 2006). In many areas of the Midwest, fields must be drained to support agriculture. Traditional drainage methods, such as agricultural tiles and agriculture drainage wells, bypass soil or shorten soil water residence times and allow nutrients to contaminate surface water and groundwater. Recent advances in drainage technology, such as wood chip bioreactors, have demonstrated that nitrate can be reduced by 95% before entering the drainage network (Greenan et al., 2009).

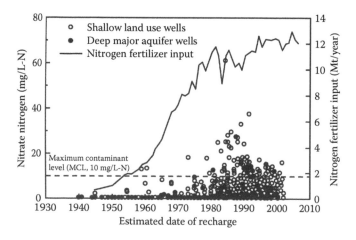

FIGURE 7.5 Relationship between nitrogen fertilizer use and nitrate concentration in groundwater in the United States. (From Dubrovsky, N.M. and Hamilton, P.A., Nutrients in the Nation's streams and groundwater: National findings and implications, U.S. Geological Survey Fact Sheet 2010-3078, 2010.)

Pesticides protect crops from interference by harmful organisms and increase productivity. Approximately 2.5 Mt of pesticides are used annually worldwide, 0.34 Mt of which are applied to fields in the United States (Kiely et al., 2004). Despite soil's remarkable ability to adsorb and degrade organic compounds, pesticides are widely distributed in both surface water and shallow groundwater. The U.S. Geological Survey reports that nearly 100% of agricultural and urban streams sampled in the United States contained detectable (>0.01 µg/L) concentrations of at least one pesticide (Larson et al., 1999). The most common pesticide was the broad-leaf herbicide atrazine, which had a median monthly concentration in streamwater that exceeded the 3 µg/L MCL in 16% of sites sampled. Detectable (>0.05 µg/L) concentrations of pesticides were reported in 48% of groundwater wells sampled nationwide in the 1990s (Kolpin et al., 2000). Pesticides were detected in shallow groundwater beneath agricultural (60%) and urban (48%) areas. Although pesticides were ubiquitous in agricultural groundwater, agrichemicals present in urban groundwater suggest that homeowners and landscapers apply significant quantities of pesticides in urban watersheds. Fortunately, pesticide concentrations in both agricultural and urban groundwater rarely exceeded current drinking water standards. Regardless, the presence of these compounds in surface-water and groundwater sources suggests that even though these compounds sorb strongly and degrade rapidly, soil is ineffective at removing them from soil water entirely.

7.3.6 Emerging Contaminants, Soil, and Water Quality

For the past 10 years, public health officials have been concerned about the widespread distribution of pharmaceuticals, hormones, surfactants, flame retardants, steroids, and other trace organic compounds referred to in the popular press as "emerging contaminants." Many of these compounds are released into the environment through wastewater and could potentially affect human health at very low concentrations. The estrogenic hormone 17β-estradiol, for example, is commonly found in municipal wastewater, and has been shown to influence aquatic organisms at concentrations below 1 ng/L (Irwin et al., 2001). Recent advances in analytical techniques have enabled analysis of hundreds of these trace organic pollutants in surface water and groundwater. Kolpin et al. (2002) reported finding organic wastewater contaminants in 80% of samples collected from 139 streams in 30 states in the United States. The most frequently detected compounds were coprostanol and cholesterol (steroids), N,N-diethyltoluamide (insect repellant), caffeine (stimulant), triclosan (disinfectant), tri(2-chloroethyl) phosphate (fire retardant), and 4-nonylphenol (detergent metabolite). Organic wastewater contaminants are also widespread in U.S. groundwater. Barnes et al. (2008) reported detecting organic wastewater contaminants in 81% of U.S. groundwater sources sampled in their study, most notably N,N-diethyltoluamide, bisphenol A (plasticizer), tri(2-chloroethyl) phosphate, sulfamethocazole (antibiotic), and 4-octylphenol monoethoxylate (detergent metabolite). Although surface-water and groundwater concentrations of these compounds were low (less than 1 µg/L), drinking water standards have not been established for many of these compounds so it is difficult to predict the actual risk to human health.

Most emerging contaminants are organic compounds that sorb strongly to soil organic carbon and biodegrade rapidly under aerobic conditions. Column studies evaluating the fate and transport of 17β-estradiol through the A-horizon of a natural soil, for example, report an instantaneous sorption coefficient (K_d) of 27.7 L/kg and a first-order degradation coefficient (μ) of 0.15/h (Casey et al., 2005). The widespread distribution of these compounds in surface water and groundwater, therefore, suggests that wastewater bypasses soil or passes through soil in such great quantities or at short timescales as to overwhelm the soil's ability to sorb or degrade contaminants. This may be a result of rapid infiltration or another mechanism such as colloid-facilitated transport that warrants further investigation, and underscores the importance of carefully managing wastewater to maximize soil's ability to remediate.

7.4   SOIL AND ON-SITE SEWAGE DISPOSAL

7.4.1   Importance of Soil and On-Site Sewage Disposal

Proper treatment and disposal of human sewage is vitally important to public health. Many of the plagues of history can be tied to improper disposal of sewage. Even today, cholera and dysentery outbreaks in undeveloped countries or in the aftermath of natural disasters are directly related to contamination of surface waters and potable water supplies with human sewage. The majority of Earth's population disposes of its sewage in a low-technology fashion, which is often soil-based. Even in developed countries such as the United States, 25% of the population disposes of their sewage using on-site sewage disposal systems, also known as septic systems (Canter, 1997). An estimated 1 trillion gal of septic-tank waste are released to the subsurface annually (Canter and Knox, 1984). The use of soil-based sewage disposal prevents pathogens from being introduced directly into surface waters or to the ground surface where contact with humans can cause disease.

7.4.2   Methods of On-Site Sewage Treatment and Disposal

The methods used in on-site sewage disposal take many forms, though most are variations on a basic theme (Figure 7.6). The oldest and most basic type of on-site sewage disposal is the privy, also known as an outhouse or pit toilet. The basic design of an outhouse is an open pit excavated into the soil with a basic open toilet mounted on top inside a structure. The liquid portion of the waste soaks into the soil while the solid portion is allowed to decompose biologically. When the pit becomes full, the outhouse is traditionally moved to a new location, the old pit is backfilled, and a new pit is dug. The advent of the outhouse in historic time was a step forward in public health in that it removed human waste from surface drainage ways and waterways and therefore from ready contact with other humans. The disadvantage of the outhouse, from an environmental and human health standpoint, is that it can allow pollutants and pathogens to leach into groundwater. In addition, some disease vectors (e.g., flying insects) can still access the human waste and transmit disease. This is not a significant problem in a low-density population, but this can present a problem as more people come to live in close proximity. One solution has been to construct outhouses over a sealed vault instead of an unlined pit. The vault needs to be emptied when it is full instead of simply abandoning it (Freedman, 1977). Although this corrects the problem of pathogens leaching into groundwater, it still allows flying disease vectors to have some access to the waste. Outhouses are widely used in developing countries and are still in use in developed countries and will probably remain in use for the foreseeable future as a low-technology method of on-site sewage disposal. Outhouses provide a low-cost, low-tech solution to a serious public health dilemma, and may be the most important safeguard of public health in developing countries.

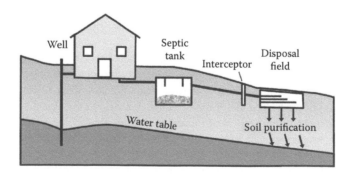

FIGURE 7.6   Typical on-site sewage treatment system. (From Environmental Enhancements, *Septic Systems [On-Site Sewage Disposal System]*, 2005. http:// environmentalenhancements.com/systems.html. Accessed 13 January 2011, verified 25 February 2012.)

The advent of the flush toilet and increasing popularity of indoor plumbing in the late 1800s created the need to dispose of larger amounts of sewage. In areas that lack central sewers, a modification of the outhouse concept was developed. This consists of an open pit lined with bricks, stones or cement blocks with open joints, known as a cesspool or a variation on this theme, an open pit filled with gravel or crushed stone, known as a seepage pit. This configuration is sometimes augmented by pretreatment with a septic tank, which is a sealed tank that allows separation of solids from liquids and a certain degree of anaerobic digestion of organic solids. Cesspools and seepage pits are essentially shallow injection features. In fact, anecdotal evidence from older excavators who installed cesspools in the United States relates that two similar pits would typically be excavated on opposite sides of a house under construction. Both pits would be excavated until water was encountered. The pit which encountered water first would be used as the potable water well and the other would become the cesspool. Conventional wisdom at the time was that the excavation had to encounter water in order to work properly. Such a configuration is focused primarily on disposal and not on treatment of the sewage. A cesspool constructed in this fashion introduces partially treated sewage directly into the groundwater.

As population density increases, the need for treatment of the sewage becomes apparent. The configuration that evolved next was the septic tank–soil absorption system (i.e., a drainfield) combination. This uses the septic tank as noted previously then disposes of the partially treated sewage by distribution over a larger area and at shallower depths where active soil microfauna and microflora exist in significant numbers. The soil acts as a physical, biological, and chemical filter, removing suspended solids, adsorbing cations and anions onto soil colloids, and allowing the endemic soil microorganisms to feed on the organic matter. This method of sewage disposal, used in the proper soils, has been shown to be one of the most cost-effective and environmentally sound methods (U.S. EPA, 1997).

This configuration and its variants have been adopted as the industry standard. Unfortunately, this system has been adopted and standardized, mostly on a state-by-state basis, based upon limited scientific data. The sizing of a soil absorption system is typically based upon a set of empirical parameters, usually an estimate of the permeability of the soil and the number of occupants a house is expected to have. The depth of unsaturated soil to be maintained underneath a drainfield varies from jurisdiction to jurisdiction, again based upon limited scientific data. For example, Pennsylvania requires 120 cm (4 ft) of unsaturated soil beneath a drainfield, neighboring Delaware requires 90 cm (3 ft) of separation from a water table, and North Carolina requires only 60 cm (2 ft). Limited research has been conducted to determine the actual depth of unsaturated soil required to adequately renovate sewage to a degree that it is no longer an environmental or public health hazard. This depth would be expected to vary depending on the composition of the soil, nature of the sewage, climate, configuration of the soil absorption system, microorganisms present, and oxygen content of the soil. Research by Hepner et al. (2006) suggests that most of the renovation of sewage within the soil column occurs within the first 30 cm (1 ft) below the soil absorption system.

7.4.3 How On-Site Sewage Treatment Systems Work

The efficiency of a soil absorption system to renovate sewage is further complicated by the formation of a biomat at the absorption system–soil interface. Research has shown that a soil absorption system develops a biomat composed of polysaccharides produced by partial digestion of organic matter by microorganisms (Beal et al., 2005, 2006). This biomat rapidly changes the biological diversity of the community of microorganisms present in the soil. Native microorganism communities in soil tend to be aerobic or facultative (i.e., a combination of aerobic and anaerobic) in nature. The biomat formed from a septic tank–soil absorption system is anaerobic. This is an artifact of the anaerobic nature of the septic tank effluent coupled with the biological oxygen demand (BOD) of the sewage. Even if a soil absorption system receives an input of oxygen, as happens following the passage of a wetting front from a rain event, which then draws atmospheric air in behind it, the balance rapidly returns to an anoxic environment. The biomat becomes a part of the renovation process, as the biomat adsorbs and filters the sewage. Studies conducted on areas with seasonal-use housing such as vacation beach houses that are uninhabited through the winter show that, upon initial use of the soil absorption system, background fecal coliform counts in the groundwater rapidly spike to high levels, which then gradually taper off as biomats form within the systems. Over the course of months the fecal coliform levels return to baseline levels as the biomats become more efficient at filtering the sewage (Postma et al., 1992).

The formation of a biomat also affects the permeability of the soil and therefore the hydraulic characteristics. Beal et al. (2005, 2006) found that biomats in soil may reduce permeability by four orders of magnitude and decrease the long-term acceptance rate (LTAR) by a factor of ten. A relationship was found between the level of organic loading and the extent of biomat development, though the endemic permeability of the soil seemed to have an influence on the development of the biomat as well. Once the biomat was established, however, the LTAR of the system was governed more by the permeability of the biomat than by the permeability of the soil absorption system. It has been concluded that, although the permeability of the native soil is important in the establishment of the biomat within the soil, it is the permeability of the biomat that ultimately dictates the LTAR of the system. This data calls into question the practice of basing the size of a soil absorption system upon an estimate of the permeability of the soil and suggests that an estimate of the permeability of the eventual biomat should be a factor in the design of such a system.

The biomat plays a crucial role in maintaining an unsaturated zone below the drainfield. As the biomat develops and permeability decreases, the sewage stops being drawn directly down through the bottom of the drainfield and begins moving laterally out of the sides. This creates an unsaturated zone directly below the drainfield and a partially saturated zone around and below the sides. In cases where saturation occurred beneath drainfields, enhanced production of nitrates from ammonia were encountered as well as enhanced transport of pathogens that could impact public health (Whelan and Barrow, 1984; Meeroff and Morin, 2005).

Most hydraulically failed soil absorption systems appear to do so because of a biological overload that leads to an overthickened biomat and prevents absorption of the sewage into the soil. Remediation of the problem by installation of a new soil absorption system and resting of the failed system has consistently shown that the rested soil absorption system will return to use over a relatively short time, consisting of months to a few years. This supports the concept that the biomat is the limiting factor in the LTAR of the system. Potts et al. (2004) found that aeration of the drainfield produced a different type of biomat dominated by aerobic and facultative microbes that did not significantly impact the LTAR yet significantly lowered the total nitrogen and fecal coliform bacteria.

The latest advances in on-lot sewage disposal technology have focused on combating the overformation of a biomat by equalizing distribution of sewage effluent throughout the soil absorption system to maximize filtration. This has been attempted for some time by using pressure distribution. Pressure distribution uses a pump or siphon to distribute the sewage evenly across the absorption field. Success using pressure distribution has been limited and may not be worth the added cost of installation (Converse, 2000).

Other attempts to overcome pollutant transport and biomat development include various methods of advanced pretreatment of sewage effluent before discharge into the soil absorption system. These include passive filters such as sand filters and peat filters as well as active filters such as aerobic treatment systems. The most prevalent problem with advanced pretreatment on systems serving single-family homes is lack of maintenance. These pretreatment systems have met with some success.

7.5 STORMWATER MANAGEMENT, SOIL, AND WATER QUALITY

Urban development produces impervious surfaces such as roofs, roads, and parking lots that intercept precipitation that would otherwise infiltrate and pass through soil. This increases runoff to rivers and streams, causing flooding and excessive erosion (Leopold, 1968). Increases in the volume of stormwater runoff, rate, and quality negatively impact the chemical, physical, and biological properties of receiving waters (Pitt et al., 2002; PADEP, 2006). A recent study conducted by the National Research Council (2008) concluded that stormwater runoff is a leading source of surface-water contamination in the United States. New stormwater management techniques and best management practices (BMPs) utilize soil infiltration to store, slow, and remediate urban storm runoff. We must proceed cautiously, however, to avoid concentrated infiltration that can exceed the soil's capacity to remediate urban pollutants.

Most watersheds impacted by anthropogenic forces respond by increasing the frequency and magnitude of floods (Leopold, 1968). Urban streams flood during minor precipitation events due to changes in land use and ineffective stormwater management strategies that increase the volume of runoff. This increase in stormwater volume is the direct result of expanding impervious surfaces and conversion of natural landscapes to commercial or residential use.

Stormwater runoff can carry high concentrations of pollutants washed or leached from altered land surfaces and impervious areas. Steep, unvegetated soils with low permeability are most likely to be eroded and add sediments and contaminants to run off (Gardiner and Miller, 2004). Typical urban pollutants include sediments, nutrients, pesticides, organic compounds, and heavy metals. In agricultural settings, runoff carries sediments, nutrients such as phosphorus, and pesticides to streams. These pollutants can cause serious degradation of water quality resulting in possible impact, either directly or indirectly, on human health (PADEP, 2006; Pitt et al., 2002; Zarriello, 1999).

In the past, stormwater management methodologies focused on slowing the runoff by detention or retention of stormwater (Figure 7.7). More recently, this focus has shifted to infiltration of stormwater in an attempt to replicate the original water budget. This has met with some success; however, this success has been limited by the proper application of engineering principles and the lack of characterization of the receiving soils and underlying bedrock (Zarriello, 1999; PADEP, 2006).

FIGURE 7.7 Infiltration basin (a) and detention basin (b) in Southeastern Pennsylvania.

In order to better apply the principle of stormwater infiltration, it is necessary to examine both the receiving mediums (i.e., soils and geology), the hydrology of the watershed, the anticipated precipitation events, and the anticipated levels of additional impervious surfaces. Soil scientists must determine the depth, drainage classification, and permeability of the soils. Subsurface infiltration of stormwater is most successful on deep, well-drained, somewhat excessively drained, or excessively drained soils with adequate permeability. Well-drained soils are defined as those that exhibit a seasonal high-water table or persistent saturation with water below 100 cm (40 in.) from the surface.

TABLE 7.1
Soil Drainage Classes Recognized by the USDA-NRCS

Soil Drainage Class	Description
Excessively drained	Water is removed from the soil very rapidly. The occurrence of internal free water is usually very rare or very deep. The soils are commonly coarse-textured and have very high hydraulic conductivity or are very shallow.
Somewhat excessively drained	Water is removed from the soil rapidly. Internal free water occurrence is usually very rare or very deep. The soils are commonly coarse-textured and have high saturated hydraulic conductivity or are very shallow.
Well drained	Water is removed from the soil readily but not rapidly. Internal free water occurrence is usually deep or very deep; annual duration is not specified. Water is available to plants throughout most of the growing season in humid regions. Wetness does not inhibit growth of roots for significant periods during most growing seasons. The soils are mainly free of the deep to redoximorphic features that are related to wetness.
Moderately well drained	Water is removed from the soil somewhat slowly during some periods of the year. Internal free water occurrence is usually moderately deep and transitory through permanent. The soils are wet for only a short time within the rooting depth during the growing season, but long enough that most mesophytic crops are affected. They commonly have a moderately low or lower saturated hydraulic conductivity in a layer within the upper 1 m, periodically receive high rainfall, or both.
Somewhat poorly drained	Water is removed slowly so that the soil is wet at a shallow depth for significant periods during the growing season. The occurrence of internal free water is usually shallow to moderately deep and transitory to permanent. Wetness markedly restricts the growth of mesophytic crops, unless artificial drainage is provided. The soils commonly have one or more of the following characteristics: low or very low saturated hydraulic conductivity, a high water table, additional water from seepage, or nearly continuous rainfall.
Poorly drained	Water is removed so slowly that the soil is wet at shallow depths periodically during the growing season or remains wet for long periods. The occurrence of internal free water is shallow or very shallow and common or persistent. Free water is commonly at or near the surface long enough during the growing season so that most mesophytic crops cannot be grown, unless the soil is artificially drained. The soil, however, is not continuously wet directly below plow-depth. Free water at shallow depth is usually present. This water table is commonly the result of low or very low saturated hydraulic conductivity of nearly continuous rainfall, or of a combination of these.
Very poorly drained	Water is removed from the soil so slowly that free water remains at or very near the ground surface during much of the growing season. The occurrence of internal free water is very shallow and persistent or permanent. Unless the soil is artificially drained, most mesophytic crops cannot be grown. The soils are commonly level or depressed and frequently ponded. If rainfall is high or nearly continuous, slope gradients may be greater.

Source: From Soil Survey Division Staff, *Soil Survey Manual*, Soil Conservation Service. U.S. Department of Agriculture Handbook 18, U.S. Government Print Office, Washington, DC, 1993.

Any attempt to infiltrate stormwater through subsurface means on moderately well-drained, somewhat poorly drained, or poorly drained soils regardless of the permeability of those soils will fail because those soils will be saturated with groundwater during the times when infiltration is most needed. This is analogous to trying to add water to a glass that is already full. Official definitions of soil drainage classes are given in Table 7.1.

Concentrated stormwater infiltration, or capture of stormwater runoff from a large area and attempting to infiltrate it into a relatively small area, can lead to some of the same problems as those associated with on-site sewage disposal. Although the organic load of stormwater is much less than that of sewage, the fraction of sediments, detritus, and hydrocarbons can form a clogging mat that limits the LTAR of the soils. As this is not a mat composed of organic matter, biological means will have little remediating effect upon it. Any clogging mat formed in a stormwater infiltration system will need to be removed mechanically in order to remediate the lost permeability; this typically provides limited renovation. For this reason, pretreatment of stormwater prior to infiltration needs to be considered. Utilization of settling ponds, forebays, or tanks prior to infiltration will greatly enhance the efficiency and longevity of these systems. Even with such pretreatment, some pollutants, such as hydrocarbons, may be introduced into the soil column, leaving the soil alone to deal with these contaminants. Maintenance of groundwater quality while infiltrating stormwater in large amounts will be a major challenge in a continually developing world.

7.6 UTILIZING SOIL TO PROTECT WATER QUALITY AND HUMAN HEALTH

The link between soil, water quality, and human health is clear. Soil processes such as biodegradation and pathogen inactivation render contaminants harmless or remove them from the system. But soil is not an unlimited sink for contaminants. We must use soil in a responsible way and seek methods to ensure sustainability.

Recent adoption of drip irrigation technology has opened up new possibilities for disposal of on-site sewage in marginal areas in an environmentally sound fashion. Previous attempts sought to flood the absorption area on a demand–dose cycle, thereby allowing sewage to then be drawn by gravity through the biomat and into the soil under saturated flow. Drip irrigation technology, metered and controlled by a microprocessor, allows distribution of the sewage effluent over a large area at a very slow rate. The sewage is then drawn into the soil by capillary action and by unsaturated flow. The advantage to this is maintenance of an aerobic or facultative environment within the soil. This in turn prevents or limits the development of a biomat and allows endemic soil organisms to act upon the organic matter in the sewage. Although this method is more complicated and therefore more costly, its promise as a long-term solution to on-lot sewage disposal merits attention (Hepner et al., 2006).

A benefit of unsaturated flow in drip irrigation is that an increased amount of the sewage effluent is retained within the root zone where water can be evapotranspired to the atmosphere. Pollutants are also retained in this aerobic zone, where they can be sorbed and degraded. An example of this is nitrate, which, as previously noted, is an anion and is not readily sorbed to the soil. In more traditional soil absorption systems, where pressure distribution results in flood dosing or gravity distribution results in a relatively steady trickle, a pollutant such as nitrate that is not treated or sorbed in the soil column will be transported downward to the water table where it may persist. Drip irrigation systems increase the ability of plants to intercept soil water and utilize or transpire nitrogen to the atmosphere. Further advances and innovations in development, such as a system using low-profile geotextile as the drainfield medium, hold equal promise (Potts et al., 2008).

The rediscovery of the use of biochar (charcoal derived from the pyrolysis of biomass, such as waste wood, brush, or agricultural residue) to amend soil opens up many new possibilities for improving soil's ability to improve water quality. Addition of biochar has been shown to increase the availability of major cations as well as increase CEC and water retention (Rondon et al., 2006;

Lehmann and Rondon, 2006; Chan et al., 2007; Van Zweiten et al., 2010). Biochar may be amended to agricultural soils and may enhance treatment of stormwater and wastewater as a filter medium. Application of biochar to the bottom of a stormwater infiltration basin, for example, would increase the soil's water retention ability as well as its ability to filter and adsorb chemical pollutants. Similarly, installing agricultural tile drains in a bed of biochar could help slow the leaching of nitrate from fertilizer applications into groundwater and surface water.

7.7 CONCLUSIONS

Throughout human history, soil has provided the majority of our food and fiber while serving as a filter medium to sustain plentiful fresh water. In the modern world, population growth has necessitated an increase in agricultural productivity coupled with urbanization, both of which stress soil and threaten water quality. The demand for food forces increased use of fertilizers, pesticides, and irrigation, thereby increasing the input of agrichemicals to groundwater. As wastewater disposal methods shift from treated sewage discharged to water bodies to land application, the possibility of contamination of groundwater increases. To meet the global demand for potable water, communities have increased their reliance on groundwater. Therefore, we have become more reliant on soil than ever before to protect the quality of the water that is so critical to human health.

Soil is an outstanding filter medium. It is capable of rendering human waste drinkable by removing enteric pathogens, the leading cause of child mortality in developing countries. Soil removes organic and heavy metal contaminants from water so effectively that they remain adsorbed to soil particles for decades. Soil is a living system whose microorganisms degrade many contaminants into harmless by-products. We must carefully manage soil, however, to ensure the protection of groundwater. Excessive concentrations of contaminants, high rates of flow, and overapplication of fertilizers and sewage can overwhelm soil's ability to remediate. We must also keep in mind that soil can be a source of contaminants and can facilitate transport through natural colloids.

Agricultural drainage wells, storm drains, and impervious surfaces cause potentially contaminated water to bypass soil and be transmitted directly to groundwater or surface water. This results in contamination of water resources and flooding, both of which threaten human health. The current initiative in the United States is to promote soil infiltration to mitigate agricultural runoff, urban stormwater, and on-lot sewage treatment. Although the motive is positive, we must be careful not to overestimate soil's ability to remediate large, concentrated volumes of wastewater.

The link between soil, the quality of water resources, and human health is inextricable. Water is a limited resource whose primary protector is the soil mantle. As population increases we will need to consider new ways of utilizing soil to manage stormwater, agriculture, and sewage. Such stresses on this finite resource require a thorough and practical understanding of the relationships between soil and water quality in order to protect the health of future generations.

REFERENCES

Azadpour-Keeley, A., B.R. Faulkner, and J. Chen. 2003. *Ground Water Issue: Movement and Longevity of Viruses in the Subsurface*. U.S. EPA National Risk Management Laboratory, Cincinatti, OH.

Bales, R.C., C.P. Gerba, R.W. Harvey, M.T. Yahya, S. Li, and K.M Maguire. 1995. Virus and bacteria transport in a sandy aquifer, Cape Cod, MA. *Ground Water* 33: 651–661.

Barnes, K.K., D.W. Kolpin, M.J. Focazio, E.T. Furlong, M.T. Meyer, S.D. Zaugg, S.K. Haak, L.B. Barber, and E.M. Thurman. 2008. Water-quality data for pharmaceuticals and other organic wastewater contaminants in groundwater and in untreated drinking water sources in the United States, 2000–01. U.S. Geological Survey, Reston, VA.

Beal, C.D., E.A.Gardner, and N.W. Menzies. 2005. Process, performance, and pollution potential: A review of septic tank-absorption systems. *Australian Journal of Soil Research* 43(7): 781–802.

Beal, C.D., E.A. Gardner, G. Kirchof, and N.W. Menzies. 2006. Long-term flow rates and biomat zone hydrology in soil columns receiving septic tank effluent. *Water Research* 40(12): 2327–2338.

Black R.E., S. Cousens, H.L. Johnson, et al. 2010. Global, regional, and national causes of child mortality in 2008: A systematic analysis. *The Lancet* 375(9730): 1969–1987.

Bolton, K.A. and L.J. Evans. 1996. Cadmium adsorption capacity of selected Ontario soils. *Canadian Journal of Soil Science* 76(2): 183–189.

Brady, N.C. and R.R. Weil. 2004. *Elements of the Nature and Properties of Soils*, 2nd edn. Pearson Prentice Hall, Upper Saddle River, NJ.

Buckingham, W.A. 1989. *Operation Ranch Hand: The USAF and Herbicides in Southeast Asia, 1961–1971*. Air Force Historical Studies Office, U.S. Government Printing Office, Washington, DC.

Burgos, W.D., C.M. Munson, and C.J. Duffy. 1999. Phenanthrene adsorption–desorption hysteresis in soil described using discrete-interval equilibrium models. *Water Resources Research* 35(7): 2043–2051.

Canter, L.W. 1997. *Nitrates in Groundwater*. CRC Lewis Publishers, Boca Raton, FL.

Canter, L.W. and R.C. Knox. 1984. *Evaluation of Septic Tank System Effects on Ground Water Quality*. U.S. Environmental Protection Agency, Washington, DC.

Casey, F.X.M., J. Simunek, J. Lee, G.L. Larsen, and H. Hakk. 2005. Sorption, mobility, and transformation of estrogenic hormones in natural soil. *Journal of Environmental Quality* 34: 1372–1379.

Chan, K.Y., L. Van Zwieten, I. Meszaros, A. Downie, and S. Joseph. 2007. Agronomic values of greenwaste biochar as a soil amendment. *Australian Journal of Soil Research* 45(8): 629–634.

Clothier, B.E., S.R. Green, and M. Deurer. 2008. Preferential flow and transport in soil: Progress and prognosis. *European Journal of Soil Science* 59: 2–13.

Converse, J.C. 2000. Pressure distribution network design. Small Scale Waste Management Project, University of Wisconsin-Madison, Madison, WI.

Craun, G.F. 1985. A summary of waterborne illness transmitted through contaminated groundwater. *Journal of Environmental Health* 48: 122–127.

Dang, T.C.H., H. Allen, B.H. Nguyen, V. Fong, T.H. Dam, N.Q. Nguyen, Q.H. Nguyen, K.H.C. Phung, and T.N.A. Dao. 2008. Joint study of bioremediation at pilot scale for detoxification of herbicide/dioxin in Da Nang hot spot, Vietnam. *In Proceedings of the 30th International Symposium on Halogenated Persistent Organic Pollutants (POPs)*, San Antonio, TX, 12–17 September 2010, 4 p.

DeNovio, N.M., J.E. Saiers, and J.N. Ryan. 2004. Colloid movement in unsaturated porous media: Recent advances and future directions. *Vadose Zone Journal* 3: 338–351.

DeSimone, L.A., P.A. Hamilton, and R.J. Gilliom. 2009. The quality of our nation's waters: Quality of water from domestic wells in principal aquifers of the United States, 1991–2004—Overview of major findings. U.S. Geological Survey Circular 1332.

Dubrovsky, N.M. and P.A. Hamilton. 2010. Nutrients in the Nation's streams and groundwater: National findings and implications. U.S. Geological Survey Fact Sheet 2010-3078.

El-Farhan, Y.H., N.M. DeNovio, J.S. Herman, and G.M. Hornberger. 2000. Mobilization and transport of soil particles during infiltration experiments in an agricultural field, Shenandoah Valley, Virginia. *Environmental Science & Technology* 34: 3555–3559.

Environmental Enhancements. 2005. *Septic Systems (On-Site Sewage Disposal System)*. http://environmental-enhancements.com/systems.html. Accessed 13 January 2011 (verified 25 February 2012).

Fetter, C.W. 2001. *Applied Hydrogeology*, 4th edn. Prentice Hall, Upper Saddle River, NJ.

Focazio, M.J., A.H. Welch, S.A. Watkins, D.R. Helsel, and M.A. Horn. 2000. A retrospective analysis of the occurrence of arsenic in ground-water resources of the United States and limitations in drinking-water-supply characterizations. Water Resources Investigations Report 99-4279, U.S. Geological Survey, Reston, VA.

Freedman, B. 1977. *Sanitarian's Handbook*. Peerless Publishing, New Orleans, LA.

Freundlich, H. 1909. Kapillarchemie: eine darstellung der chemie der kolloide und verwanter gebiete. Academische Bibliothek, Leipzig, Germany.

Gardiner, D.T. and R.W. Miller. 2004. *Soils in Our Environment*, 10th edn. Pearson Prentice Hall, Upper Saddle River, NJ.

Greenan, C.M., T.B. Moorman, T.B. Parkin, T.C. Kasper, and D.B. Jaynes. 2009. Denitrification in wood chip bioreactors at different water flows. *Journal of Environmental Quality* 38: 1664–1671.

Harter, T., S. Wagner, and E.R. Atwill. 2000. Colloid transport and filtration of *Cryptosporidium parvum* in sandy soils and aquifer sediments. *Environmental Science and Technology* 34: 62–70.

Harvey, C.F., K.N. Ashfaque, W. Yu, A.B.M. Badruzzaman, M.A. Ali, P.M. Oates, H.A. Michael, et al. 2006. Groundwater dynamics and arsenic contamination in Bangladesh. *Chemical Geology* 228: 112–136.

Hatfield Consultants. 2009. Comprehensive assessment of dioxin contamination in Da Nang Airport, Viet Nam: Environmental levels, human exposure and options for mitigating impacts—Final report. Hatfield Consultants, North Vancouver, BC.

Hemond, H.F., and E.J. Fechner. 1994. *Chemical Fate and Transport in the Environment.* Academic Press, San Diego, CA.

Hepner, L., D. Linde, C. Weber and D. Smith. 2006. Alternative on-lot technology research. Soil based treatment systems, Delaware Valley College, Doylestown, PA.

Hofmann, T. and A. Wendelborn. 2007. Colloid facilitated transport of polychlorinated dibenzo-p-dioxins and dibenzofurans (PCDD/Fs) to the groundwater at Ma Da Area, Vietnam. *Environmental Science and Pollution Research* 14: 223–224.

Hutson, S.S., N.L. Barber, J.F. Kenney, K.S. Linsey, D.S. Lumia, and M.A. Maupin. 2004. Estimated use of water in the United States in 2000. U.S. Geological Survey Circular 1268.

Irwin, L.K., S. Gray, and E. Oberdorster. 2001. Vitellogenin induction in painted turtle, *Chrysemys picta*, as a biomarker of exposure to environmental levels of estradiol. *Aquatic Toxicology* 55: 49–60.

Jamieson, R.C., R.J. Gordon, K.E. Sharples, G.W. Stratton, and A. Madani. 2002. Movement and persistence of fecal bacteria in agricultural soils and subsurface drainage water: A review. *Canadian Biosystems Engineering* 44: 1.1–1.9.

Kato, S., M.B. Jenkins, E.A. Fogarty, and D.D. Bowman. 2003. *Cryptosporidium parvum* oocyst inactivation in field soil and its relation to soil characteristics: Analysis using the geographic information system. *Science of the Total Environment* 321: 47–58.

Kiely, T., D. Donaldson, and A. Grube. 2004. *Pesticides Industry Sales and Usage: 2000 and 2001 Market Estimates.* U.S. Environmental Protection Agency, Washington, DC.

Kinley, D. and Z. Hossain. 2003. Poisoned waters. *World Watch* 16: 22–27.

Kinniburgh, D.G. and P.L. Smedley. 2001. *Arsenic Contamination of Groundwater in Bangladesh, Vol. 2: Final Report.* British Geological Survey, Nottingham.

Kolpin, D.W., J.E. Barbush, and R.J. Gilliom. 2000. Pesticides in ground water of the United States, 1992–1996. *Ground Water* 38: 858–863.

Kolpin, D.W., E.T. Furlong, M.T. Meyer, E.M. Thurman, S.D. Zaugg, L.B. Barber, and H.T. Buxton. 2002. Pharmaceuticals, hormones, and other organic wastewater contaminants in U.S. streams, 1999–2000: A national reconnaissance. *Environmental Science and Technology* 36: 1202–1211.

Kretzschmar, R., M. Borkovec, D. Grolimund, and M. Elimelech. 1999. Mobile subsurface colloids and their role in contaminant transport. *Advances in Agronomy* 66: 121–193.

Lal, R. and M. Shukla. 2004. *Principles of Soil Physics.* Marcel Dekker, New York, NY.

Langmuir, I. 1916. The constitution and fundamental properties of solids and liquids. Part I: Solids. *Journal of the American Chemical Society* 38: 2221–2295.

Larson, S.J., R.J. Gilliom, and P.D. Capel. 1999. Pesticides in streams of the United States—Initial results from the national water-quality assessment program. Water-Resources Investigations Report 98-4222, U.S. Geological Survey, Sacramento, CA.

Lehmann, J. and M. Rondon. 2006. Bio-char soil management on highly weathered soils in the humid tropics. *In* N. Uphoff, A.S. Ball, E. Fernandes, et al. (eds). *Biological Approaches to Sustainable Soil Systems*, pp. 517–530. Taylor & Francis Group, Boca Raton, FL.

Leopold, L.B. 1968. Hydrology for urban planning—A guidebook on the hydrologic effects of urban land use. Geological Survey Circular 554, U.S. Geological Survey, Washington, DC.

Meeroff, D.E. and F.J. Morin. 2005. Contribution of on-site treatment and disposal systems on coastal pollutant loading. *Proceedings of the Water Environment Federation*, Session 101 through Session 110, 17: 8391–8407.

Mubiru, D.N., M.S. Coyne, and J.H. Grove. 2000. Mortality of *Escherichia coli* O157:H7 in two soils with different physical and chemical properties. *Journal of Environmental Quality* 29: 1821–1825.

Nasser, A.M., Z. Huberman, A. Zilberman, and S. Greenfeld. 2003. Die-off and retardation of *Cryptosporidium* spp. oocyst in loamy soil saturated with secondary effluent. *Water Science & Technology: Water Supply* 3: 253–259.

National Research Council. 2008. *Urban Stormwater Management in the United States.* The National Academies Press, Washington, DC.

PADEP. 2006. *Pennsylvania Stormwater Best Management Practices Manual.* Pennsylvania Department of Environmental Protection Document 363-0300-002. http://www.elibrary.dep.state.pa.us/dsweb/View/Collection-8305 (verified 25 February 2012).

Pitt, R., S. Chen, and S. Clark. 2002. Compacted urban soils effects on infiltration and bioretention stormwater control designs. *In Proceedings of the 9th International Conference on Urban Drainage*, Portland, OR, 8–13 September 2002. doi:http://dx.doi.org/10.1061/40644(2002)14.

Postma, F.B., A.J. Gold, and G.W. Loomis. 1992. Nutrient and microbial movement from seasonally-used septic systems. *Journal of Environmental Health* 55(2): 5–10.

Potts, D.A., J.H. Görres, E.L. Nicosia, and J.A. Amador. 2004. Effects of aeration on water quality from septic tank leachfields. *Journal of Environmental Quality* 33: 1828–1838.

Potts, D.A., E.L. Patenaude, J.H. Görres, and J.A. Amador. 2008. Wastewater renovation and hydraulic performance of a low profile leaching system. *In Proceedings of the 17th National Onsite Wastewater Recycling Association Conference*, Memphis, TN, 7–10 April 2008.

Reddy, K.R., R. Khaleel, and M.R. Overcash. 1981. Behavior and transport of microbial pathogens and indicator organisms in soils treated with organic wastes. *Journal of Environmental Quality* 10: 255–266.

Rondon, M.A., D. Molina, M. Hurtado, J. Ramirez, J. Lehmann, J. Major, and E. Amezquita. 2006. Enhancing the productivity of crops and grasses while reducing greenhouse gas emissions through bio-char amendments to unfertile tropical soils. *In Proceedings of the 18th World Congress of Soil Science*, Philadelphia, PA, 9–14 July 2006. Session 113-68.

Sawyer, J., E. Nafziger, G. Randall, L. Bundy, G. Rehm, and B. Joern. 2006. *Concepts and Rationale for Regional Nitrogen Rate Guidelines for Corn*. PM 2015, Iowa State University Extension Office, Ames, IA.

Shiklomanov, I.A. 2000. Appraisal and assessment of world water resources. *Water International* 25: 11–32.

Sobsey, M.D., R.M. Hall, and R.L. Hazard. 1995. Comparative reduction of Hepatitis A virus, Enterovirus, and Coliphage MS2 in miniature soil columns. *Water Science and Technology* 31: 203–209.

Soil Survey Division Staff. 1993. *Soil Survey Manual*. Soil Conservation Service. U.S. Department of Agriculture Handbook 18, U.S. Government Print Office, Washington, DC.

Solley, W.B., R.R. Pierce, and H.A. Perlman. 1993. Estimated use of water in the United States in 1990. U.S. Geological Survey Circular 1081.

Stellman, J.M., S.D. Stellman, R. Christians, T. Weber, and C. Tomasallo. 2004. The extent and patterns of usage of Agent Orange and other herbicides in Vietnam. *Nature* 422: 681–687.

Stramer, S.L. 1984. Fates of poliovirus and enteric indicator bacteria during treatment in a septic tank system including septage disinfection. Unpublished Master's thesis, University of Wisconsin-Madison, Madison, WI.

Toccalino, P.L. and J.A. Hopple. 2010. The quality of our nation's waters: Quality of water from public-supply wells in the United States, 1993–2007—Overview of major findings. U.S. Geological Survey Circular 1346.

UNICEF. 2010. Trends in infant mortality rates (1960–2009). www.childinfo.org/mortality_imrcountrydata.php. Accessed 16 January 2011 (verified 25 February 2012).

UNICEF and WHO. 2008. *Progress on Drinking-Water and Sanitation: Special Focus on Sanitation*. United Nations Children's Fund, New York, NY.

U.S. Census Bureau. 2008. Current housing reports, Series H150/07, American housing survey for the United States: 2007. U.S. Government Printing Office, Washington, DC.

USDA. 2008. *Agricultural Chemical Usage 2007 Field Crops Summary*. National Agricultural Statistics Service, U.S. Government Printing Office, Washington, DC.

U.S. EPA. 1996. *Soil Screening Guidance: Technical Background Document*, 2nd edn. Office of Emergency and Remedial Response Publication 9355.4-17A. Environmental Protection Agency, Washington, DC.

U.S. EPA. 1997. Response to Congress on use of decentralized wastewater treatment systems. EPA 832-R-97-001b.

U.S. EPA. 2000. National primary drinking water regulations: Ground water rule; proposed rules. *Federal Register* 65(91): 30202.

U.S. EPA. 2009. *National Water Quality Inventory: 2004 Report to Congress*. Office of Water, Environmental Protection Agency, Washington, DC.

U.S. EPA. 2010. Mid-Atlantic risk assessment risk-based concentration table. http://www.epa.gov/reg3hwmd/risk/human/index.htm. Accessed 31 December 2010 (verified 25 February 2012).

Van Zweiten, L., S. Kimber, S. Morris, K.Y. Chan, A. Downie, J. Rust, S. Joseph and A. Cowie. 2010. Effects of biochar from slow pyrolysis of papermill waste on agronomic performance and soil fertility. *Plant and Soil* 327(1–2): 235–246.

Whelan, B.R. and N.J. Barrow. 1984. The movement of septic tank effluent through sandy soils near Perth. I: Movement of nitrogen. *Australian Journal of Soil Research* 22(3): 283–292.

WHO. 2008a. *Global Burden of Disease, 2004 Update*. World Health Organization, Geneva, Switzerland.

WHO. 2008b. *Guidelines for Drinking Water Quality*, 3rd edn. World Health Organization, Geneva, Switzerland.

Wise, D.L., D.J. Trantolo, E.J. Cichon, H.I. Inyang, and U. Stottmeister (eds). 2000. *Bioremediation of Contaminated Soils*. Marcel Dekker, New York, NY.

Young, A.L., W.J. Van Houten, and W.B. Andrews. 2008. State-of-the-art: Agent Orange and dioxin remediation. *Environmental Science and Pollution Research* 15: 113–118.

Yu, W.H., C.M. Harvey, and C.F. Harvey. 2003. Arsenic in the groundwater in Bangladesh: A geostatistical and epidemiological framework for evaluating health effects and potential remedies. *Water Resources Research* 39: 1146–1163.

Zarriello, P.J. 1999. Watershed modeling approach to assessing the hydrologic effects of future development in the Ninemile Creek Basin, Onondaga County, New York. U.S. Geological Survey Fact Sheet FS 112-99.

Zhang, H. and H.M. Selim. 2005. Kinetics of arsenate adsorption–desorption in soils. *Environmental Science and Technology* 39(16): 6101–6108.

Section III

Human Use of and
Interactions with Soil

8 Geophagy
An Anthropological Perspective

Jacques M. Henry and F. Daniel Cring

CONTENTS

8.1 STUDY OF GEOPHAGY

The finding is clear: geophagy remains relatively mysterious. Although its existence has now been documented worldwide for at least two millennia and studied for several centuries, the reasons why humans consume soil are still a matter of puzzlement for many and a subject of research for some. Indeed, researchers from fields as diverse as chemistry and geography, medicine and geology, sociology and pediatrics, history of religion and pedology, nutrition and primatology have investigated the practice, the practitioners, and the product consumed. Myths and falsehoods have been debunked and much has been learned: people do not eat clay because of mental or physical sickness, and soil really does fulfill physical and emotional needs.

The liminal status of geophagy has a lot to do with this unsatisfactory situation. As noted by some, the practice provides a vivid example of human behavior where culture not only meets nature but eats it too. It is a "well-grounded diet," a serious scholar of food and medicine writes half-jokingly (Johns, 2000), while the jacket of *Salt: A World History* hypes the historical examination of the "only rock humans eat!" (Kurlansky, 2002), even though salt consumption is typically not included among geophagic behaviors. Geophagy uncannily links the centrality of earth to the existence of mankind with the unique thinking ability of *Homo sapiens*. Gravity binds us to the ground and it has shaped the evolution of our physical constitution. From the ground, humans have dug, collected, or grown the means of their subsistence. Meanwhile, human thought has made soil a core element of our being. Truly universally, from the Dogon to the Hopis, from Antique Greece to the Melanesian island of Tikopia, the matter on which humans stand has entered their mind. Geophagy brings it into the body.

Not surprisingly, explanations of geophagy, both ancient and recent, strive to integrate the natural and cultural dimensions of the phenomenon in an attempt to overcome the dichotomy long-central to Western thought. Individually, stand-alone scientific disciplines have ultimately been shown to be insufficient in apprehending the rationale and mechanisms of geophagy. Early on, physicians tended not to concern themselves with the cultural variables of clay consumption and geographers could only skim the surface of the chemical process of nutrient absorption. Dual approaches combining biological and cultural dimensions have yielded important findings and opened new vistas (Vermeer and Frate, 1979; Reid, 1992; Young, 2011). Lately, collaborative efforts have developed (e.g., this volume) and calls for even deeper multidisciplinarity have been issued (MacClancy et al., 2007; Young et al., 2008).

8.2   FOUR-FIELD ANTHROPOLOGICAL APPROACH TO GEOPHAGY

Notwithstanding the enlightening and ongoing efforts to bring together different fields of study, we would advance that anthropology's holistic approach to human behavior is well suited to investigate geophagy. The combination of its archeological, biological, cultural, and linguistic branches offers an integrated paradigm apt to capture the multiple dimensions of geophagy and their interrelations.* The anthropology of geophagy also allows full focus on adaptation, now a widely recognized feature of the phenomenon.

The consumption of *muktuk* provides a brief illustration of how anthropology may contribute to the understanding of geophagy. *Muktuk* is the skin and fat of marine mammals consumed raw by people of the Arctic. As one consequence of biological evolution, the frugivorous primates, including humans, lost the ability to synthesize vitamin C (ascorbic acid), which then became one of many essential nutrients that had to be obtained from the diet. For most of human history, people consumed a diversified diet consisting of animals (including insects) and plants (fruits and seeds of grasses), and thus a vitamin C deficiency was unlikely.

However, as some prehistoric populations migrated into the Arctic regions, acquiring plants and other sources of vitamin C became difficult and vitamin C deficiencies became a potential problem. The people of the Arctic adapted culturally by eating *muktuk*, a food rich in vitamin C.

Muktuk is a cultural adaptation (cultural anthropology) to a potential vitamin C deficiency caused by evolution (biological anthropology) and the prehistoric migrations of ancient peoples into the Arctic (archaeological anthropology). What about linguistic anthropology? These peoples of the high Arctic were called *Eskimo* by their Cree neighbors to the south, which means "eaters of raw meat."

So, let us start with issues of terminology. The term *geophagy* (or *geophagia*) has long been applied to the practice by humans and animals of consuming soil. This blanket term has been used regardless of the circumstances of the act, which show significant variations in the material consumed, participants, rationale, and consequences of the ingestion.

In 2000, a panel convened by the Agency for Toxic Substances and Disease Registry (ATSDR) of the Centers for Disease Control (Atlanta) attempted to clarify the definition of geophagy. It dealt with several of the issues that have been debated in the literature: Is soil-pica intentional? Is it abnormal behavior? Is it necessarily recurrent? What is the consumed material (surface soil or known sources)? Is age a risk factor? Is sex a factor? Is geophagy adaptive or pathological? Is it a relevant example of where nature and culture meet? But first it proposed the following distinctions:

> *Soil ingestion* is the consumption of soil. This may result from various behaviors including, but not limited to, mouthing, contacting dirty hands, eating dropped food, or consuming soil directly. *Soil-pica* is the recurrent ingestion of unusually high amounts of soil (i.e., on the order of 1,000–5,000 milligrams per day). Groups at risk of soil-pica behavior include children aged 6 years and younger and individuals who are developmentally delayed. *Geophagy* is the intentional ingestion of earths and is usually associated with cultural practices. (ATSDR, 2001, p. 3)

---

* Formalized by Franz Boas, the founder of modern American anthropology, the four-field approach is arguably culture-bound, faces epistemological issues (few if any anthropologists are fluent in all disciplines), and has received its share of postmodern critique. It nevertheless remains the framework recognized by the American Anthropological Association and the one organizing most anthropology departments in American universities.

This precise yet wide-ranging definition is helpful for this chapter, primarily concerned with geophagy defined as the deliberate and culturally sanctioned consumption of limited amounts of soil. Accordingly, the unintentional or involuntary ingestion of soil is not included and neither is pica, a craving and deliberate ingestion of high amounts of nonfood items such as ice, chalk, ash, and soil over significant periods of time (see Young, 2011, for a recent comprehensive treatment of pica). We propose that the examination of its archeological, biological, cultural, and linguistic dimensions may offer a valid paradigm to better understand this sporadic, puzzling, yet human, behavior. We will begin by looking at the cultural and linguistic aspects before addressing the archeological and biological dimensions.

8.3 CLAY-MADE CONSUMABLE: CULTURAL ASPECTS OF GEOPHAGY

In this section, we focus on the mechanisms used in different cultures to select and prepare soil to make it consumable by humans. We also examine the symbolic meanings associated with soil and its consumption, as well as its socioeconomic dimensions. The cultural variability of the phenomenon is documented, thus helping with its explanation and understanding.

It is telling that some of the early historical reports of geophagy are those of *terra sigillata* or "sealed earth," lozenges of "clay bearing little images" from the Greek island of Lemnos (Laufer, 1930, p. 164) whose consumption and curative powers were documented in the second century AD and continued to be subjects of sporadic reports until the nineteenth century. To enter the body, soil must go through the human hand and thought. As anthropologist Claude Levi-Strauss (1963, p. 89) famously offered about the consumption of natural species, they "are chosen not because they are 'good to eat' but because they are 'good to think of'." This view can be extended to minerals. Culture is ubiquitous in geophagy. Detailed reports of geophagy unequivocally demonstrate that humans make clay consumable by making it human, which is by bringing it into culture. People select the clay, process it, cook it, shape it, share it, and name it. Human agency is further evidenced in the distribution of geophagy in societies, the symbolic meanings attributed to the practice, and the categorization of consumers.

Not all soil is considered good to eat; it is carefully selected on the basis of appearance, texture, and taste. Von Humboldt (1872, p. 495) reported that "the Ottomacs do not eat every kind of clay indifferently; they choose the alluvial beds or strata, which contain the most unctuous earth and the smoothest to the touch." Decades later and thousands of miles to the north, the same attention to provenance was noted among African Americans in North Carolina (Hertz, 1947) and rural Mississippi (Vermeer and Frate, 1979). Frate (1984, p. 35) observed residents in the process of selecting their material: "Not just any soil will do; sandy loams and other coarse varieties are rejected in favor of fine-grained clays, prized for their distinctively sour taste. Choice soils, like choice wines, acquire a reputation over the years and are sometimes known by their area of origin." More recently, extensive research by Geissler (2000) among the Luo in Western Kenya showed that children prefer material collected from termite mounds, a well-known and frequent source of material for humans and animals alike, with a marked liking for fine, dark red clay collected from the inside walls where it is considered to be purer.

The processing of some soils begins with cleaning. It was noted that in the nineteenth-century Dutch Indies, "the Javanese would first remove sand and other hard substances from the edible clay and then reduce it to a paste by kneading it with water" (Laufer, 1930, p. 130). In 1880, Rutherford reported from Batanga in West Africa (likely modern-day Cameroon) that there was "evidently a large quantity of organic matter" in the clay. However, after it had been baked in 5 in. balls, "a small portion is broken off, and placed in the hollow of any smooth leaf and reduced to powder between the finger and thumb. The leaf is then gently shaken in order to cause the harder and grittier particles to fall aside. These are carefully removed, and the residue, consisting of a fine powder, is transferred to the mouth, masticated, and swallowed" (Rutherford, 1881, p. 467). A century later attention to cleanliness continues; for instance, Luo children remove unwanted debris such as pieces of wood and stones from their clay before consuming it (Geissler, 2000). Such concerns about purity echo the importance of the concept in nutritional anthropology, especially as framed by the analysis of Mary Douglas (1966), who argued that foodstuff must be construed as pure and nonpolluting to be consumed, regardless of its nutritional content.

Sometimes the material is further prepared for consumption. The preparation can be as simple as cutting the clay into "the shape of thin tiles of small size and simply dried in the sun" as reported in Northern Vietnam around 1900 (Laufer, 1930, p. 128). In Mississippi, "the clay is usually not eaten at the clay bank although there is some mention of people having eaten the clay at the 'dirt-hole.' Generally the clay is first baked in the oven to dry it thoroughly" (Hertz, 1947, p. 343). Hunter (1973, p. 174) reported how in Ghana "the raw clay is mixed with water and pounded with a heavy pestle on a board (not a mortar). Next, pieces are kneaded by hand into uniform size and shape, smoothed with further application of water, often distinctively marked and then sun-dried until hard." The accompanying picture shows a woman sitting on a low stool by neatly arranged rows of clay cakes bearing some resemblance to bread rolls. In some cases, the preparation can be as sophisticated as "ready-made pottery" as Mitra reported from India where the clay was shaped in the forms of cups and baked in a kiln (Laufer, 1930, p. 140). Frate (1984, p. 37) reports how in 1971 in Central Mississippi "some women season their clay with salt and vinegar and more than two-thirds of the geophagous women we surveyed cooked their clay in pans atop a stove or in ovens sometimes for as long as three hours" in order to dry it. Some use kitchen utensils to extract it, a practice also noted recently in Chimayo, New Mexico (Callahan, 2003).

So, soil is selected, cleaned, baked, shaped, and cooked by the human hand. All this contributes to making geophagic material a commodity. Clay was sold and bought in markets in India, Africa, and the American South and continues to be in countries such as Kenya, Ghana, Guatemala, Britain, and the United States. One writer recently found some in a West African grocery store in Newark, New Jersey (Stokes, 2006), and in 2010 we simply acquired our sample of California calcium montmorillonite from an online store dedicated to selling geophagic material. There is no word on how Galen carried the twenty thousand lozenges of Lemnos earth in the second century AD back to Rome, but we received our clay via UPS after due credit card payment.

The commodification of edible soil is not a surprising development given the existence of the communal and private property of land, and the commerce of mineral building and landscaping material. Also predictable is the accompanying process of stratification. Access to geophagic material is then a function of resources. For instance, Geissler (2000, p. 658) describes how only Luo women of the elite, such as the wives of educated men and female teachers, could afford to consume the good earth and by doing so "displayed their wealth and distinguished themselves ... from the uneducated labouring women in the poorer households, who ate ordinary earth."

The distribution of geophagy worldwide and within societies brings additional evidence of the social nature of the phenomenon. Nineteenth-century accounts suggested that, in addition to Ancient Greece, soil was consumed in South America, Java, New Caledonia, Gambia, and the Caribbean, as well as Germany, Portugal, and Russia (Von Humboldt, 1872). It was not until Berthold Laufer's 1930 bibliographical essay that the near worldwide presence of geophagy became apparent. Even though he documented the practice under its many guises on all continents (except Antarctica), he contended that "geophagy is not a universal phenomenon" and neither is it "general in any particular tribal or social groups," which cannot "positively be labeled with a clear distinction as geophagists or non-geophagists" (Laufer, 1930, p. 103). Indeed, his and later evidence show that geophagy is practiced and socially sanctioned in only some cultures (e.g., the Luo, the Ewe, the Tiv, etc.) and within these societies by only some categories of people. In the present and in the past, these categories include pregnant women; young children; African slaves; African American women in the United States; and immigrants from Africa, India, and Pakistan to Great Britain and the United States. In addition, the greater propensity to consume soil by members of larger categories (women, rural dwellers, members of ethnic and racial minorities in Western societies, people in non-Western societies) has also been asserted but hardly well documented.

Estimates of geophagy in large populations are few and quite imprecise. Vermeer (1971) recorded rates of geophagy of 14% of men and 46% of women among the Ewe of Ghana. Halsted (1968) estimated pica prevalence among African American women at approximately 40%–50%, while others reported rates of geophagy as high as 57% (Vermeer and Frate, 1979).

Laufer presented convincing evidence from Melanesia, India, Africa, and Mexico that points to a relationship between pregnancy and geophagy. Since then research has confirmed that link. Drawing from a variety of ethnographic sources on Africa, Wiley and Katz (1998) identified 60 African populations where geophagy had been recorded in some manner (Figure 8.1). In most of these African populations, geophagy was virtually universal (10), common (28) or occasional (6) among pregnant women while it was inexistent (9) or rare (7) in others. For his part, Hunter (1993) estimated that 90% of pregnant women in southern Africa regularly snacked on clay, a finding similar to that of Vermeer (1966) among the Tiv of Ghana. More precise and accurate figures are hard to obtain either because the data are old and unreliable, samples of a studied population are not necessarily representative, or pica behavior beyond geophagy is recorded. Collating a number of studies conducted in northern and southern, urban and rural areas of the United States from the 1940s to the 1960s, Hunter (1973) reported percentages of pregnant women consuming clay ranging from 27% to 55%. Vermeer and Frate (1979) reported a geophagia prevalence rate of 28% among pregnant women in rural Mississippi. A meta-analysis of pica behavior in the United States showed rates ranging from 68% to 12% for African American pregnant women and 12% to 0% for white women (Horner et al., 1991). A 1995 study of pica in a rural obstetric population of 125 women in Georgia offered a strikingly different finding in that 14.4% engaged in pica, with only a handful reporting geophagia (Smulian and Motiwala, 1995).

The extent of geophagy among children is not much better documented because studies are fraught with the same methodological impairments. While a study of school-age children in Mississippi reported that 25% of the 207 African American children interviewed had eaten clay within 2 weeks of the interview (Dickins and Ford, 1942), another later study, also in rural Mississippi, found a rate of 16% (Vermeer and Frate, 1979). Rates of pica behavior among children in the United States, which includes clay-eating, tend to be higher and show a consistent pattern of greater incidence among African Americans (between 25% and 33%) than among white Americans and in rural areas more than in cities (Hunter, 1973; Lacey, 1990; Reid, 1992). More recently, panelists gathered by the ATSDR noted that a team of researchers has estimated that 33% of children ingest more than 10 g of soil 1 or 2 days a year (ATSDR, 2001). These figures would suggest that about a third of children in the United States are likely geophagists.

Interestingly, rates tend to be much higher in other regions of the world. Data from India in the 1980s suggest that prevalence among children varies between 29% and 75% (Abrahams and Parsons, 1996) and Geissler (2000, p. 653) reports that "previous studies among the Luo of Western Kenya have found that about three quarters of primary schoolchildren eat earth every day."

Despite limited statistical data, all point to a phenomenon limited in scope to a significant minority of people in societies and, within these, clearly bounded by social, cultural, and biological features. This is what emerges from the latest and still limited data available. Young et al. (2010) report than on the island of Pemba (Tanzania), 7.1% of pregnant women and 4.5% of young children consumed soil, something that men and nonpregnant women nearly never do.

Finally, geophagy has a symbolic dimension. Eating soils means more than simply fulfilling a need or a craving. We have seen evidence of variation according to culture, social class, sex, and age. There is also an association between geophagy and religion. It ranges from the antique lozenges of *terra sigilatta*, extracted by a priestess and mixed with goat's blood, to the clay tablets marked with Roman Catholic symbols and images of the cult of Esquipulas in Guatemala (Frate, 1984). Consumption of material from sacred sites for its expected healing properties has also been noted in India (Laufer, 1930) and New Mexico (Callahan, 2003).

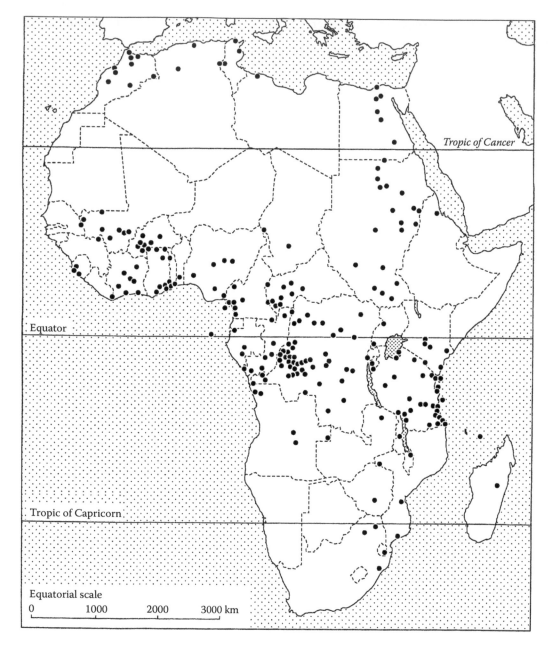

FIGURE 8.1 Geophagy in Africa (shown by the dots). The distribution of geophagy in Africa shows the practice to be ubiquitous but not universal, since it is virtually absent from large areas (Sahara, Madagascar) and concentrated in others. (From Anell, B. and Lagercrantz, S., *Geophagical Customs*, Studia Ethnographica Upsaliensia 17, Humanistiska Fonden, Stockholm, Sweden, 1958; Abrahams, P.W. and Parsons, J.A., *Geographical Journal*, 162, 63–72, 1996.)

8.4 LANGUAGE OF GEOPHAGY

In fact, a most obvious sign of the symbolic dimension of geophagy is its linguistic dimension. The naming of the practice and of geophagic material is the unmistakable sign of its enculturation. Reinbacher (2003, p. 53) claims, tongue in cheek, that the term "geophagy" was coined "around 1850 to make ethnologists sound more sophisticated than when they say dirt gobblers, mud feeders, or clay eaters." The process reflects the varying cultural meaning of "earth." While most languages do not have an "earth/land split," in languages such as Latin, German, French, and English, different words are used to refer to the different connotations of earth (the planet, the land under our feet, a mineral different from water and air).

Indeed, the conceptualization of soil by cultures is interesting. In Western culture elements from Greek philosophy and the Judeo-Christian tradition view soil as the constitutive product of human existence. Earth is one of the four basic elements in Empedocles' metaphysics; in the Bible, God fashions human and animal life from the dust of the ground and human life follows a cycle from dust to dust. The political and economic significance of land and landowning in agricultural then industrial economies amply testifies to the positive connotations earth has had at at least some points in time. Recently, environmental activists have been promoting greater harmony between humans and Mother Earth (Henry and Kwong, 2003).

However, symbolic meanings associated with soil change through time and across cultures. Earth or soil is not granted very high status in contemporary Western cultures. In American English, expressions such as "dirty," "dirt poor," "dirt cheap," and "dirt bag" refer to undesirable things or people, as do other earth-related terms, "soiled," "muddy," "mudslinging," and "muck," which convey a similar meaning (see also Brevik and Hartemink, 2010, p. 21). Dirt is considered a contaminant to be mopped and dusted out of homes, washed out of clothing and off bodies, and purified out of water and the air because, after all, "cleanliness is next to godliness." If the "salt of the earth" is the best, the worst food "tastes like dirt."

Occupations that require people to work closely with the earth, land, or dirt, such as ditch diggers, grave diggers, farmers, and miners, are rarely high status or associated with high pay. The French and Italian languages share a similar attitude: a *cul-terreux* is a pejorative term for peasants, farmers, or people living in the countryside, while in Italian a *terrone* is a pejorative term used by urban northerners to denigrate southerners as peasants.

Non-Western and naturalist cultures tend to place a greater value on soil as a basic formative element and a symbol of life and fertility. Ethnographic examples abound, so for the sake of brevity, a couple of references limited to the African continent will suffice. Creation myths, ideologies, and economic practices among the Igbo in Nigeria, Gikuyu of Kenya, and Dogon of Mali place soil at the beginning of humanity in combination with another element (water, sky), a spirit, or a creature, and practically view earth as the provider for human and animal life (Henry and Kwong, 2003).

The naming of geophagic material further documents the cultural boundedness of geophagy. The etymology, referent, and number of terms vary from culture to culture. Eighteenth-century African slaves in the Caribbean ate *aboo*, *caioo*, or *trieng*. Colin Chisholm, a nineteenth-century English physician, speculated that "*Aboo* and *Caioo* [were] probably corruptions of the two French words *boue*, signifying dirt or clay; and *craie*, chalk" while *triend* was "an Ebo word, signifying a purer kind of pipe-clay, much used with food by most of the inhabitants of the coast of Guinea" (cited in Higman, 1984, p. 296). African American women in Mississippi distinguished between *Franklin dirt*, *hill dirt*, and *gumbo* (not a good dirt) (Vermeer and Frate, 1975); Luo children prefer *odongo* but consume mostly *loo* (earth) (Geissler, 2000); and Tanzanians differentiate between *udongo*, *mchanga*, *vitango pepeta*, and *ufue* (Young et al., 2010). U.S. merchants tend to name their material with its geological moniker such as *bentonite* or with a suggestive brand name. We bought some TerraMin "100% calcium montmorillonite," with the label showing the chemical composition of the material as per the format established by the U.S. Department of Agriculture, along with the symbol of a duly protected registered brand (Figure 8.2), as expected in a market economy. The message is clear: soil is a food and a commodity and it is packaged to look like one.

At this juncture, the cultural treatment of geophagy becomes a bit more complicated. As the discussion moves from the materiality of the practice to its immateriality, it is necessary to distinguish between the facts that are reported or examined and the discourse in which they are embedded. For example, when historians collate instances of geophagy drawing from antique medical treatises, medieval court cases, memoirs by eighteenth-century explorers, or Caribbean slave-owners'

FIGURE 8.2 A bag and a sample of geophagic material purchased online in 2010. The packaging provides instructions for consumption, and the label mentions the two leading hypotheses for geophagy: nutritional supplementation and detoxification. (Photo by J. Henry.)

journals, they do not simply relay information on the physical qualities of consumed clay or the social identities of the consumers. They convey the views of geophagy by the original writers and interpret the primary data from their own cultural perspectives. They carry not only the descriptions of the practice but also the connotations of geophagy and its practitioners.

Historically, two perspectives have been particularly powerful in shaping the popular understanding and scientific examination of geophagy. The medical gaze and the colonial gaze have cast the practice and the participants as pathology and patients for the former and as an uncivilized aberration and the "other" for the latter. These underlying narratives are still active today, which is why their influence needs to brought into the open.

The medical gaze, a term coined by French philosopher Michel Foucault (1973), refers to the medicalization process "by which nonmedical problems become defined and treated as medical problems, usually in terms of illness or disorders" (Tischler, 1999, p. 550). As physicians and scientists sought to understand and cure painful events and conditions thought to be socially undesirable, human reproduction, hygiene, diet, alcoholism, even poverty, and, in our case, geophagy became medical conditions, and their sufferers became patients.

Physicians in antiquity appear to have been the first to record the behavior, at least according to the records at hand. Hippocrates (460–370 BC) was reportedly the first to write about geophagy. Galen, the great second-century AD Greek physician, a prolific and meticulous writer, and later Avicenna added their own accounts.

These physicians seem to have tolerated geophagy as long as its assumed therapeutic functions were deemed legitimate, helping with nutrient supplementation especially during pregnancy and food shortages, with the adsorption of toxins in cases of poisoning, and with the relief of gastrointestinal discomfort. The propensity of pregnant women to eat clay provided in part the basis for the medical acceptance of geophagy, as did the belief that clay fills the stomach in times of famine, a hopeless but understandable remedy, and that it may also act as a poison antidote.

However, tolerance of geophagy quickly abated as these circumstances ceased to exist. Around AD 1000, the Arab physician and philosopher Avicenna suggested the use of whips, restraints, and prison for young and old geophagists and consumers of nonedible items. He even advocated letting death occur for hopeless sufferers of the condition since an "appetite for sour, sharp things is most effectively changed while that for drying and dry things such as clay and charcoal is of much greater corruption" (quoted from Halsted, 1968, p. 1385; Frate, 1984).

In fact, as Avicenna's comment indicates, geophagy was associated with pica. The term, believed to be derived from the Latin word for magpie, a bird known to collect various edible and nonedible items, had been coined in the sixth century AD by Aetius of Amida (Young, 2011). The term endured; the syndrome was further described in the sixteenth century by French physician Ambroise Paré, among others (Danford, 1982), and is still widely used today with the same meaning (Young, 2010).

Most of the historical accounts of geophagy that emerged in the eighteenth and nineteenth centuries consist of cases recorded by physicians outside of Europe, many involving African slaves in the Americas and Caribbean, and native populations in areas colonized by Western European nations. Physicians give informative remarks about the material consumed—its provenance, appearance, and taste—and about the consumers—their sex, age, and state of health. They describe the symptoms: sluggishness, inertia, skin coloration, depression, shortness of breath, abdominal swelling, and melancholy. *Mal d'estomac* and *cachexia africana* were the most frequent names given to this "condition." They also propose hypotheses about the causes and consequences of geophagy.

For instance, James Maxwell, a surgeon in Jamaica, remarked in 1835 that "the moderate use of this [clayey] earth is considered by the negroes neither dangerous nor disgraceful; and those who eat it, take it as much to gratify an acquired taste, similar to that of chewing tobacco or opium, as to satisfy any morbid desire" (quoted in Higman, 1984, p. 295). Another physician, James Mason, suggested a novel approach. Noting "the strong propensity which pregnant females have to take alkaline earths is generally well-known" (quoted in Kiple, 1984, p. 101), he proposed "that dirt-eating, instead of being a disease, or the cause of a disease, is actually a remedy; and probably the various earths, marl and ashes made use of, really contain useful ingredients mixed up with much hurtful matter" (quoted from Higman, 1984, p. 297).

By the nineteenth century, the medical gaze still paints geophagy as a pathology. It is still seen this way today. At best, geophagy is considered an eating disorder, a form of pica, or a public health problem. Horner and his colleagues (1991) recommend that the practice should not be considered benign and should be stopped as an indication and possibly a cause of anemia and malnutrition. At worst, it kills in cases where deaths of clay-eaters have been linked to excessive consumption or ingestion of polluted soil (Lacey, 1990).

Current medical research on soil-pica or geophagy continues to emphasize the pathological dimension. Geophagic material is analyzed for levels of toxicity due to the contents of lead (Abrahams et al., 2006), arsenic, and mercury (U.S. Department of Health and Human Services, 2009) and the presence of human and animal feces (Shigova and Moturi, 2009). Maybe an extreme example of the association of geophagy and pathology, Dissanayake and Chandrajith (2009, p. 223) deal with geophagy in a chapter dedicated to the "the geological basis of podoconiosis, geophagy and other diseases."

The colonial gaze can be defined as the self-given right by European colonizers to order, name, and control reality in colonized areas (Pratt, 1992). In addition to the writings of plantation physicians, where the medical and colonial gazes meet, it is found in the accounts of geophagy produced by colonial administrators (Debien, 1974); slave owners; and the multitude of travelers, explorers, and scientists who investigated the ways of life of non-Western people and cultures. These accounts show that despite their careful attention to description and valiant attempts to explain native behaviors, colonial writers could not overcome their cultural bounds. Hence, already framed by Western medical discourse, the examination of geophagy is further framed by moralizing and the social construction of the "other."

The colonial gaze on geophagy emerging from accounts by colonial administrators, slave owners, and other Western writers offers a narrative parallel to that of the medical gaze. The differences between both—later emergence of colonialism, and replacement of physiological and therapeutic concerns by cultural, economic, and political issues—do not change the evaluation of geophagy. It is widely considered a depraved appetite that affects the uncivilized natives of Africa and South America and results in various maladies and economic losses. Seventeenth-century slave owners, for instance, were more interested in preventing the behavior from happening than in understanding it. They also strived to eradicate it from their properties by various means including facial restraints, flogging, and imprisonment (see MacClancy et al., 2007 for striking drawings of geophagy masks from this period). From their perspective, geophagy was detrimental not only to health but profits too: a 1687 report from Jamaica attributed approximately half of the deaths of slaves to geophagy as a form of suicide (Abrahams and Parsons, 1996).

A pioneer in the presentation of geophagy, and a founder of the modern study of geography, Alexander Von Humboldt wrote about his contact with clay-eating Ottomacs encountered during his travel in South America in the early 1800s. His account provides a vivid illustration of the power of an ethnocentric worldview in shaping his understanding of the phenomenon:

> We found heaps of earth-balls in their huts, piled up in pyramids three or four feet high. These balls were five or six inches in diameter. The earth which the Ottomacs eat is a very fine and unctuous clay, of a yellowish grey color; and, when being slightly baked at the fire, the hardened crust has a tint inclining to red, owing to the oxide of iron which is mingled with it ... If necessity force the Indians to eat earth during two months (and from three quarters to five quarters of a pound in twenty-four hours), he eats it from choice during the rest of the year ... It is most surprising that the Ottomacs do not become lean by swallowing such quantities of earth: they are, on the contrary, extremely robust. (Von Humboldt, 1872, pp. 495–496)

This excerpt is remarkable for the details it provides on the physical qualities, chemical composition, and location of the material consumed, as well as those on its consumption, its native explanation, and its assumed consequences for the health of the consumers. From our perspective, however, problems arise in the subsequent passages. Von Humbold continues:

> I observed everywhere within the torrid zone, in a great number of individuals, children, women, and sometimes even full-grown men, an inordinate and almost irresistible desire of swallowing earth. (Von Humboldt, 1872, p. 496)

A behavior that had just been described as willful and beneficial to alleviate hunger has become "inordinate" and surprising among adult males. In addition to soothing hunger pains, the "mania of eating earth" may in fact be a consequence of a "regimen purely vegetable and deprived of spirituous liquors" (Von Humboldt, 1872, p. 497). Ultimately, the Ottomacs, "a restless, turbulent people, with unbridled passions" are deemed to consume earth like they consume fermented drinks, snuff, and tobacco, so they can throw "themselves into a peculiar state of intoxication, we might say of madness" (Von Humboldt, 1872, p. 504). What started as a reliable ethnographic account of geophagy has turned into an ethnocentric indictment of "man in a savage state" (Von Humboldt, 1872, p. 502) who cannot control his taste for intoxication, whatever the source.

The colonial gaze continues to exert significant influence. Berthold Laufer's 1930 groundbreaking review of the literature on the topic provides an interesting example. He begins by admonishing researchers that "the days are gone when the discussion of a problem started with the Greeks and Romans" (Laufer, 1930, p. 101) and then proceeds to focus on the practice in Asia. He criticizes the labeling of geophagy as evil, vicious, disgusting, or depraved because "such characterizations are subjective and meaningless, and do not help us in understanding the phenomenon" (Laufer, 1930, p. 102). Noting the flimsiness of the evidence, he challenges "the sweeping assertion that geophagy is a universal phenomenon and was practiced in antiquity. Neither of these statements is true." It is merely occurring "sporadically almost anywhere" (Laufer, 1930, p. 103).

However, right after this laudable call for methodological and epistemological rigor, Laufer goes on to state that since it "may spring up anywhere spontaneously and independently ... the habit is more or less individual." Any regularity "can be explained only through contact and diffusion" (Laufer, 1930, p. 104). His rejection of a cultural basis to geophagy and the individualization of the behavior rests on several tenets of anthropological orthodoxy of the time—rejection of racial or tribal explanations of human behavior—and reliance on imitation and diffusion to explain the social location of practices, now all quite obsolete paradigms.

What precedes should not be interpreted as the indictment of an archaic ethnocentric scholar. His pioneering work still informs the field in content (despite its stance he still managed to identify the major social functions of geophagy) and in method (the meta-analysis of secondary data is a rather frequent occurrence in the study of geophagy, this chapter included) (see also Cooper, 1957; Halsted, 1968). In fact, some argue (Wiley and Katz, 1998; Henry and Kwong, 2003) that the ideological flaws that irked Laufer are still hindering the understanding of geophagy by studying it as deviant behavior, especially when it occurs in Western societies.

It would be wrong to believe, however, that the stigmatization of geophagy is uniquely a Western male attitude. In his remarkable analysis of the cultural aspects of geophagy among the contemporary Luo, Geissler (2000) argues that for the Luo earth-eating belongs to the "lower" part of society where one finds the young, the weak, and the female. In the "negative male discourse against earth-eating," the practice is associated with the sick, the dirty, the traditional, the poor, and the uneducated versus the healthy, the clean, the modern, the wealthy, and the educated elite. His work calls our attention to the fact that, notwithstanding the Western stigmatization of geophagy, social and cultural tensions in societies require a greater scrutiny of its characterization. Whereas Luo women and children see nothing wrong with geophagy, their husbands, fathers, and brothers do not see it as appropriate behavior for themselves.

In sum, the examination of the culture and language of geophagy has confirmed the relevance of its sociocultural dimensions, but it has also shown the limitation of its explanation. However, the survey of sociocultural dimensions is insufficient because it does not fully account for the materiality of the practice. It is imperative to understand biological change, or biological evolution (by means of natural selection) and how a species adapts to its environment. Change, whether cultural or biological, does not always result in an adaptation; sometimes consequences occur, resulting in maladaptations.

8.5   ARCHEOLOGY OF GEOPHAGY

Archaeological anthropology is the subfield that studies the human cultural past. The excavation of archaeological sites and the study of artifacts and ecofacts (animal bones and plant remains) show that human paleodiets have been quite diverse through time and space and included the consumption of soil.

Findings from archeological research leave little doubt that geophagy has long been taking place, but they yield few details about the practice. Clark (2001) has reported on possible prehistoric evidence for geophagy at Kalambo Falls, Zambia. This Paleolithic site is exceptional for its preservation of a living floor where ancient humans processed their food. In addition to butchered animal bones, well-preserved botanical remains of edible tree nuts were recovered by archaeologists, who believe that these early humans were exploiting savanna and forest habitats between 100,000 and 200,000 years ago (Clark, 1970). Johns and Duquette (1991) analyzed soil samples from there and found that Kalambo Falls clay samples were similar to the modern clay samples consumed in other parts of Africa today. It is possible that the ancient peoples of Kalambo Falls were using clay soils in order to detoxify the tree nuts collected from the nearby forests. As noted by Johns (1990, p. 98),

> If early human foraging was analogous to the feeding behavior of modern nonhuman primates, then food procurement by humans must have been subject to the constraints imposed by plant defensive compounds. Geophagy is a behavior with antecedents that are certainly prehominoid, and higher primates, including humans, have apparently maintained it as a mechanism for dealing with naturally occurring toxins.

Indeed, ethnographers have documented a putatively similar practice among the Pomo people of California and Sardinians, who used clay soil to detoxify their acorn meal.

Brady and Rissolo (2006) argue that the ancient Mayans extracted clay from multiple cave sites in the Americas, which was then used in rituals involving geophagy. Their evidence is indirect, arguing that it was a well-documented behavior in the prehistoric Americas and it is still practiced in Guatemala. However, like Kalambo Falls, this evidence is, again, inferential.

Aside from extraction sites like the Mayan caves and the clay soils found out of their natural context in archaeological sites like Kalambo Falls, evidence for prehistoric geophagy may depend upon finding clay "artifacts" like the *terra sigillata* of ancient Greece. Reinbacher (2003) documents these clay "coins" from ancient Bronze Age Lemnos in the Aegean Sea continuing in Europe to the nineteenth century. These clay "coins" were used for medicinal purposes.

Geophagy is not easily recognized in the archaeological record. A case in point may be the enigmatic Poverty Point objects found in the Late Archaic archaeological site (1730–1350 BCE) near the Mississippi River in northeastern Louisiana. The hand-formed baked clay objects (Figure 8.3) are thought to have been used in cooking and they are remarkably similar to the clay "balls" collected in contemporary Ghanian markets (Hunter, 1973). If the Poverty Point peoples were primarily eating acorns, as Gibson (1996) states, then they may have been using clay in order to make the acorns more palatable, as noted earlier.

Furthermore, Turpin (2004) reports on a prehistoric archaeological site in a braided floodplain in south central Texas consisting of earth ovens with the weathered remains of clay balls (Burned Clay Objects) that were used in cooking. This could perhaps even be a processing site for the clay balls themselves. Again, the ethnographic evidence of the Washo people is consistent with the hypothesis that these clay balls (Poverty Point Objects or Burned Clay Objects) were used in cooking. The Washo, a Great Basin tribe of Native Americans living in California and Nevada, would heat clay balls or stones and then drop them into tightly woven baskets for cooking purposes (Downs, 1966). The indigenous Australians would also use clay cooking balls, but in earth ovens in the floodplains of southern Australia. Rowland (2002) questions whether this Australian plant processing technology, using earth ovens and clay cooking balls, might be a cultural adaptation in order to detoxify the rather toxic and starchy *Typha* plant.

Not only does archaeology document food processing technologies and paleodiets of ancient peoples, but it also endeavors to understand the dynamic process by which ancient cultures adapt to their natural and sociocultural environments. Sometimes, humans had to adapt to the unforeseen consequences of culture change. The archaeological classic is the dilemma created by the transition from a foraging lifestyle (hunting and gathering) to plant and animal domestication, known as the Neolithic Revolution. The dietary diversity of the foraging lifestyle was sufficient to the extent that malnutrition was seldom a problem. However, when some human cultures transitioned to plant domestication, dietary diversity decreased to the extent that malnutrition became a problem. Perhaps plant cultivation in the tropical areas of the world presented the greatest challenge to

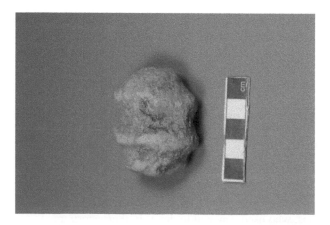

FIGURE 8.3 A clay ball from the Poverty Point (Louisiana) archeological site with visible finger prints dated 1650–700 BCE. (Photo by J. Henry.)

human nutrition. The major root crops like manioc, taro, and yam are low in nutritional value. These starchy tuberous crops provided food for energy, but nutritional deficiencies were a likely problem. If geophagy was triggered by nutritional deficiencies, then geophagy would be more commonly found in the Old and New World Tropics.

When some prehistoric North American populations focused on maize cultivation, niacin deficiency (resulting in pellagra) and amino acid (lysine and tryptophan) deficiencies became a problem (Katz et al., 1974). Native Americans avoided these deficiencies by using lye, lime, or wood ashes during food preparation. This alkali treatment of the maize prior to consumption increased the bioavailability of both niacin and the amino acids lysine and tryptophan. For those cultures depending on wheat cultivation, certain amino acid deficiencies (like lysine) and deficiencies in the micronutrient zinc became a problem. In this context, geophagy may again be seen as one of the many ways that humans adapted to dietary deficiencies caused by dietary change.

So, what does archeology tell us about geophagy? It ascertains that geophagy is indeed an ancient behavior that is temporally and culturally variable. Interestingly, the analyses of archeological data contribute to building a framework for a better understanding of geophagy since they suggest a diversity of explanations, namely a medicinal function, nutrient supplementation function, and detoxification function.

8.6   BIOLOGICAL ASPECTS OF GEOPHAGY

Biological anthropology is the most basic subfield of anthropology. It is the study of the origins of our species as well as contemporary human biological diversity and human adaptation. Subfields include medical and nutritional anthropology. Central to this field is the cross-species technique, which compares the biology and behavior of humans to that of other species.

8.6.1   AMONG ANIMALS

Even though this chapter is dedicated to human geophagy, the widely reported occurrence of the practice in the animal kingdom calls for attention. Many animals do indeed consume soil in some manner and for some purpose. In fact, geophagy is documented among all the chordate orders (amphibians, birds, fish, mammals, and reptiles) and some invertebrates such as the appropriately named earthworms and termites. It appears that small and large creatures alike consume soil, but much research has focused on mammals, from bats to zebras. Primates, especially monkeys, macaques, and chimpanzees, as well as a variety of ungulates are the subjects of many studies (Krishnamani and Mahaney, 2000).

A common form of mineral consumption by animals involves salt licks. These have been reported among African forest elephants, deer, pigs, macaws and parrots in the Amazon, pigeons in South Africa, large ungulates such as elk, moose, Stone's sheep, and mountain goats (British Columbia), mountain tapir (Peru), Bongo antelope, Greenland muskoxen, bighorn sheep, spider monkeys, and feral horses in Alberta (Canada). Again, mammals dominate the list, as this observation by researchers in the Deramakot forest reserve in the province of Sabah (Malaysia) illustrates:

> All food-habit types of mammals (carnivore, herbivore-frugivore, insectivore, and omnivore), which included 29 (78.4%) of 37 species known in Deramakot, were recorded at the natural licks. The sambar deer, followed by the bearded pig, the lesser mouse-deer, the Malay badger, and the orangutan were the most commonly recorded species and represented 77.5% in terms of the frequency of appearance in all photographs taken throughout the year. (Matsubayashi et al., 2007, p. 742)

Analyses of soil and water samples collected at salt licks and clay licks show high levels of elements such as calcium (Ca), cobalt (Co), magnesium (Mg), manganese (Mn), potassium (K), phosphorus (P), selenium (Se), sodium (Na), and sulfur (S). The presence of these components in consumed material, especially Na and Ca, leads some researchers to hypothesize that animal geophagy is a behavioral adaptation to nutritional deficiencies by providing the required nutrients in bioavailable form. Nutritional deficiencies can be caused not only by dietary deficiencies but also by increased demand for certain nutrients during pregnancy, lactation, or even during disease stress. However, a consensus emerges among researchers that geophagy likely serves other functions, especially adsorbing secondary plant compounds such as alkaloids (Klaus et al., 1998) and providing agents such as kaolinite which act as pharmaceutical agents to remedy gastrointestinal upset (Abrahams, 1999).

8.6.2 Among Humans

In the investigation of a behavioral phenomenon such as geophagy, it is important to consider all the factors that determine its frequency and distribution. This includes primarily the composition of the material consumed, its contribution to the consuming organism, and the overall nutritional context in which the behavior takes place. It is interesting to note that, faced with the limitations of the cultural and medical approaches, which have long dominated the analyses of geophagy, the examination of the biological, physical, and chemical properties of edible soil and digestive processes has recently increasingly yielded significant, informative findings. The point here is not to surrender the field to physical and chemical analyses of geophagy and understand the practice as some form of mechanical adaptation to innate nutritional requirement, but rather to integrate the information with cultural dimensions within the holistic framework of anthropology.

Early descriptions and chemical analyses of geophagic soils identified major elements of their composition, especially iron, calcium, and magnesium (Laufer, 1930). Modern research has confirmed the presence of these mineral nutrients and has revealed many others in various concentrations. A review of works that present a detailed analysis of consumed soils show that the human nutrients calcium, iron, magnesium, manganese, potassium, silicon, and sodium are almost universally present. The presence of other elements, some human nutrients and some not, such as aluminum, arsenic, barium, cerium, cobalt, chromium, copper, gallium, iodine, lanthanum, lead, nickel, niobium, titanium, rubidium, strontium, thorium, uranium, vanadium, and zinc, has been reported (Mahaney et al., 2000; Reilly and Henry, 2000; Vermeer and Frate, 1975; Wilson, 2003; Young et al., 2010), but researchers still need to establish their role if any. Interestingly, this nonexhaustive list contains most of the natural and nonradioactive metals identified in the periodic table of elements.

Soils contain more than these core elements. "Soil is a considerable biological sink," writes an immunologist (Callahan, 2003, p. 1019). He refers in part to the estimated 4600 species of prokaryotic microorganisms identified by DNA-hybridization analyses in 1 g of natural soil and the 700–7000 g of biomass per cubic meter of soil found by other analyses.

We can now clearly understand why physicians and others have long considered geophagy to be detrimental to the health of the individuals, even to the extent of causing disease such as anemia or helminthic infestation, and that many continue to do so (Horner et al.,1991). For one, soil is universally considered a nonfood and thus not fit for human consumption, even if earth is positively conceptualized. Recent chemical analyses of geophagous material clearly show the presence of harmful components. For instance, lead and uranium make no known useful contribution to the human diet. Moreover, the possible excessive presence of useful components may have negative consequences; for instance, too much iron may result in health problems such as hemosiderosis. Finally, soil, and particularly surface soil, may contain external elements such as feces and helminth ova, causing diarrhea, helminth infections, or just plain poisoning (Shigova and Moturi, 2009).

However, others, including physicians and researchers, have considered the possibility that some ingestive behaviors, while not a normal pattern, may in fact be an adaptation by certain individuals to cope with three main conditions: dietary deficiencies, plant toxins, or gastrointestinal distress (Laufer, 1930). These explanations are still framing the research today, although it can be proposed

TABLE 8.1

Major Hypotheses on Geophagy

Sources	Nutrient Hypotheses	Detoxification Hypotheses
Abrahams and Parsons (1996)	Minerals—especially iron	Detoxification (bimodal)
Brevik (2009)	Enhanced mineral nutrition	
Hooda and Henry (2007)	Calcium deficiency	Detoxification
Johns and Duquette (1991)	Calcium, sodium	Acorn tannins
Johns (1999)		Detoxification
Vermeer (1966)	Calcium deficiency	Plant toxins
Young (2007)	Calcium deficiency	Detoxification
Wiley and Katz (1998)	Calcium and other minerals	Detoxification (bimodal)

that the field is now divided between two major hypotheses, nutrient supplementation and detoxification (Table 8.1).

Johns (1999) considers that some nondietary ingestive behaviors may be maladaptive (aberrant), but then considers the detoxification function of geophagy as possibly a behavioral adaptation to the ingestion of plant toxins in specific diets. He lists cases of clay consumption with unpalatable plants by natives in Western Australia, among the Indians in Florida in 1527, and Ainu of the North Asian Pacific. He also describes how the Pomo of California baked the so-called "black bread" from ground unbleached acorns and a specific type of red clay (Johns, 1990, pp. 85–86). For their part, Wiley and Katz (1998) hypothesize that geophagy practiced by pregnant women in nonpastoral African cultures is an adaptation that serves to supplement their diet with minerals, notably calcium, and also serves to reduce pregnancy sickness and to detoxify secondary plant compounds in their diet.

It must be made clear, however, that while some studies emphasize one hypothesis over the other because of research design, theoretical outlook, place of study, or scientific specialty, the bulk of research considers both as likely explanations of geophagy. Indeed, Wiley and Katz (1998, p. 541) conclude that "geophagy is clearly multifactorial, deriving from multiple causes and having multiple benefits that vary across species and populations."

Symptomatic of the recent directions taken by researchers is a 2007 edited volume (MacClancy et al., 2007) that approaches geophagy from a nutritional perspective (see also Young, 2011). In one chapter, Young (2007) reviews six mechanisms by which geophagic material may act on the body. Four can be grouped together as variations of the detoxification hypothesis. The first hypothesis deals with the high cation exchange capacity of many clay soils (good at holding to positively charged ions and even microbes), which makes them efficient at binding with toxins, thus reducing the effects of noxious plant chemicals such as tannins and glycoalkaloids. However, this faculty may also bind useful micronutrients such as iron and zinc or macronutrients such as potassium. The second detoxification mechanism considered is the way that clays affect digestion by reinforcing the intestinal mucosa and blocking the gut so digestion is slowed to allow better absorption. Pivotal here is the general presence of smectite, a clay known to help binding, but constipation and intestinal blockage are significant risks. The third mechanism involves quelling gastrointestinal upset since kaolin and smectite are known to reduce nausea and diarrhea. The fourth mechanism consists of increasing the pH level, thus making any ingested content less acid than the stomach's hydrochloric acid and in doing so, potentially allaying heartburn.

Young (2007) also examines the nutrient supplementation hypothesis, especially the role of iron, zinc, and calcium. She notes that few clinical studies indicate that geophagists were deficient in these elements prior to consuming soil and concludes that only calcium supplementation would be a possible reason for geophagy. In another chapter, Hooda and Henry (2007) indeed report that, after conducting *in vitro* experiments that simulate digestion by the human stomach and small intestine, geophagy might actually reduce iron and zinc absorption while potentially increasing calcium supplementation.

In addition to the clarification on the actual biological effects of geophagy, another important research direction is evidenced: the reliance on *in vitro* experiments. Interestingly, findings tend to coalesce around both leading hypotheses. For instance, using the TNO Intestinal Model to test the capacity of kaolin, a commonly ingested clay, to adsorb quinine (an alkaloid) and two types of tannin (digestion inhibitors), a group of researchers found that kaolin reduced the bioavailability of each compound by ≤30% and thus concluded that "gastrointestinal adsorption is the most plausible function of human geophagy" (Dominy et al., 2004, p. 319). However, another research group, using a comparable *in vitro* procedure (the physiologically based extraction test or PBET) on commercial clays consumed in Britain by young pregnant Asian women, concluded that "despite the low bio-accessibility of Fe, with the quantity of soil consumed, one sample can provide 41%–54% of this mineral nutrient required by a 15–18 year old female, with the other sample providing 90%–119%" (Abrahams et al., 2006, p. 98).

In sum, when it comes to the contribution of geophagy to human health, the jury is still out. While the hunger hypothesis has now been rejected, people do not eat soil because they are hungry, and the nutrient supplementation theory is being increasingly challenged, the detoxification hypothesis appears to stand as "the best supported explanation" (Young, 2011, p. 137). However, evidence and definitive proof remain elusive. Still, undoubtedly future research will continue to unveil the chemical mechanisms at play in the digestion of soil by humans and clarify the role of the minerals, and possibly organisms, on the human organism.[*]

8.7   CURRENT AND FUTURE STATES OF GEOPHAGY

In conclusion of this survey of the biological, physical, cultural, and symbolic dimensions of geophagy, several findings emerge about the basis and effects of the behavior and about the research conducted to shed light on aspects still unclear or unknown. It is strongly established that the consumption of geophagic soil fulfills some physiological and cultural functions, whether it is detoxification or whether it is by custom or association with religious purposes or social status. These functions are not all adaptive or positive since there are serious physical downsides and cultural stigma attached to the practice. Then, the concentration of geophagists among specific and universally quite stable categories of people defined by gender (women), age (young children), physical status (pregnancy), social status (people exposed to significant nutritional deficiencies), and culture suggests that physical and cultural environmental factors play a pivotal role. Eighteenth-century African slaves in the Caribbean, African American women in the American South, Luo children in Kenya, Guatemalan Catholics, and Asian immigrants in the United Kingdom do not consume clay for the same reason and for the same effect. There may be several reasons for geophagy. This is increasingly confirmed by the physical and chemical diversity of geophagic material and its differential treatment by the human digestive system, the analyses of which now comprise the most advanced and informative research on the topic. The physical condition of individual geophagists, framed in part by their ecological environment, their culturally sanctioned diet, their access to nutrients, and their nutritional needs, is clearly related to the propensity to consume soil.

Research has adapted to this evolution of our understanding of geophagy. Once conducted in separate fields and shaped by different paradigms (Reid, 1992), it is now generally pluridisciplinary,

---

[*] One possible avenue would be the consideration of organisms found in geophagic soil and pica material in addition to the analysis of elements. For example, cobalamin (vitamin B12) is an essential nutrient that cannot be synthesized by humans and thus must be obtained from a diet rich in cobalamin such as insects (like termites) or directly from the bacteria that synthesize this complex vitamin (Oxnard, 2008). Cobalamin deficiencies may result in pernicious anemia and various mental disorders including depression. We hypothesize that dietary anomalies such as coprophagy in some primates such as gorillas as well as geophagy in humans could be behavioral adaptations to directly obtain cobalamin from the bacteria in faeces or from the bacteria in soil.

where cultural and biological explanations are brought, sometimes wrought, together rather than opposed to one another. The six-step methodology put forth by Young et al. (2008) to study pica provides an excellent example of possible avenues for future research. Young and her colleagues propose to combine data drawn from oral interviews with consumers, the careful collection of soil samples, and the analysis of their physical (color, particle size distribution, consistency, dispersion/flocculation characteristics), mineralogical, and chemical traits (elemental analysis, NaCl content, measurement of pH, cation exchange capacity, etc.), as well as the analysis of biological interventions through both *in vitro* (checking for bioaccessibility) and *in vivo* experiments (bioavailability, adsorption of harmful chemicals and pathogens, parasitology).

Geophagy is still practiced in twenty-first-century societies, but it is not a widely discussed topic. A Lexis/Nexis search conducted in September 2010 yielded 277 English language mentions over an 18-year period, from 1984 to 2006. Many were mere repeats of the same news articles or announcements of research findings. As a matter of example, of the 139 mentions in newsletters, which comprised the largest category of reports, 106 came from one single source, NewsRx, an Internet-based company dedicated to distributing health and medical information, and 64 were repeats of research findings about pregnant women and geophagy in Kenya in 2005.

There are, however, episodic appearances of geophagy in the popular media, even if our perusal was mostly limited to English-language published and Internet sources. Over the past two decades, rare but lively Internet discussions have argued about the merits and perils of geophagy and picas, relayed scientific information and analyses along with limited bibliographies, and even provided contact information for providers of edible clays (Warshall, 1999). The tone of press articles is generally quite positive, ranging from being informative to borderline facetious. For instance, a 2009 *New York Times* headline, "Babies Know: A Little Dirt Is Good for You" was placed over a picture of a bowl of dirt with a fork planted in it. The article went on to report physicians' views that "the millions of bacteria, viruses, and especially worms that enter the body along with 'dirt' spur the development a healthy immune system" (Brody, 2009). A 2010 CNN no-budget diners' guide featured dirt alongside leather, bugs, and tree bark as inexpensive sources of nutrients in hard economic times (Clark, 2010). Reporting on the increased use of specialty salts in upscale restaurants, a 2000 *New York Times* article appeared to nudge geophagy toward upper-class luxury. In a stark contrast to the general image of geophagy usually associated with possible pathologies and disadvantaged individuals, a chef, apparently smitten with his $32-a-pound Hawaiian red clay salt, noted how "the clay adds a real earthiness to the foie gras that the other salts don't contribute" (Clark, 2000).

Media reports are not all whimsical though. Echoing Depression-Era accounts of geophagy in the Deep South, an Associated Press dispatch noted that the practice continued "especially among poor folks in rural middle Georgia—where most of the world's kaolin comes from." It featured a shameful 51-year-old nurse depicted as addicted to her white clay fix, "lethargic, isolated and chronically constipated" (Smith, 1999). A recent *Maclean's* article about poor Haitians reduced to eating mud cakes to alleviate hunger pain conveyed a similar message (Kennedy, 2008).

Although it is difficult to evaluate, the practice of geophagy is likely waning (Parry-Jones and Parry-Jones, 1992). Increased access to medical care and the medicalization of modern life allow for better diagnosis and treatment of some of the conditions associated with geophagy. Also, the growing urbanization in modern societies, almost complete in some nations, has reduced access to edible dirt sites and increased the distance to their consumers.

Geophagy has also taken new guises, especially in the proliferation of nutritional mineral supplements and additives. When packaged, advertised, and sold as scientifically balanced and doctor-recommended, the ingestion of clay components, a form of clay eating, becomes acceptable. This is certainly demonstrated by the wide use of over-the-counter drugs such as Kaopectate and multivitamins. Pills containing minerals "from A to zinc," digestive aids, high doses of calcium added to milk and orange juice, and mineral-loaded vegetables can be viewed as acceptable forms of geophagic regimens. Calcite, a mineral that is the main constituent of limestone, is found in products such as Di-Gel, Rolaids, Mylanta, and Maalox, which are used to relieve indigestion and acid stomach. Kaolinite and attapulgite are clays that can be found in many over-the-counter antidiarrheal medications.

Finally, the commercialization of geophagy is not to be neglected. While most clay eaters, especially in Africa, continue to personally avail themselves of their material, evidence from the United States and Great Britain suggests that store-bought material is becoming the norm, although there are currently efforts in Great Britain to stop the sale of geophagic materials. Immigrants and internal migrants can thus maintain their habit even when removed from their preferred sites. The existence of ethnic grocery stores in many Western cities and the growth of Internet shopping can counter if not eliminate the effect of displacement.

Arguably, such a deconstructed geophagy, soil separated into mineral components, virtually available *urbi et orbi*, may bear little relation to the habits of African pregnant women or the hungry Ottomacs (and no longer fits the definition by the ATSDR given at the beginning of this chapter). Ironically, as geophagy is now better understood and may finally be yielding its secrets, the practice also seems to be abating.

REFERENCES

Abrahams, P.W. 1999. The chemistry and mineralogy of three savanna lick soils. *Journal of Chemical Ecology* 25(10): 2215–2228.

Abrahams, P.W. and J.A. Parsons. 1996. Geophagy in the tropics: A literature review. *Geographical Journal* 162(1): 63–72.

Abrahams, P., M. Follansbee, A. Hunt, B. Smith, and J. Wragg. 2006. Iron nutrition and possible lead toxicity: An appraisal of geophagy undertaken by pregnant women of UK. *Applied Geochemistry* 21: 98–108.

Anell, B. and S. Lagercrantz. 1958. *Geophagical Customs*. Studia Ethnographica Upsaliensia 17. Humanistiska Fonden, Stockholm, Sweden.

ATSDR (Agency for Toxic Substances and Disease Registry). 2001. *Summary Report for the ATSDR Soil-Pica Workshop*, Atlanta, GA. Prepared by Eastern Research Group, Lexington, MA.

Brady, J. and D. Rissolo. 2006. A reappraisal of ancient Maya cave mining. *Journal of Anthropological Research* 62: 471–490.

Brevik, E.C. 2009. Soil, food security, and human health. *In* W. Verheye (ed.), *Soils, Plant Growth and Crop Production*. Encyclopedia of Life Support Systems (EOLSS), Developed under the Auspices of the UNESCO, EOLSS Publishers, Oxford. http://www.eolss.net (verified 7 January 2012).

Brevik, E. and A.E. Hartemink. 2010. History, philosophy, and sociology of soil science. *In* W. Verheye (ed.), *Soils, Plant Growth and Crop Production*. Encyclopedia of Life Support Systems (EOLSS), Developed under the Auspices of the UNESCO, EOLSS Publishers, Oxford. http://www.eolss.net (verified 7 January 2012).

Brody, J. 2009. Babies know: A little dirt is good for you. *The New York Times*, January 27: D7.

Callahan, G. 2003. Eating dirt. *Emerging Infectious Diseases* 9(8): 1016–1021.

Clark, D. 2010. The no-budget diners' guide. http://www.cnn.com/2010/LIVING/03/08/mf.eat.dirt.other.objects/index.html?iref=allsearch. Accessed 14 October 2010 (verified 7 January 2012).

Clark, J.D. 1970. *The Prehistory of Africa. Ancient Peoples and Places*. Praeger, Santa Barbara, CA.

Clark, J.D. 2001. *Kalambo Falls Prehistoric Site*, Vol. 3. Cambridge University Press, London.

Clark, M. 2000. Chefs' new obsession: Grains from the wild. *The New York Times*, August 30: B1.

Cooper, M. 1957. *Pica*. Charles C. Thomas Publisher, Springfield, IL.

Danford, D. 1982. Pica and nutrition. *Annual Review of Nutrition* 2: 303–322.

Debien, G. 1974. *Les esclaves aux Antilles françaises (XVIIe–XVIIIe siècles)*. Société d'histoire de la Guadeloupe/Société d'histoire de la Martinique, Fort de France, Martinique.

Dickins, D. and R. Ford. 1942. Geophagy (dirt eating) among Mississippi negro school children, *American Sociological Review* 7: 59–65.

Dissanayake, C.B. and R. Chandrajith. 2009. *Introduction to Medical Geology*. Springer, Dordrecht, The Netherlands.

Dominy, N., E. Davoust, and M. Minekus. 2004. Adaptive function of soil consumption: An *in vitro* study modeling the human stomach and small intestine. *Journal of Experimental Biology* 207: 319–324.

Douglas, M. 1966. *Purity and Danger*. Routledge, London.

Downs, J. 1966. *The Two Worlds of the Washo: An Indian Tribe of California and Nevada*. Holt, Rinehart and Winston, New York, NY.

Foucault, M. 1973. *The Birth of the Clinic: An Archeology of the Human Sciences*. Tavistock Publication, London.

Frate, D.A. 1984. Last of the earth eaters. *Sciences* 24(6): 34–38.

Geissler, P.W. 2000. The significance of earth-eating: Social and cultural aspects of geophagy among Luo children. *Africa* 70(4): 653–682.

Gibson, J. 1996. *Poverty Point: A Terminal Archaic Culture of the Lower Mississippi Valley*. Anthropological Study Series Number 7. Louisiana Archaeological Survey and Antiquities Commission, Baton Rouge, LA.

Halsted, J.A. 1968. Geophagia in man: Its nature and nutritional effects. *American Journal of Clinical Nutrition* 21: 1384–1393.

Henry, J. and A.M. Kwong. 2003. Why is geophagy treated like dirt? *Deviant Behavior* 24(4): 353–371.

Hertz, H. 1947. Notes on clay and starch eating among Negroes in a southern urban community. *Social Forces* 25: 343–344.

Higman, B.W. 1984. *Slave Population of the British Caribbean 1807–1834*. The John Hopkins University Press, Baltimore, MD.

Horner, R.D., C.J. Lackey, K. Kolasa, and K. Warren. 1991. Pica practices of pregnant women. *Journal of the American Dietetic Association* 91: 34–38.

Hooda, P. and J. Henry. 2007. Geophagia and human nutrition. *In* J. MacClancy, J. Henry, and H. MacBeth (eds), *Consuming the Inedible. Neglected Dimensions of Food Choice*, pp. 89–98. Berghahn Books, New York, NY.

Hunter, J.M. 1973. Geophagy in Africa and in the United States: A culture-nutrition hypothesis. *Geographical Review* 63: 170–195.

Hunter, J.M. 1993. Macroterm geophagy and pregnancy clays in southern Africa. *Journal of Cultural Geography* 14: 69–92.

Johns, T. 1990. *The Origins of Human Diet and Medicine*. University of Arizona Press, Tucson, AZ.

Johns, T. 1999. The chemical ecology of human ingestive behaviors. *Annual Review of Anthropology* 28: 27–50.

Johns, T. 2000. Well-grounded diet. *In* A. Goodman, D. Dufour, and G. Pelto (eds), *Nutritional Anthropology: Biocultural Perspectives on Food and Nutrition*, pp. 122–126. McGraw-Hill, San Francisco, CA.

Johns, T. and M. Duquette. 1991. Detoxification and mineral supplementation as functions of geophagy. *The American Journal of Clinical Nutrition* 53: 448–456.

Katz, S., M.L. Hediger, and L.A. Valleroy. 1974. Traditional maize processing techniques in the New World. *Science* 184: 765–773.

Kennedy, K. 2008. Dirt poor: Eating mud to survive. *Maclean's* 121(32): 30.

Kiple, K. 1984. *The Caribbean Slave: A Biological History*. Cambridge University Press, Cambridge.

Klaus, G., C. Klaus-Hügia, and B. Schmid. 1998. Geophagy by large mammals at natural licks in the rain forest of the Dzanga National Park, Central African Republic. *Journal of Tropical Ecology* 14: 829–839.

Krishnamani, R. and W.C. Mahaney. 2000. Geophagy among primates: Adaptive significance and ecological consequences. *Animal Behaviour* 59: 899–915.

Kurlansky, M. 2002. *Salt: A World History*. Walker Publishing, New York, NY.

Lacey, E.P. 1990. Broadening the perspective of pica: Literature review. *Public Health Reports* 105: 29–35.

Laufer, B. 1930. *Geophagy*. Field Museum Press, Chicago, IL.

Levi-Strauss, C. 1963. *Totemism*. Beacon Press, Boston, MA.

MacClancy, J., J. Henry, and H. MacBeth (eds). 2007. *Consuming the Inedible: Neglected Dimensions of Food Choice*. Berghahn Books, New York, NY.

Mahaney, W.C., M.W. Milner, M. Hs, R.G.V. Hancock, S. Aufreiter, M. Reich, and M. Wink. 2000. Mineral and chemical analyses of soils eaten by humans in Indonesia. *International Journal of Environmental Health Research* 10: 93–109.

Matsubayashi, H., P. Lagan, N. Majalap, J. Tangah, J.R.A. Sukor, and K. Kitayama. 2007. Importance of natural licks for the mammals in Bornean inland tropical rain forests. *Ecological Research* 22: 742–748.

Oxnard, C. 2008. *Ghostly Muscles, Wrinkled Brains, Heresies and Hobbits*. World Scientific Publishing, Hackensack, NJ.

Parry-Jones, B. and W.L. Parry-Jones. 1992. Pica: Symptom or eating disorder? A historical assessment. *British Journal of Psychiatry* 160: 341–354.

Pratt, M.L. 1992. *Imperial Eyes: Travel Writing and Transculturation*. Routledge, London.

Reid, R.M. 1992. Cultural and medical perspectives on geophagia. *Medical Anthropology* 13: 1337–1351.

Reilly C. and J. Henry. 2000. Geophagia: Why do humans consume soil? *Nutrition Bulletin* 25: 141–144.

Reinbacher, W.R. 2003. *Healing Earths: The Third Leg of Medicine*. 1st Books Library, Bloomington, IN.

Rowland, M.J. 2002. Geophagy: An assessment of implications for the development of Australian Indigenous plant processing technologies. *Australian Aboriginal Studies* 2: 51–66.

Rutherford, D. 1881. Notes on the people of Batanga. *The Journal of Anthropological Institute of Great Britain and Ireland* 10: 463–470.

Shigova, W. and W. Moturi. 2009. Geophagia as a risk factor for diarrhea. *Journal of Infection in Developing Countries* 3: 94–98.

Smith, E.N. 1999. Some middle Georgians don't like to talk about 'dirty' habit. *The Associated Press*, April 19.

Smulian, J. and S. Motiwala. 1995. Pica in a rural obstetric population. *Southern Medical Journal* 88: 1236–1240.

Stokes, T. 2006. The earth eaters. *Nature* 444(30): 543–544.

Tischler, H. 1999. *Introduction to Sociology*, 6th edn. Harcourt Press, Fort Worth, TX.

Turpin, J. 2004. Baked clay balls and earth ovens in south central Texas. *Current Archeology in Texas* 6(2): 9–14. www.thc.state.tx.us/archeology/aapdfs/CAT_Nov_04.pdf. Accessed 15 May 2011 (verified 8 January 2012).

U.S. Department of Health and Human Services. 2009. *Evaluation of Arsenic and Mercury in Soil along BIA 120. Elem Indian Colony, Sulphur Bank Mercury Mine, Clearlake Oaks, California.* U.S. Department of Health and Human Services Agency for Toxic Substances and Disease Registry, Division of Health Assessment and Consultation, Atlanta, GA. http://www.atsdr.cdc.gov/hac/pha/ElemIndianColony/ElemIndianColonyHC06-09-2009.pdf (verified 8 January 2012).

Vermeer, D. 1966. Geophagy among the Tiv. *Annals of the Association of American Geographers* 56: 197–204.

Vermeer, D. 1971. Geophagy among the Ewe of Ghana. *Ethnology* 10: 56–72.

Vermeer, D.E. and D.A. Frate. 1975. Geophagy in a Mississippi County. *Annals of the Association of American Geographers* 65(3): 414–424.

Vermeer, D. and D.A. Frate. 1979. Geophagia in rural Mississippi: Environmental and cultural context and nutritional implications. *American Journal of Clinical Nutrition* 32: 2129–2135.

Von Humboldt, A. 1872. *Personal Narrative of Travels to the Equinoctial Regions of America during the Years 1799–1804*, Vol. II. Belle and Daldy, London.

Warshall, P. 1999. Eating Earth. Whole Earth. http://www.wholeearth.com/issue/2096/article/75/eating.earth (verified 8 January 2012).

Wiley, A.S. and S.H. Katz. 1998. Geophagy in pregnancy: A test of a hypothesis. *Current Anthropology* 39(4): 532–545.

Wilson, M.J. 2003. Clay mineralogical and related characteristics of geophagic material. *Journal of Chemical Ecology* 29(7): 1525–1547.

Young, S. 2007. Evidence for the consumption of the inedible. Who, What, When, Where and Why? *In* J. MacClancy, J. Henry, and H. MacBeth (eds), *Consuming the Inedible. Neglected Dimensions of Food Choice*, pp. 17–30. Berghahn Books, New York, NY.

Young, S. 2010. Pica in pregnancy: New ideas about an old condition. *Annual Review of Nutrition* 30: 403–422.

Young, S. 2011. *Craving Earth: Understanding Pica*. Columbia University Press, New York, NY.

Young, S., M.J. Wilson, D. Miller, and S. Hillier. 2008. Toward a comprehensive approach to the collection and analysis of pica substances, with emphasis on geophagic material. *PloS One* 3(9): e3147. doi:10.1371/journal.pone.0003147. http://www.plosone.org/article/info%3Adoi%2F10.1371%2Fjournal.pone.0003147 (verified 8 January 2012).

Young, S., M.J. Wilson, S. Hillier, E. Delbos, S.M. Ali, and R.J. Stoltzfus. 2010. Differences and commonalities in physical, chemical and mineralogical properties of Zanzibari geophagic soils. *Journal of Chemical Ecology* 36: 129–140.

9 Soil Minerals, Organisms, and Human Health
Medicinal Uses of Soils and Soil Materials

Monday Mbila

CONTENTS

9.1 INTRODUCTION

Soil is a complex assemblage of primary and secondary minerals as well as nonmineral organic matter. The nature, composition, and interaction of the minerals and nonmineral components determine the behavior and use of soils. Throughout the early stages of the development of soil science as a discipline, the major focus was on the use of soils to grow crops (Brevik and Hartemink, 2010). However, some important soil-related advances of the twentieth century brought the realization that soil serves many functions, such as regulating water supplies, recycling raw materials, providing a habitat for soil organisms, and serving as a foundation material for engineering (Brevik and Hartemink, 2010; Brevik, 2009a; Brady and Weil, 2008), and can also be a source of compounds that can be used to combat human diseases and as health supplements (Brevik, 2009b; Deckers and Steinnes, 2004; Abrahams, 2002). During the same time, many studies showed that some human illnesses could be attributed to soil constituents such as aluminum, arsenic, cadmium, copper, fluorine, iodine, iron, lead, selenium, thallium, and zinc being either deficient or present in toxic amounts in food or water (Brevik, 2009b; Hough, 2007; Deckers and Steinnes, 2004). Knowing that these elements originate from soils, where they are bound to minerals in different forms, relating human health and nutrition problems to soil conditions gained some prominence in soil science and human health research (Hough, 2007; Abrahams, 2005; Fuge, 2005; Henry and Kwong, 2003; Hunter, 1973) and has become a subject of major interest in soil mineralogy (Velazquez et al., 2010; Sing and Sing, 2010; Haydel et al., 2008; Tong et al., 2005).

The increased interest in studying the uses of soil for purposes other than food and fiber production came at a time of immense interest in understanding the conscious ingestion of clay by humans, a practice referred to as geophagy. Many cultures ingest soils as part of their traditional medicine, for dietary supplements, and as a general pastime (Henry and Cring, Chapter 8, this volume; Abrahams, 2005; Abrahams and Parsons, 1996; Vermeer, 1985; Hunter and de Kleine, 1984). The objectives of this chapter are to (a) describe the origin, nature, and composition of soil clay minerals in relation to human health; (b) discuss the chemical and mineralogical properties of soils that make them suitable for nutritional, medicinal, and other uses; (c) identify medicines from soils; and (d) discuss soil-related human health issues and diseases.

9.2 ORIGIN, COMPOSITION, AND STRUCTURE OF SOIL CLAYS

9.2.1 SOIL CLAYS

Different soils have different types of clays that, when combined with organic matter in soils, determine the behavior of those soils. It has long been the view of soil scientists that these soil clays, with their small sizes, thin flat shapes, high surface areas, and electrically charged surfaces, allow the soil to serve as nature's chemical reactor (Brady and Weil, 2008). To gain a better understanding of the capacity of soil to supply some of the ingredients that serve as natural medicines or nutrient supplements needed for human health, a brief and general description of the origin, structure, and composition of soil clays is in order. For a comprehensive discussion of the subject, refer to Shultze (1989).

9.2.2 FORMATION, TYPES, AND STRUCTURE OF SOIL CLAYS

Clay minerals are crystalline materials that have formed from the partial or complete dissolution and precipitation or recombination of the original (primary) minerals, such as feldspars, micas, or hornblende, that were derived from the parent rock. Due to their formation from the alteration of primary minerals, clay minerals are often referred to as secondary minerals.

The two main types of clay minerals in nature are the 1:1 and the 2:1 type minerals. A third, but less common and less understood, type is the 2:1:1 mineral. These different types of clay minerals in soils have different properties and behaviors that arise from their different structures, but nearly all the clays contain two basic components (building blocks) that are made up of the elements silicon (Si), aluminum (Al), and oxygen (O). These two basic building blocks of clay minerals are the silica tetrahedron (Figure 9.1) and the aluminum octahedron (Figure 9.2). The structures of the clay

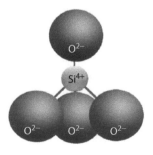

FIGURE 9.1 Silicon tetrahedron structure showing a silicon ion in coordination with four oxygen ions to form a tetrahedral structure. (Redrawn from Harter, R., Building the phyllosilicates, 1998. http://pubpages.unh.edu/~harter/crystal.htm, verified 3 March 2012.)

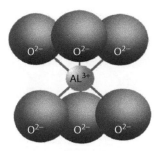

FIGURE 9.2 Aluminum octahedral structure showing aluminum in coordination with six oxygen ions. (Redrawn from Harter, R., Building the phyllosilicates, 1998. http://pubpages.unh.edu/~harter/crystal.htm, verified 3 March 2012.)

mineral building blocks will be briefly described to show the occurrence of important elements within the structure of the soil minerals.

In a silica tetrahedron structure, a single silicon ion is surrounded by four oxygen ions to form a four-sided structural unit, the orthosilicate anion (Figure 9.1). In the tetrahedron structure, the silicon ion shares its charge equally between the four oxygen ions, leaving each oxygen ion with an excess negative charge. In the orthosilicate anion, the negative charge is neutralized when the oxygen ions with excess negative charges bond with other Si ions. For electron coordination chemistry purposes, sharing of the corner oxygen ions is the most likely arrangement to neutralize the charge. The bonding of the corner oxygen ions with two silicon ions results in one plane of oxygen ions bonded to two silicon ions that extend in two directions to form a sheet of silicon tetrahedrons (Figure 9.3).

For the aluminum octahedral building block unit, aluminum coordinates with six oxygen ions in an octahedral coordination (Figure 9.2). The result of this coordination is that each aluminum shares +0.5 of its charge with each of the surrounding oxygen ions, leaving each oxygen ion with a negative 1.5 charge to be neutralized by bonding to other aluminum ions. With each oxygen bonded to two aluminum ions, each oxygen ion is left with a −1 charge that is then neutralized by bonding with a hydrogen cation (H^+) to create a hydroxide anion (OH^-). In clay minerals, the remaining excess negative charge (−1) of the oxygen ions is neutralized by the apical oxygen ions of the tetrahedral layer that are in the octahedral layer through bonding to one silicon ion and two aluminum ions. The result of this arrangement is the basic structure of kaolins, a 1:1 mineral group that has one sheet of silica tetrahedrons combined with one sheet of aluminum octahedrons (Figure 9.4).

Other groups of clay minerals result from different arrangements of the aluminum octahedral and silica tetrahedral sheets. When the bonding of the sheets is such that an octahedral sheet

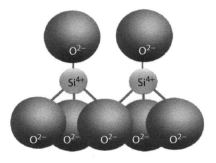

FIGURE 9.3 Silicon tetrahedron sheet showing one plane of oxygen ions bonded to two silicon ions in two directions to form a sheet of silicon tetrahedrons with unbalanced charges on the apical O ions. (Redrawn from Harter, R., Building the phyllosilicates, 1998. http://pubpages.unh.edu/~harter/crystal.htm, verified 3 March 2012.)

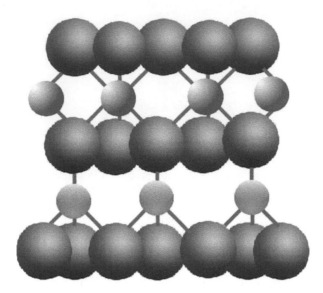

FIGURE 9.4   The structure of kaolinite (~0.7 nm thick from the bottom oxygen to the top oxygen) showing apical oxygen ions of a silicon tetrahedral sheet bonded with the octahedral sheet to form a 1:1 layer mineral. (Redrawn from Harter, R., Building the phyllosilicates, 1998. http://pubpages.unh.edu/~harter/crystal.htm, verified 3 March 2012.)

coordinated by Al or Mg is bonded to tetrahedral sheets coordinated by silica on both sides of the octahedral sheet, the basic structure of the 2:1 mineral group results (Figure 9.5). Minerals in the 2:1 group include the micas and clays such as illite, smectite, and vermiculite. In some cases, a third group of minerals based on the arrangements of the building sheets arises due to the bonding of an additional Al octahedral sheet to a 2:1 mineral structure, resulting in an extra aluminum octahedral layer forming between and binding the 2:1 layers. Minerals having this arrangement are known as the chlorite group.

During soil weathering, some of the original (primary) minerals partially dissolve and recrystallize to become secondary minerals. In that process, some ions tend to substitute for other ions of similar size and charge in the tetrahedral or octahedral structure without changing the shape of the structures in a process termed isomorphous substitution. If the two ions do not have the same charge, then an unsatisfied net charge remains at that point, leaving a net negative charge on the crystal. Common substitutions are aluminum cations (Al^{3+}) for silicon cations (Si^{4+}), magnesium cations (Mg^{2+}) for aluminum cations (Al^{3+}), and iron cations (Fe^{2+}) for aluminum cations (Al^{3+}).

9.2.3   Elemental Composition of Soil Minerals

Almost all soil clay minerals are phyllosilicates that utilize the two building blocks described earlier to form sheet-like components that are stacked in different arrangements to form layers of clays (Figures 9.4 and 9.5). The degree of isomorphous substitution within the building blocks of the clays causes a certain negative electrical charge to develop at different sites within the layers. Such negative charge sites must be balanced by attracting positive ions such as calcium (Ca^{2+}), potassium (K^+), sodium (Na^+), and magnesium (Mg^{2+}) from the soil solution. Some of these attracted cations play roles in nutrition and medicine. The concentrations of the cations and their availability depend to a large extent on the location and magnitude of the negative charge within the clay mineral structure (Dahlgren et al., 2004).

Thus, elements such as oxygen (O), silicon (Si), aluminum (Al), calcium (Ca), magnesium (Mg), potassium (K), sodium (Na), and others are naturally present as a part of and/or in association with

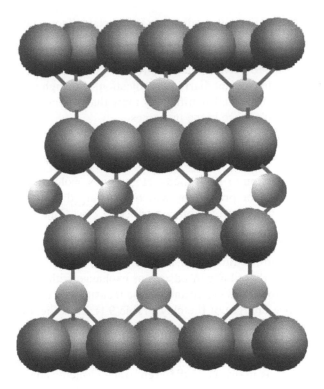

FIGURE 9.5 The basic structure of 2:1 clay minerals showing two silicon tetrahedral layers on top and bottom and one aluminum octahedral layer in the middle of the structure. (Redrawn from Harter, R., Building the phyllosilicates, 1998. http://pubpages.unh.edu/~harter/crystal.htm, verified 3 March 2012.)

the clay minerals. Oxygen, Si, and Al occur mainly as structural constituents of minerals and as oxides. Iron (Fe) occurs mainly in the form of oxides and ferromagnesian minerals. Calcium occurs mainly in primary minerals such as calcite, gypsum, apatite, and dolomite. Magnesium is present mainly in primary minerals such as dolomite and hornblende. Potassium occurs mainly in microcline (a feldspar) and mica minerals. Micronutrients like molybdenum (Mo), manganese (Mn), zinc (Zn), copper (Cu), and boron (B) occur in inorganic forms in soils. In general, most soil clay minerals contain high levels of Al, Fe, Ca, Cu, and Mg. Iron, Ca, Cu, K, Zn, and Mg are among the essential minerals for the human diet (Combs, 2005) and are commonly used in health supplements. Sodium, Si, Mo, and Mn are also all essential minerals in the human diet (Combs, 2005).

9.3 MEDICINAL PROPERTIES OF CLAYS

9.3.1 ABSORPTION OF WATER BY CLAY

Phyllosilicate clays are by definition very small in size, being less than 2 μm in diameter. Due to their small sizes, their specific surfaces (surface area per unit mass) are far larger than those of silt and sand. For instance, mineralogists have discovered that a single gram of phyllosilicate clay has a surface area of up to about 800 m^2, as compared with only about 0.009 m^2/g for fine sand (Brevik, Chapter 1, this volume). The high specific surface of clay is the reason for its capacity for absorption, adsorption, and the development of surface charge, all of which have enhanced the medicinal properties of clays.

Absorption by clays is the physical capacity of the clay structural lattice to be filled with liquids or gases, like a sponge. Some clays, especially the smectites, can hold so much water from their environment that swelling of the clays occurs. This water-absorbing characteristic of clays is the basis for their use in the antidiarrhea medication Kaopectate. Kaopectate originally contained purified kaolin (Allport, 2002; Tobey and Covington, 1975), which is a type of clay that is capable of absorbing large quantities of water without much swelling. Thus, when kaolin in Kaopectate got into the intestines, it absorbed excess water and relieved diarrhea. Smectites and other swelling clays are not great candidates for antidiarrhea medication. Although they have the capacity to absorb large amounts of water, they swell significantly more than kaolin and could thus cause other medical problems related with blockage in the intestines.

9.3.2 Adsorption of Elements by Clay

Adsorption describes the adhesion of substances (atoms, ions, molecules of gas, liquid, or dissolved solids) to a surface. It is the process by which the charged particles of other substances combine with the charged particles on the outer surface of clays. Due to the fact that soil clay particles are negatively charged, they have the capacity to attract and hold onto, or adsorb, cations, and at times exchange the adsorbed cations with other positively charged ions such as potassium, calcium, magnesium, sodium, iron, and other cations that can be of medicinal importance. The number of cations that clay can adsorb or exchange with its surroundings depends upon the clay structure described previously and can be measured by the cation exchange capacity (CEC, amount of negative charge) of the clay (Table 9.1). Thus the high adsorptive and ion exchange capacity of clays are due to the clays' uniquely large surface areas compared with other components of soil such as sand and silt. Those capacities explain why clays contain high amounts of elements necessary for medicinal purposes. For example, montmorillonite clays (a type of smectite) can have as many as 75 different trace elements adsorbed to their surfaces (U.S. Patent 6962718). Clays holding adsorbed substances with the right elemental content, when consumed, as is common in some cultures (geophagy), may release nutrients to the body (Henry and Cring, Chapter 8, this volume; Abrahams, 2005).

Additionally, clay adsorption capacity has been shown to have direct application in medicines. Studies at Texas A&M University and other places have shown that the adsorption of smectite clays to aflatoxins, which are naturally occurring, toxic, carcinogenic mycotoxins that are produced by many species of *Aspergillus* in both animals and humans, rendered the mycotoxin less toxic (Velazquez et al., 2010; Dixon et al., 2009). The addition of bentonite clay to an aflatoxin-contaminated diet diminishes most of the deleterious effects of the aflatoxin due to its strong adsorption capacity for the toxin (Abdel-Wahhab et al., 1998).

TABLE 9.1
Select Properties Important to the Medicinal Properties of Clay

| | Colloid | | | | | |
| | 2:1 Clays | | | | 1:1 Clay | Organic |
Property	Smectite	Vermiculite	Illite	Chlorite	Kaolinite	Humus
External surface (m²/g)	80–140	70–120	70–100	70–100	10–30	20–800
Internal surface (m²/g)	570–660	600–700	None	None	None	None
Total surface (m²/g)	650–800	670–820	70–100	70–100	10–30	20–800
Net negative charge (cmol/kg)	80–120	100–180	15–40	15–40	2–5	100–550
Swelling	Yes	Partial	No	No	No	No

Source: Data from Brady, N.C. and Weil, R.R., *The Nature and Properties of Soils*, 14th edn, Pearson Prentice Hall, Upper Saddle River, NJ, 2008.

MDR and extensively-drug-resistant (XDR) strains of *Mycobacterium tuberculosis* (MDR-TB and XDR-TB) have been identified as major emerging threats to public health (Fischbach and Walsh, 2009). But as antibiotic resistance spreads to different types of bacteria, more attention will be focused on soil by scientists who are already at work trying to find the microbe that will provide the next remedy to antibiotic-resistant organisms.

9.4.2 OTHER MEDICINES FROM SOIL

Kaolinite has a wide range of medical uses and may be the most used soil material for medical purposes other than antibiotics. It is an adsorbent that has been widely used to treat diarrhea. The absorptive capacity of kaolinite is utilized in medicine to adsorb some of the bacterial toxins that often cause diarrhea. Kaolinite was an ingredient in Kaopectate in the United States up until the 1990s (Allport, 2002; Tobey and Covington, 1975). The original kaolin–pectin formula for Kaopectate was then replaced by attapulgite, another soil mineral, before both were phased out in the United States in the early 2000s (Brevik, 2009b). However, in many other countries around the world kaolin products are still used to treat diarrhea and kaolin tablets can still be purchased for medicinal use (Brevik, 2009b; Abrahams, 2005). Other medical uses of kaolin include treatment of diaper rash and as an emollient and drying agent in treating poison ivy, poison oak, and poison sumac cases (AJN, 1989). In these cases the absorbent properties of kaolinite are used to dry the area with the rash. Kaolin is also used in some toothpaste formulas (EPA, 1999), where it plays a role in dental health.

The 2:1 clays also find use as medicines, although their use is not nearly as extensive as kaolinite due to the possibility that they could swell within the digestive tract, leading to other problems. Fuller's earth and bentonite both contain the clay mineral montmorillonite, which is used as an antipoison drug (Abrahams, 2005; Clark, 1971). In fact, one of the prevailing theories concerning the practice of geophagy is that the consumed soil acts as a food detoxifier (Henry and Cring, Chapter 8, this volume; Young, 2011; Abrahams, 2005). There are other therapeutic and cosmetic uses of soil clay and nonclay minerals that are orally or topically administered to patients (Table 9.2; Carretero and Pozo, 2010).

9.5 SOIL-RELATED DISEASES

Soils are not just crucial to human health through their impact on human nutrition and as a source of medicines, but are also potentially a source of human pathogens, which can result in many diseases (Abrahams, 2006). Soil is a complex habitat that supports a diversity of organisms and is home to, by some estimates, about 25% of Earth's species (Jeffery and van der Putten, 2011). For instance, one teaspoon of soil contains several thousand microbial species, several hundred meters of fungal hyphae, and more than 1 million individuals (Wardle et al., 2004; Schaefer and Schauermann, 1990). The overwhelming majority of soil organisms are involved in ecosystem processes that make l possible, but some of the organisms are capable of causing diseases in humans and animals. J and van der Putten (2011) classified infectious organisms capable of causing disease into fi phylogenetic groups: viruses, bacteria, protozoa, fungi, and helminths (nematodes); an diseases into two groups: euedaphic pathogenic organisms (EPOs) and soil-transmitt (STPs) (Table 9.3). EPOs are potential pathogens that are true soil organisms, tat is the soil. STPs are organisms that, while they may be able to survive i of time, are not true soil organisms, but rather are obligate pathogens th to complete their life cycles (Jeffery and van der Putten, 2011).

The list of soilborne diseases in Table 9.3 is by no means exhausti (2009b), and Loynachan (Chapter 5, this volume) have reported on a are associated with the soil. Other soil-related diseases include a (Finegold, 1976), goiters from iodine deficiency (WHO, 2007), le itai from cadmium toxicity (Nordberg and Cherian, 2005; Abrah

TABLE 9.2

Soil Minerals and Their Therapeutic and Cosmetic Actions When Orally or Topically Administered to Patients

Mineral	Chemical Formula	Therapeutic Activity or Cosmetic Action
Rutile	TiO_2	Dermatological protector, solar protector
Periclase	MgO	Antacid, osmotic oral laxative, mineral supplement
Zincite	ZnO	Antiseptic and disinfectant, dermatological protector, solar protector
Carbonates		
Calcite	$CaCO_3$	Antacid, antidiarrheal, mineral supplement, abrasive, and polishing agent in toothpaste
Magnesite	$MgCO_3$	Antacid, osmotic oral laxative, mineral supplement
Hydrozincite	$Zn_5(CO_3)_2(OH)_6$	Dermatological protector
Smithsonite	$ZnCO_3$	Dermatological protector
Sulphates		
Epsomite	$MgSO_4 \times 7H_2O$	Osmotic oral laxative, mineral supplement, bathroom salts
Mirabilite	$Na_2SO_4 \times 10H_2O$	Osmotic oral laxative, bathroom salts
Melanterite	$FeSO_4 \times 7H_2O$	Antianemic, mineral supplement
Chalcanthite	$CuSO_4 \times 5H_2O$	Direct emetic, antiseptic, and disinfectant
Zincosite	$ZnSO_4$	Direct emetic, antiseptic, and disinfectant
Goslarite	$ZnSO_4 \times 7H_2O$	Direct emetic, antiseptic, and disinfectant
Alum	$KAl(SO_4)_2 \times 12H_2O$	Antiseptic and disinfectant, deodorant
Chlorides		
Halite	$NaCl$	Homeostatic, mineral supplement, decongestive eye drops, bathroom salts
Sylvite	KCl	Homeostatic, mineral supplement, bathroom salts
Hydroxides		
Brucite	$Mg(OH)_2$	Antacid, osmotic oral laxative, mineral supplement
Gibbsite	$Al(OH)_3$	Antacid, gastrointestinal protector, antidiarrheal
Hydrotalcite	$Mg_6Al_2(CO_3)(OH)_{16} \times 4H_2O$	Antacid
Others		
Sulfur	S	Antiseptic and disinfectant, keratolytic reducer
Greenockite	CdS	Keratolytic reducer
Borax	$Na_2B_4O_7 \times 10H_2O$	Antiseptic and disinfectant
Hydroxyapatite	$Ca_5(PO_4)_3(OH)$	Mineral supplement
Niter	KNO_3	Anesthetizer in toothpastes
Phyllosilicates		
Smectites	Montmorillonite: $(Al_{1.67}Mg_{0.33})Si_4O_{10}(OH)_2M_{0.33}^+$ Saponite: $Mg_3(Si_{3.67}Al_{0.33})O_{10}(OH)_2M_{0.33}^+$ Hectorite: $(Mg_{2.67}Li_{0.33})Si_4O_{10}(OH)_2M_{0.33}^+$	Antacid, gastrointestinal protector, antidiarrheal, dermatological protector, cosmetic creams, powders, and emulsions
Palygorskite	$(Mg, Al, Fe^{3+})_5(Si,Al)_8O_{20}(OH)_2(OH_2)_4 \times 4H_2O$	Antacid, gastrointestinal protector, antidiarrheal, cosmetic creams, powders, and emulsions
Sepiolite	$Mg_8Si_{12}O_{30}(OH)_4(OH_2)_4 \times 8H_2O$	Antacid, gastrointestinal protector, antidiarrheal, cosmetic creams, powders, and emulsions
Kaolinite	$Al_2Si_2O_5(OH)_4$	Gastrointestinal protector, antidiarrheal, dermatological protector, anti-inflammatory and local anesthetic, cosmetic creams, powders, and emulsions
Talc	$Mg_3Si_4O_{10}(OH)_2$	Dermatological protector, cosmetic creams, powders, and emulsions
Mica	Muscovite: $KAl_2(Si_3Al)O_{10}(OH)_2$	Cosmetic creams, powders, and emulsions

Source: Adapted from Carretero, M.I. and Pozo, M., *Applied Clay Science,* 47, 171–181, 2010.

TABLE 9.3

Soilborne Infectious Diseases and Their Causative Agents (italics)

Euedaphic Pathogenic Organisms	Soil-Transmitted Pathogens
Actinomycetoma: e.g., *Actinomyces israelii*	Poliovirus
Anthrax: *Bacillus anthracis*	Hantavirus
Botulism: *Clostridium botulinium*	Q Fever: *Coxiella burnetii*
Campylobacteriosis: e.g., *Campylobacter jejuni*	Lyme disease: *Borrelia* spp.
Leptospirosis: e.g., *Leptospira interrogans*	Ascariasis: *Ascaris lumbricoides*
Listeriosis: *Listeria monocytogenes*	Hookworm: e.g., *Ancylostoma duodenale*
Tetanus: *Clostridium tetani*	Enterobiasis (pinworm)
Tularemia: *Francisella tularensis*	Strongyloidiasis: e.g., *Strongyloides stercoralis*
Gas gangrene: *Clostridium perferingens*	Trichuriasis (whipworm): *Trichuris trichiura*
Yersiniosis: *Yersinia enterocolitica*	Echinococcosis: e.g., *Echinococcus multicularis*
Aspergillosis: *Aspergillus* spp.	Trichinellosis: *Trichinella spiralis*
Blastomycosis: e.g., *Blastomyces dermatitidis*	Amoebiasis: *Entamoeba histolytica*
Coccidioidomycosis: e.g., *Coccidiodes immitis*	Balantidiasis: *Balantidium coli*
Histoplasmosis: *Histoplasma capsulatum*	Cryptosporidiosis: e.g., *Cryptosporidium parvum*
Sporotrichosis: *Sporothrix schenckii*	Cyclosporiasis: *Cyclospora cayetanensis*
Mucormycosis: e.g., *Rhizopus* spp.	Giardiasis: *Giardia lamblia*
Mycetoma: e.g., *Nocardia* spp.	Isosporiasis: *Isospora belli*
Strongyloidiasis: e.g., *Strongyloides stercoralis*	Toxoplasmosis: *Toxoplasma gondii*
	Shigellosis: e.g., *Shigella dyseneriae*
	Pseudomonas aeruginosa
	Escherichia coli
	Salmonellosis: e.g., *Salmonella enterica*

Source: Adapted from Jeffery, S. and van der Putten, W.H., *Soil Borne Human Diseases*, European Commission Joint Research Centre, Institute for Environment and Sustainability, EUR 24893 EN, 2011.

disease from selenium deficiency in soils (VitaminsDiary.com, 2011; Beck et al., 2003; Ellis and Salt, 2003), and other diseases. While the literature is somewhat scanty on the various soilborne diseases that can affect humans, and the basis for considering each disease to be soilborne, good overviews of the subject can be found in Jeffery and van der Putten (2011) and Loynachan (Chapter 5, this volume).

9.6   CONCLUSIONS

Soils perform unique functions that are basic to life. Soils can also be a source of compounds that can be used as health supplements and to combat human diseases. Exploring the use of soil materials to alleviate problems has become a major part of soil mineralogy and soil microbiology.

To gain a better understanding of the capacity of soil to supply some of the ingredients that serve as natural medicines or nutrient supplements needed for human health, a brief and general description of the origin, structure, and composition of soil clays is needed. The different types of clay minerals in soils have different properties and behaviors that arise from their different structures (containing silicon-tetrahedral and aluminum-octahedral sheets), which are made up of the elements silicon (Si), aluminum (Al), and oxygen. Such arrangements result in the development of clay structural charge for the 1:1 type minerals, 2:1 type minerals, and 2:1:1 type minerals that must be balanced by attracting positive ions such as calcium (Ca^{2+}), potassium (K^+), sodium (Na^+), and magnesium (Mg^{2+}), some of which play roles in nutrition and medicine. The concentrations of the cations and their availability depend to a large extent on the location and magnitude of the negative charge within the mineral structure.

Other properties of clays that make them suitable for medicinal purposes are their absorption of water, adsorption of ions, CEC and availability of cations, antimicrobial properties, and trace element content. Due to these properties of soil, soil is the major source of antibiotics and other important medicines that enhance human health.

Soils as complex habitats are not just crucial to human health through their impact on human nutrition and as a source of medicines, but are also potentially a source of human pathogens, which can result in many diseases. Also, since soil bacteria are constantly exposed to antibiotics produced by other soil microorganisms, they have the tendency to develop resistance to antibiotics. But as antibiotic resistance spreads to different types of bacteria, more attention will be focused on soil by scientists who are already at work trying to find the microbe that will provide the next remedy to antibiotic-resistant organisms and save more lives.

REFERENCES

Abdel-Wahhab, M.A., S.A. Nada, I.M. Farag, N.F. Abbas, and H.A. Amra. 1998. Potential protective effect of HSCAS and bentonite against dietary aflatoxicosis in rat: With special reference to chromosomal aberrations. *Natural Toxins* 6: 211–218.

Abrahams, P.W. 2002. Soils: Their implications to human health. *Science of the Total Environment* 291: 1–32.

Abrahams, P.W. 2005. Geophagy and the involuntary ingestion of soil. *In* O. Selinus, B. Alloway, J.A. Centeno, R.B. Finkelman, R. Fuge, U. Lindh, and P. Smedley (eds), *Essentials of Medical Geology*, pp. 435–458. Elsevier, Amsterdam, The Netherlands.

Abrahams, P.W. 2006. Soil, geography and human disease: A critical review of the importance of medical cartography. *Progress in Physical Geography* 30: 490–512.

Abrahams, P.W. and J.A. Parsons. 1996. Geophagy in the tropics: A literature review. *The Geographical Journal* 162: 63–72.

Adriano, D.C. 2001. *Trace Elements in Terrestrial Environments: Biogeochemistry, Bioavailability, and Risks of Metals*. Springer-Verlag, New York, NY.

AJN. 1989. Nurses' drug alert. *American Journal of Nursing* 89(8): 1063–1070.

Allen, L., B. de Benoist, O. Dary, and R. Hurrell. 2006. *Guidelines on Food Fortification with Micronutrients*. World Health Organization, Geneva, Switzerland.

Alloway, B.J. 2005. Bioavailability of elements in soil. *In* O. Selinus, B. Alloway, J.A. Centeno, R.B. Finkelman, R. Fuge, U. Lindh, and P. Smedley (eds), *Essentials of Medical Geology*, pp. 347–372. Elsevier, Amsterdam, The Netherlands.

Allport, S. 2002. Women who eat dirt. *Gastronomica: The Journal of Food and Culture* 2(2): 28–37.

ATSDR. 1988. *The Nature and Extent of Lead Poisoning in Children in the United States: A Report to Congress*. Centers for Disease Control and Prevention, Agency for Toxic Substances and Disease Registry, Atlanta, GA.

Aubert, H. and M. Pinta. 1980. *Trace Elements in Soils*. Elsevier, Amsterdam, The Netherlands.

Beck, M.A., O. Levander, and J. Handy. 2003. Selenium deficiency and viral infection. *Journal of Nutrition* 133: 1463–1467.

Berdy, J. 1974. Recent developments of antibiotic research and classification of antibiotics according to chemical structure. *Advances in Applied Microbiology* 18: 309–406.

Brady, N.C. and R.R. Weil. 2008. *The Nature and Properties of Soils*, 14th edn. Pearson Prentice Hall, Upper Saddle River, NJ.

Brevik, E.C. 2009a. Soil health and productivity. *In* W. Verheye (ed.), *Soils, Plant Growth and Crop Production*. Encyclopedia of Life Support Systems (EOLSS), Developed under the Auspices of the UNESCO, EOLSS Publishers, Oxford. http://www.eolss.net (verified 3 March 2012).

Brevik, E.C. 2009b. Soil, food security, and human health. *In* W. Verheye (ed.), *Soils, Plant Growth and Crop Production*. Encyclopedia of Life Support Systems (EOLSS), Developed under the Auspices of the UNESCO, EOLSS Publishers, Oxford. http://www.eolss.net (verified 3 March 2012).

Brevik, E.C. and A.E. Hartemink. 2010. Early soil knowledge and the birth and development of soil science. *Catena* 83: 23–33.

Campbell, D.M. 1988. Trace element needs in human pregnancy. *Proceedings of the Nutrition Society* 47: 45–53.

Carretero, M.I. and M. Pozo. 2010. Clay and non-clay minerals in the pharmaceutical and cosmetic industries Part II: Active ingredients. *Applied Clay Science* 47(3–4): 171–181.

Chan, S., B. Gerson, and S. Subramaniam. 1998. The role of copper, molybdenum, selenium, and zinc in nutrition and health. *Clinics in Laboratory Medicine* 18: 673–685.

Clardy, J., M.A. Fischbach, and C.T. Walsh. 2006. New antibiotics from bacterial natural products. *Nature Biotechnology* 24: 1541–1550.

Clark, D.G. 1971. Inhibition of the absorption of paraquat from the gastrointestinal tract by adsorbents. *British Journal of Industrial Medicine* 28: 186–188.

Combs Jr, G.F. 2005. Geological impacts on nutrition. *In* O. Selinus, B. Alloway, J.A. Centeno, R.B. Finkelman, R. Fuge, U. Lindh, and P. Smedley (eds), *Essentials of Medical Geology*, pp. 161–177. Elsevier, Amsterdam, The Netherlands.

Dahlgren, R.A., M. Saigusa, and F.C. Ugolini. 2004. The nature, properties and management of volcanic soils. *Advances in Agronomy* 82: 113–182.

Dantas, G., M.O.A. Sommer, R.D. Oluwasegun, and G.M. Church. 2008. Bacteria subsisting on antibiotics. *Science* 320(5872): 100–103.

Deckers, J. and E. Steinnes. 2004. State of the art on soil-related geo-medical issues in the world. *In* D.L. Sparks (ed.), *Advances in Agronomy*, Vol. 84, pp. 1–35. Elsevier, Amsterdam, The Netherlands.

Dixon, J.B., A.L.B. Velazquez, and Y. Deng. 2009. Smectite clay adsorption of aflatoxin illustrates applied clay science to the public. *In XIV International Clay Conference*, Castellaneta Marina, Italy, 14–20 June 2009, p. 552.

Dreosit, I.E. 1978. Trace elements in human nutrition with special reference to zinc. *Proceedings of the Nutrition Society of Australia* 3: 25–31.

Ellis, D.R. and D.E. Salt. 2003. Plants, selenium and human health. *Current Opinion in Plant Biology* 6: 273–279.

EPA. 1999. *Kaolin (100104) Fact Sheet.* Environmental Protection Agency, Washington, DC.

Falkinham, J., T. Wall, J. Tanner, K. Tawaha, F. Alali, C. Li, and N. Oberlies, 2009. Proliferation of antibiotic-producing bacteria and concomitant antibiotic production as the basis for the antibiotic activity of Jordan's Red Soils. *Applied and Environmental Microbiology* 75: 2735–2741.

Finegold, I. 1976. Allergic reactions to molds. *Cutis* 17: 1080–1084.

Fischbach, M.A. and C.T. Walsh. 2009. Antibiotics for emerging pathogens. *Science* 325(5944): 1089–1093.

Franklin, R.B. and A.L. Mills. 2009. Importance of spatially structured environmental heterogeneity in controlling microbial community composition at small spatial scales in an agricultural field. *Soil Biology and Biochemistry* 41: 1833–1840.

Fuge, R. 2005. Soils and iodine deficiency. *In* O. Selinus, B. Alloway, J.A. Centeno, R.B. Finkelman, R. Fuge, U. Lindh, and P. Smedley (eds), *Essentials of Medical Geology*, pp. 417–433. Elsevier, Amsterdam, The Netherlands.

Geissler, P.W., C.E. Shulman, R.J. Prince, C. Mutemi, C. Mnazi, H. Friis, and B. Lowe. 1988. Geophagy, iron status and anaemia among pregnant women on the coast of Kenya. *Transactions of the Royal Society of Tropical Medicine and Hygiene* 92: 549–553.

Gibson, R.S. 1991. Trace element deficiencies in humans. *Canadian Medical Association Journal* 145(3): 231.

Graham, T.W. 1991. Trace element deficiencies in cattle. *Veterinary Clinics of North America: Food Animal Practice* 7: 153–215.

Harter, R. 1998. Building the phyllosilicates. http://pubpages.unh.edu/~harter/crystal.htm (verified 3 March 2012).

Haydel, S., C. Remenih, and L. Willams. 2008. Broad-spectrum *in vitro* antibacterial activities of clay minerals against antibiotic-susceptible and antibiotic-resistant bacterial pathogens. *Journal of Antimicrobial Chemotherapy* 61: 353–361.

Henry, J. and A.M. Kwong. 2003. Why is geophagy treated like dirt? *Deviant Behavior* 24(4): 353–371.

Hooda, P. 2010. *Trace Elements in Soils.* Blackwell, Oxford.

Hough, R.L. 2007. Soil and human health: An epidemiological review. *European Journal of Soil Science* 58: 1200–1212.

Hunter, J.M. 1973. Geophagy in Africa and in the United States. *Geographical Review* 63: 170–195.

Hunter, J.M. and R. de Kleine. 1984. Geophagy in Central America. *Geographical Review* 74: 157–169.

Jeffery, S. and W.H. van der Putten. 2011. *Soil Borne Human Diseases.* European Commission Joint Research Centre, Institute for Environment and Sustainability. EUR 24893 EN, ISBN 978-92-79-20797-6.

Kabata-Pendias, A. and H.K. Pendias. 2001. *Trace Elements in Soils and Plants.* CRC Press, Boca Raton, FL.

Kawai, K., E. Saathoff, G. Antelman, G., Msamanga, and W. Fawzi. 2009. Geophagy (soil-eating) in relation to anemia and helminth infection among HIV-infected pregnant women in Tanzania. *American Journal of Tropical Medicine and Hygiene* 80: 36–43.

Neuman, M. 1987. Relationship between chemical structure of antibiotics and pharmacokinetics. *Drugs Under Experimental and Clinical Research* 3: 115–124.

Nicolaou, K.C., J.S. Chen, D.J. Edmonds, and A.A. Estrada. 2009. Recent advances in the chemistry and biology of naturally occurring antibiotics. *Angewandte Chemie International Edition* 48: 660–719.

Njiru, H., U. Elchalal, and O. Paltiel. 2011. Geophagy during pregnancy in Africa: A literature review. *Obstetrical & Gynecological Survey* 66: 452–459.

Nordberg, M. and M.G. Cherian. 2005. Biological responses of elements. *In* O. Selinus, B. Alloway, J.A. Centeno, R.B. Finkelman, R. Fuge, U. Lindh, and P. Smedley (eds), *Essentials of Medical Geology*, pp. 179–200. Elsevier, Amsterdam, The Netherlands.

Pepper, I. and C. Rice. 2010. The influence of soil on public health. *In 19th World Congress of Soil Science*, Soil Solutions for a Changing World, Brisbane, Australia, 1–6 August 2010. Published on DVD.

Pepper, I.L., C.P. Gerba, D.T. Newby, and C.W. Rice. 2009. Soil: A public health threat or savior? *Critical Reviews in Environmental Science and Technology* 39: 416–432.

Ranjard, L. and A. Richaume. 2001. Quantitative and qualitative microscale distribution of bacteria in soil. *Research in Microbiology* 152: 707–716.

Sanchez, P.A. and M.S. Swaminathan. 2005. Hunger in Africa: The link between unhealthy people and unhealthy soils. *The Lancet* 365: 442–444.

Sandberg, A.S. 1994. Food processing influencing iron and zinc bioavailability. *In Ninth International Symposium on Trace Elements in Man and Animals*, P. Wilhelm, et al. (eds). National Research Council Canada, 19–24 May 1994.

Schaefer, M. and J. Schauermann. 1990. The soil fauna of beech forests: Comparison between a mull and a moder soil. *Pedobiologia* 34: 299–314.

Shultze, D.G. 1989. An introduction to soil mineralogy. *In* J.B. Dixon and S.B. Weed (eds), *Minerals in Soil Environments*, 2nd edn, pp. 1–34. SSSA Book Series: 1, Soil Science Society of America, Madison, WI.

Sing, D. and C. Sing. 2010. Impact of direct soil exposures from airborne dust and geophagy on human health. *International Journal of Environmental Research and Public Health* 7: 1205–1223.

Steinnes, E. 2010. Human health problems related to trace element deficiencies in soil. *In 19th World Congress of Soil Science, Soil Solutions for a Changing World*, Brisbane, Australia, 1–6 August 2010. Published on DVD.

Steinnes, E. 2011. Soils and human health. *In* T.J. Sauer, J.M. Norman, and M.V.K. Sivakumar (eds), *Sustaining Soil Productivity in Response to Global Climate Change: Science, Policy, and Ethics*, pp. 79–86. Wiley, Oxford.

Strobel, G. and B. Daisy. 2003. Bioprospecting for microbial endophytes and their natural products. *Microbiology and Molecular Biology Reviews* 67: 491–502.

Tiedje, J.M., J.C. Cho, A. Murray, D. Treves, B. Xia, and J. Zhou. 2001. Soil teeming with life: New frontiers for soil science. *In* R.M. Rees, B. Ball, C. Watson, and C. Campbell (eds), *Sustainable Management of Soil Organic Matter*, pp. 393–412. CABI, Wallingford.

Tobey, L.E. and T.R. Covington. 1975. Antimicrobial drug interactions. *The American Journal of Nursing* 75: 1470–1473.

Tong, G., M. Yulong, G. Peng, and X. Zirong. 2005. Antibacterial effects of the Cu(II)-exchanged montmorillonite on *Escherichia coli* K88 and *Salmonella choleraesuis*. *Veterinary Microbiology* 105: 113–122.

Torsvik, V. and L. Ovreas. 2002. Microbial diversity and function in soil: From genes to ecosystems. *Current Opinion in Microbiology* 5: 240–245.

U.S. Patent 6962718: Montmorillonite Components; Average Nutrient Content. Accessed 23 September 2011.

Velazquez, A.L.B., J. Fowler, R. Kakani, A. Haq, C.A. Bailey, M.G.T. Arvide, Y. Deng, and J. Dixon. 2010. Texas bentonites as an amendment of aflatoxins in poultry feed. *In Soil Science Society of America Annual Meeting Abstracts*. Published on CD-ROM.

Vermeer, D.E. 1985. Geophagy among the Ewe of Ghana. *Ethnology* 9: 56–72.

VitaminsDiary.com. 2011. *Keshan Disease: Causes, Symptom and Treatment*. http://www.vitaminsdiary.com/disorders/keshan.html. Accessed 10 January 2012 (verified 4 March 2012).

Waksman, S.A. 1945. *Microbial Antagonisms and Antibiotic Substances*. The Commonwealth Fund, New York, NY.

Waksman, S.A. 1952. Streptomycin: Background, isolation, properties, and utilization. Nobel Lecture, 12 December 1952. http://www.nobelprize.org/nobel_prizes/medicine/laureates/1952/waksman-lecture.pdf (verified 4 March 2012).

Walsh, C. 2003. *Antibiotics: Actions, Origins, Resistance*. ASM Press, Washington, DC.

Wardle, D.A., R.D. Bardgett, J.N. Klironomos, H. Setälä, W.H. van der Putten, and D.H. Wall. 2004. Ecological linkages between aboveground and belowground biota. *Science* 304(5677): 1629–1633.

Williams, L., D. Metge, D. Eberl, R. Harvey, A. Turner, P. Prapaipong, and A. Poret-Peterson. 2011. What makes a natural clay antibacterial? *Environmental Science & Technology* 45: 3768–3773.

WHO. 2007. *Assessment of Iodine Deficiency Disorders and Monitoring Their Elimination*, 3rd edn. World Health Organization, Geneva, Switzerland.

Young, S. 2011. *Craving Earth: Understanding Pica*. Columbia University Press, New York, NY.

10 Soils, Human Health, and Wealth

A Complicated Relationship

C. Lee Burras, Mary Nyasimi, and Lorna Butler

CONTENTS

> Good health is essential to human welfare and to sustained economic and social development.
>
> **World Health Organization, 2011**

10.1 INTRODUCTION

Most people do not routinely discuss "soil," yet the mention of "health" and/or "wealth" immediately captures people's attention, whether it is a discussion between two individuals or an evaluation of a nation. Human health and wealth are global human rights, according to the Universal Declaration of Human Rights from 1947 (United Nations, 2011). Yet, soil is often an underlying component to both, given human history of wars fought over access to land. Soil access continues to have direct application to even highly food-secure nations such as the United States. Demand for land in Africa, Asia, and possibly Latin America is projected to escalate such that the U.S. armed forces become involved (National Intelligence Council, 2008). Two activities exacerbating these tensions are the conversion of arable land from food crops to biofuel crops and the "global landgrabs" occurring as wealthy investors buy large tracts of farmland in Africa and beyond (Zoomers, 2010; Robertson and Pinstrup-Andersen, 2010). A less obvious but equally real concern is a growing perception that people are living on soil that is poisoning them (Weber et al., 2001).

Paradoxically, given health and wealth's incredible hold on the thinking of individuals and nations, research examining the role of soil in human well-being is limited (Steinnes, 2009; Deckers and Steinnes, 2004; Oliver, 1997). It is widely recognized that high-quality soils are crucial to food production, food quality, and agricultural economies—just as soil processes and properties underlie many aspects of water quality, waste decomposition, flood mitigation, and a wide host of other issues that directly affect human well-being (Brevik, 2009; Abrahams, 2002). Likewise, health risks are clear and well documented, involving soils with toxic levels of elements and/or pathogens (Brevik, 2009; Oliver, 1997; Abrahams, 2002). Even wind erosion is a known health risk when it results in inhalation or ingestion of harmful materials (Brevik, 2009).

Soil, health, and wealth are independently amazingly complex phenomena (e.g., see Hopkins et al., 1966). As a result, traditional research, which strives to isolate cause–effect relationships by examining one detail at a time, is only slowly coming to understand the interplay between them. Attempts to create a broader understanding are limited by the lack of collaboration between scientists in relevant fields, such as economics, health science, and soil science. Finally, scientific testing of these relationships normally has a low funding priority. The impact of soils on (poor) human health and wealth is mostly in impoverished nations, yet those nations have limited public infrastructure, which creates a need to ruthlessly focus on obvious pressing problems, especially hunger, malnourishment, and poverty. Discussion with individuals and policy makers from governments and NGOs indicates that they recognize the complex interplay of issues, yet, and commendably, when faced with dying people, they focus on rapid solutions such as food aid, increased availability of fertilizers, and other solutions with clear cost/benefit ratios.

This chapter very briefly reviews health and wealth issues that directly tie to soils, followed by an example illustrating the intertwined complexities of aforementioned three issues at the local scale. For further discussion on soils and health, interested readers are encouraged to refer to the very useful papers by Brevik (2009), Steinnes (2009, 2011), Deckers and Steinnes (2004), Oliver (1997), and Abrahams (2002), as well as other chapters in this volume.

10.2 IS THERE A DIRECT RELATIONSHIP BETWEEN SOILS AND THE HEALTH OF HUMANS?

Three direct relationships exist between soils and human health. They are (a) food availability and quality, (b) select chemical constituents, and (c) pathogens. In discussing these, it is useful to remember that "good health" itself is a vague concept tied to vitality and other qualitative concepts. Risk and toxicity are more quantitatively assessed. Risk is proportional to (a) exposure pathway, (b) exposure duration, and (c) exposure intensity. An individual's perception of risk is often very different to the reality of the risk.

The importance of adequate food to human health is so well established that it does not need discussion here. Hungry people have poor health and, generally, low wealth. Given human diets are overwhelmingly dependent on terrestrial crops, it is clear that human health is directly dependent on the soil's capacity to produce foodstuffs and that any risk to soil is a risk to human health (Sears, 1949; National Intelligence Council, 2008; Lal, 2010; FAO, 2011; Brevik, Chapter 2, this volume).

The risks to human health posed by soil constituents are often well-known and scientifically accepted direct hazards. A few common risks that are easy to explain are shown in Figure 10.1 and discussed here:

- Fluorine content in food correlates with soil type, with moderate fluoride intake improving health (especially dental health), while high content causes skeletal fluorosis (Steinnes, 2009; Ludwig et al., 1962).
- Anemia caused by iron deficiency is common in humans (Steinnes, 2011). It often results from low Fe content in foods. Iron content is typically controlled by the pH–Eh conditions of the soil that grew the food plant. Some cultures address Fe deficiency through geophagia, that is, intentional eating of Fe-rich soil. Such soil is even sold in some markets of the world, for example, the senior author purchased "herbal flavored soil" in Namasagali, Uganda, with the vendor recommending he give it to his wife for better vitality. Geophagia is most common in menstruating and pregnant women, both populations especially susceptible to anemia. Interestingly, geophagia can cause a number of health problems including hepatosplenomegaly, dwarfism, hypogonadism, and Zn deficiencies (Oliver, 1997).
- Lead poisoning in young children from contaminated soil and/or food grown on contaminated soil is recognized by the U.S. Environmental Protection Agency as a significant risk. It has been documented in many areas, for example, Mielke et al. (1999) found lead poisoned

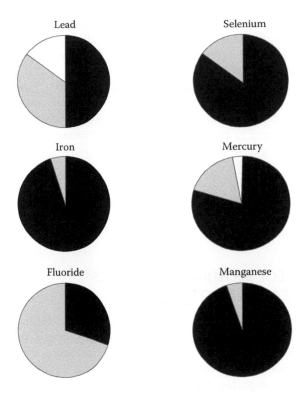

FIGURE 10.1 Proportions of food and soil (black in the book [blue in the CD-ROM figure]), water (gray in the book [red in the CD-ROM figure]), and air (white in the book [green in the CD-ROM figure]) exposed to six elements that have direct impacts on human health (lead, selenium, mercury, and fluoride are modified from Abrahams [2002]; iron and manganese are estimated by the authors).

children in Baltimore, Minneapolis/St. Paul, and New Orleans. These same children tend to have elevated concentrations of zinc and cadmium in their blood (Mielke et al., 1999). The health impacts of elevated blood Pb are wide-ranging and include decreased IQ, reproductive failure, loss of motor functions, and damage to the nervous system.

- Manganese content in soils correlates, sometimes positively but more often negatively, with cancer rates (Marjanen and Soini, 1972; Deckers and Steinnes, 2004).
- Mercury from soil that is released into surface waters where methylation occurs, followed by uptake in fish and other aquatic organisms that are subsequently eaten, causes chronic illness or even death (Dumont and Kosatsky, 1990; Steinnes, 2009).
- Selenium-deficient soils and selenium-excessive soils can damage health. Tan et al. (2002) reported that selenium deficiencies in eroded Entisols in China are associated with Keshan disease and Kashin–Beck disease. A variety of other studies have reported that selenium and iodine deficiencies are often related within humans. Excess soil Se results in selenoisis. Problematically high concentrations of Se have been found in the Mollisol region of the Great Plains in the United States (Steinnes, 2009, 2011) and Mollisols and Aridisols of China (Tan et al., 2002), with each of these areas being important to their respective nation's food supply.
- Low soil zinc is a common problem and results in Zn deficiencies in crops, which in turn can result in Zn deficiencies in humans who eat little meat (Steinnes, 2009, 2011). Low Zn in humans has a number of health impacts vis-à-vis pregnancy, brain development, dwarfism, etc. Alloway (2005) estimates Zn deficiency affects perhaps one-third of all humans. Soils with high Zn contents experience increased wind erosion since high Zn is toxic to plants.

- Tetanus occurs when unimmunized individuals are infected by toxins from spores of *Clostridium tetani*, an anaerobic bacteria that is especially abundant in manured fields of the tropics (Abrahams, 2002). Infections occur when wounds are exposed to the soil, making subsidence farmers and soldiers especially at risk.
- Hookworms affect about 25% of all humans (Abrahams, 2002), most of whom become infected when their skin comes into direct contact with contaminated soil. Given that hookworm larvae need damp, warm (24°C–32°C) soil, this is seen as a tropical problem, although hookworms used to be common even in the rural south of the United States (Keller et al., 1934).

Each of the preceding soil-health relationships is preventable and/or treatable. Each is recognized by public health and environmental quality agencies worldwide. As a result, it is likely that each will be solved in the twenty-first century, at least in those nations with effective public health and environmental quality infrastructures.

10.3   IS THERE A DIRECT RELATIONSHIP BETWEEN SOILS AND THE WEALTH OF HUMANS?

On a broad historic scale, ownership of land has led to considerable wealth directly and, especially, indirectly since many cultures tied and still tie land ownership to societal leadership and the corresponding access to labor and privilege (Shenk et al., 2010). The soil acts as a palate for creating wealth via management, a possession that has value as real estate, and a means to merit additional social resources. Soil is currently an important source of greater wealth as commodity and farmland prices reach record levels in many countries, with wealth gains going to farmers and owners of farmland across all sizes of farms (Jack et al., 2009). Since 2007, the economic return from investing in farmland has exceeded that of stocks and bonds in the United States. However, arguing for a direct relationship between wealth and soils is tenuous. There are incredible examples of wealth being lost and even civilizations failing because of mismanagement of soil or erroneous speculation about the worth of soil (McNeill and Winiwarter, 2006; Diamond, 2005; Troeh et al., 2004). This issue is especially relevant today as global change, population growth, and land use intensification cause increased soil degradation, salinization, and desertification (Lal, 2010). Finally, wealth from soil depends on a stable, agreed-upon land tenure system, which is often not the case when competing cultures vie for the same land. Illustrations of this are replacement of Native American hunter-gatherers by European farmers on the U.S. Great Plains, nationalization of private lands by the former Soviet Union, and modern land investment corporations purchasing tribal lands in sub-Saharan Africa.

10.4   WHAT COMPLICATES THE RELATIONSHIP BETWEEN SOILS, HUMAN HEALTH, AND WEALTH?

The relationship between soils, health, and wealth continues to evolve in response to human enterprise and global change. As a result there are only weak correlations connecting the three when comparing across nations. A brief analysis of the 28 nations in Table 10.1 can illustrate this. The strongest correlation ($r = -0.77$) is between the Global Hunger Index (GHI) and life expectancy, although the second strongest correlation is between income and percentage of urban population (0.74). These indicate the obvious: healthier people live longer than less healthy people and there is more wealth in cities than in rural areas. Multiple linear regression analysis indicates that 49% of these 28 nations' GHIs are predicted by knowing soil quality index (SQI) and per capita arable land (CAR):

$$GHI = 3.6\ SQI - 4.3\ CAR - 4.1,\ n = 28,\ R^2 = 0.49,\ p = .002.$$

TABLE 10.1

Selected Data of Human Health and Wealth and Soil Health of 28 Countries

Nation	Years Independent (YI) (Years)	Population Growth (%)	Percentage Urban Population (%)	Arable Land Proportion (%)	Life Expectancy (Years)	Population Density (PD) (people/km²)	Per Capita Arable Land (CAR) (ha/person)	Per Capita Rural Dweller Arable Land (ha/person)	Global Health Index (GHI)	Per Capita Income ($)	Common Soil Orders Used for Cropping	Soil Quality Index (SQI)
Afghanistan	22	2.4	23	12.1	45	45.7	0.3	0.3	30	900	Entisols	6
Algeria	49	1.2	66	3.2	75	14.7	0.2	0.6	4	7,300	Inceptisols, Alfisols	3
Argentina	195	1.0	92	10.0	77	15.3	0.7	8.2	3	14,700	Mollisols, Alfisols	2
Australia	110	1.2	89	6.2	82	2.8	2.2	19.7	1	41,000	Alfisols, Mollisols	5
Bangladesh	40	1.6	28	55.9	70	1,218.2	0.0	0.1	24	1,700	Inceptisols	4
Belarus	20	−0.4	75	26.8	71	47.2	0.6	2.3	3	13,600	Alfisols	2
Bermuda	0	0.6	100	20.0	81	1,271.8	0.0	15.7	2	69,900	Inceptisols	5
Bhutan	62	1.0	35	2.3	67	18.5	0.1	0.2	24	5,500	Ultisols	7
Bolivia	186	1.7	67	2.8	68	9.2	0.3	0.9	11	4,800	Ultisols, Alfisols, Mollisols	5
Brazil	189	1.1	87	6.9	73	24.0	0.3	2.2	5	10,800	Oxisols, Ultisols, Alfisols	5
Canada	144	0.8	81	4.6	81	3.7	1.2	6.4	1	39,400	Mollisols, Alfisols	3
China	62	0.5	47	14.9	75	139.7	0.1	0.2	6	7,600	Inceptisols, Mollisols, Ultisols	4
El Salvador	190	0.3	64	31.4	73	293.0	0.1	0.3	6	7,200	Andisols, Ultisols	5
France (metro)	500	0.5	85	33.5	81	118.8	0.3	1.9	1	33,100	Alfisols	3
Germany	500	−0.2	74	33.1	80	233.7	0.1	0.5	1	35,700	Alfisols	4
Hungary	500	−0.2	68	49.6	75	111.3	0.4	1.4	1	18,800	Mollisols	1

(continued)

TABLE 1C.1 (Continued)
Selected Data of Human Health and Wealth and Soil Health of 28 Countries

Nation	Years Independent (YI) (Years)	Population Growth (%)	Percentage Urban Population (%)	Arable Land Proportion (%)	Life Expectancy (Years)	Population Density (PD) (people/km⁻²)	Per Capita Arable Land (CAR) (ha/person)	Per Capita Rural Dweller Arable Land (ha/person)	Global Health Index (GHI)	Per Capita Income ($)	Common Soil Orders Used for Cropping	Soil Quality Index (SQI)
Indonesia	62	1.1	44	11.0	71	135.6	0.1	0.1	13	4,200	Ultisols	5
India	64	1.3	30	48.8	67	400.0	0.1	0.2	24	3,500	Vertisols, Inceptisols, Alfisols	4
Japan	500	−0.3	67	11.6	82	347.0	0.0	0.1	1	34,000	Andisols	5
Kazakhstan	20	0.4	59	8.3	69	5.7	1.4	3.5	3	12,700	Mollisols, Aridisols	2
Kenya	48	2.5	22	8.0	59	72.2	0.1	0.1	20	1,600	Oxisols, Ultisols	6
Nigeria	51	1.9	50	33.0	48	170.4	0.2	0.4	18	2,500	Oxisols	6
Pakistan	64	1.6	36	24.4	66	243.0	0.1	0.2	19	2,500	Entisols	6
Paraguay	200	1.3	61	7.5	76	16.3	0.5	1.2	6	5,200	Mollisols, Alfisols	3
Russia	500	−0.5	73	7.2	66	8.5	0.8	3.1	2	15,900	Alfisols, Mollisols	4
Rwanda	49	2.8	19	45.6	58	460.9	0.1	0.1	27	1,100	Oxisols, Ultisols	6
Uganda	49	3.6	13	21.6	53	175.6	0.1	0.1	15	1,300	Oxisols, Ultisols	6
United States	235	1.0	82	18.0	78	34.2	0.5	2.9	1	47,200	Mollisols, Alfisols	3

Sixty-two percent of GHI is predicted just by knowing SQI, CAR, population density (PD), and years of national independence (YI):

$$GHI = 3.9 + 2.7\ SQI - 5.6\ CAR - 0.0\ 2YI - 0.01\ PD, n = 28, R^2 = 0.62, p = .0001.$$

SQI correlations with GHI and income are 0.62 and −0.23, respectively. These indicate nations having high-quality soil have better health, although better soils and higher incomes only marginally trend together. Given that a cause–effect relationship is not routinely accepted between the income and soils of a nation, no regression analyses are reported.

Table 10.1 is included as a reminder that while it is very tempting to use databases to statistically find links between soils and health or between health and wealth, the reality is that those links can only be distilled by in-depth understanding and reflection about the soils, health, and wealth of a given population. Fortunately, there are a variety of well-known and respected scientific approaches to establish this. The authors of this chapter prefer the sustainable rural livelihood method, which uses eight "capitals" and their interplay and feedback loops to explain human well-being. This approach was pioneered by Chambers and Conway (1992) and refined by Ashley and Carney (1999). The eight capitals are natural resources, physical infrastructure, social resources, financial resources, cultural resources, political resources, information access, and human capital (Flora and Flora, 2008). The value of the livelihood framework is in establishing how a shift in one capital affects the other capitals and, ultimately, a people's well-being. In essence the livelihood framework is a combined biophysical-social analog to Le Chatelier's Principle.

A useful example of a livelihood's analysis is given by Nyasimi (2007), who used it to explain the connections between soils, human health, and wealth in the Awach River watershed in Western Kenya. The watershed covers 1050 km² and empties into eastern Lake Victoria. It is home to 250,000 people, most of who are engaged in subsistence agriculture. The population is growing rapidly, with many young women having their first children by age 15. Members of the politically connected Kipsigis tribe own and farm the uplands mantled with volcanic ash, and members of the politically disenfranchised Luo tribe own and farm the western lowlands. Those lowlands are part of the Lake Victoria basin. A 400-m tall escarpment of Precambrian granite that was exposed during rifting that also created Lake Victoria separates the uplands and lowlands. Fifty-eight percent of the area has experienced significant water erosion (Shepherd and Walsh, 2002).

Nyasimi (2007; also see Nyasimi et al., 2007) documented the diverging experiences of the Kipsigis and Luo people in terms of soils, health, and wealth. The people in her study villages had fairly similar health and wealth in 1980, even though their soils, climates, and elevations were very different. As a note of caution, Nyasimi's (2007) health and wealth data was self-reported and the two groups define wealth a little differently. Nyasimi's (2007) critical finding is that upland Kipsigis in Ainamoi were able to maintain their wealth while the lowland Luo in Kanyibana experienced a dramatic increase in terribly impoverished households, from 45% to 78%, over a 25-year period (Table 10.2). Furthermore, the Kipsigis are on track to improve their health even with HIV-AIDS being endemic. The Luo are not. The diverging experiences of these people, who literally often live within walking distance of one another, reflect how they adjusted to changing capitals—some of which are biophysical and some of which are human-induced.

The Kipsigis realized their future well-being required simultaneous adjustments in culture and use of soils in order to survive uncertainties in the economy, the climate, land fragmentation, population growth, and a major health risk, HIV-AIDS. Culturally they changed from mandating the inheritance of a widow by her dead husband's brother and they began allowing a widowed daughter—really any at-risk adult daughter—to be able to return to and even inherit some of her father's land. This helped control the spread of HIV-AIDS, since husbands were often dying from it and if a wife was inherited she would infect not only her new husband but also his other wives. Since each of those wives would also eventually have been inherited, this was culturally catastrophic. Nyasimi's (2007) findings indicate the Kipsigis were able to make this cultural adjustment because

TABLE 10.2
Selected Features of Two Representative Villages in the Awach Watershed in Western Kenya

Case Study Village	Ainamoi (Uplands)	Kanyibana (Lowlands)
Physical Environment		
MAP (mm)	1800	1200
MAT (°C)	24	32
Elevation (m)	2000	1200
Surficial geology	Holocene ash over pre-Cambrian basement rock	Holocene alluvium, colluvium, and lacustrine sediments
Local relief	Flat to gently rolling	Flat to gently undulating
Common soil suborders	Udands, Udalfs, Udoxs	Usterts, Ustalfs, Psamments
Soil fertility (20 cm)	Low to medium	Very low to low
Soil quality (20 cm)	Good	Poor and becoming poorer
Postindependence erosion	Low to moderate sheet and rill erosion	Moderate to severe gully erosion; moderate wind erosion
Access to clean water	Moderate to good	Poor to terrible
Flood risk	None	Very high
Drought risk	Low	High
Agriculture		
Typical farm size	1–3 ha	1–2 ha
Common crops	Maize, beans, tea, coffee, pineapples, sweet potato, bananas, finger millet, vegetables, pumpkins, Irish potato	Maize, beans, cassava, sorghum, vegetables
Livestock	Grade dairy cattle, Zebu bulls, goats, sheep, donkeys, chickens	Zebu cattle, goats, chickens
Weed pressure	Little to moderate	Moderate to strong, especially by *Striga hermonthica*
Agriculture knowledge	Moderate but increasing	Moderate but decreasing
Cropping decisions	High independence	Moderate independence
Planting date	High independence	Low independence
Culture		
Tribe	Kipsigis	Luo
Descent identity	Through males	Through males
Postmarriage home	Husband's family	Husband's family
Marriage type	Polygamous and polygny	Polygamous and levirate
Adaptability of mores	Moderately adaptable	Not adaptable
Main foods	Ugali from finger millet, milk, meat, vegetables	Ugali from maize, vegetables
Daughters inherit land and can return home?	Yes, as of the early twenty-first century	No
Typical meals/day	2–3	1–2
Poor/not poor, 2004	60/40	78/22
Poor/not poor, 1994	55/45	62/38
Poor/not poor, 1979	59/41	45/55

Source: Adapted from Nyasimi, M.K., Transforming lands and livelihoods in the Awach River Basin of Lake Victoria, western Kenya, Unpublished PhD dissertation, Iowa State University, Ames, IA, 2007; Nyasimi, M.K., Butler, L.M., Burras, L., Ilahiane, H., Schultz, R., and Flora, J., *Journal of Ecological Anthropology*, 11, 43–57, 2007; and unpublished data.

Note: MAP, mean annual precipitation; MAT, mean annual temperature.

they owned high-quality soils (especially Andisols), which remained highly productive even when moderately eroded (Table 10.2). They also had adequate land area to allow for returning individuals to be able to have their own fields. The Kipsigis did not stop with simply changing their cultural practices. They also exploited the change in the region's economy by diversifying their crops. They aggressively developed a model of identifying and producing high-value crops. Their approach was highly cooperative and inclusive, which let them take advantage of each other's knowledge and resources. Their political connections provided government extension help including the acquisition of safe drinking water from clean aquifers. Finally, their proximity to the increasingly impoverished Luo meant they, the Kipsigis, had a growing pool of cheap labor to facilitate their expanding economy.

The Luo story is less happy. Nyasimi's (2007) livelihood analyses show a critical feature is their location downstream from the Kipsigis along the River Awach. The intensification of land use in the headwaters by the Kipsigis has resulted in more flooding and more intense floods being experienced by the Luo, over the past 25 years. It has also resulted in reduced water quality, with a concurrent deterioration of the health of the Kanyibana villagers, since most of them get their drinking water directly from the river (Figure 10.2). These floods have also caused incredible soil degradation (Figure 10.3), which has greatly reduced their soils' production capability (Table 10.2). The Luo in Kanyibana village have responded to these increasing biophysical limitations by rigorously following their traditional cultural mores. They continue to mandate inheritance of widows even though it increases the risk of HIV-AIDS for the new husband and his other wives. They continue to grow traditional crops, especially maize, even though the soils and weather have changed to being better suited for sorghum or other less-water-demanding crops. They are becoming increasingly isolated and exploited. Part of this is due to rapidly growing gullies literally making it impossible for neighbors to visit each other (Figure 10.4). This is also due to their need to find off-farm work since their own crops are inadequate to feed a family. The major avenues of off-farm work are, for the women, as low-paid migrant farm workers for the Kipsigis upstream or, for the men, as low-paid manual laborers in a city. Given manual labor in cities is almost exclusively limited to healthy young men, they are the most common group to leave. Many of these men are husbands and fathers, who periodically return home. Some are infected with HIV-AIDS and infect their wives. This means children and grandparents are maintaining the farms. The former is the least knowledgeable group and the latter is the least likely to adopt new practices better suited to their changing world. Collectively these biophysical and familial changes are causing a breakdown in social support and increased feelings of embarrassment and powerlessness. Finally, the Luo in Kanyibana have no political clout, which means they

FIGURE 10.2 A Kanyibana villager carrying water from the River Awach for use in her house.

FIGURE 10.3   A pedestal of soil in Kanyibana village, which illustrates the extent of general soil degradation that had occurred in the past 50 years.

FIGURE 10.4   A gully within Kanyibana village that had formed in the past 30 years. It continues to downcut and extend headward, destroying roads, houses, and fields.

have—for all practical purposes—no access to government services and knowledge that might be able to teach them ways out of this depressing cycle.

The authors have seen similar complex interplay between soils, health, and wealth in other places, most notably in Uganda. The details differ but the ramifications are the same. The future of a given population depends on how they balance and respond to the eight capitals identified by the livelihoods framework. All parameters have a level of codependence. This might suggest it would be wise for at-risk populations to seek out well-educated people for assistance, although it is equally likely that a group undergoing declining health and wealth is unlikely to be receptive to outsiders.

10.5   CONCLUSION

Soils are a critical component of human health and wealth; however, the relationships between these three topics range from being very straightforward to deceptively complex. Increased food security and international economic stability will require improved understanding of the relationship.

REFERENCES

Abrahams, P.W. 2002. Soils: Their implications to human health. *Science of the Total Environment* 291: 1–32.

Alloway, B.J. 2005. Bioavailability of elements in soil. *In* O. Selinus, B. Alloway, J.A. Centeno, R.B. Finkelman, R. Fuge, U. Lindh, and P. Smedley (eds), *Essentials of Medical Geology*, pp. 347–372. Elsevier, Amsterdam, The Netherlands.

Ashley, C. and D. Carney. 1999. *Sustainable Livelihoods: Lessons from Early Experience*. Department for International Development, London.

Brevik, E.C. 2009. Soil, food security, and human health. *In* W. Verheye (ed.), *Soils, Plant Growth and Crop Production*. Encyclopedia of Life Support Systems (EOLSS), developed under the Auspices of the UNESCO, EOLSS Publishers, Oxford. http://www.eolss.net (verified 21 February 2012).

Chambers, R. and G. Conway. 1992. *Sustainable Livelihoods: Practical Concepts for the 21st Century*. Institute of Development Studies, U.K. http://opendocs.ids.ac.uk/opendocs/handle/123456789/775. Accessed 3 January 2012 (verified 21 February 2012).

Deckers, J. and E. Steinnes. 2004. State of the art on soil-related geo-medical issues in the world. *In* D.L. Sparks (ed.), *Advances in Agronomy*, Vol. 84, pp. 1–35. Elsevier, Amsterdam, The Netherlands.

Diamond, J. 2005. *Collapse—How Societies Choose to Fail or Succeed*. Viking, New York, NY.

Dumont, C. and T. Kosatsky. 1990. Methylmercury in northern Canada. *In* J. Lag (ed.), *Excess and Deficiency of Race Elements in Relation to Human and Animal Health in Arctic and Subarctic Regions*, pp. 109–133. The Norwegian Academy of Science and Letters, Oslo.

Flora, C. and J. Flora. 2008. Community capitals summary. North Central Regional Center for Rural Development. http://ncrcrd.msu.edu/uploads/files/133/Community%20Capitals.pdf. Accessed 3 January 2012 (verified 21 February 2012).

FAO. 2011. Global soil partnership for food security launched at FAO. http://www.fao.org/news/story/en/item/89277/icode/. Accessed 30 December 2011 (verified 21 February 2012).

Hopkins, H.T., E.H. Stevenson, and P.L. Harris. 1966. Soil factors and food composition. *American Journal of Clinical Nutrition* 18: 390–395.

Jack, C.G., J.E. Moss, and M.T. Wallace. 2009. Is growth in land-based wealth sustaining part-time farming? *EuroChoices* 8: 29–35.

Keller, A.E., W.S. Leathers, and H.C. Ricks. 1934. An investigation of the incidence and intensity of infestation of hookworm in Mississippi. *American Journal of Epidemiology* 19: 629–656.

Lal, R. 2010. Managing soils for a warming earth in a food-insecure and energy-starved world. *Journal of Plant Nutrition and Soil Science* 173: 4–15.

Ludwig, T.G., W.B. Healy, and R.S. Malthus. 1962. Dental caries prevalence in specific soil areas at Napier and Hastings. *In* G.J. Neale (ed.), *Transactions of Joint Meeting of Commissions IV and V, International Society of Soil Science*, pp. 895–903. Soil Bureau, P.B., Lower Hutt, New Zealand.

Marjanen, H. and S. Soini. 1972. Possible relationship between nutrient imbalances, especially manganese deficiency, and susceptibility to cancer in Finland. *Annales Agriculturae Fenniae* 11: 391–406.

McNeill, J.R. and V. Winiwarter (eds). 2006. *Soils and Societies—Perspectives from Environmental History*. The White Horse Press, Isle of Harris.

Mielke, H.W., C.R. Gonzales, M.K. Smith, and P.W. Mielke. 1999. The urban environment and children's health: Soils as an integrator of lead, zinc and cadmium in New Orleans, Louisiana, USA. *Environmental Research* 81: 117–129.

National Intelligence Council (NIC). 2008. *Global Trends 2025: A Transformed World*. U.S. Government Printing Office, Washington, DC. http://www.dni.gov/nic/NIC_2025_project.html. Accessed 18 April 2011 (verified 21 February 2012).

Nyasimi, M.K. 2007. Transforming lands and livelihoods in the Awach River Basin of Lake Victoria, western Kenya. Unpublished PhD dissertation, Iowa State University, Ames, IA. Proquest document ID 1320943501.

Nyasimi, M.K., L.M. Butler, L. Burras, H. Ilahiane, R. Schultz, and J. Flora. 2007. Differentiating livelihood strategies among the Luo and Kipsigis people of western Kenya. *Journal of Ecological Anthropology* 11: 43–57.

Oliver, M.A. 1997. Soil and human health: A review. *European Journal of Soil Science* 48: 573–592.

Robertson, B. and P. Pinstrup-Andersen. 2010. Global land acquisition: Neo-colonialism or development opportunity? *Food Security* 2: 271–283.

Sears, P.B. 1949. Soils and health. *Canadian Medical Association Journal* 60: 88–92.

Shenk, M.K., M.B. Mulder, J. Beise, et al. 2010. Intergenerational wealth transmission among agriculturalists: Foundation of agrarian inequality. *Current Anthropology* 51: 65–83.

Shepherd, K.D. and M.G. Walsh. 2002. Development of reflectance spectral libraries for characterization of soil properties. *Soil Science Society of America Journal* 66: 988–998.

Steinnes, E. 2009. Soils and geomedicine. *Environmental Geochemistry and Health* 31: 523–535.

Steinnes, E. 2011. Soil and human health. *In* T.J. Sauer, J.M. Norman, and M.V.K. Sivakumar (eds), *Sustaining Soil Productivity in Response to Global Climate Change: Science, Policy, and Ethics*, pp. 79–86. Wiley, Oxford.

Tan, J., W. Zhu, W. Wang, R. Li, S. Hou, D. Wang, and L. Yang. 2002. Selenium in soil and endemic diseases in China. *Science of the Total Environment* 284: 227–235.

Troeh, F.R., J.A. Hobbs, and R.L. Donahue. 2004. *Soil and Water Conservation for Productivity and Environmental Protection*, 4th edn. Prentice Hall, Upper Saddle River, NJ.

United Nations. 2011. The universal declaration of human rights. http://www.un.org/en/documents/udhr/. Accessed 28 August 2011 (verified 21 February 2012).

Weber, O., R.W. Scholz, R. Buhlmann, and D. Grasmuck. 2001. Risk perception of heavy metal soil contamination and attitudes toward decontamination strategies. *Risk Analysis* 21: 967–977.

World Health Organization (WHO). 2011. *The World Health Report: Health Systems Financing: The Path to Universal Coverage*. World Health Organization, Rome.

Zoomers, A. 2010. Globalization and the foreignisation of space: Seven processes driving the current global land grab. *Journal of Peasant Studies* 37: 429–447.

11 Human Contact with Plants and Soils for Health and Well-Being

Joseph R. Heckman

CONTENTS

11.1 INTRODUCTION

Many studies reveal that in addition to needing food and good nutrition from fertile soils, people also need contact with nature for health and well-being. While active work with plants and soils engenders a sense of well-being and resultant stress reduction in the gardening public, even the passive viewing of landscapes by nongardeners can improve human health. A classic study conducted in a hospital showed that patients recovered more quickly from surgery and required less pain medication if they had a view of a green landscape rather than a view of a brick wall. Other studies reveal that walking in a garden helps cancer patients recover and reduces the number of violent outbursts for Alzheimer's patients. Also, children with learning difficulties improve in their ability to concentrate when they are moved from areas with little or no green spaces to areas of increased greenery. Beneficial effects of viewing scenes of nature, as compared with urban scenes or hardscape, can result in physiological changes such as lowered blood pressure and reduced muscle tension (Figure 11.1). Some studies suggest that levels of greenness may influence the incidence of crime. The practice of using plants as tools to rehabilitate people with disabilities has matured into a discipline known as horticultural therapy, and the concept of designing healing landscapes at hospitals, hospices, nursing homes, and urban environments has matured into the allied discipline known as therapy landscapes. Yet, the healing factor of healthy landscapes has not gained the broader attention of soil scientists, agronomists, and farmers in managing much of Earth's land resource, even though healing landscapes are rooted in soil. Beyond food production and the usual environmental issues, soil scientists and agronomists may find new opportunities by asking and researching questions about landscapes that appear to be lifeless as a result of tillage, herbicides, and cropping regimes, the aesthetic value of these landscapes, and their potential impact on the physiological and psychological health of the viewing public.

FIGURE 11.1 The aesthetic quality and emotional impact of vegetation and landscape are a living expression of soil quality, and this subjective relationship has objective consequences for human health and well-being. (a) Garden scene, (b) degraded soil and landscape, and (c) hardscape. ([a, c] Photos by Joseph Heckman and [b] courtesy of Roy Flannery.)

11.2 HISTORICAL CONTEXT FOR SOIL–PLANT–PEOPLE RELATIONSHIPS

There has been little discussion on the relationship between human health and well-being and human contact with nature in the discipline of soil science, especially in comparison to the study of food quality and nutrition. However, horticultural therapy—an outgrowth of a subdiscipline of horticulture often referred to as plant–people relationships—has not only made this relationship a focus of investigation but has become a part of the health-care system itself (Relf and Lohr, 2003). The purpose of this chapter is to provide an overview of plant–people relationships and to illustrate ways in which these relationships are in fact rooted in soil. My main thesis is based on the premise that the aesthetic quality and emotional impact of vegetation and landscape are a living expression of soil quality and that this subjective relationship has objective consequences for human health and well-being.

Well before horticultural therapy developed into a formal discipline, its rudiments were present in folk wisdom (Louv, 2008). In 1699, the *English Gardener* declared, "spare time in the garden, either digging, or setting out, or weeding: there is no better way to improve your health." In the 1700s, Dr. Benjamin Rush prescribed "digging in the soil" as a curative for the mentally ill. Frederick Law Olmsted (1870) wrote about the value of human contact with natural scenery in parks and landscapes within urban settings and about how they bring "tranquility and rest to the mind." In the 1800s, asylums for the mentally ill were frequently associated with farms or market gardens, providing these institutions with fresh food as well as an occupation for inmates. Within these institutions, farmwork and gardening activities were believed to contribute to mental health (Sempik and Aldridge, 2006).

By the mid-1900s, the horticulture therapy movement was developing into a discipline with degree curriculums. Graduates of these programs would then use knowledge of plant–people relationships to manage distressing social situations and a wide range of mental and physical disorders.

The list of plant benefits to human health and well-being is now extensive. Over the last three decades, an accumulation of research-based information on people–plant interactions—including the presence of, the viewing of, and activity with plants and their associated health effects—has developed into a broad body of knowledge that has become the subject of several more recent reviews (e.g., Flagler and Poincelot, 1994; Relf and Lohr, 2003; Hine et al., 2009). Being in the presence of plants or the passive viewing of them has been documented to improve recovery from stress, lower blood pressure, reduce muscle tension, improve student attention and concentration, reduce perception of pain, and speed up healing from surgery.

A study by Ulrich (1984) on the recovery of patients from gall bladder surgery is considered a classic study on the positive influence of viewing greenery. Hospital patients with a view of trees rather than a view of a brick wall recovered more quickly, had shorter postoperative hospital stays, used less pain medication, and exhibited fewer negative evaluative comments as documented by nurses. Other studies have documented benefits for a more active approach. Walking through a garden, for example, helps women recovering from breast cancer restore mental clarity (Cimprich, 1993). Active participation in gardening itself also has a long list of associated benefits. Surveys found that participants in gardening activities felt calmer and more relaxed; felt nature was essential to their well-being; and had increased self-esteem, enhanced personal satisfaction, and improved quality of life (Lohr and Pearson-Mins, 2002). In short, gardening activity raises community spirit (Figure 11.2).

(a)

(b)

FIGURE 11.2 Exposure to garden scenes has been shown to provide many health benefits. (a) The garden scene at the home of Pearl S. Buck, author of *The Good Earth*, in Bucks County, Pennsylvania, United States. (b) A community garden. ([a] Photo by Ann Heckman-Kreitzer and [b] photo by Joseph Heckman.)

(a)   (b)

FIGURE 11.3 Many desert plants, including the Joshua tree (*Yucca brevifolia* Engelm) (a) and various varieties of cacti (*Cactaceae* family) (b) are covered in spines. These examples are found in Death Valley National Park, California, United States. (Courtesy of Eric Brevik.)

It is interesting to note that the idea that the quality of landscapes, particularly deserts, has an influence on human behavior and social interaction was put forth in 1955 by biopsychiatrist and natural scientist Dr. Wilhelm Reich (1979). Reich associated the desert environment with a deadness of emotion, considering crime and psychosis as consequences of the "emotional desert." Reich states: "The whole plant is covered with bristles, reminding us, in analogy, of the prickly outer behavior of human beings who are empty and desert-like inside. This is not a mere analogy. The simile goes very far, indeed." (Figure 11.3). And: "With the spreading of the global desert, civilizations go under, life perishes completely in the affected realm, man either tries to escape or he too adjusts to the life in the desert on rare spots of green called oases."

More recently, geographer James DeMeo (2006) built upon Reich's hypothesis. DeMeo conducted a comprehensive, cross-cultural analysis of human behavior, and settlement and migration patterns, examining and mapping the geographical distribution of 6000 years of human warfare and social violence against the backdrop of soils, climate, and landscape quality. Using archaeological, historical, and anthropological data, the dichotomous behaviors, attitudes, and social institutions associated with either deserts or green landscapes are documented in DeMeo's book titled *Saharasia: The 4000 BCE Origins of Child Abuse, Sex-Repression, Warfare and Social Violence in the Deserts of the Old World* (DeMeo, 2006). His work indicates that male dominance, social hierarchy, and violence are pervasive human behaviors in deserts. DeMeo argues that patriarchal social behaviors that may have originated with desertification are subject to diffusion beyond desert regions as they became institutionalized and ingrained in human character structure. His findings support Reich's view that harsh desert landscapes play a role in shaping human social behavior. An important question remains about to what extent emotional health in individuals and social conditions in general can be restored by improving landscape ecology.

11.3 HUMAN CRAVING FOR PERCEPTUAL PLEASURE

Neuroscientists that study cognition and perception have taken a serious interest in the human craving for perceptual pleasure (Biederman and Vessel, 2006). They would probably agree with this

FIGURE 11.4 Humans tend to respond positively to views of attractive landscapes, such as this view from Olympic National Park in Washington, United States. (Courtesy of Eric Brevik.)

statement by Deming (1994): "At times I feel green deprived and would not be surprised to learn that there exists a psychic malady that can be cured only by the visual ingestion of green wildness." Neurobiologists, too, are investigating what makes pleasurable experiences, such as viewing a dramatic landscape, so profoundly gratifying. Evidence suggests that opioid receptors are involved (Biederman and Vessel, 2006).

Subjective impressions about how people respond to certain types of stimuli can also be objectified with functional magnetic resonance imaging. This tool enables observations on how the brain responds to various types of visual stimulus. Findings tend to support our subjective impression that people generally prefer broad, rather than restricted, views of landscapes; novel, rather than mundane, visual stimuli; and natural, rather than man-made, environments (Biederman and Vessel, 2006). Humans apparently have a strong preference for certain types of experiences and stimuli and they are not limited to vision. They also involve hearing, touch, taste, and smell, all of which may accompany the total landscape experience.

A straightforward approach to further investigation of not only visual pleasure but the entire range of perceptual pleasure may be to consider how total landscape experience influences the autonomic nervous function (Reich, 1982). Attractive landscape experiences associated with pleasure may be expected to stimulate the parasympathetic nervous system whereas unattractive, degraded landscapes may stimulate the sympathetic nervous system. A sympathetic response is associated with anxiety, stress, difficulty relaxing, and high blood pressure. Chronic overstimulation of the sympathetic nervous system is unhealthy. Many studies (Ulrich, 1986; Cimprich, 1993; Tennessen and Cimprich, 1995; Lohr and Pearson-Mims, 2006; Barton and Pretty, 2010) have shown positive human responses to attractive landscapes, such as reductions in stress and restoration from mental fatigue (Figure 11.4). These findings seem to suggest that attractive landscapes stimulate the parasympathetic nervous system. Perhaps this basic reaction to landscape quality is the underlying, common functioning principle that applies to understanding soil–plant–people relationships in general.

11.4   VIEWS OF SOIL SCIENTISTS

Although the design and benefit of healing landscapes in hospitals, hospices, nursing homes, and urban environments have captured the focus and attention of horticulturalists and public health specialists, they have not gained much attention among soil scientists, even though healing landscapes are rooted in soil. At the same time, horticulturalists interested in plant–people relationships are so heavily focused on plants that they rarely even mention the soils the plants grow in. As a soil scientist reading the literature on plant–people relationships, it seems almost as if plants exist and survive independently from soil.

A recent article in *CSA News* (Pepper, 2010) discussing the many economic and social values of soil failed to mention the vital role of healthy soil in supporting healthy landscapes. However, in the previous century, a distinguished soil scientist named Firman E. Bear recognized the therapeutic value of landscapes, noting that the starting point was healthy soil:

> The starting point in the conservation on natural beauty is the soil itself. Bare soil is a blot on the land-scape … Yet there are much larger and more disturbing aspects to the problems than these. People are greatly influenced by their surroundings. If they are careless about the land on which they live they tend to become even more careless about themselves. Ignorance, immorality, and crime tend to grow and thrive in areas where the natural beauty of the surroundings has been destroyed; this applies to country and city alike. When attention is given to correcting such conditions, the people living in these areas and participating in their renovation tend to take on greatly improved attitudes as well. (Bear, 1962)

Although this was written by Bear (1962) before there was much research to back these ideas, the data that has since been collected lends considerable support to his argument, as already discussed.

FIGURE 11.5 Different soils, and different layers within a given soil, have different colors, textures, and other properties that make working with soils a multisensory experience. (a) The surfaces of three Mollisols in Iowa, United States. (b) The profile of a Spodosol in New Jersey, United States. ([a] Photo by Lynn Betts, USDA NRCS and [b] photo by Joseph Heckman.)

An article in *Soil Science* by Bidwell and Hole (1965) discusses how human activities factor into soil formation and also suggests that every major soil landscape leaves an imprint on human culture. Today, however, the role of soil in what may be called soil–plant–people relationships remains largely unexplored.

There is a need for collaborative research among soil scientists, horticulturalists, climate scientists, and social scientists to identify the role soil and climate play in plant–people relationships. The use of gardening activities in horticultural therapy should attempt to elucidate the role of soil in the personal experience of the gardener. It could, for example, investigate the multisensory experience of people working and digging in qualitatively different soils (Figure 11.5). Anyone who has ever dug into soil quickly develops an awareness of soil texture, color, smell, and perhaps even taste.

11.5 BIOPHILIA AND SOILS

The love of life or of living systems has been referred to as "biophilia" by E. O. Wilson (1993), who hypothesized that human beings subconsciously seek connections with other life forms. The subjective urge to affiliate with other life forms and the positive reaction to open verdant landscapes (Biederman and Vessel, 2006) is not limited to the above ground. People, gardeners in particular, may respond positively to soil as a living system (Figure 11.6). Digging opens up the soil, creating opportunities for humans to encounter earthworms and other soil fauna. Gardeners and farmers alike have a feeling for the soil and know firsthand the restorative experience of working with it. Although the psychological value of working or making contact with soil is beyond our subjective impressions, largely unknown, the considerable subjective evidence of this value is a good place to start.

For example, compacted and degraded soils are the bane of the urban gardener. This is evident from numerous questions asked and concerns expressed by gardeners. When digging is attempted in an urban soil that has been compacted and degraded in quality as a result of heavy machinery traffic, it is not likely to be a positive gardening experience. Extension professionals have an important role in teaching the public how to improve soil quality, alleviate compaction, and build soil fertility and organic matter content. Rejuvenation of degraded urban soils should enhance gardening as a pleasurable and therapeutic experience. In 1965, Bidwell and Hole wrote: "we await the pleasure of the soil and its environment."

The organic system of gardening and farming emphasizes the living nature of soils (Balfour, 1976). Most importantly, it uses cultural practices that build the soil's organic matter content, thus increasing the numbers of living organisms within it. The relationship between organic farming

FIGURE 11.6   Digging in soil may provide health benefits. (Photo by Joseph Heckman.)

and soil quality is discussed in Carr et al. (Chapter 12, this volume). Pioneers of the organic movement wrote with great enthusiasm and fondness about the profound influence of the living aspects of humus-rich soil on plants and people. To give an example, Howard (1943) wrote: "The effect of humus on the crop is nothing short of profound. The farmers and peasants who live in close touch with Nature can tell by a glance at the crop whether or not the soil is rich in humus. The habit of the plant then develops something approaching personality: the leaves acquire depth of colour: the minute morphological characters of the whole of the plant organs become clearer and sharper. Root development is profuse; the active roots exhibit not only turgidity but bloom." Perhaps this passage may be interpreted as a subconscious display of Howard's "biophilia." My own personal experience in interacting with numerous organic farmers and gardeners suggests to me that "biophilia" for healthy, humus-rich soils is not only common, but a motivation for adopting organic farming methods. Thus, creating a healthy living soil, and the pleasurable soil–plant–people relationships it engenders, produces its own rewards. It may even be one of the motivating forces behind the organic movement itself (Heckman, 2006).

Soils are experienced in a tactile way when people walk over or dig into them. Healthy soils are resilient; they absorb and cushion the human step and influence the pleasure of leisurely walks over landscapes. Cushioning impact is of special significance on athletic playing fields and the prevention of sports injuries.

In *The Good Earth*, Pearl S. Buck described walking barefoot on soil this way: "The earth lay rich and dark and fell apart lightly under the points of their toes" (Buck, 1931). Contrast this with people in the modern world walking mostly over paved surfaces. Within the United States, the total area of impervious surfaces—buildings, roads, parking lots, roofs, etc.—is estimated to be almost equivalent in size to Ohio (Elvidge et al., 2004). This major alteration of the land surface not only changes local hydrological patterns, climate, and nutrient cycling, but it also greatly reduces the potential for human skin and eye contact with soil.

Reconnecting with the earth by walking barefoot outside is an emerging trend referred to as "Earthing." People who have adopted the routine of Earthing report numerous health benefits, such as better sleep, reductions in pain and stress, more rapid wound healing, and synchronization of internal biological clocks (Ober et al., 2010). All of these subjective claims deserve further scientific investigation.

11.6 HEALING AGRICULTURAL LANDSCAPES

As hunter–gatherers, humans evolved in places of natural beauty and with much contact with soil. This contrasts markedly with the modern world, where most people procure food in areas of hardscape. Some attempts are being made to remedy this situation by encouraging people to frequent farmers' markets or take advantage of direct on-farm sales. One such initiative is the USDA program "Know Your Farmer, Know Your Food." This program encourages people to visit farms or farm markets. While occasional travel to farms to know the source of our food can be a valuable experience, distance makes it impractical for most people in cities. Nevertheless, agricultural landscapes that have been designed with nature in mind provide another alternative to the "over-crowded park" as a place for people to make contact with plants and soils.

Unfortunately, advocates for another Green Revolution perpetuate the myth that farms are "ecological sacrifice zones" that are incompatible with natural habitats (Jackson and Jackson, 2002). More intensive farming, it is argued, is necessary to increase production, and space for wilderness should be conserved elsewhere (Cassman, 2007). This approach assumes that farms and wilderness are mutually exclusive. Here, even farms that are intensively managed to produce only a few commodities take on the appearance of rural factories and offer little in the way of a natural experience.

An alternative system, known as "Farming with Nature in Mind," has been referred to as "Farmscaping" by Av Singh (2004). Farmscaping goes beyond food production to providing ecological services, where wild plants and animals coexist in natural habitats along with domesticated crops and livestock. Traditional small organic farms that foster an abundance of biodiversity easily lend themselves to farmscaping.

Another alternative farming system is biodynamic agriculture, which was founded by Rudolf Steiner (BFGA, 2011). In biodynamic agriculture, biodiversity and sustainable cultural practices serve to create an "individual farm organism." Organic farming also ideally includes this "farm organism" concept by embracing biodiversity and relying on complex crop rotation cycles and integration of plant and animal agriculture. By design these farming systems avoid monoculture and tend to naturally comprise many visually interesting farmscape features (Hansen et al., 2006).

Pasture-based feeding of livestock is an important cultural practice in biodynamic and organic farming and is now mandated in USDA-NOP standards for organic dairy. In contrast to row crops, pastures typically support a diversity of plant–animal ecosystems as well as wildlife. Protecting public health requires the protection of soil quality. Pastures are an important part of crop rotation for building soil organic matter content and improving soil quality because they provide perennial soil coverage and thereby protect soil from erosion (Wedin and Fales, 2009; Heckman, 2010).

Cows and other livestock on pasture offer pastoral views that are particularly attractive, especially when presented in contrast to views of confined animal feeding operations (CAFOs) (Figure 11.7). Interestingly, just as humans in urban settings are deprived of contact with plants and soils by increasing amounts of hardscape, industrialized agriculture has placed livestock in an analogous

(a)

(b)

FIGURE 11.7 Cattle grazing on pastures with green, lush backgrounds (a) provide a more appealing view than cattle in CAFO (b). ([a] Photo by Gary Kramer, USDA NRCS and [b] photo by Jeff Vanuga, USDA NRCS.)

environment—animals stand on concrete or in manure-covered feedlots without a blade of green grass. Livestock are typically healthier on pasture compared to those in CAFOs. For example, "Decreased incidence of lameness and teat injuries, fewer abnormal behaviors, and longer bouts of resting" are associated with dairy cows with access to pasture (Campbell-Arvai, 2009). In organic farming systems, the mandate for pasture feeding is generally regarded as beneficial to animal health and welfare (Clark, 2009). While dietary differences and environmental factors may influence animal health, one should also consider that livestock may benefit from the same perceptual pleasure that humans derive from contact with natural settings, in this case direct contact with soils and grass. Nevertheless, the dietary benefits from pasture-based farming for both the animal producing and the human eating the meat, milk, or eggs comes along with the psychological health benefits derived from the aesthetic value of pastoral landscapes (Karsten et al., 2010).

The aesthetic value of open spaces, provided by farms in highly urbanized states like New Jersey, is recognized and supported by public policy. The almost universal appeal of pastoral landscapes provided by traditional livestock farms where animals are pastured appears to be deeply rooted in human behavior. For example, the human preference for a savanna-like environment is essentially duplicated in the form of a "lawn," found surrounding nearly every home (Kaufman and Lohr, 2002) but usually with pets instead of livestock. A great deal of time, money, and effort is devoted to the re-creation of this "pastoral landscape." Conventional lawns do little to feed the growing human population, but the movement toward keeping poultry or even a family goat or cow demonstrates the potential to provide both a pastoral scene and food.

Even though adults and children from urban and suburban areas only occasionally visit or encounter agricultural landscapes (as in shopping for food at on-farm markets) or view farms or scenes along a highway as places for recreation and entertainment (Ulrich, 1974), these experiences still may have long-term value. Surveys of adult attitudes and sensitivities toward nature and the environment indicate that they are shaped by their surroundings and experiences during childhood (Lohr and Pearson-Mins, 2002). Although active involvement with nature has a stronger influence than passive involvement, Lohr and Pearson-Mins (2002) argue that "any involvement with nature during childhood" has positive value.

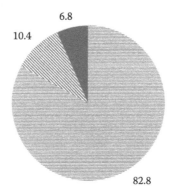

≣ Metropolitan statistical areas

≋ Micropolitan statistical areas

■ Outside metropolitan and micropolitan statistical areas

FIGURE 11.8 Population distribution in the United States during the 2010 census. Numbers are the percent of the U.S. population that lives in each area. Metropolitan Statistical Areas have a core urban area population of 50,000 or more, and Micropolitan Statistical Areas have a core urban area population of 10,000–49,999. (Data from U.S. Census Bureau, 2011, *Population Distribution and Change: 2000 to 2010*. http://www. census.gov/prod/cen2010/briefs/c2010br-01.pdf, verified 3 February 2012.)

The opening of increasing numbers of farmers' markets is a positive trend occurring at a time when over 80% of the U.S. population now lives in Metropolitan Statistical Areas (Figure 11.8) (U.S. Census Bureau, 2011). Perhaps consumers are seeking not just access to authentic farm-fresh foods but also a chance to reconnect with nature in ways that are possible when farm-to-consumer relationships are respected and even encouraged. Unfortunately, direct farm-to-plate relationships have increasingly come under attack under the pretense of food safety (Gumpert, 2009). The need for protection of farm-to-consumer relationships has even spurred the founding of a legal defense fund (Farm-to-Consumer Legal Fund, 2007).

Agriculture does more to alter the surface character and appearance of planet Earth than any other human activity. A critic of modern industrial farming systems has cogently remarked: "Agriculture is the dilemma of civilization because it both separates us from nature and connects us to it." Fortunately, highly productive, alternative agricultural systems are possible and can both feed people and satisfy the human need for a pleasant landscape experience (Kirschenmann, 2007). A prominent example is Joel Salatin's Polyface Farm where rolling green pastures are dotted with herds of beef cattle and flocks of chickens (Purdum, 2005). Such pasture-based farming systems significantly decrease plowing, tillage, erosion, and herbicide use, all of which can mar landscape quality.

11.7 CONCLUSIONS

In summary, beyond providing food and shelter, soil scientists and agronomists, in following the lead of horticulturalists, may find new opportunities by asking and researching questions about the destructive influence of landscapes that appear lifeless because of commonly used cultural practices. Examples include how tillage exposes soils to erosion (Figure 11.9), how the use of burndown herbicides deadens green landscapes (Figure 11.10), and how monocropping versus diverse complex crop rotations changes the aesthetic value of landscapes (Figure 11.11). The potential impact on the physiological and psychological health of the viewing public has not yet reached the consciousness of many agriculturalists, but there are many potential benefits to applying the same functioning principles of healing landscapes at hospitals, schools, or parks to healing landscapes on farms. Widespread adoption of pasture-based farming systems would appear to be one healthy step toward providing healthy, green, healing farmscapes.

FIGURE 11.9 This tilled field has experienced severe soil erosion, as evidenced by the light colors on hilltops due to topsoil loss and the numerous rills and gullies cut into the field's surface. (Photo by Lynn Betts, USDA NRCS.)

FIGURE 11.10   In contrast to the surrounding green areas, the herbicide burndown has left this field with a dead look. (Photo by Joseph Heckman.)

FIGURE 11.11   A diverse assemblage of crops within a field provides an array of colors and textures that create an aesthetically pleasing landscape. (Photo by Joseph Heckman.)

REFERENCES

Balfour, E.B. 1975. *The Living Soil and the Haughley Experiment*. Universe Books, New York, NY.

Barton, J. and J. Pretty. 2010. What is the best dose of nature and green exercise for improving mental health? A multi-study analysis. *Environmental Science and Technology* 44: 3947–3955.

Bear, F.E. 1962. *Earth: The Stuff of Life*. University of Oklahoma Press, Norman, OK.

BFGA. 2011. *Who Was Rudolf Steiner?* http://www.biodynamics.com/steiner.html. Accessed 12 May 2011 (verified 5 January 2012).

Bidwell, O.W. and F.D. Hole. 1965. Man as a factor of soil formation. *Soil Science* 99: 65–72.

Biederman, I. and E.A. Vessel. 2006. Perceptual pleasure and the brain. *American Scientist* 94: 247–253.

Buck, P.S. 1931. *The Good Earth*. John Day, New York, NY.

Campbell-Arvai, V. 2009. Literature Review: A comparison of dairy production systems. Department of CARRS, Michigan State University. http://www.mottgroup.msu.edu/uploads/files/59/Pasture-basedDairyLiteratureReview.pdf. Accessed 5 April 2011 (verified 5 January 2012).

Cassman, K. 2007. Editorial response by Kenneth Cassman: Can organic agriculture feed the world—Science to the rescue? *Renewable Agriculture and Food Systems* 22: 83–84.

Cimprich, B. 1993. Development of an intervention to restore attention in cancer patients. *Cancer Nursing* 16: 83–92.

Clark, E.A. 2009. Forages in organic crop-livestock systems. *In* C. Francis (ed.), *Organic Farming: The Ecological System*, pp. 85–112. Agronomy Monograph 54. ASA, CSSA, SSSA, Madison, WI.

DeMeo, J. 2006. *Saharasia: The 4000 BCE Origins of Child Abuse, Sex-Repression, Warfare and Social Violence in the Deserts of the Old World*. Natural Energy Works, Ashland, OR.

Demming, A.H. 1994. Claiming the yard—Desert gardening. *Sierra Magazine*, July/August. http://findarticles. com/p/articles/mi_m1525/is_n4_v79/ai_15518160/ (verified 5 January 2012).

Elvidge, C.D., C. Milesi, J.B. Dietz, B.T. Tuttle, P.C. Sutton, R. Nemani, and J.E. Vogelmann. 2004. U.S. constructed area approaches the size of Ohio. *Transactions, American Geophysical Union* 85: 233–240.

Farm-to-Consumer Legal Fund. 2007. Mission statement. http://www.farmtoconsumer.org/mission-statement. html. Accessed 15 December 2010 (verified 5 January 2012).

Flagler, J. and R.P. Poincelot (eds). 1994. *People–Plant Relationships: Setting Research Priorities*. Food Products Press, Binghamton, NY.

Gumpert, D.E. 2009. *The Raw Milk Revolution: Behind America's Emerging Battle Over Food Rights*. Chelsea Green Publishing, White River Junction, VT.

Hansen, L., E. Noe, and K. Højring. 2006. Nature and values in organic agriculture: An analysis of contested concepts and values among different actors in organic farming. *Journal of Agricultural and Environmental Ethics* 2: 147–168.

Heckman, J. 2006. A history of organic farming: Transitions from Sir Albert Howard's war in the soil to USDA National Organic Program. *Renewable Agriculture and Food Systems* 21: 143–150.

Heckman, J.R., N. Bhashyam, and K. Hair. 2010. Soil organic matter levels under pastures and row-crops. *Soil Science Society of America Annual Meeting Abstracts*. Published on CD-ROM.

Hine, R., J. Pretty, and J. Barton. 2009. *Research Project: Social, Psychological and Cultural Benefits of Large Natural Habitat & Wilderness Experience*. University of Essex. http://www.essex.ac.uk/ces/occasionalpapers/ Kerry/Literature%20Review%20for%20WF.pdf (verified 6 February 2012).

Howard, A. 1943. *An Agricultural Testament*. Oxford University Press, New York, NY.

Jackson, D.L. and L.L. Jackson (eds). 2002. *The Farm as Natural Habitat: Reconnecting Food Systems with Ecosystems*. Island Press, Washington, DC.

Karsten, H.D., P.H. Patterson, R. Stout, and G. Crews. 2010. Vitamins A, E and fatty acid composition of the eggs of caged hens and pastured hens. *Renewable Agriculture and Food Systems* 25: 45–54.

Kaufman, A.J. and V.I. Lohr. 2002. Where the lawn mower stops: The social construction of alternative front yard ideologies. *In* C.A. Shoemaker (ed.), *International Interaction by Design: Bringing People and Plants Together for Health and Well-Being: An International Symposium*, pp. 291–300. Iowa State Press, Ames, IA.

Kirschenmann, F.L. 2007. Potential for a new generation of biodiversity in agroecosystems of the future. *Agronomy Journal* 99(2): 373–376.

Lohr, V.I. and C.H. Pearson-Mins. 2002. Childhood contact with nature influences adult attitudes and actions towards trees and gardening. *In* C.A. Shoemaker (ed.), *International Interaction by Design: Bringing People and Plants Together for Health and Well-Being—An International Symposium*. Iowa State Press, Ames, IA.

Lohr, V.I. and C.H. Pearson-Mims. 2006. Responses to scenes with spreading, rounded, and conical tree forms. *Environment and Behavior* 5: 667–688.

Louv, R. 2008. *Last Child in the Woods: Saving Our Children from Nature-Deficit Disorder*. Algonquin Books, Chapel Hill, NC.

Ober, C., S.T. Sinatra, and M. Zucker. 2010. *Earthing*. Basic Health Publications, Laguna Beach, CA.

Olmsted, F.L. 1870. *Public Parks and the Enlargement of Towns*. Riverside Press, Cambridge, MA.

Pepper, I. 2010. Soils the good the bad and the beautiful. *CSA News* 55: 4–7.

Purdum, T.S. 2005. High priest of the pasture. *The New York Times Style Magazine*, 1 May: 76–79.

Reich, W. 1979. *Selected Writings: An Introduction to Orgonomy*. Farrar, Straus, and Giroux, New York, NY.

Reich, W. 1982. *The Bioelectrical Investigation of Sexuality and Anxiety*. Farrar, Straus, and Giroux, New York, NY.

Relf, P.D. and V.I. Lohr. 2003. Human issues in horticulture. *HortScience* 38: 984–993.

Sempik, J. and J. Aldridge. 2006. Care farms and care gardens. *In* J. Hassink and M. van Dijk (eds), *Farming for Health*, pp. 147–161. Springer, Dordrecht, The Netherlands.

Sing, A. 2004. Farmscaping, farming with nature in mind. *The Canadian Organic Grower* Fall: 56–58.

Tennessen, C.M. and B. Cimprich. 1995. Views to nature: Effects on attention. *Journal of Environmental Psychology* 15: 77–85.

Ulrich, R.S. 1974. *Scenery and the Shopping Trip: The Roadside Environment as a Factor in Route Choice.* Michigan Geographical Publications No. 12. Department of Geography, University of Michigan, Ann Arbor, MI.

Ulrich, R.S. 1984. View through a window may influence recovery from surgery. *Science* 224: 420–421.

Ulrich, R.S. 1986. Human responses to vegetation and landscapes. *Landscape and Urban Planning* 13: 29–44.

U.S. Census Bureau. 2011. *Population Distribution and Change: 2000 to 2010.* http://www.census.gov/prod/cen2010/briefs/c2010br-01.pdf (verified 3 February 2012).

Wendin, W.F. and S.L. Fales (eds). 2009. *Grassland, Quietness and Strength for a New American Agriculture.* ASA, CSSA, and SSSA, Madison, WI.

Wilson, E.O. 1993. Biophilia and the conservation ethic. *In* S.R. Kellert and E.O. Wilson (eds), *The Biophillia Hypothesis*, pp. 31–41. Island Press, Washington, DC.

12 Organic Farming
Impacts on Soil, Food, and Human Health

Patrick M. Carr, Kathleen Delate, Xin Zhao,
Cynthia A. Cambardella, Pattie L. Carr, and
Joseph R. Heckman

CONTENTS

12.1 INTRODUCTION

The importance of responsible stewardship in managing soil is a central tenet of organic farming. Organic farmers believe that practices that stimulate biology and overall quality of soil enhance production of healthy and nutritious crops. Few involved in agriculture would argue this point. Nevertheless, disagreement exists among agriculturists about the relative importance that should be placed on organic farming for meeting global food needs in the twenty-first century. Organic farming proponents insist that careful management of on-farm ecological processes creates soil capable of supplying adequate amounts of nutritious food for a growing world population, with reduced need for off-farm inputs. Critics contend that organic farming can degrade soils and will contribute to the underproduction of food crops that are no more nutritious than foods grown using synthetic agrichemicals (i.e., conventional farming), if adopted on a large scale. This chapter was not written to settle the debate but to provide a historical context for the belief that organic methods promote soil quality, to compare organic and conventional farming systems regarding impacts on soil quality and crop production, and to summarize comparisons in food quality between the two farming systems. While these three topics cannot be discussed in detail in a single chapter, enough information is provided to give readers with limited knowledge of organic farming a better understanding of why consideration of this farming method is appropriate in a discussion about soil and its relation to human health. Although organic farming is practiced globally, space limitations restrict most of the focus in this chapter to organic farming within the United States. In addition, while it is important to emphasize the tremendous impact that animals have on soil and food quality in integrated organic systems that include crop and livestock enterprises, space constraints virtually limit the discussion in this chapter to organic crop/plant production systems.

12.2 HISTORICAL CONTEXT OF ORGANIC FARMING: SOIL HEALTH CONNECTION

12.2.1 Introduction

The International Federation of Organic Agriculture Movements (IFOAM) is considered one of the leading global organic farming organizations, comprising organic farmers, processors, and marketing groups (Sansavini and Wollesen, 1992). The high value placed on soil and good stewardship of this resource are included in the definition of organic farming on the IFOAM web page: "Organic agriculture is a production system that sustains the health of soils, ecosystems and people. It relies on ecological processes, biodiversity and cycles adapted to local conditions, rather than the use of inputs with adverse effects" (IFOAM, 2011). This definition reflects a persistent emphasis given to farming practices that maintain or enhance soil health by organic farming proponents from the early days of the "modern" organic farming movement. Others have extended the positive impacts of organic farming methods to soil quality (Liebig and Doran, 1999), which has been defined as the capacity of a soil to function, within natural or managed ecosystem boundaries, to sustain plant and animal productivity, maintain or enhance water and air quality, and support human health and habitation (Doran and Parkin, 1994; Karlen et al., 1997).

12.2.2 Humus and Mineralist Theories: Eighteenth and Nineteenth Centuries

Francis and Van Wart (2009) and others (e.g., Korcak, 1992) provide a brief history of some of the earliest recorded organic production methods in Europe, Asia, and the Americas. The evolution of these practices, along with discoveries about the soil–crop interface, created diverging schools of thought on proper soil management for optimum food crop production. One of the prevailing theories by the eighteenth century was that crops derived nutrients, including carbon (C), from the soil or "humus," as the term was then applied (Korcak, 1992). This explains the emphasis placed on tillage by Jethro Tull in the 1700s, since small soil particles could be extracted by plant roots more easily than larger particles (Wicker, 1957). The German agronomist Abrecht Thaer emphasized the central role that soil played in plant nutrition in *Principles of Rational Agriculture*, published in 1809. As summarized by Manlay et al. (2007), Thaer concluded that "the majority of plant dry matter derives from the soil nutritive juices contained in the fraction of soil organic matter that is soluble in hot water." Treadwell et al. (2003), along with Waksman (1942) in Korcak (1992), credited Thaer with being responsible for formulating the humus theory of plant nutrition.

An alternative to the humus theory was the mineral theory of plant nutrition, where plants obtained nutrients contained in mineral compounds and not directly from soil organic matter. Justus von Liebig was a vocal proponent of the mineral theory and is often credited with being the first scientist to argue that plant dry matter is derived from inorganic rather than organic compounds. In fact, Liebig largely synthesized earlier ideas developed by Carl Sprengel and others about processes governing plant–nutrient relationships (van der Ploeg et al., 1999; Heckman, 2006). Liebig's persuasive arguments convinced many agricultural scientists to abandon the humus theory, in part because the mineral theory explained plant response when fertilizer inputs were supplied in nutrient-deficient environments (Manlay et al., 2007). By the late 1800s, most agricultural scientists had adopted the mineral theory to guide their understanding of plant nutrient uptake. This is reflected in modern soil fertility textbooks (e.g., Tisdale et al., 1993), where the mineral theory of plant nutrition is presented in considerable detail. Interestingly, research suggests that plants can extract organic forms of nutrients, in addition to inorganic forms, in some environments (Owen and Jones, 2001; Jones and Darrah, 1994).

Adoption of the mineral theory had significant implications for farming systems. Emphasizing form rather than source of nutrients elevated mineral compounds to equal status with soil organic matter from a plant nutrition standpoint. Recognizing this, Liebig, along with John Lawes and Joseph Gilbert, at what became known as the Rothamsted Experiment Station in England, became early advocates of mineral fertilizer additions to correct nutrient deficiencies in agricultural soils (Manlay et al., 2007). As reported by Conford (2001), the "Broadbalk" field experiment at Rothamsted demonstrated that adequate fertility could be provided to maintain wheat yield for over 100 years by the application of mineral fertilizer alone, eliminating the need to supply the soil with animal manure or other organic amendments. While it took decades before the manufacture and use of synthetic fertilizer became widespread, the rationale for replacing regular applications of organic materials (e.g., animal manure applications) with mineral fertilizers was established.

12.2.3 Birth of the Modern Era of Organic Farming: Twentieth Century

The influence of the mineral theory on plant nutrient management became widespread during the twentieth century. Mechanization in agriculture also occurred during this period, as did reliance on synthetic pesticides. Not all agriculturists were enamored with these new developments, particularly those that seemed to diminish the importance of soil organic matter in agricultural systems designed for long-term sustainability. A countermovement developed that was and still is opposed to many practices adopted by the majority of farmers in conventional agricultural systems. One of the most influential individuals of the movement was Sir Albert Howard (1873–1947), considered to be one of the pioneers of the modern organic farming movement, which Conford (2001) suggested had begun by the late 1920s. An excellent summary of Howard's impact on modern organic farming practices is provided by Heckman (2006). Chief among them was Howard's insistence on the "Law of Return" or the need to recycle organic waste materials back onto fields so that soil organic matter levels were replenished. Howard knew firsthand the positive impact that recycled waste had on crop response from his experience in developing the Indore method of composting while director of an agricultural research center in India. As retold by Heckman (2006), Howard described the mixture of plant, animal, ash, and urine-soaked soil, along with pile size, heat, moisture, and aeration, that was used to create compost using this method in *An Agricultural Testament* (Howard, 1943), and he encouraged the application of all composted organic wastes onto fields. He even advocated returning sewage sludge onto farmland, a practice that is prohibited under National Organic Program standards developed by the U.S. Department of Agriculture (USDA) (USDA-AMS, 2011) because of the potential for toxins in today's municipal sewage.

Albert Howard was as critical of the use of synthetic fertilizer as he was an advocate of compost. Similarly, he condemned modern agricultural, small-plot research methods, many of the statistical tools used to analyze crop experiments, and "the unsoundness" of research at the Rothamsted Experiment Station, including the Broadbalk field experiment (Howard, 1947). He believed that Liebig was a qualified chemist but lacked an appreciation of the field of soil biology and its importance in soil fertility. He decried the age of the "laboratory hermit" where scientists applied fertilizer to plants potted in a sand medium and then extrapolated results to guide nutrient management in farm fields. Howard noted rightfully that the controlled-environment studies failed to account for the huge impact that macrofauna and microfauna have in nutrient cycling and other processes in soil. In many ways, Howard was ahead of his time in recognizing the importance of soil organisms in sustainable crop production.

Howard made important contributions to our understanding of the soil–plant interface, most notably in India. However, his incessant criticism of many others in the scientific community cost Howard much of his reputation among peers. Eventually, Howard lost virtually all of his credibility among fellow scientists because of his extreme views, even though Hershey (1992) suggested that many of Howard's criticisms were justified to a point. Some of his beliefs seem extreme even today, such as a firm conviction that the source of virtually all human disease can be traced to the soil and an inability to supply specific proteins to plants (Howard, 1947).

A more controversial figure than Albert Howard in the area of alternative farming methods was Rudolf Steiner (1861–1925). Like Howard, Steiner is considered an important figure in the modern organic farming movement (Treadwell et al., 2003). Steiner was the founder of the school of anthroposophy in 1913, which includes what is today known as the biodynamic method of farming (Biodynamics, 2010). Steiner had a deep appreciation for traditional methods of farming and was critical of the use of mineral fertilizers, specialization, and other practices that were being promoted by most agricultural scientists (Conford, 2001). Steiner also linked human health to plants and farming practices, stating that the use of synthetic fertilizer invariably leads to a loss in the nutritive value of crops. However, as noted by Treadwell et al. (2003), unlike Howard, Steiner also incorporated "astral and zodiacal forces" into his soil and crop management philosophies. Howard rejected many of Steiner's theories outright, as do many agricultural scientists today (Hershey, 1992). Nevertheless, Steiner and his teachings greatly affected the thinking of some members of the emerging organic farming community in the 1920s, both in Europe and abroad.

Jerome Rodale (1898–1971) was a very influential figure in popularizing organic farming concepts in the United States. Rodale grew up in New York City and had no practical farm experience. However, after reading *An Agricultural Testament* by Sir Albert Howard in 1940, Rodale became a tireless advocate of Howard's teachings on agricultural sustainability (Kelly, 1992). Rodale purchased a farm in Pennsylvania to put into practice many of Howard's teachings and began publishing the magazine *Organic Farming and Gardening* in 1942. He popularized many of Howard's concepts about the relationship between soil and human health in his magazine, even convincing Howard to serve as an associate editor until his death. Initially, the magazine attracted very few subscribers, but hard work and promotion by Rodale resulted in consistent growth in the number of readers, particularly among gardening enthusiasts. By 1980, *Organic Gardening*, as the magazine then was known, had over 1 million subscriptions. Readership has declined some since that time, but there still were over 350,000 subscribers in June 2010 (G. Horsford, personal communication).

There are many other pioneers of the modern organic farming era. Lady Eve Balfour (1899–1990) was so influenced by the writing of Sir Albert Howard that she and others established the Haughley Experiment in 1939 in England, with the goal of demonstrating the advantages of "ecological" farming over conventional methods on large tracts of land (Balfour, 1977). William Albrecht (1888–1974), like Howard and many others, linked soil quality to human health, even suggesting that social problems in some regions may be linked to poor soil quality (Conford, 2001). Lord Northbourne was the first to use the term "organic" in the context of farming as described by Steiner, Howard, and others (Scofield, 1986). Readers are referred to Conford (2001), Treadwell et al. (2003), and others referenced in this section (e.g., Heckman, 2006) for thorough discussions of the key figures and their contributions to the development of organic farming concepts during the modern era.

The relationship between organic farming proponents and most agricultural scientists was contentious by the 1940s. Scientists at U.S. land grant institutions—the universities where the majority of agricultural research occurred—were among the most critical of organic farming methods. The dean of agriculture at one institution even referred to organic farming proponents as members of a "cult" and believed that their activities could stymie the progress of "important agricultural research" (Treadwell et al., 2003). This harsh criticism continued through the 1960s, with agricultural scientists frequently making reference to the "organic cult" and ridiculing advocates for their unwillingness to accept theories about soil–plant interactions that were widely accepted and being taught at most U.S. land grant institutions (Truog, 1963).

A shift in view within the scientific community was signaled by a report on organic farming published by the USDA in 1980 (USDA, 1980). The report indicated that there was increasing concern about how soils could be mismanaged when using conventional farming practices and, conversely, how many soil and crop management practices used by organic farmers were considered "best management practices." This publication was followed by a symposium during the annual meeting of the American Society of Agronomy in 1981, where scientists discussed some of the opportunities, but also challenges, offered by organic farming methods (Bezdicek and Power, 1984). Two organic farms were included among 11 farms highlighted in *Alternative Agriculture* by the National Research Council, published in 1989, where the need for and advantages of adopting alternatives to conventional farming methods were discussed (National Research Council, 1989).

12.2.4 ORGANIC FARMING CERTIFICATION AND CURRENT STATUS

The importance of regulating organic farming practices led to the first certification body of Demeter International, a biodynamic association, in 1928 (Demeter, 2010), which was followed by IFOAM in 1972 (IFOAM, 2011) and California Certified Organic Farmers (CCOF) in 1973 (CCOF, 2010). In 1990, the U.S. government passed the Organic Food Production Act (OFPA), which codified certified organic practices. In 2002, U.S. organic standards went into effect, creating a uniform set of regulations across all certifiers in the United States. At the heart of the regulations is the protection or enhancement of soil organic matter, which has been shown to be critical for maintaining soil quality in sustainable systems (Manlay et al., 2007).

Interest in research into organic farming methods began to increase among scientists in the 1980s, reflecting the growth in popularity of organic farming methods among commercial producers and consumers. Still, distrust in organic farming methods remains among some agricultural scientists because of a perceived inability of organic farming to supply adequate food for a growing world population (Trewavas, 2001; Cassman, 2007). Critics also suggest that organic farming is inferior to no-tillage conventional farming methods for enhancing soil health and soil quality, and organic food has nutrition equivalent to that of conventional food and, in some instances, actually may pose a greater health risk when consumed, based on unproven beliefs about manure contamination of organic foods (Trewavas, 2004). The remainder of this chapter discusses the recent literature comparing organic farming and conventional farming methods regarding their impact on soil quality and subsequent effects on food crop quality.

12.3 COMPARISONS OF ORGANIC AND CONVENTIONAL FARMING

12.3.1 THE EARLY YEARS: 1940s THROUGH 1980s

Comparisons of organic and conventional farming methods started prior to the 1970s. Lady Eve Balfour began the Haughley Experiment in England in 1939 (Balfour, 1977). Jerome Rodale established the Soil and Health Foundation 8 years later in Pennsylvania, which included a laboratory and a grant program that was established to provide funding for scientists working on organic farming methods. These early efforts to demonstrate the benefits of adopting organic farming methods were generally unsuccessful. As summarized by Conford (2001), results generated by the Haughley Experiment were inconclusive, and the laboratory and grants program established by Rodale were "... an almost total failure" for several years. Part of the challenge faced by Rodale was an inability to attract widely respected members of the scientific community to conduct research on organic farming systems. This unwillingness to study organic farming methods began to change as a younger generation of scientists emerged in the 1960s and 1970s, with some coming from nonagricultural backgrounds.

A group of scientists at Washington University in St. Louis began a 5-year study on organic farming in 1974. The motivation for this effort was the 1973 oil embargo and concerns that a long-term disruption in the fossil fuel supply could impact conventional U.S. farming methods (Lockeretz et al., 1984). The group used surveys and other research tools to compare agronomic, economic, and energetic performances of organic and conventional farms in five U.S. states; also soil, insect, and crop yield data were collected from 26 pairs of fields managed organically and conventionally and analyzed in side-by-side comparisons (Lockeretz et al., 1980). Differences in crop yield were not detected between the organic and conventional fields; however, there was a trend for crop yields to be higher for conventional fields under favorable growing conditions and lower under adverse growing conditions. Soils comprising the organic fields were higher in organic C and total nitrogen (N), even though grain protein content was higher when harvested from the conventional fields. However, soil phosphorus levels were lower in the organic fields. No differences were detected between soils under the two contrasting management systems for exchangeable potassium, cation exchange capacity, or pH. Results of this study suggested some potential soil quality advantages when soils were managed organically, with the exception of a soil phosphorus limitation if levels were not replenished with manure or compost over time.

A study was established in 1975 at the University of Nebraska Agricultural Research and Development Center at Mead to determine if farming system strategy (organic vs. conventional) affected soil microbial populations, as well as selected soil chemical and physical properties, in the upper 30 cm of the soil profile. By 1982, soil organic C, Kjeldahl N, and potentially mineralizable N were 20%–40% greater in organic plots compared with conventional plots (Fraser et al., 1988). Microbial populations and activity mirrored soil chemical differences between organic and conventional systems. The researchers attributed the enhancements in soil quality to manure applications in the organic plots, in contrast to the synthetic fertilizer applications in the conventional plots. A USDA study team relied heavily on case study analyses of 69 organic farms in developing their Report and Recommendations on Organic Farming (USDA, 1980). The study team observed that "[m]uch of the research conducted to date that relates to organic farming has been somewhat piecemeal and fragmentary." The study team urged that organic farming researchers become more systematic in their approach and include well-designed studies so that the relative performance of organic and conventional farming methods could be assessed quantitatively.

In 1981, the Rodale Institute, then the Rodale Research Center, began a field study that compared organic cash grain, organic cash grain and livestock, and conventional cash grain cropping systems (Harwood, 1984). The three systems were established in an experimental design that was valid scientifically and subjected data to analyses of variance. Results indicated that maize (*Zea mays* L.) yields were lower under organic management from 1981–1985, while soybean (*Glycine max* L.) yields were equal or higher (Liebhardt et al., 1989). However, differences in maize yield between the organic and conventional systems narrowed over time. This tendency for crop yield to decline over a 3- to 5-year transition phase when moving from conventional to organic management, and then to rebound, is a common experience (Power and Doran, 1984). However, the reason for the "organic transition effect" is debated; Lotter (2003) suggested that some research implicated soil biology changes as explaining the decrease and then increase in crop yield, while results of other studies did not. Martini et al. (2004) suggested that improvements in management skill rather than soil quality considerations may explain the change in crop yield during and shortly after the transition from conventional to organic management.

Soil quality was improved by organic management in the Rodale Farming Systems Trial. There were no differences in soil quality parameters (organic matter, C, N, phosphorus, potassium, calcium, and magnesium contents; cation exchange capacity; and pH; Liebhardt et al., 1989) across the organic and conventional cropping systems when the study began. By 1985, soils under organic management during particular crop phases had more water-stable aggregates, faster water infiltration rates, and greater microbial biomass and activity than soils under conventional management did (Peters, 2004). By 1995, soil C levels had increased by 7%–12% in plots managed organically, whereas no change in soil C levels was detected in conventional plots (Drinkwater et al., 1998). By 2002, the soils farmed organically had greater populations of arbuscular mycorrhizae fungi spores

(Pimentel et al., 2005). The presence of arbuscular mycorrhizae fungi in agricultural soils has been suggested as an important indicator of a biologically active and healthy soil (Hamel and Strullu, 2006).

The Sustainable Agriculture Farming Systems (SAFS) project was established in 1988 at the University of California at Davis Agronomy Farm. Four-year rotations of grain, oilseed, and vegetable crops were compared under organic, "low-input" (i.e., fertilizer and pesticide inputs were reduced, with greater reliance on legume cover crops to supply N needs and mechanical cultivation for weed control), and conventional management, along with a 2-year vegetable–grain crop rotation. Soil C, along with soluble phosphorus, exchangeable potassium, and pH, was higher in organic and low-input systems than in conventional systems at both 4 and 8 years after the study began (Clark et al., 1998). The differences in soil chemical properties between cropping system strategies generally were predicted by crop nutrient budgets and reflected the application of manure to organic and low-input systems, and the use of synthetic fertilizer in the conventional systems. Consistent differences in crop yields were not detected between cropping systems in the study (Poudel et al., 2002).

A side-by-side comparison of two fields, with one having never received synthetic fertilizer inputs and only limited pesticide applications, while the other received synthetic fertilizer and pesticide inputs regularly since the early 1950s, was conducted in the Palouse region of Washington. Soil samples collected in 1981 and 1982 indicated that the field managed "organically" had greater levels of urease, phosphatase, and dehydrogenase activity than the field managed conventionally, indicating higher levels of soil microbial activity in the field under organic management (Bolton et al., 1985). The researchers were unable to detect differences in soil microorganism numbers between the two fields using plate counts, but microbial biomass was greater in mid-July as winter wheat (*Triticum* spp.) approached maturity and postharvest after late-season tillage in November in the organic field. Subsequent research indicated that more erosion had occurred in the field managed conventionally, which Reganold (1988) attributed to differences in cropping practices used on the two fields, most notably the inclusion of a green manure crop in the organic rotation. Overall crop yield averaged 8% less on the organic farm than on the conventional farm, based on estimates provided by farm managers and USDA Agricultural Stabilization and Conservation Service personnel (Reganold et al., 1987). However, crop yield on the organic farm was comparable to a second conventional farm located nearby.

The impact of organic and conventional management was compared on commercial farms in central California involved in fresh market tomato (*Lycopersicon esculentum* Mill.) production during 1989 and 1990. Differences in tomato fruit yield were not detected between the two farm management types, but potentially mineralizable N, microbial biomass and diversity, and parasitoid number and species diversity were greater in fields managed organically (Drinkwater et al., 1995). The researchers noted that, in soils managed organically, soil organic matter levels tended to be greater than in conventionally managed soils and attributed this enhancement to the more frequent vegetative cover from leguminous green manure crops for soil nutrient supplementation on organic farms, compared with conventional farms where generally only synthetic fertilizers were used to provide nutrients to crops.

12.3.2 THE LATER YEARS: 1990S TO PRESENT

Several long-term studies comparing organic and conventional farming systems took place during the 1990s. The Long-Term Research on Agricultural Systems comparison was established near Davis, California, in 1993, as a companion study to the 5-year-old SAFS project. Ten 2-year rotations were compared for crop yield and impact on soil quality, along with other environmental factors (e.g., water use). An organic farming system consisting of tomato, maize, and winter legume cover crops was included along with nine other cropping systems managed conventionally. Maize yields were 34% lower following a 3-year transition period in the organic system than in the conventional tomato–maize system between 1996 and 2002 (Denison et al., 2004). The researchers speculated that the yield depression in the organic system could have resulted from a delay in planting maize to prevent buildup of seed corn maggot (*Hylema cilicrura*) after the winter legume cover crop was incorporated, since fresh organic matter additions attract seed corn maggots. Tomato yield was comparable between the two systems during that same period. Martini et al. (2004) reported that composted poultry litter was applied to plots managed organically, and soil C and N content was greater in these plots than in conventional plots. Results of fatty acid methyl ester (FAME) analyses indicated that organic and conventional plots differed in soil microbial community composition.

The Wisconsin Integrated Cropping Systems Trial began in 1989, although crop rotations included in the study were not established until 1990. Crop yields in organic and conventional grain-based and forage-based cropping systems were compared for a 6- to 9-year span, depending on the crop. Maize yield averaged 9% less in the organic grain-based system and 12% less in the organic forage-base system compared with the respective conventional counterparts (Posner et al., 2008). Similarly, soybean managed organically yielded 8% less than soybean managed conventionally. In contrast, alfalfa (*Medicago* spp.) managed organically yielded 22% more during the establishment year and 10% more during the second year than alfalfa managed conventionally. Other researchers likewise reported that alfalfa and other forages managed organically were higher yielding than when managed conventionally (Entz et al., 2001). More total hydrolysable N and amino acid N occurred in the top 8 in. of soil under the organic grain-based cropping system than under a maize monoculture managed conventionally in the Wisconsin study, but there was no difference detected in organic C, potentially mineralizable C and N, or amino sugar N content between soils under the two different management systems (Wander et al., 2007). The lack of soil quality differences between soils under organic and conventional management systems was attributed to a history of manuring at the site prior to establishing the study, and the unusually high levels of labile N and associated soil factors that were detected, as manure applications can affect soil chemical properties for decades after being discontinued (Jenkinson, 1991).

The Farming Systems Project was established in 1996 at the USDA-ARS Agricultural Research Center at Beltsville, Maryland, to compare relative performances of organic and conventional farming systems in Coastal Plains soils in the mid-Atlantic region. Maize yield averaged 20%–40% less under organic management compared with conventional management over a 9-year period, depending on the organic and conventional systems compared (Cavigelli et al., 2008). Maize yield increased as crop rotations became more diverse and included forages, particularly legumes or crop mixtures including legumes. Low N availability explained over 70% of the variability in organic maize yields. Yields of soybean were an average of 19% lower under organic compared with conventional management over the 9-year period, with weed competition explaining the yield depression under organic management.

The Neely-Kinyon Long-Term Agricultural Research site in Greenfield, Iowa, was established in 1998 to determine if organic farming systems could produce competitive yields, while maintaining soil quality and plant protection, compared with conventional systems. A 2-year maize–soybean rotation managed conventionally was compared with three organic systems: a 2-year rotation of soybean and rye (*Secale cereale* L.) or wheat; a 3-year rotation of maize, soybean, and alfalfa+oat (*Avena sativa* L.) mixture; and a 4-year maize–soybean–alfalfa+oat mixture–alfalfa rotation. Maize and soybean yields in the organic systems were equal to or greater than those produced in the conventional system during the first 4 years of the study (Delate and Cambardella, 2004). Likewise, differences in insect pest damage and numbers were not detected for either maize or soybean across organic and conventional cropping systems. There was no consistent effect of management system type (i.e., organic and conventional) on weed population, unlike other long-term studies (e.g., Teasdale et al., 2007), where weed competition was greater under organic management.

Results of several other long-term studies comparing organic and conventional farming systems have been reported in the literature, and readers are referred elsewhere for discussion of these (e.g., Benitez et al., 2007; Paul et al., 1999; Porter et al., 2003). Marriott and Wander (2006a,b) compared the impact of organic and conventional farming systems on selected soil quality parameters in nine long-term studies, some of which have been discussed. The two researchers concluded from soil analyses across the studies that particulate organic matter content was greater in soils managed organically than conventionally. Organic farming practices increased surface soil C and total N concentrations by an average of 14%, and that increase was irrespective of the legume-based organic systems including manure applications or not. Results of this research indicated the propensity for soil quality to be enhanced by organic farming compared with conventional farming in long-term studies across the United States.

12.3.2.1 Tillage Effects on Soil Quality

The effect of tillage on soil quality is a common theme in many cropping system studies comparing organic and conventional farming systems (e.g., Reganold, 1988). The deleterious effects that tillage can have on soil quality are well known, including decreases in water-stable aggregates (Green et al., 2005), organic C and N (Ismail et al., 1994), and microbial populations (Karlen et al., 1994) in surface soils. Crop yields are often greater under conventional no-till than in conventional tilled systems (Carr et al., 2006; Tanaka et al. 2002; Triplett and Dick, 2008). Experiments were conducted where tillage was eliminated during specific crop phases of a crop rotation in systems managed organically (Creamer and Dabney, 2002; Ashford and Reeves, 2003), but not throughout an entire cropping system, as in conventional no-till. Triplett and Dick (2008) point out that the benefits of no-till farming tend to be greatest when tillage is eliminated entirely from a cropping system. Trewavas (2004) suggested that conventional no-till farming may be superior to organic farming for enhancing soil quality since tillage can be reduced but, so far, not eliminated entirely from organic farming systems. Robertson et al. (2000) warned that organic farming and similar food production systems relying on tillage contributed more to greenhouse gas emissions than conventional no-till systems, even though the greenhouse gas data for the organic system were not at problematic levels.

The Sustainable Agriculture Demonstration Project, established in 1994 by USDA scientists at Beltsville, Maryland, to compare three conventional no-till farming systems with an organic system where some tillage was used, included cover crops in two of the conventional no-till systems and the organic system. Results over 9 years indicated that maize yields were depressed by 16%–28% in the organic system compared with the conventional systems, depending on the systems being compared (Teasdale et al., 2007). No difference was detected in the yields of wheat between the organic system and the conventional no-till system that excluded cover crops, while organic wheat yield was 20% lower compared with the conventional system that included wheat and cover crops. The yield reduction probably resulted from excessive weed pressure in the organic system, which worsened over time. However, soil C and N levels were higher under organic than conventional management, presumably because of regular additions of animal manure and extensive use of green manure crops in the organic system. The researchers suggested that organic systems might offer "improved soil quality and yield-enhancing benefits" compared with conventional no-till systems if weeds in organic systems were controlled. They conceded that similar benefits might result if organic soil amendments and perennial crops were incorporated into conventional no-till systems.

A cropping system that transitioned to certified organic management was compared to four conventional no-till systems between 2000 and 2003 in Montana. In contrast to the study by Teasdale et al. (2007), where crop yields typically were lower under organic compared with conventional no-till management, wheat yield in the organic system was equal to or greater than that in the conventional systems in the Montana study (Miller et al., 2008). Grain protein content was lower in the organic than the conventional no-till systems, but this depression probably resulted from a dilution of N from the higher organic yield, since total grain N yield (the product of grain yield and N content) was comparable between organic and conventional no-till systems. Potentially mineralizable N was 23% higher in the organic system at the end of the 4-year period, presumably because of the inclusion of a leguminous green manure crop in the organic system since no compost or manure was applied. Results of the Montana study suggested that selected soil quality parameters were enhanced by organic farming compared with conventional no-till systems, while weed pressure was increasing in the organic system by the end of the study. This was also the case in the study by Teasdale et al. (2007).

12.3.2.2 Cropping Practices That Promote Soil Quality

Considerable research indicates that soil quality and microbial processes can be enhanced by organic farming compared with conventional farming. However, other studies suggest that there are no consistent benefits when organic farming is compared with conventional farming methods. For example, Shannon et al. (2002) failed to detect a consistent difference in total soil or microbial C between soils under organic and conventional management in England, and Gardner and Clancy (1996) generally were unable to detect differences in topsoil depth, bulk density, and organic matter levels between soils managed organically and conventionally in the north central USA. This apparent inconsistency in the literature might be explained by differences in specific cropping practices between the organic and conventional farming systems, and not, *per se*, in the management strategy, that is, organic versus conventional.

Organic crop rotations are typically more diverse than conventional crop rotations, and organic farmers use green manure crops and organic inputs to a much greater extent than their conventional counterparts (Buttel and Gillespie, 1988; McCann et al., 1997; Taylor et al., 1992). Forages are grown on many organic farms as a rotational crop and less frequently on conventional farms. Many researchers incorporate these differences in their comparisons of the farming systems so that the research has relevance to farmers' actual situations. Unfortunately, the impact that different crops, rotations, and inputs have on soil quality can be profound and can confound attempts to separate management systems (i.e., organic vs. conventional) from cropping practices (e.g., presence or absence of green manure crops).

The benefits of crop rotation diversity on agricultural system resiliency and performance have been recognized for millennia (Higgs et al., 1990) and extend to soil quality factors (Karlen et al., 2006). Thus, crop rotations that are more diverse—as usually is the case for organic systems in comparison studies with conventional systems—will benefit from an obvious rotation effect that would not exist if the same rotation was used across both systems. Similarly, inclusion of forages, particularly perennial species, seems to provide additional soil quality benefits that are not achieved when the crop rotation is limited to annual grain and seed crops (Entz et al., 1995, 2002; Karlen et al., 2006; Stanger and Lauer, 2008). Likewise, inclusion of green manure crops in cropping systems is widely recognized as a strategy for improving soil quality (Cherr et al., 2006; Snapp et al., 2005), as is the addition of organic inputs, particularly manure (Russelle et al., 2007). Liebig and Doran (1999) attributed improvements in soil quality to the use of more diverse rotations and the addition of organic inputs in their comparison of organic and conventional farms in the central United States. These two researchers also noted less tillage was used when crop rotations included perennial forages, and perennial forages were not uncommon on organic farms. Diverse rotations incorporating forages and even organic inputs are being used on a growing number of conventional farms. Hence, studies comparing organic and conventional systems at a range of management levels are needed to accurately reflect the differences among integrated systems.

12.4   ORGANIC FARMING AND HUMAN HEALTH

12.4.1   Soil and Food Quality

One of the four pillars upon which organic agriculture is based is the principle of health, and it includes a belief that, "[H]ealthy soils produce healthy crops that foster the health of animals and people" (IFOAM, 2011). A link between human health and soil quality was a central belief among organic farming advocates during the early period of organic agriculture history, as was discussed previously. Readers are reminded that soil also can be a source of human illness in instances where deficiencies of essential nutrients or excesses of heavy metals occur, or when foodborne pathogens are present (Oliver, 1997). Clearly, farming practices affecting soil quality can have a profound effect on human health directly or indirectly through the crops that are produced in the soil and subsequently consumed. Farming practices also can affect human health through impacts on air and water quality, particularly in rural areas.

In general, food quality assessment encompasses sensory attributes (e.g., external appearance, texture, taste, and flavor), nutritive values (e.g., proteins, carbohydrates, minerals, vitamins), and nonnutritive bioactive compounds with health-promoting effects (e.g., phenolics, glucosinolates) as well as undesirable attributes (e.g., microbiological contamination, pesticide residues). Soil affects crop and food quality as a result of physical (e.g., texture, structure, and water-holding capacity), chemical (e.g., pH and its effect on plant nutrient solubility/availability, nutrient supply and its effect on plant uptake of essential nutrients), and biological (e.g., arbuscular mycorrhizae populations and plant uptake of phosphorus, microbial community in the rhizosphere, and plant root function) characteristics. While food quality also is affected by the genetic makeup of crops; local growing conditions; production practices; and postharvest handling, processing, and distribution systems (Hornick, 1992), soil quality remains an important and unique influence and may exhibit complex interactions with other preharvest and postharvest factors. The impact of soil quality on food quality, as well as on other ecosystem services that are provided by high-quality soils, is reflected by Karlen et al. (1997) in their insistence on a multifaceted approach when studying soil quality: "A systems approach is important for soil quality assessments because numerous trade-offs must be considered when trying to meet diverse societal goals such as enhancing water quality, sustaining productivity, ensuring food quality, increasing biodiversity, and improving recreational opportunities."

Several studies have been discussed that indicate that soil quality is improved following the adoption of organic farming practices. The causal reason(s) for this enhancement in soil quality warrants additional research. More work also is needed to elucidate why organic farming has failed to improve soil quality compared with conventional farming in some studies where no advantage to organic farming has been demonstrated. Nevertheless, since most long-term research suggests improvements in soil quality following adoption of organic farming practices and management systems, and soil quality improvements are expected to promote nutritional and health benefits in food, it follows that food produced organically may be more nutritious than food grown conventionally.

12.4.2 Organic Farming and Food Quality

Consumer demand for food produced organically is fueled by increased confidence in the health benefits of organic food, coupled with growing concern about the safety and nutrition of food grown conventionally (Winter and Davis, 2006; Hoefkens et al., 2009a; Zhao et al., 2006). Interestingly, there is disagreement in the literature on the benefits that result when organic food is consumed compared to conventional food. Reganold et al. (2010) noted that out of ten review articles published between 2000 and 2010, eight of the studies found some evidence that food produced organically was more nutritious, whereas two reports found no consistent trends. Reganold's team found in their own study that strawberries (*Framaria* × *amanassa* Duch.) had a longer shelf life, enhanced antioxidant activity, higher ascorbic acid concentration, and more total phenolic compounds when grown organically than conventionally, while phosphorus and potassium concentrations were higher in conventional strawberries. In the same study, sensory panels judged one strawberry cultivar to be sweeter and have better flavor when grown organically. In an earlier experiment comparing organic and conventional apple production, Reganold et al. (2001) found that enhancement of soil quality in organic apple production systems can lead to measurable improvements in fruit nutritional quality, taste, and storability.

Two recent studies compared the effects of organic and conventional production methods on nutritional quality of tomato fruits. Dried tomato samples preserved over a 10-year period from the Long-Term Research on Agricultural Systems project in Davis, California, were analyzed for flavonoid content. Quercetin and kaempferol aglycone contents were higher in tomatoes grown organically than in those grown conventionally (Mitchell et al., 2007). Moreover, within the organic system the flavonoid level in tomato fruits was elevated in the last 4 years of the study when lowering the compost application rate, which suggested that overfertilization, resulting in high levels of readily available N, may not favor phytochemical enhancement. Juice was higher in soluble solids and consistency when prepared from organic tomatoes on four Californian farms in a separate study (Barrett et al., 2007). However, the conventional tomato juice was brighter in red color, contained more ascorbic acid, and had higher total phenolic content than the organic juice. Zhao et al. (2009) found that organic fertilization methods resulted in significantly higher phenolic concentrations in pac choi (*Brassica rapa* L. *chinensis*) compared with synthetic fertilizer applications. The researchers cautioned that differences in insect feeding and crop yield between the two fertilization methods confounded interpretation of the research results, indicating that additional work was needed.

Several scientists reviewed previous research to clarify what is known about the impact of organic farming methods on food quality. Worthington (1998) reported in a review of the literature published over a 50-year period that crops receiving "organic fertilization" were higher in protein and lower in nitrate than crops grown conventionally, and that feeding studies showed better growth and reproduction when animals were provided with grain grown organically. However, she acknowledged challenges in arriving at firm conclusions because of differences in the treatments imposed across studies, the research methods used, and other factors.

Williams (2002) summarized results from a previous review indicating that organic farming methods enhanced, had no effect, and decreased several nutritional factors compared with conventional farming methods. For example, vitamin C content of crops grown organically was greater in 21, the same in 12, and less in 3 studies, while zinc concentration was higher in 4, the same in 9, and lower in 3 instances. The author concluded that few well-controlled, scientific studies have been conducted, making it difficult to determine how, or even if, organic farming methods affect food quality consistently compared with conventional methods. After reviewing approximately 50 previous studies, Bourn and Prescott (2002) also pointed out the need for more scientifically sound research in comparing organically and conventionally produced foods for sensory and nutritional qualities and food safety. They noted inconsistent findings on comparison of sensory attributes and a lack of evidence indicating high microbiological contamination in organic produce. The authors called for attention to the influence of organic production systems on nutrient bioavailability and phytochemical levels as opposed to traditional studies of nutritive values.

Recent reviews of the literature on the effect of organic farming methods on food quality have been mixed. The Organic Center's 2005 "State of Science Review" on antioxidants concluded that organic farming and food processing increased average antioxidant content by about 30% (Benbrook, 2005). Lairon (2010), in a recent review, also reported higher levels of antioxidant compounds in organic plant products such as phenolics. However, meta-analyses by Hoefkens et al. (2009b), involving data from 74 studies comparing organic and conventional farming effects on food quality of several vegetables and potato (*Solanum tuberosum* L.), generated contrasting results. Comparisons of nutritional data from 39 of these studies failed to indicate a consistent superiority in food quality when crops were grown organically. For example, β-carotene content was greater in several vegetables, with the exception of lettuce (*Lactuca* spp.), where differences were not detected ($p = .06$) between plants grown organically and conventionally. Conversely, vitamin C content was higher in carrots (*Daucus carota* L., subsp. *sativus* [Hoffm.] Archang) and potatoes when grown conventionally, but greater in organic tomatoes. The researchers pointed out several shortcomings in the data used in the meta-analyses and emphasized that well-designed, paired studies were needed before definitive conclusions could be made about the impacts of organic practices on food quality.

Additional meta-analyses were attempted on studies conducted over a 50-year period by Dangour et al. (2010) to identify the nutritional benefits that could be attributed to organic rather than conventional food consumption. Heterogeneity of study designs between the 12 different studies that met the initial screening criteria prevented statistical meta-analyses from being performed, but an alternative method was used for a systematic evaluation of the results from the studies. In general, the studies failed to demonstrate a consistent positive effect of consuming organic food products, with the exception of one study that suggested a possible association of reduced eczema risk in infants when consuming strictly organic dairy products. The researchers emphasized the need for additional research comparing organic and conventional food products, and encouraged these studies to include appropriate research design and adequate sample size so that valid conclusions can be made.

A consistent observation made by several researchers is a general lack of well-designed studies upon which to base definitive statements about the impact of adopting organic farming methods on food quality (Dangour et al., 2010; Hoefkens et al., 2009a,b; Reganold et al., 2010). Similar concerns were articulated by other scientists earlier (e.g., Hornick, 1992). A concerted effort is needed to ensure that future research aimed at determining the impact of organic farming on food quality and human nutrition follows rigorous scientific protocols that include appropriate experimental design, an adequate sample size, inclusion of a control treatment, same or similar cultivars of organic and conventional crops, elimination of other confounding factors, and valid statistical analyses, among other things. Research also is needed on how organic livestock systems, which emphasize feeding animals outdoors on pasture, influence the nutritional quality of foods. Efforts should be made to include members from nonagricultural (e.g., medicine) as well as agricultural (i.e., agronomy and food science) disciplines when forming the research team so that results can be targeted to providing public health benefits by solving major health issues. In addition, future research needs to elucidate the multidirectional linkages between organic farming systems, change of soil quality characteristics, and modification of food quality attributes, with the goal of providing recommendations.

12.4.3 ORGANIC FARMING AND PESTICIDES

Most synthetic pesticides are not permitted in organic farming (USDA-AMS, 2011). Natural pesticides are allowed, but some fall under a restricted status because of dangers posed to humans and/or the environment from long-term use (e.g., copper). A logical outcome of this is an expectation that pesticide residues, if even detectable, are much lower in foods produced by organic methods than in those produced conventionally. Meta-analyses were used to test the hypothesis that pesticide concentration is lower in organic than conventional foods. Baker et al. (2002) used data from three test programs to demonstrate that food grown organically had pesticide residue concentrations that were approximately 67% lower than those of food grown conventionally. Further, the meta-analyses indicated an ability to detect pesticide residue in a much smaller percentage in organic than in conventional foods. The researchers acknowledged that even food grown organically may contain traces of pesticides in some instances, but that the likelihood of consuming food with pesticide residues is minimized by choosing organic foods. Hoefkens et al. (2009b) conducted meta-analyses of food contaminants that included data from 35 studies: 10 of the studies were reported in the peer-reviewed literature, 23 were reports or databases of governments and research institutions, and 2 were the result of personal communication. Synthetic pesticide levels were lower in foods grown organically and there were several instances where pesticide levels were undetectable in the food samples. In cases where pesticide residues were detected on foods from organic farming systems, levels were much lower than what is considered unsafe and a health risk. Results of the meta-analyses by Hoefkens et al. (2009b) and Baker et al. (2002) demonstrate that lower levels of pesticide residues occur in organic as compared to conventional foods. Exposure of children to organophosphate pesticides through food intake was significantly reduced when organic foods instead of conventional ones were consumed (Curl et al., 2003; Lu et al., 2006). Even researchers generally critical of organic farming methods (e.g., Trewavas, 2001) acknowledge that pesticide residue is lower in organic than in conventional foods (Trewavas, 2004).

12.4.4 ORGANIC FARMING AND OTHER FOOD CONTAMINANTS

Regarding nitrate contamination derived from N-containing fertilizers, or animal wastes, Hoefkens et al. (2009b) concluded from their meta-analyses that lettuce did contain higher nitrate content when grown conventionally than organically, but nitrate levels were lower in conventional spinach (*Spinacia oleracea* L.). Nitrates can form carcinogenic compounds (nitrosamines) in the GI tract and can limit the oxygen-carrying capacity of hemoglobin. A trend of lower nitrate levels in organic produce has been indicated in other reviews (Bourn and Prescott, 2002; Lairon, 2010), which is primarily related to the slower release of N under organic production in contrast to the conventional system. Hoefkens et al. (2009b) also reported a trend of equivalent levels of the heavy metals lead and cadmium in organic and conventional production methods in their meta-analyses. Similarly, the amounts of mycotoxins present in crops and in their food products, particularly cereals, were similar when grown organically and conventionally (Lairon, 2010).

Thus far, there is sufficient research to conclude that organic farming methods can reduce pesticide residue in food, but the challenge remains to understand the mechanistic processes between soil biological communities and food nutrients. According to Lundegardh and Martensson (2003), the possible difference in health benefits between organic and conventionally grown food crops may result from differences between production systems in plant protection and health, supply and availability of nutrients, and soil chemical and biological processes. Additional in-depth interdisciplinary research is needed to provide science-based information to organic farmers for improving their production systems to promote improved soil quality and crop nutritional quality.

REFERENCES

Ashford, D.L. and D.W. Reeves. 2003. Use of a mechanical roller-crimper as an alternative kill method for cover crops. *American Journal of Alternative Agriculture* 18: 37–45.

Baker, B.P., C.M. Benbrook, E. Groth III, and K.L. Benbrook. 2002. Pesticide residues in conventional, integrated pest management (IPM)-grown and organic foods: Insights from three US data sets. *Food Additives and Contaminants* 19: 427–446.

Balfour, L.E. 1977. Towards a sustainable agriculture—The living soil. Address given at the 4th IFOAM General Assembly, Sissach, Switzerland. http://journeytoforever.org/farm_library/balfour_sustag.html. Accessed 23 July 2011 (verified 5 January 2012).

Barrett, D.M., C. Weakley, J.V. Diaz, and M. Watnik. 2007. Qualitative and nutritional differences in processing tomatoes grown under commercial organic and conventional production systems. *Journal of Food Science* 72: C441–C451.

Benbrook, C.M. 2005. Elevating antioxidant levels in food through organic farming and food processing. An Organic Center State of Science Review. http://www.organic-center.org/reportfiles/AntioxidantReport.pdf. Accessed 25 July 2011 (verified 5 January 2012).

Benitez, M.S., F.B. Tustas, D. Rotenberg, M.D. Kleinhenz, J. Cardina, D. Stinner, S.A. Miller, and B.B. McSpadden Gardener. 2007. Multiple statistical approaches of community fingerprint data reveal bacterial populations associated with general disease suppression arising from the application of different organic field management strategies. *Soil Biology & Biochemistry* 39: 2289–2301.

Bezdicek, D.F. and J.F. Power (eds). 1984. *Organic Farming: Current Technology and Its Role in Sustainable Agriculture*. ASA Special Publication 46. American Society of Agronomy, Madison, WI.

Biodynamics. 2010. What is biodynamic agriculture? Biodynamics Farming and Gardening Association. https://www.biodynamics.com/biodynamics.html. Accessed 23 July 2011 (verified 5 January 2012).

Bolton Jr, H., L.F. Elliott, R.I. Papendick, and D.F. Brzdicek. 1985. Soil microbial biomass and selected soil enzyme activities: Effect of fertilization and cropping practices. *Soil Biology & Biochemistry* 17: 297–302.

Bourn, D. and J. Prescott. 2002. Comparison of the nutritional value, sensory qualities, and food safety of organically and conventionally produced foods. *Critical Reviews in Food Science and Nutrition* 42: 1–34.

Buttel, F.H. and G.W. Gillespie Jr. 1988. Preferences for crop production practices among conventional and alternative farmers. *American Journal of Alternative Agriculture* 3(1): 11–16.

Carr, P.M., R.D. Horsley, and G.B. Martin. 2006. Impact of tillage and crop rotation on grain yield of spring wheat I: Tillage effect. *Crop Management*. http://www.plantmanagementnetwork.org/pub/cm/research/2006/wheat1/. Accessed 23 July 2011 (verified 5 January 2012).

Cavigelli, M.A., J.R. Teasdale, and A.E. Conklin. 2008. Long-term agronomic performance of organic and conventional field crops in the mid-Atlantic region. *Agronomy Journal* 100: 785–794.

Cassman, K. 2007. Editorial response by Kenneth Cassman: Can organic agriculture feed the world—Science to the rescue? *Renewable Agriculture and Food Systems* 22: 83–84.

CCOF. 2010. Structure of CCOF. http://www.ccof.org/about.php. Accessed 23 July 2011 (verified 5 January 2012).

Cherr, C.M., J.M.S. Scholberg, and R. McSorley. 2006. Green manure approaches to crop production: A synthesis. *Agronomy Journal* 98: 302–319.

Clark, M.S., W.R. Horwath, C. Shennan, and K.M. Scow. 1998. Changes in soil chemical properties resulting from organic and low-input farming practices. *Agronomy Journal* 90: 662–671.

Conford, P. 2001. *The Origin of the Organic Movement*. Bell & Blain, Glasgow.

Creamer, N.G. and S.M. Dabney. 2002. Killing cover crops mechanically: Review of recent literature and assessment of new research results. *American Journal of Alternative Agriculture* 17: 32–40.

Curl, C.L., R.A. Fenske, and K. Elgethun. 2003. Organophosphorus pesticide exposure of urban and suburban preschool children with organic and conventional diets. *Environmental Health Perspectives* 111: 377–382.

Dangour, A.D., K. Lock, A. Hayter, A. Aikenhead, E. Allen, and R. Uauy. 2010. Nutrition-related health effects of organic foods: A systematic review. *American Journal of Clinical Nutrition* 92: 203–210.

Delate, K. and C.A. Cambardella. 2004. Agroecosystem performance during transition to certified organic grain production. *Agronomy Journal* 96: 1288–1298.

Demeter. 2010. History of the biodynamic movement. http://www.demeter.net. Accessed 23 July 2011 (verified 5 January 2012).

Denison, R.F., D.C. Bryant, and T.E. Kearney. 2004. Crop yields over the first nine years of LTRAS, a long-term comparison of field crop systems in a Mediterranean climate. *Field Crops Research* 86: 267–277.

Doran, J.W. and T.B. Parkin. 1994. Defining and assessing soil quality. *In* J.W. Doran, D.C. Coleman, D.F. Bezdicek, and B.A. Stewart (eds), *Defining Soil Quality for a Sustainable Environment*, pp. 3–21. SSSA Special Publication 35. Soil Science Society of America, Madison, WI.

Drinkwater, L.E., D.K. Letourneau, F. Workneh, and A.H.C. van Bruggen. 1995. Fundamental differences between conventional and organic tomato agroecosystems in California. *Ecological Applications* 4: 1098–1112.

Drinkwater, L.E., P. Wagoner, and M. Sarrantionio. 1998. Legume-based cropping systems have reduced carbon and nitrogen losses. *Nature* 396: 262–265.

Entz, M.H., W.J. Bullied, and F. Katepa-Mupondwa. 1995. Rotational benefits of forage crops in the Canadian prairie cropping systems. *Journal of Production Agriculture* 8: 521–528.

Entz, M.H., R. Guilford, and R. Gulden. 2001. Crop yield and soil nutrient status on 14 organic farms in the eastern portion of the northern Great Plains. *Canadian Journal of Plant Science* 81: 351–354.

Entz, M.H., V.S. Baron, P.M Carr, D.W. Meyer, S.R. Smith Jr, and W.P. McCaughey. 2002. Potential of forages to diversify northern Great Plains cropping systems. *Agronomy Journal* 94: 240–250.

Francis, C. and J. Van Wart. 2009. History of organic farming and certification. *In* C. Francis (ed.), *Organic Farming: The Ecological System*, pp. 3–17. American Society of Agronomy, Madison, WI.

Fraser, D.G., J.W. Doran, W.W. Sahs, and G.W. Lesoing. 1988. Soil microbial populations and activities under conventional and organic management. *Journal of Environmental Quality* 17: 585–590.

Gardner, J.C. and S.A. Clancy. 1996. Impact of farming practices on soil quality in North Dakota. *In* J.W. Doran and A.J. Jones (eds), *Methods of Assessing Soil Quality*, pp. 337–343. SSSA Special Publication 49. Soil Science Society of America, Madison, WI.

Green, V.S., M.A. Cavigelli, T.H. Dao, and D.C. Flanagan. 2005. Soil physical properties and aggregate associated C, N, and P distributions in organic and conventional cropping systems. *Soil Science* 170: 822–831.

Hamel, C. and D.G. Strullu. 2006. Arbuscular mycorrhizal fungi in field crop production: Potential and new direction. *Canadian Journal of Plant Science* 86: 941–950.

Harwood, R.R. 1984. Organic farming research at the Rodale Research Center. *In* D.F. Bezdicek, et al. (eds), *Organic Farming: Current Technology and Its Role in a Sustainable Agriculture*, pp. 1–17. ASA Special Publication 46. American Society of Agronomy, Madison, WI.

Heckman, J. 2006. A history of organic farming: Transitions from Sir Albert Howard's war in the soil to USDA National Organic Program. *Renewable Agriculture and Food Systems* 21: 143–150.

Hershey, D. 1992. Sir Albert Howard and the Indore process. *HortTechnology* 2: 267–269.

Higgs, R.L., A.E. Peterson, and W.H. Paulson. 1990. Crop rotation: Sustainable and profitable. *Journal of Soil and Water Conservation* 45: 68–70.

Hoefkens, C., I. Vandekinderen, B. De Meulenaer, F. Devlieghere, K. Baert, I. Sioen, S. De Henauw, W. Verbeke, and J. Van Camp. 2009a. A literature-based comparison of nutrient and contaminant contents between organic and conventional vegetables and potatoes. *British Food Journal* 111: 1078–1097.

Hoefkens, C., W. Verbeke, J. Aertsens, K. Mondelaers, and J. Van Camp. 2009b. The nutritional and toxicological value of organic vegetables: Consumer perception versus scientific evidence. *British Food Journal* 111: 1062–1077.

Hornick, S.B. 1992. Factors affecting the nutritional quality of crops. *American Journal of Alternative Agriculture* 7: 63–67.

Howard, A. 1943. *An Agricultural Testament*. Oxford University Press, New York, NY.

Howard, A. 1947. *The Soil and Health*. Shocken Books, New York, NY (reprinted 1972).

IFOAM. 2011. Definition of Organic Farming. http://ifoam.org/growing_organic/definitions/doa/index.html. Accessed 23 July 2011 (verified 5 January 2012).

Ismail, I., R.L. Blevins, and W.W. Frye. 1994. Long-term no-tillage effects on soil properties and continuous corn yields. *Soil Science Society of America Journal* 58: 193–198.

Jenkinson, D.S. 1991. The Rothamsted long-term experiments: Are they still of use? *Agronomy Journal* 83: 2–10.

Jones, D.L. and P.R. Darrah. 1994. Amino-acid influx at the soil-root interface of *Zea mays* L. and its implications in the rhizosphere. *Plant and Soil* 163: 1–12.

Karlen, D.L., N.C. Wollenhaupt, D.C. Erbach, E.C. Berry, J.B. Swan, N.S. Eash, and J.L. Jordahl. 1994. Long-term tillage effects on soil quality. *Soil and Tillage Research* 32: 313–327.

Karlen, D.L., M.J. Mausbach, J.W. Doran, R.G. Cline, R.F. Harris, and G.E. Schuman. 1997. Soil quality: A concept, definition, and framework for evaluation. *Soil Science Society of America Journal* 61: 4–10.

Karlen, D.L., E.G. Hurley, S.S. Andrews, C.A. Cambardella, D.W. Meek, M.D. Duffy, and A.P. Mallarino. 2006. Crop rotation effects on soil quality at three northern corn/soybean belt locations. *Agronomy Journal* 98: 484–495.

Kelly, W.C. 1992. Rodale Press and organic gardening. *HortTechnology* 2: 270–271.

Korcak, R.F. 1992. Early roots of the organic movement: A plant nutrition perspective. *HortTechnology* 2: 263–267.

Lairon, D. 2010. Nutritional quality and safety of organic food. A review. *Agronomy for Sustainable Development* 30: 33–41.

Liebhardt, W.C., R.W. Andrews, M.N. Culik, R.R. Harwood, R.R. Janke, J.K. Radke, and S.L. Rieger-Schwartz. 1989. Crop production during conversion from conventional to low-input methods. *Agronomy Journal* 81: 150–159.

Liebig, M.A. and J.W. Doran. 1999. Impact of organic production practices on soil quality indicators. *Journal of Environmental Quality* 28: 1601–1609.

Lockeretz, W., G. Shearer, S. Sweeney, G. Kuepper, D. Wanner, and D.H. Kohl. 1980. Maize yields and soil nutrient levels with and without pesticides and standard commercial fertilizers. *Agronomy Journal* 72: 65–72.

Lockeretz, W., G. Shearer, D.H. Kohl, and R.W. Klepper. 1984. Comparison of organic and conventional farming in the Corn Belt. *In* D.F. Bedzicek, et al. (eds), *Organic Farming: Current Technology and Its Role in a Sustainable Agriculture*, pp. 37–49. ASA Special Publication 46. American Society of Agronomy, Madison, WI.

Lotter, D.W. 2003. Organic agriculture. *Journal of Sustainable Agriculture* 21: 59–128.

Lu, C., K. Toepel, R. Irish, R.A. Fenske, D.B. Barr, and R. Bravo. 2006. Organic diets significantly lower children's dietary exposure to organophosphorus pesticides. *Environmental Health Perspectives* 114: 260–263.

Lundegardh, B. and A. Martensson. 2003. Organically produced plant foods—Evidence of health benefits. *Acta Agriculturae Scandinavica, Section B—Soil & Plant Science* 53: 3–15.

Manlay, R.J., C. Feller, and M.J. Swift. 2007. Historical evolution of soil organic matter concepts and their relationships with the fertility and sustainability of cropping systems. *Agriculture, Ecosystems & Environment* 119: 217–233.

Marriott, E.E. and M.M. Wander. 2006a. Qualitative and quantitative differences in particulate organic matter fractions in organic and conventional farming systems. *Soil Biology & Biochemistry* 38: 1527–1536.

Marriott, E.E. and M.M. Wander. 2006b. Total and labile soil organic matter in organic and conventional farming systems. *Soil Science Society of America Journal* 70: 950–959.

Martini, E.A., J.S. Buyer, D.C. Bryant, T.K. Hartz, and R. Ford Denison. 2004. Yield increases during the organic transition: Improving soil quality or increasing experience? *Field Crops Research* 86: 255–266.

McCann, E., S. Sullivan, D. Erikson, and R. De Young. 1997. Environmental awareness, economic orientation, and farming practices: A comparison of organic and conventional farmers. *Environmental Management* 21: 747–758.

Miller, P.R., D.E. Buschena, C.A. Jones, and J.A. Holmes. 2008. Transition from intensive tillage to no-tillage and organic diversified annual cropping systems. *Agronomy Journal* 100: 591–599.

Mitchell, A.E., Y.J. Hong, E. Koh, D.M. Barrett, D.E. Bryant, R.F. Denison, and S. Kaffka. 2007. Ten-year comparison of the influence of organic and conventional crop management practices on the content of flavonoids in tomatoes. *Journal of Agriculture and Food Chemistry* 55: 6154–6159.

National Research Council. 1989. *Alternative Agriculture*. National Academy Press, Washington, DC.

Oliver, M.A. 1997. Soil and human health: A review. *European Journal of Soil Science* 48: 573–592.

Owen, A.G. and D.L. Jones. 2001. Competition for amino acids between wheat roots and rhizosphere microorganisms and the role of amino acids in plant N acquisition. *Soil Biology & Biochemistry* 33: 651–657.

Paul, E.A., D. Harris, H.P. Collins, U. Schulthess, and G.P. Robertson. 1999. Evolution of CO_2 and soil carbon dynamics in biologically managed, row-crop agroecosystems. *Applied Soil Ecology* 11: 53–65.

Peters, S. 2004. Organic and conventional beyond transition. *Organic Farmer* 11: 1–5.

Pimentel, D., P. Hepperly, J. Hanson, D. Douds, and R. Seidel. 2005. Environmental, energetic, and economic comparisons of organic and conventional farming systems. *BioScience* 55: 573–582.

Porter, P.M., D.R. Huggins, C.A. Perillo, S.R. Quiring, and R.K. Crookston. 2003. Organic and other management strategies with two and four year crop rotations in Minnesota. *Agronomy Journal* 95: 233–244.

Posner, J.L., J.O. Baldock, and J.L. Hedtcke. 2008. Organic and conventional production systems in the Wisconsin Integrated Cropping Systems Trials. I: Productivity 1990–2002. *Agronomy Journal* 100: 253–260.

Poudel, D.D., W.R. Horwath, W.T. Lanini, S.R. Temple, and A.H.C. van Bruggen. 2002. Comparison of soil N availability and leaching potential, crop yields and weeds in organic, low-input, and conventional farming systems in northern California. *Agriculture, Ecosystems, & Environment* 90: 125–137.

Power, J.F. and J.W. Doran. 1984. Nitrogen use in organic farming. *In* R.D. Hauck (ed.), *Nitrogen in Crop Production*, pp. 585–598. American Society of Agronomy, Madison, WI.

Reganold, J.P. 1988. Comparison of soil properties as influenced by organic and conventional farming systems. *American Journal of Alternative Agriculture* 3: 144–155.

Reganold, J.P., L.F. Elliott, and Y.L. Unger. 1987. Long-term effects of organic and conventional farming on soil erosion. *Nature* 330: 370–372.

Reganold, J.P., J.D. Glover, P.K. Andrews, and H.R. Hinman. 2001. Sustainability of three apple production systems. *Nature* 410: 926–930.

Reganold, J.P., P.K. Andrews, J.R. Reeve, L. Carpenter-Boggs, C.W. Shadt, J.R. Alldredge, C.F. Ross, N.M. Davies, and J. Zhou. 2010. Fruit and soil quality of organic and conventional strawberry agroecosystems. *PloS ONE* 5(9): e12346. doi:10.1371/journal.pone.0012346.

Robertson, G.P., A.P. Eldor, and R.R. Harwood. 2000. Greenhouse gases in intensive agriculture: Contributions of individual gases to the radiative forcing of the atmosphere. *Science* 289: 1922–1925.

Russelle, M.P., M.H. Entz, and A.J. Franzluebbers. 2007. Reconsidering integrated crop-livestock systems in North America. *Agronomy Journal* 99: 325–334.

Sansavini, S. and J. Wollesen. 1992. The organic farming movement in Europe. *HortTechnology* 2: 276–280.

Scofield, A.M. 1986. Organic farming—The origin of the name. *Biological Agriculture and Fertilizers* 4: 1–5.

Shannon, D., A.M. Sen, and D.B. Johnson. 2002. A comparative study of the microbiology of soils managed under organic and conventional regimes. *Soil Use and Management* 18: 274–283.

Snapp, S., S.M. Swinton, R. Labarta, D. Mutch, J.R. Black, R. Leep, J. Nyiraneza, and K. O'Neil. 2005. Evaluating cover crops for benefits, costs, and performance within cropping system niches. *Agronomy Journal* 97: 322–332.

Stanger, T.F. and J.G. Lauer. 2008. Corn grain yield response to crop rotation and nitrogen over 35 years. *Agronomy Journal* 100: 645–650.

Tanaka, D.L., J.M. Krupinsky, M.A. Liebig, S.D. Merrill, R.E. Ries, J.R. Hendrickson, H.A. Johnson, and J.D. Hanson. 2002. Dynamic cropping systems: An adaptable approach to crop production in the Great Plains. *Agronomy Journal* 94: 957–961.

Taylor, D.C., T.L. Dobbs, and J.D. Smolik. 1992. Beliefs and practices of sustainable farmers in South Dakota. *Journal of Production Agriculture* 5: 545–550.

Teasdale, J.R., C.B. Coffman, and R.W. Mangum. 2007. Potential long-term benefits of no-tillage and organic cropping systems for grain production and soil improvement. *Agronomy Journal* 99: 1297–1305.

Tisdale, S.L., W.L. Nelson, J.D. Beaton, and J.L. Havlin. 1993. *Soil Fertility and Fertilizers*, 5th edn. Macmillan, New York, NY.

Treadwell, D.D., D.E McKinney, and N.G. Creamer. 2003. From philosophy to science: A brief history of organic horticulture in the United States. *HortScience* 38: 1009–1014.

Trewavas, A. 2001. Urban myths about organic farming. *Nature* 410: 409–410.

Trewavas, A. 2004. A critical assessment of organic farming-and-food assertions with particular respect to the UK and the potential environmental benefits of no-till agriculture. *Crop Protection* 23: 757–781.

Triplett Jr, G.B. and W.A. Dick. 2008. No-tillage crop production: A revolution in agriculture! *Agronomy Journal* 100: S153–S165.

Truog, E. 1963. The organic gardening myth. *Soil Survey Horizons* 4(4): 12–18.

USDA. 1980. *Report and Recommendations on Organic Farming*. U.S. Department of Agriculture. U.S. Government Print Office, Washington, DC.

USDA-AMS. 2011. National organic program overview 7 CFR Part 205 [Docket Number: TMD-00-02-FR] RIN: 0581-AA40. http://www.ams.usda.gov/nop/. Accessed 26 July 2011 (verified 5 January 2012).

van der Ploeg, R.R., W.W. Bohm, and M.B. Kirkham. 1999. On the origin of the theory of mineral nutrition of plants and the law of the minimum. *Soil Science Society of America Journal* 63: 1055–1062.

Waksman, S.A. 1942. Liebig: The humus theory and the role of humus in plant nutrition. *In* F.R. Moulton (ed.). *Liebig and After Liebig*, pp. 56–63. AAAS Special Publication 16. American Association for the Advancement of Science, Washington, DC.

Wander, M.M., W. Yun, W.A. Goldstein, S. Aref, and S.A. Khan. 2007. Organic N and particulate organic matter fractions in organic and conventional farming systems with a history of manure application. *Plant Soil* 291: 311–321.

Wicker, E.R. 1957. A note on Jethro Tull: Innovator or crank? *Agricultural History* 31: 46–48.

Williams, C.M. 2002. Nutritional quality of organic food: Shades of grey or shades of green? *Proceedings of the Nutrition Society* 61: 19–24.

Winter, C.K. and S.F. Davis. 2006. Organic foods. *Journal of Food Science* 71: R117–R124.

Worthington, V. 1998. Effect of agricultural methods on nutritional quality: A comparison of organic with conventional crops. *Alternative Therapies* 4: 58–69.

Zhao, X., E.E. Carey, W. Wang, and C.B. Rajashekar. 2006. Does organic production enhance phytochemical content of fruit and vegetables? Current knowledge and prospects for research. *HortTechnology* 16: 449–456.

Zhao, X., J.R. Nechols, K.A. Williams, W. Wang, and E.E. Casey. 2009. Comparison of phenolic acids in organically and conventionally grown pac choi (*Brassica rapa* L. *chinensis*). *Journal of the Science of Food and Agriculture* 89: 940–946.

13 Addressing Soil Impacts on Public Health

Issues and Recommendations

Roni A. Neff, Cheryl Carmona, and Rebecca Kanter

CONTENTS

13.1 INTRODUCTION

The word "human" derives from the Latin "humus," meaning soil, perhaps reflecting an ancient understanding of soil's importance to human existence (Merriam-Webster, 2008). Soil health and human health intertwine in multiple and powerful ways. As will be described, soil is fundamental to our food security. However, soil degradation may be threatening the ability to feed populations, both in the United States and globally. Soil is also an important vector for human exposure to pathogens and toxic contaminants, which contribute to an as yet unquantifiable portion of the burden of cancer and other chronic diseases, as well as to health conditions resulting from acute exposures. Additionally, poor soil health contributes to climate change, water degradation, and other ecological conditions with human health effects including infections, respiratory conditions, heat stress, injury, and the stresses of social dislocation (Helmke and Losco, Chapter 7; Brevik, Chapter 16, this volume). Many of these soil-related impacts are inequitably distributed across populations, both geographically and demographically. Contemporary public health has afforded little attention to soil or its impact on food security and ecological threats, although a range of public health activities do address soil contamination.

This chapter is premised on the idea that, as the Institute of Medicine states, the field of public health has the mission of "fulfilling society's interest in assuring *conditions* in which people can be healthy" [emphasis added] (Institute of Medicine, 1988). This mission appropriately includes work to provide soil conditions that protect and advance human health, including promoting soil quality, working to evaluate and reduce human exposures to soil-related hazards, supporting remediation of degraded soils, and preventing the use of toxic chemicals or substances that lead to downstream soil contamination. We provide a roadmap for those tasks, identifying ways to bring the public health field's strengths to bear, in partnership with other experts and stakeholders.

We begin by discussing major soil-related human health concerns and how soil contributes to them (referred to herein as "soil risk factors"). Section 13.2 then presents examples of "direct interventions" needed to address these soil risk factors. While implementing these direct interventions is outside the traditional purview of public health, we will describe how the field can contribute to efforts to advance their implementation. In Section 13.3, we describe the strategies and tools needed to promote the direct interventions, such as policy and education activities. Finally, Section 13.4 recommends a set of roles for public health in applying these strategies and tools. These roles are organized around the "ten essential services of public health" framework (Figure 13.1), which articulates the field's central responsibilities (Public Health Functions, 1994). The ten essential services can be applied to define a clear role and responsibility for public health professionals to continue addressing soil contamination and to engage far more actively in addressing the food security and ecological threats associated with damage to soil quality (Figure 13.1).

Soils vary widely based on both inherent properties that do not change on human timescales, and dynamic properties shaped by natural events, human management, and land use decisions. Examples of inherent properties are soil texture (from sand to clay loam), depth, and parent material. Dynamic properties include soil organic matter (SOM), soil structure, and water-holding capacity.

This chapter's discussion is rooted in the concept of soil quality. Soil quality is measured both by how well soil provides "ecosystem services" (benefits provided by nature with often quantifiable positive impacts on human, animal, or plant health—such as food production, water filtration, or carbon sequestration) and by the soil's ability to perform other functions with less direct human benefit, such as regulating water and air, sustaining plant and animal life, cycling nutrients, and supporting structures within developed areas (Brevik, 2009a; Karlen, 2003; Doran, 2000; Rapport et al., 1998). The term *soil degradation* refers to changes in soil quality resulting in lost capacity to perform these functions (Eswaran et al., 2001).

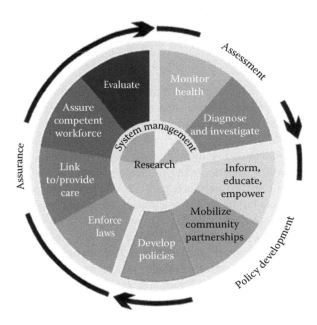

FIGURE 13.1 Ten essential services of Public Health. (From Public Health Functions, Steering Committee. Public health in America, 1994. http://www.health.gov/phfunctions/public.htm, verified 9 February 2012.)

This chapter focuses on the United States, although many of the findings are relevant elsewhere. We place particular emphasis on agriculture, due to its centrality in U.S. land use. About half of the land in the United States supports agriculture including crop land, pasture, and rangelands. Of agricultural land, 27.5% is designated as prime agricultural land or "Class I or II land capability class" (Lubowski et al., 2006).

When discussing the impacts of agriculture on soil quality, we distinguish between high-input and low-input agricultural systems. High-input systems rely heavily on fossil fuel–based inputs and nonrenewable resources (e.g., fertilizers, pesticides, mechanized equipment) and natural resources such as irrigation water. While these methods have increased short-term crop and livestock yields, they have also exacted a significant environmental toll including, in many areas, long-term soil damage (Jackson-Smith, 2010). For example, substantially as a result of damage to soils from mechanized and high-input agricultural methods, U.S. waterways have been greatly impacted over time; 44% of rivers and streams and 64% of lakes and reservoirs were impaired in 2004 (EPA, 2009b). Additionally, an estimated 30% of the crops grown on irrigated land in the United States experience yield reductions due to soil salinity (USDA Agricultural Research Service, 2005). In another example, 1.73 billion tons of soil eroded from U.S. croplands in 2007, and 28% of U.S. cropland is experiencing faster soil erosion than is considered sustainable (USDA, 2009). Studies have indicated that erosion is occurring at two to three times the sustainable rate in Iowa (Veenstra, 2010; Konen, 1999). While soil damage can occur due to both natural forces and human activities, as will be discussed, intensified high-input agriculture has amplified the extent of degradation. High-input agricultural methods also depend on continued availability and affordability of inputs, which are not assured.

Low-input agricultural systems, by contrast, take a systems approach, accounting for relationships and interactions among soils, climate, plants, and animals, thus enabling them to work more efficiently with natural resources and require fewer inputs. In these systems, land management and land use are oriented around building and maintaining soil health and biodiversity, thus reducing the need for fertilizers, pesticides, fossil fuels, and irrigation.

Most agricultural production exists somewhere on the continuum from high- to low-input, using practices from each, though higher-input industrial production dominates in the United States.

13.2   SOIL AND HEALTH

Our analysis focuses on three broad categories of human health concerns: food security, exposures to soil contaminants, and ecological threats with human health implications. We do not discuss the potential role of soil health in mediating food's nutritional content, as much more research is needed to clarify this issue. We also do not touch on soil's role as a vector for parasites and infectious diseases; information on this can be found in Loynachan (Chapter 5, this volume). Figure 13.2 summarizes the key relationships of concern in this chapter: human activities and natural processes affect soil quality and dynamic soil properties (e.g., SOM, soil bulk density, or chemical contamination). In turn, soil quality affects human health via impacts on ecosystem services such as soil's ability to support food production and water quality (Brevik, 2009b) or to sequester greenhouse gases (Brevik, Chapter 16, this volume) as well as via human exposures to contaminated soil (Burgess, Chapter 4; Helmke and Losco, Chapter 7, this volume). As Handschumacher and Schwartz (2010) note, in most cases, soil is one element of a broader system mediating the relationships between these natural processes/human activities and human health, but it is not the sole or root cause of the health impacts (Figure 13.2).

In the following subsections and in Table 13.1, we summarize how soil can contribute to health concerns. It is beyond the scope of this chapter to describe these mechanisms in detail; that information is covered in other sections of this book, as well as in stand-alone articles (i.e., Handschumacher and Schwartz, 2010; Brevik, 2009b; Pepper et al., 2009; Sing and Sing, 2008; Abrahams, 2002).

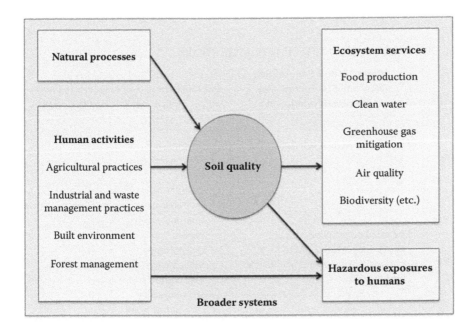

Natural processes

Human activities

Agricultural practices

Industrial and waste
management practices

Built environment

Forest management

Soil quality

Ecosystem services

Food production

Clean water

Greenhouse gas
mitigation

Air quality

Biodiversity (etc.)

Hazardous exposures
to humans

Broader systems

FIGURE 13.2 Human activities affecting soil, and exposure pathways to human health.

13.2.1 Food Security

Food security and hunger are public health concerns of the highest importance. In 2010, 925 million people worldwide were considered undernourished (FAO, 2010) and in the United States, 14.7% of the population was categorized as food insecure (Nord et al., 2010). Undernutrition is responsible for approximately 35% of global child mortality and 11% of disease burden (Black et al., 2008). Malnutrition's consequences go beyond physical health to impact educational outcomes, economic status, and even offspring birth weights (Victora et al., 2008).

Global food security pits agricultural production against rapidly depleting natural resources, with the challenge continually increasing due to ecological threats from soil degradation, climate change, water overuse and contamination, biodiversity loss, and peak oil. Growing populations compound the crisis; the U.S. Census Bureau projects a 47% increase in U.S. population between 2010 and 2050 (U.S. Census Bureau, 2009), while the global population is expected to surpass 9 billion by 2050 (U.S. Census Bureau, 2010). Demand-side factors add to the pressure on limited resources: increasing global affluence, migration and population dynamics, and crop diversion to animal feed and biofuels (Cribb, 2010). Experts project that global food demand will rise by 70%–100% or more by 2050 (Godfray et al., 2010). While food insecurity today is a phenomenon of misdistribution rather than inadequate production, over the long term there may be significant domestic and global food supply shortages. Population disparities in food access and affordability may become even starker.

Soil health may impact food security primarily in the following three ways. (1) Degraded soils reduce crop and pasture yields and quality, reduce drought resistance, and challenge other ecosystem services. Pimentel (2006) cites estimates suggesting that agricultural land degradation could reduce world food production by 30% over 25–50 years. (2) Land development (particularly near population centers), soil degradation, and crop production for biofuels and animal feeds reduce available land for human food production, even as rising petroleum prices are expected to eventually increase the impetus for regional food production (Neff et al., 2011; Clancy and Ruhf, 2010; Kirschenmann, 2010a; Peters et al., 2007, 2008; Kloppenburg et al., 1996). Development claimed over 23 million acres of agricultural land in the United States between 1982 and 2007, 38% of which

TABLE 13.1

Examples of "Soil Risk Factors" Affecting Human Health

Health Concern	Pathway of Soil Contribution (How Soil Risk Factors May Affect Health Concern)	Soil Risk Factors (Soil Practices and Problems That Can Affect This Health Concern)
A. Food Insecurity		
Rising prices, inadequate food	Decreased drought resilience	• Depletion of SOM • Soil compaction • Erosion • Desertification • Crop production on vulnerable agricultural land • Crop production in arid areas
	Decreased crop productivity (particularly when compensatory chemical inputs become unavailable/ unaffordable)	• Erosion • Desertification • Salinization • Acidification • Soil compaction • Depletion of SOM • Reduced soil microbial biodiversity • Crop production on vulnerable agricultural land
	Dependence on soil inputs (especially phosphorus, nitrogen) with limited global supply	*Example: Phosphorus (P)* • P necessary in agriculture; few alternate sources • Inadequate recycling of P (e.g., in food or human wastes) • Inefficient usage of manure and P fertilizers • Pesticide/fungicide damage to mycorrhizal fungi (which increase root system P access and uptake) leading to excessive P usage
Inadequate food, distribution issues	Lost agricultural land for food production	• Urban development on fertile land • Producing biofuels, animal feed grains on arable land
B. Soil Contamination (Human Exposure via Soil/Dust, Water, Air, and Food)		
Exposure to pathogens (including antimicrobial-resistant pathogens)	Pathogen contamination of soils, leading to contamination of surface water, groundwater, crops, dust	• Animal manure: spreading, storage, spills, direct dropping • Antimicrobials in animal feeds, resistant pathogens in manure • Biosolids and municipal waste applied to soil • Wastewater used for irrigation • Septic system surfacing sewage from soil absorption field
Exposure to hazardous chemicals	Industrial and other chemical contamination of soils, leading to contamination of surface water, groundwater, crops, dust	• Air emissions deposited on soils • Spills, storage tank leakage, leaching, dumping, buried materials • Flooding or other natural events leading to spills or leaching • Contaminated water applied to soils for irrigation or other purposes • Inadequate remediation
	Metal contamination of soils, leading to contamination of surface water, groundwater, crops, dust	• Industrial, power plant emissions • Manure from poultry fed arsenical drugs: spreading, storage, spills, direct dropping • Biosolids and municipal waste applied to soil

TABLE 13.1   (Continued)
Examples of "Soil Risk Factors" Affecting Human Health

Health Concern	Pathway of Soil Contribution (How Soil Risk Factors May Affect Health Concern)	Soil Risk Factors (Soil Practices and Problems That Can Affect This Health Concern)
	Pesticide, fungicide, herbicide contamination of soils, leading to contamination of surface water, groundwater, crops, dust	• Wastewater used for irrigation • Soil acidification (increases metal movement to water supply, crops) • Use and excess application of chemicals on crop and developed land • Soil accumulation of persistent pesticides, herbicides, and fungicides from previous land applications
C. Ecological Impacts That Affect Health		
Climate change (and associated health threats)	Reduced soil carbon sequestration and increased soil greenhouse gas emissions; emissions from nitrogen fertilizer manufacture and use	• Depletion of SOM • Soil erosion • Soil desertification • Soil compaction • Change from native cover to other land use • Leaving soil bare • Nitrogen fertilizer application
Water	*Example: Harmful algal blooms, nitrates* Soil runoff leads to overnutrification of water, feeding algae populations, including harmful ones; excess nitrogen contributes to nitrates in water supplies	• Erosion • Soil compaction • Excessive nutrient applications on crop, pasture, or developed land (e.g., manure, N and P fertilizer, compost, wastewater) • Untreated on-site sewage system discharge of nutrients

was prime agricultural land (Dempsey, 2010). (3) Most agriculture today depends on soil inputs for which long-term availability and affordability are not guaranteed. For example, the tremendous yield increases in industrial agriculture over the past 50 years were made possible with the use of synthetic fertilizers providing nitrogen, phosphorus, and potassium (N, P, and K). These fertilizers face supply challenges, although phosphorus is of greatest concern. There is no substitute for phosphorus in crop growth, for reasons including its role in photosynthesis and as a component of DNA and adenosine triphosphate (ATP), the plant's key energy source. Yet, by some estimates, global reserves of rock phosphate may only last another 50–100 years (Cordell et al., 2009). Food security issues are discussed in greater detail in Blum and Nortcliff (Chapter 14, this volume).

13.2.2   EXPOSURES TO SOIL CONTAMINANTS

Soil contamination can occur in a variety of ways, including spills of hazardous substances, deposition of airborne pollutants, land application of chemicals or organic materials (e.g., pesticides, manure, surfacing septic tank effluents, biosolids), leaching of metals, toxins, or pathogens into soil from waste storage or contaminated water (e.g., rainfall or irrigation), and dumping (both legal and illegal). A soil's capacity to filter out or suppress contaminants is based on inherent properties and soil quality factors. As soil health declines or as the level of contamination exceeds a soil's capacity to adsorb or adequately digest contaminants, the movement of these substances into water, air, and plants can occur more readily. The threshold level of a contaminant that a soil can attenuate varies for each contaminant and each soil. An inorganic contaminant such as a metal can be immobilized within a soil and although the soil has reduced the movement of the contaminant into air or water, it still exists within the soil and may require remediation depending on land use (Harter, 1983). Lead, for example, is a toxic metal that can enter soils through multiple pathways including land application of amendments containing lead (e.g., sewage sludge), deposition to land from industrial emissions (e.g., lead smelter), or through contaminated irrigation water. The primary factors that determine the mobility or bioavailability of lead are pH, organic matter content, cation exchange capacity (CEC), soil mineralogy, and the amount of metal applied or deposited. Soils that are non-calcareous and have a high CEC or clay content with strong adsorption capacity will more strongly adsorb or immobilize lead on soil particles. Soil quality factors such as pH and organic matter content impact the solubility of lead in soils. As these factors change due to land use and management they have the ability to affect lead accumulation in plants and movement of lead into water (Martinez and Motte, 2000). Further, land development with impervious surfaces allows runoff water carrying sediment, nutrients, and potential contaminants to bypass soil filtration.

Humans can be exposed to soil contaminants through respiration, ingestion, and dermal contact with soil and dust in occupational, environmental, and recreational settings. Crops for human consumption may be contaminated via surface deposition of water and plant uptake. Soil contaminants can also leach or run off into water supplies used for drinking or irrigation. These environmental and occupational hazards are not evenly distributed in the population. An extensive literature documents that contaminated sites are more frequently proximate to African American and lower-income communities than others (Handschumacher and Schwartz, 2010; Martuzzi et al., 2010; Brown, 1995). Work exposures also differ by socioeconomic status and ethnicity (Dembe et al., 2004; Murray, 2003; Robinson, 1989). For example, Hispanics and other immigrants disproportionately perform agricultural labor that exposes them to pesticides and other soil contaminants (USDL, 2004).

It is difficult to quantify how much chronic illness may be attributed to soil contamination, given gaps in exposure information, exposures to multiple media containing the same contaminant, lag time to health outcomes, and the many hazardous exposures most people encounter. Further, for many contaminants, inadequate risk assessment data exist and information may be lacking about bioavailability from soil exposures versus those from other media, such as air or water. Nonetheless, given the unquantified but substantial extent of human exposure to soil contaminants, there is reason to believe that soil may play some significant role in outcomes such as the estimated 19% of all cancers globally that are associated with occupational or environmental causes (WHO, 2011), the rising rates of antibiotic-resistant infections, and the estimated 19 million waterborne illnesses in the United States annually (Reynolds et al., 2008).

13.2.3 Ecological Threats with Human Health Implications

Soil is central to Earth's ecosystems, and thus its health is connected to many of the top ecological and health threats we face today. For example, *The Lancet* describes climate change as "the biggest global health threat of the 21st century." Climate change poses serious challenges to food and water security, and contributes both directly and indirectly to infectious disease, heat stress, injury, respiratory conditions, and other health problems (Brevik, Chapter 16, this volume; Costello et al., 2009). Soil health is connected to climate change in two main ways. First, soil is one of the world's most important "carbon sinks." Soils, by one estimate, offset up to 15% of world fossil fuel emissions (Lal, 2007), and soils with high organic matter and biomass sequester more CO_2 (Johnson et al., 2007). Second, nitrogen fertilizers, when applied to soil, lead to emissions of the greenhouse gas nitrous oxide. Nitrogen in agricultural soil management accounted for about 3% of U.S. greenhouse gas emissions from 1990 to 2009 (EPA, 2011a); this relatively small percentage is magnified in impact because the greenhouse potential of nitrous oxide is 300 times that of CO_2. Further, these fertilizers require large amounts of energy to produce—by one estimate half the energy used in commercial agriculture (Woods et al., 2010).

Another example of an ecological link between soil health and human health is water. Degraded soils and those with low levels of organic matter are less effective at retaining water, and thus are more vulnerable to drought and can require far more irrigation than healthy soils, contributing to problems of water depletion. Healthy soils also perform an important filtering role, preventing and/or reducing contaminants from reaching water supplies (Helmke and Losco, Chapter 7, this volume). Finally, nonpoint sources such as soil runoff are the main contributors of nutrients in U.S. surface waters. These nutrients contribute to eutrophication, hazardous nitrate levels in water supplies, and harmful algal blooms (Heisler et al., 2008; Carpenter et al., 1998). High water nitrate levels are associated with methemoglobinemia ("blue baby syndrome") and livestock toxicity, while acute exposure to harmful algal blooms may cause neurological, gastrointestinal, and other symptoms, with possible severe illness (NOAA, 2009).

In sum, soil health and human health intersect in numerous ways. A single "soil risk factor," such as erosion, can impact health through pathways including reduced crop yields and food insecurity, reduced carbon sequestration, groundwater contamination, contribution to harmful algal blooms and eutrophication, and respiratory conditions due to blowing dust. Accordingly, the impacts of any one soil health intervention can yield multiple benefits for human health.

13.3   DIRECT INTERVENTIONS

We describe "direct interventions" to address soil risk factors and thus prevent or mitigate the related health concerns. Direct interventions are typically implemented by farmers, engineers, and others working directly with the soil. Public health practitioners are typically not involved in performing these interventions themselves but rather can take steps (described later) to improve and increase performance of these interventions. In the following subsections, we organize examples of interventions into three categories: (A) improve agricultural practices, (B) reduce and manage industrial and other soil contamination, and (C) manage land use and built environment. Most interventions are multipurpose, addressing several soil risk factors at once. The impacts of these interventions may vary significantly based on factors such as soil type, hydrology, and climate conditions; elaborating that variation is beyond the scope of this chapter. We emphasize that while these direct interventions have clear benefits for soil health and human health, decisions about their implementation will necessarily also take account of economic and other considerations. Table 13.2 provides a list of examples of direct interventions.

13.3.1   IMPROVE AGRICULTURAL PRACTICES

Most agriculture in the United States involves high-input methods, although lower-input methods that improve soil health are increasingly practiced, and some, such as no-till agriculture, have become fully mainstream (Horowitz et al., 2010). There is a need both to accelerate farm transitions toward more broadly sustainable, low-input agricultural systems and to accelerate the mainstream shift toward adopting basic soil conservation interventions. Transitions to sustainable methods increase soil resilience to drought and other ecological stressors and protect natural resources in the long term. Yet, the sheer extent of high-input agriculture means even small shifts in the mainstream can have great overall impacts on soil quality at the national level.

13.3.2   CROP PRODUCTION

13.3.2.1   Reduce Manufactured Pesticide Usage

Insecticides, herbicides, and fungicides (herein referred to as pesticides) are used to manage pests, control weed growth, or kill fungi that may be damaging to crops. They also disrupt soil biological activity (Hussain et al., 2009). Soil microorganisms, earthworms, and arthropods are essential to healthy soil

TABLE 13.2

Examples of "Direct Interventions" to Address Soil Risk Factors

Direct Intervention	Soil Risk Factors Addressed	
	Improve	**Reduce**
A. Improve Agricultural Practice		
1. Crop Production		
Reduce usage of manufactured pesticides, herbicides, and fungicides.	SOM, soil biomass.	Greenhouse gas emissions (GHG), soil contamination, chemical use.
Use crop rotation.	Soil biodiversity.	Chemical use.
Use nutrient management plans to minimize fertilizer overuse; reduce dependence on manufactured fertilizers.	SOM, soil biomass, sustainability of phosphorus supply.	Acidification, GHG, eutrophication.
Incorporate organic matter into soil annually (as part of a nutrient management plan).	SOM, soil biomass, drought resilience, nutrient cycling, sustainability of phosphorus supply.	Acidification, GHG.
Use no-till or conservation tillage and leave plant residues on soil.	SOM, soil biomass, nutrient cycling, biodiversity.	Compaction, erosion, eutrophication.
Use cover crops.	SOM.	Erosion, GHG, eutrophication.
Adapt agricultural production practices and plant selection to site- and region-specific soil and other conditions.		Desertification, erosion, compaction.
Use lighter-weight farm vehicles; restrict traffic through crop beds.		Compaction, GHG, eutrophication, pathogens.
Control water and wind erosion through land management, other interventions.		Erosion, eutrophication.
Diversify crop and livestock production on farms; reduce monocropping.	SOM, soil biomass, nutrient cycling.	Chemical use, GHG, compaction.
2. Animal Agriculture		
Shift away from CAFO production.	SOM, soil biomass.	Pathogens, AMR, metals, erosion, use and excess application of chemicals.
Eliminate antimicrobial amendments, arsenical drugs from animal feeds.		AMR, metals.
Improve manure treatment prior to land application.		Pathogens, AMR, metals, pharmaceutical residues.
Implement nutrient management planning for manure.		Pathogens, AMR, metals, eutrophication.
Avoid use of phosphorus in livestock feeds. Use phytase in feed to improve P efficiency (phytase is an enzyme that breaks apart molecules to make inorganic P bioavailable to animals).	P supply.	Eutrophication.
Improve grazing management to allow grass/pasture rejuvenation.		Erosion, desertification, GHG.
For grazing, provide buffer zones near riparian areas and watering areas for animals.		Compaction, erosion, SOM, eutrophication, pathogens.

TABLE 13.2   (Continued)

Examples of "Direct Interventions" to Address Soil Risk Factors

Direct Intervention	Soil Risk Factors Addressed	
	Improve	**Reduce**
B. Reduce and Manage Industrial Soil Contamination		
Eliminate or minimize use of the most hazardous substances and those with inadequate safety data. Seek substitutes.		Exposure to hazardous chemicals.
Engineer physical environment and work practices to prevent soil, water, and air emissions at existing facilities, and to minimize exposures.		Exposure to hazardous chemicals.
Maintain records of historical land use, chemical use, and spills, to enable improved identification of potential hazards.		Exposure to hazardous chemicals.
Remediate contaminated soils.		Exposure to hazardous chemicals.
Improve treatment methods and monitor water treatment facilities, septic systems, wastewater treatment plants.	Soil evaluation and treatment specific to site needs.	Exposure to metals, pathogens, AMR, pharmaceutical residues, other chemical contaminants.
Perform health-based soil testing for agriculture/gardening.		Exposures to metals, chemicals.
C. Manage Land Use and Built Environment, Particularly in Vulnerable Areas		
1. Pervious Surfaces		
Ensure that roads, parking lots, and other built structures are planned to minimize water runoff.		Erosion, eutrophication, water contamination.
Plant natives and perennials, and reduce grass monocultures; reduce chemical use in landscaping.	SOM, soil biomass.	Erosion, acidification, desertification, GHG, eutrophication.
2. Agricultural Land		
Make agricultural land use decisions based on soil qualities, soil morphology, and other regional factors (e.g., climate).	Food security.	Erosion, salinization, acidification, desertification.
Take fragile or sensitive land out of agricultural production, restore land.	SOM, biomass.	Erosion, eutrophication, salinization.
Protect agricultural land from development.	Adequacy of available agricultural land.	Compaction, erosion, eutrophication.
Continue to develop agricultural production in urban, periurban areas.	Adequacy of available agricultural land.	

ecosystems that cycle nutrients, mitigate plant pathogens, and develop healthy plant–soil relationships. Thus, curtailing pesticide use can improve long-term food security by protecting soil biological health, including total biomass and species diversity of soil biota (Pimentel, 2005). Reducing pesticide usage also decreases the likelihood of human exposure to pesticides in soil and other media, thus preventing known and potential pesticide health effects, including nervous system damage, endocrine disruption, birth defects, and several types of cancers (Burgess, Chapter 4, this volume; EPA and OPP, 2011; Rezg et al., 2010; Weichenthal et al., 2010; Bassil et al., 2007; Sanborn et al., 2007).

Two main agriculture systems substantially reduce pesticide use. Organic agriculture avoids all manufactured pesticides, using an array of nonchemical interventions including attracting beneficial insects, early action with insecticidal soaps and sprays, physical destruction of weeds and affected plants, and diversified agriculture with intercropping and crop rotations (Carr et al., Chapter 12, this volume). The second system, integrated pest management, uses similar techniques but reserves the option to use pesticides as a last resort.

13.3.2.2   Reduce Manufactured Fertilizer Usage

Most U.S. agriculture relies on manufactured fertilizers, yet, particularly when applied in excess quantities, these can negatively impact soil in several ways. Manufactured nitrogen fertilizers may increase the risk of acidification, which can reduce a soil's food-producing capacity and require costly lime inputs to address. They can also contribute to soil nutrient imbalances, which in turn can reduce SOM, lead to release of sequestered soil carbon (a greenhouse gas), and alter soil microbial populations (Francis et al., 1990; Tisdale et al., 1985). Conversely, soil testing and proper fertilizer application can be used intentionally to address micronutrient and macronutrient imbalances, which will improve crop yields and hence build SOM through increased crop returns of organic matter. Years of nitrogen fertilizer application on agricultural crop land without additions of organic matter (e.g., cover cropping, compost application) reduce a soil's capacity to support healthy crop growth due to reduced quality and quantity of SOM (Manna et al., 2007; Melero et al., 2007). Reduced quality and quantity of organic matter also lead to changes in soil structure, thereby leading to reduced aggregation, infiltration capacity, hydraulic conductivity, and water-holding capacity, which may also hinder healthy crop growth (Haynes and Naidu, 1998). As described earlier, fertilizers can contribute to eutrophication, risk of methemoglobemia, and emissions of the potent greenhouse gas N_2O, especially when overapplied or applied without careful attention to timing, placement, and weather conditions; their production uses much energy.

In addition to nitrogen fertilizers, most U.S. agriculture also relies on mined phosphate to supply the macronutrient phosphorus. Yet, as noted earlier, global phosphate reserves are limited. Demand for mined phosphorus could be greatly reduced with improved recovery of phosphorus from animal manure and human excreta. As will be discussed, however, for this to be a feasible option, critical improvements are needed in waste treatment processes and livestock feeds (Graham and Nachman, 2010; Cordell et al., 2009).

There are alternative fertilization methods that can ensure adequate soil nutrients while avoiding the threats from manufactured fertilizers, mined phosphorus, and treated human/animal waste. Some alternative methods include applying composted biomass from food waste and manure from animals that have not received antimicrobial feed supplements, soil biomass inoculants (e.g., arbuscular mycorrhizal fungi), blood and bone meals, and fish emulsions, as well as using crop rotation and legumes for cover crops. Additionally, SOM can be increased by leaving plant residuals on the surface and mulching, thus providing both short- and long-term nutrient supplies.

Regardless of the source of soil nutrients, nutrient management planning is essential to estimate the existing soil nutrient levels, in order to determine the remaining amounts of nitrogen, phosphorus, potassium, and other micronutrients a crop needs, and how to best meet these needs (NRCS, 2006). Nutrient management plans prevent excessive nutrient applications, which can lead to soil nutrient imbalance, nutrient leaching or runoff to surface or groundwater (Ribaudo et al., 2003), and methemoglobemia.

13.3.2.3   Continue Reducing Tillage (without Substantially Increasing Chemical Use)

Soil tillage, or plowing, is commonly used to increase soil aeration, manage weeds, and incorporate nutrients before planting. It also has several negative effects. It disrupts soil biomass and soil structure, creating important macro and micro pore spaces. Following tillage soils can become compacted, contributing to nutrient runoff and erosion (Karlen, 2004), and further reducing soil biomass. Soil aeration also promotes the breakdown of organic matter, which contributes to the release of the greenhouse gas CO_2, although the impact of this on total greenhouse gas emissions is debated (Baker et al. 2007).

Conservation tillage (minimal tillage) and no-till techniques respond to these concerns by minimizing topsoil disruption, leaving a percentage of plant residues on the soil surface and leaving root systems intact to decompose into SOM. In 2010, the USDA reported that no-till methods are practiced on 35% of cropland for eight key crops (Horowitz et al., 2010); the increase in no-till methods likely contributed to the USDA's reported 43% drop in soil erosion from 1982 to 2007 (NRCS, 2010a). Unfortunately, despite the benefits, conservation tillage may also pose threats to soil because crop residues and undisturbed soils can provide a continued habitat for weeds and other pests, potentially leading to increased pesticide use. Evidence is mixed regarding actual impacts on levels of pesticide use (Fernandez-Cornejo and Caswell, 2006). The increase in conservation tillage in the United States was partly enabled by the availability of genetically engineered crops that can withstand herbicides, particularly Roundup. As Roundup has accordingly been used extensively, resistant weeds have developed, driving many farmers in the United States to begin using more toxic herbicides than they used previously (Neumann and Pollack, 2010; Powles, 2008; Pimentel, 2005), with potential soil and human health effects. Alternative no-till techniques are available, such as using cover crops to suppress weed growth and applying manure or fertilizer in bands along rows or with injectors to incorporate nutrients in soils. Further research is needed to improve these methods and enable widespread use.

13.3.2.4 Transition from the Industrial Model of Food Animal Production to More Sustainable Methods

Most of the over 9 billion animals produced in the United States each year are raised in industrial facilities containing thousands of animals. Industrial food animal production (IFAP) has significant implications for soil quality, in addition to other negative environmental effects (Pew Commission on Industrial Farm Animal Production, 2008; Silbergeld et al., 2008; Walker et al., 2007). In IFAP, animals are fed grain grown in large-scale industrial monoculture systems; such grain occupies a significant portion of U.S. cropland and poses threats to soil quality, as discussed earlier.

Additionally, to help animals survive the stressful setting and to speed growth, animal feeds typically contain antimicrobial drugs at subtherapeutic doses. Indeed, 80% of antibiotics consumed in the United States are fed to food animals, with 74% of these administered as feed additives (Office of Congresswoman Louise Slaughter, 2011). These drugs contribute to the development and promotion of antimicrobial-resistant bacteria, which can result in infections in humans that are challenging and costly to treat (Silbergeld et al., 2008). Broiler chickens, turkeys, and swine are also fed roxarsone, a growth-promoting drug that contains arsenic (Silbergeld and Nachman, 2008). When animal waste is applied to land, residual drugs, pathogens, including those exhibiting antimicrobial resistance (AMR), arsenic, and other contaminants can enter soils, water supplies, and crops, with potential ramifications for human health (Silbergeld et al., 2008; Nachman et al., 2008). The United States does not require manure treatment prior to land application. Further, manure is often overapplied to soils, leading to soil runoff that contributes to eutrophication (Xue and Landis, 2010; Carpenter, 2008; Heisler et al., 2008; Ribaudo et al., 2003).

Given these impacts on soil, a longer-term recommendation is to transition away from the IFAP model to more environmentally sustainable methods. Sound livestock management for healthy soils includes raising cattle on perennial-based pastures, using intensive rotational grazing practices, and mixing animal and crop agriculture on diversified farms. These methods allow for grass and pasture rejuvenation and prevent erosion, runoff, and desertification (Herrero et al., 2010; Kemp and Michalk, 2007). Manure in appropriate quantities from animals raised in such systems provides soil benefits for fertility and SOM. For the short term, however, Graham and Nachman (2010) present a three-part categorization of strategies for managing manure, which will contribute to improved soil quality. (1) Improve waste quality; in particular, eliminate antimicrobial drugs, arsenic, and other hazardous substances from feed regimens (Stingone and Wing, 2011; Nachman et al., 2005, 2008). (2) Decrease waste quantities to reduce the volume needing treatment and reduce the risk of spills (e.g., by not adding water). (3) Reduce geographic concentration of waste generation and application, for example, by addressing zoning policy relevant to IFAP and transporting waste. Additionally, as discussed earlier, until hazardous substances are removed from feeds, manure should be treated before application. Further research and development are needed to improve manure treatment methods in order to protect soils.

13.3.3 REDUCE AND MANAGE INDUSTRIAL AND OTHER SOIL CONTAMINATION

13.3.3.1 Protect and Manage Soils with Current and Former Industrial Land Use

Preventive action is needed to protect soils in areas with current and former industrial land use, including manufacturing and waste management facilities. The industrial hygiene hierarchy of controls is a useful framework for protecting workers and preventing environmental entry of these chemicals (CDC and NIOSH, 2010). This hierarchy of exposure control actions prioritizes steps based on protectiveness, while recognizing that controls lower on the list may be most feasible and affordable. (1) *Eliminate*: redesign processes to avoid using the most hazardous substances and those for which little safety data exist. (2) *Substitute*: substitute less hazardous substances. (3) *Engineering controls*: engineer systems that reduce the likelihood of exposure to contaminated soils, including use of durable physical barriers. (4) *Administrative controls*: engineer work practices to minimize exposure risks. (5) *Personal protective equipment*: provide protective equipment and tools to the potentially exposed. Beyond these activities, data tracking of chemical use and possible spills by area is essential to support future investigations. Finally, at brownfields (contaminated sites being redeveloped) and other current and former industrial sites, there is a need to assure risk assessment and adequate cleanup of contaminated soils with controls to prevent future human exposures and community engagement (Litt et al., 2002).

Policies and programs exist to manage site restoration (such as Superfund and the Department of Defense Environmental Restoration Program) (Department of Defense, 2011), regulate allowable substances used in industry (Environmental Protection Agency, EPA), and track some spill information (such as the Hazardous Substances Emergency Event Program) (ATSDR, 2009).

13.3.3.2 Perform Health-Based Agriculture/Gardening Soil Testing and Surveillance

Many agricultural land managers monitor soil nutrients and metals, and many gardeners perform basic soil tests, sometimes including lead testing. Other health-based soil testing, such as for industrial chemicals, historical pesticide use, or pathogens, is relatively rare. Tests can identify benefits of protective practices including placing barriers at the soil surface and importing soil. Such tests can be costly and require identifying contaminants of concern in advance. The EPA and other sources provide grants for soil testing on sites like brownfields and community gardens, as well as providing informational tools for site assessment (EPA, 2009a). Further government funding is needed for expanded soil testing, record keeping on historical land use, and outreach and education to make health-based soil tests more available and affordable.

A related issue is the need to perform proactive surveillance of soil amendments for sale, both natural (e.g., manure) and synthetic (e.g., industrial by-products sold as soil amendments) (Brevik, Chapter 2, this volume; Wilson, 2001). For example, pelletized poultry manure is widely sold for agricultural and home garden uses; tests find that it often contains arsenic from the growth-promoting drug roxarsone at levels above those that trigger state notification and action (Nachman et al., 2008). Manufacturers or source originators could be required to perform testing at their expense and report results to a central registry.

13.3.4 Manage Land Use and Built Environment: Target Vulnerable Areas

13.3.4.1 Use Pervious Surfaces

As described earlier, soils serve as a filter, providing an important ecosystem service by preventing contaminants from going directly into groundwater or surface water. In developed areas, built structures and pavement cover much of the soil, preventing this filtration. Further, instead of being absorbed into soils across the landscape, stormwater falling on impervious surfaces gathers, forming into streams that carry sediments, nutrients, and toxins directly to surface waters. Where possible, pervious surfaces that allow water infiltration into underlying soils should be used instead of impervious ones. Other techniques to reduce stormwater runoff include collecting rainwater from rooftops, and planting trees, rain gardens, or grass swales. Stormwater issues are discussed in greater detail in Helmke and Losco (Chapter 7, this volume).

13.3.4.2 Make Agricultural Land Use Decisions Based on Soil Characteristics

Agricultural land use decisions should be made with strong attention to soil characteristics and other regional factors such as climate conditions. For example, perennial-based pastures are not as taxing on the health of soil ecosystems as crop agriculture and can be sustained well on lower-quality soils. Crop production requiring intense water use should be sited in higher-quality soils with better moisture retention and in areas with adequate rainfall and available irrigation water. Cropland with reparable soil degradation should be treated with appropriate interventions prior to production. Fragile or sensitive lands such as wetlands, deserts, or riparian areas should be taken out of agricultural production and restored with native habitats that will increase soil biodiversity, helping to restore SOM and preventing erosion, eutrophication, and further soil degradation.

13.3.4.3 Protect and Advance Periurban and Urban Agriculture

Today most food travels great distances from production to table. In the future several factors, including population growth and peak oil, may converge, forcing change. There may be a significant rise in regional food production and distribution, including near population centers, in order to provide for long-term food security. Trade will likely continue, but regions that can provide more of their own food may benefit economically (Neff et al. 2011; Clancy and Ruhf, 2010; Kirschenmann, 2010a; Peters et al., 2007; Kloppenburg et al., 1996).

Larger spaces in and near urban areas (periurban) can be converted for food production purposes, while infill land (vacant or underutilized space between existing construction) can be reclaimed for community gardens and green spaces. Frequently, urban and periurban land is among the most fertile land available, which led to historical choices to settle in these areas. That land, however, is often damaged by building structures and paving. Development can compact and alter soil composition or texture for foreseeable time frames, making the land unsuitable for agricultural production. Such losses may significantly affect future food production and security. Accordingly, zoning and other policies to maintain agricultural use of land (especially Classes 1 and 2, which represent prime agricultural land) near urban centers should be pursued.

13.3.4.4 Properly Design, Maintain, and Use On-Site Sewage Systems

On-site sewage systems are utilized where centralized sewage systems are not available. In 2007, septic systems served an estimated 20% (26.1 million) of all U.S. housing units (EPA, 2008). A conventional on-site system consists of a septic tank and soil absorption field. The septic tank allows primary treatment (i.e., settling of solids) and the effluent must go to unsaturated soil to have secondary treatment (aerobic biodegradation) of the organic materials by soil bacteria and to allow soil to filter out other contaminants such as pathogens (EPA, 2010a). The soil must also provide good drainage to allow recycling of the water portion of the wastewater to the environment. Septic systems in the United States discharge more than 1 trillion gallons of wastewater to the soil each year. A number of factors can cause on-site systems to fail, including unsuitable soil conditions, improper design and installation, and inadequate maintenance practices. Failing on-site systems are often defined as those with surfacing wastewater. Failing systems are often associated with soils that lack drainage and tend to remain wet such as those soils high in clay content and/or with water tables near the surface. They contribute significant amounts of pollutants to the soil, especially nitrogen and microbiological pathogens. Soils that stay saturated with water do not allow aerobic biodegradation and will allow contaminants to flow off-site. Even a properly operating septic system can release more than 10 pounds of nitrogen per person per year to the environment. Nutrients (nitrogen and phosphorus) from failing septic systems contribute to the eutrophication of water bodies. Surfacing septic systems can directly expose humans to pathogens as the system is often near the dwelling (i.e., in the yard). Bacteria and viruses from failing septic systems are public health risks that can contaminate water supplies and contribute to waterborne illness (EPA, 2006).

Selecting the right type of on-site system to handle the site-specific conditions (e.g., soil type, soil depth, and water table depth) will decrease the likelihood of septic system failure. A thorough site and soil evaluation (such as soil morphology description by a soil scientist) should be conducted when the septic system is still in the planning stages (National Environmental Services Center, 2004). Soil morphology evaluations can identify seasonal high water tables in the soil profile using redoximorphic features such as mottling. Septic system absorption trench bottoms need vertical separation from the water table to allow the soil around and below the absorption line to remain aerobic and percolating. Systems can be designed to have additional advanced treatment components such as aerobic treatment units and filters to help further reduce pollutants such as organics and nitrate. Standards for advanced-treatment on-site systems have been developed by the National Sanitation Foundation to allow these systems to demonstrate that they can provide at least a 50% nitrogen reduction in addition to significantly reducing organic pollutants (National Sanitation Foundation, 2011). However, the greater performance in the reduction of pollutants comes at a price, as advanced treatment systems easily cost $10,000 or more over conventional septic systems (EPA, 2002). Laws and regulations for the construction, design, and maintenance of on-site systems vary by local authority (i.e., county, state) across the United States. Homeowners need to be aware of their specific local requirements prior to construction of septic systems.

Septic systems must be used and maintained properly to prevent soil and water contamination. Toxic substances, such as paints, thinners, pesticides, waste oils, and other hazardous chemicals cannot be treated by soil bacteria and in fact will disrupt the natural treatment process of the septic system and should never be disposed of in a septic system (National Environmental Services Center, 2004). After a few years, the solids that accumulate in the septic tank should be pumped out and disposed of at an approved location. If septic tank solids are not removed, they will eventually accumulate in the absorption field and clog the soil pores, preventing treatment of the wastewater (Tyler et al., 1977). Additional discussion of septic systems can be found in Helmke and Losco (Chapter 7, this volume).

13.3.5 Strategies and Tools

The direct interventions discussed earlier and in Table 13.2 are, to varying degrees, already taking place. All need immediate and decisive action. We next present a set of strategies and tools designed to encourage such action. By strategies, we mean approaches to leveraging social, economic, and political forces in order to achieve goals. The strategies we propose include: (A) policy strategies, (B) multisectoral and multistakeholder strategies, (C) education, and (D) research. What we refer to as tools are more specific instruments within these general strategies. Just as most individual direct interventions can target multiple soil risk factors, so too many of these strategies and tools can simultaneously address multiple direct interventions. Table 13.3 summaries some of these strategies and tools.

TABLE 13.3

Examples of "Strategies and Tools" to Advance Action on the Direct Interventions

Strategies	Example(s)	Tools
A. Policy Strategies		
Incentives		
Provide economic incentives for soil stewardship in agriculture.	Fund conservation reserve programs, on-farm environmental interventions, and sustainable agriculture (including transitions).	Farm Bill.
Transition (carefully) away from farm policy incentives that support soil-damaging crop and livestock production, toward strengthened incentives for soil stewardship.	Shift from commodity subsidies to "green" payments. Enforce environmental requirements for farms and other businesses receiving federal aid.	Farm Bill.
Improve access to credit and insurance for farmers producing in soil-conserving ways.	Mandate access to loans and insurance coverage for organic and sustainable farms. Facilitate actuarial projections via improved tracking of production yields from such agriculture.	Farm Bill, state agricultural policy.
Use taxes, tax breaks, and other government strategies to alter relative prices of detrimental vs. desired activities	Increase taxes on manufactured fertilizers and pesticides, or provide tax breaks for producers who meet certain criteria.	Federal, state taxation policy.
"Polluter pays"	Reinstate Superfund tax for industrial chemical users.	Federal lawmaking for EPA.
Regulations		
Use regulatory policy or legal challenge to limit allowable animal feed ingredients.	Ban nontherapeutic antimicrobial drugs and use of arsenic in animal feeds.	FDA regulations, federal lawmaking, legal suit.
Update and increase enforcement of regulations addressing practices with negative impacts on soil.	Enforce practices that prohibit manure runoff or soil erosion. Enforce practices leading to on-site sewage systems with successful site-specific treatment capabilities.	Regulations by EPA, FDA, USDA, and state/local counterparts. Federal, state, local lawmaking, budget appropriations.
Update and increase enforcement of regulations addressing occupational health concerns relevant to soil.	Expand applicability of OSHA regulations to agriculture and all government workers. Update OSHA standards for common soil hazards of concern. Strengthen enforcement and penalties.	Federal, state OSHA regulations. Federal, state, local lawmaking, budget appropriations.
Use legal system to challenge practices that cause or exacerbate soil risk factors.	Regulate false advertising. (e.g., Tyson claimed their chicken was raised "antibiotic free" when they were still feeding the chickens ionophores, a type of antibiotic that selectively decreases some types of gut microbes.)	Class action or other lawsuit.
Zoning		
Use zoning policies to preserve prime agricultural land and to prevent soil-damaging land uses.	Zoning to preserve agricultural land; zoning to allow urban agriculture; requirements that new structures, roads are built to minimize water runoff.	State and local zoning policy.

(continued)

TABLE 13.3   (Continued)
Examples of "Strategies and Tools" to Advance Action on the Direct Interventions

Strategies	Example(s)	Tools
Infrastructure		
Develop and improve infrastructure for collecting, treating, and distributing composted food waste/yard waste, manure, and biosolids. Develop improved on-site sewage systems.	Infrastructure put in place to collect municipal food and yard waste and generate compost. Use of on-site sewage systems that fit specific site needs (i.e., keep a vertical separation between absorption lines and a seasonal water table).	Farm Bill, state and municipal bonds, and other policies.
B. Multisectoral and Multistakeholder Strategies		
Engage stakeholders in developing and supporting plans and policies relevant to soil health.	Engage community members near sites with contaminated soils to shape site reuse plans and safety strategies. Design policies to incentivize sustainable agriculture methods, with input and buy-in from potential users.	Community organizations, farmer organizations, unions, and other worker organizations.
Develop and use planning infrastructure to help regions manage soil-related threats and plan for future scenarios.	Convene stakeholders, gather information and projections, set priorities, develop plans with stakeholder input.	Food policy councils, regional planning bodies.
Enhance interagency cooperation within governments.	Public health agencies meet regularly with environmental and agricultural agencies to share input and ideas, develop protocols for information-sharing, and design joint efforts to tackle interdisciplinary problems.	Interagency workgroups, interagency agreements, committee membership.
C. Education, Communication, and Behavior Change Strategies		
Provide safety training and materials to those working with soil in occupational and nonoccupational settings.	Develop linguistically and culturally appropriate training and written materials to explain hazards to which trainees may be exposed, and strategies to reduce exposures.	Worker education programs, federal and state OSHA and EPA, employers, industry associations.
Provide skills for sustainable agriculture and soil stewardship to both new and current agricultural workers.	Develop public training programs in sustainable agriculture methods through the Agricultural Extension Service; provide improved access to credit.	State education systems, Farm Bill, student loan programs.
Make information and technical assistance available to farmers.	Expand emphasis on soil conservation and preservation in Cooperative Extension System activities.	Farm Bill, agricultural extension programs.
Educate homeowners, gardeners and landscapers about soil stewardship.	Master Gardener programs, Cooperative Extension services.	Present targeted messaging through "new" and "old" media, and through relevant organizations.
Educate the public about the linkages between soil and human health (i.e., soil risk factors) to motivate and increase pressure for action of all kinds.	Educate about soil degradation, its impacts on food security, and connections to other environmental threats in the food system. Educate about the need for on-site sewage systems that provide adequate treatment of contaminants.	Present consumer messaging through "new" and "old" media, as well as K-12 education system.
Educate consumers about how to shop for sustainably produced foods (i.e., those produced in soil-sustaining ways).	Provide targeted messages to explain how to identify foods produced with methods beneficial to soil. Consider developing or modifying labeling programs.	Present consumer messaging through "new" and "old" media, and in-store information.

TABLE 13.3   (Continued)
Examples of "Strategies and Tools" to Advance Action on the Direct Interventions

Strategies	Example(s)	Tools
D. Research		
Provide public research funds for studies relevant to soil and health, including pilot studies and research training.	Develop RFAs and identify areas of interest relevant to soil and health. Develop methods to provide better on-site sewage treatment in regard to specific soil and site conditions.	USDA, EPA, National Institutes of Health, National Science Foundation, Centers for Disease Control and Prevention. Federal lawmaking for these agencies, appropriations.
Provide incentives to increase corporate R&D for sustainable agriculture interventions.	Tax breaks can support such activities, with appropriate safeguards to minimize "greenwashing"—overstating or falsifying a product's environmental benefits.	Federal, state taxation policy, Farm Bill.
Commission authoritative report to prioritize integrated recommendations for public health action on soil health issues.	Institute of Medicine report.	USDA, CDC, EPA, or private funders could commission report.

Note: These strategies and tools vary in feasibility, cost, and effectiveness.

13.3.6   POLICY STRATEGIES

13.3.6.1   Financial Incentives

The U.S. Farm Bill is a policy tool providing financial incentives that play a central role in shaping agriculture and food systems in the United States. It includes policies with both significant positive and negative impacts on soil. Even as recognition grows regarding the Farm Bill's public health relevance, there has been little attention to how the Farm Bill can impact health via soil.

The Conservation Title is the section in the Farm Bill with the most direct relevance to soil (USDA Economic Research Service, 2008a). Through two programs, it provides funds to working farms for implementing environmentally sound practices such as cover cropping and no-till agriculture. Demand typically exceeds available funds. For example, in the fiscal year (FY) 2009, the Environmental Quality Incentives Program (EQIP) funded only 29% of applications (NRCS, 2010b). The Conservation Title also includes land reserve programs that pay farmers to keep vulnerable land out of production and protect soil health with vegetative cover and/or other restorative activities. Despite the potential improvement in soil quality by restoring such lands, in times of high food demand and prices many farmers let their land reserve contracts expire. In 2011 and other years, advocates have requested the option for farmers to escape their contracts altogether (Hellerstein and Malcolm, 2011; Looker, 2011; Multiple Authors, 2011; Streitfeld, 2008).

Conservation programs represent a valuable strategy for protecting and improving soils while advancing farmer interests. To improve effectiveness, more funding, improved geographic targeting, and increased enforcement are recommended (Batie, 2009; Eubanks, 2009; Claassen and Morehart, 2006; Frisvold, 2004). Unfortunately, in times of tight budgets these programs are quickly targeted for budget cuts (Clayton, 2011; Monke, 2010; National Sustainable Agriculture Coalition, 2010).

While the Conservation Title protects soil, Farm Bill commodity policy tools including subsidies and marketing funds promote maximizing yield. Much of the agriculture these policies support uses methods that contribute to degraded soil health, including intensive planting, large-scale monocultures, the use of large quantities of fertilizers and pesticides, extensive mechanization, and lack of buffers to prevent erosion (Batie, 2009; Eubanks, 2009; Wallinga, 2009; Frisvold, 2004). As a result, soils, particularly in the Midwest, have experienced loss of SOM, pesticide contamination, and erosion, among other threats (Veenstra, 2010; USDA, 2009). Erosion and overnutrification in these soils also contributes to hypoxia in the Gulf of Mexico (Rabalais et al., 2002). In the Midwestern states known as the "Corn Belt," Farm Bill income, production, and insurance subsidies provided 7.3 times as much money as conservation subsidies from 1997 to 2009 (Cox et al., 2011). To reduce farming intensity, a contrasting policy strategy would aim to stabilize commodity supply (and price) through means such as maintaining food reserves (Ray et al., 2003).

Receipt of commodity subsidies requires soil conservation compliance when producing on "highly erodible land" and wetlands (NRCS, 2011). A USDA analysis credits the conservation compliance program with contributing 25% of the reduction in soil erosion between 1982 and 1997 (Claassen et al., 2004). However, the Environmental Working Group reports that enforcement resources were substantially cut after the 1996 Farm Bill and that the program has made little additional progress in reducing soil erosion since then (Cox et al., 2011; Perez, 2007). The program's impacts would be stronger with increased enforcement resources and broadened geographic eligibility.

Economists debate the extent to which commodity policies drive practice versus providing safety nets for farmers who might perform these practices regardless of policy incentive (Ray et al., 2003). Soil health, human health, and the environment would broadly benefit from shifting from a farm safety net that at best reinforces existing practice to one based on incentivizing conservation. We note that any change to subsidy policy should aim to minimize dislocation in the farm sector. As farm ownership declines, and as conventional family farmers are pushed to ever more intensive practices to break even financially, soil health can be further degraded.

13.3.6.2 Regulations

While the Farm Bill is primarily a funding bill, other U.S. government policies may improve soil conditions by regulating allowable farming practices. Regulatory tools can drive change and innovation, but can also create resentment and resistance among the regulated, resulting in political challenges, delays, and reduced enforcement (Neff and Goldman, 2005).

An example of a regulatory tool affecting soil is the Clean Water Act (CWA). One of the ways the CWA affects soil is through its Concentrated Animal Feeding Operation[*] (CAFO) rule that requires CAFOs to submit nutrient management plans as part of the permit process. These must undergo public review (EPA, 2010b). There is a need for enforcement. Some permitted CAFOs do not adhere to rules, while others avoid permits due to loopholes (EPA, 2011b; Natural Resources Defense Council, 2010). Additionally, a significant number of animal feeding operations (AFOs) slightly smaller than the CAFO classifying criteria (e.g., 37,500 chickens, 3,000 hogs, 300 cows) are unregulated. To improve U.S. soil and water quality, enforcement of waste storage and disposal and nutrient management should be increased, and all AFOs should be required to comply with regulations (Ribaudo et al., 2003).

We highlight one additional set of regulations: those promulgated by the U.S. Occupational Safety and Health Administration (OSHA) for worker protection. The Occupational Safety and Health Act of 1970 requires OSHA to make rules to protect workers, including those with ongoing exposure to soil contaminants (a few relevant contaminants, such as pesticides, are regulated instead by the EPA). From 1970 to 1972 the agency adopted standards for exposure to over 400 chemicals, some based on science from the 1940s and 1950s. Most have not been updated. Further, despite the over 60,000 chemicals used in the United States, OSHA has only been able to pass two chemical standards since 1997 due to political and other challenges (Feldman, 2011; Howard, 2010; McGarity et al., 2010). The only occupational health exposure regulated for agriculture is cadmium (OSHA,

---

[*] CAFO is a formal EPA term based on number of animals at a facility. There are also many IFAP facilities that, while not having enough animals to qualify as CAFOs, exhibit many of the same problems (EPA, 2011b).

2012a). A chemical-by-chemical approach may be outdated in the modern era, but improved regulatory approaches by OSHA are needed to protect agricultural and other workers. One example is OSHA's HAZWOPER standard, governing exposures to soil and other hazards encountered in hazardous waste and emergency response work. This process-based standard requires provision of training and personal protective equipment, rather than setting exposure limits by chemical (OSHA, 2012b).

Even where OSHA standards exist, enforcement is weak (AFL-CIO, 2011; Feldman, 2011; Howard, 2010; McGarity et al., 2010; Silverstein, 2008; National Coalition for Agricultural Safety, 1989). Federal and state OSHA regulations also do not apply to many of those at risk from soil-related hazards, including public sector workers in most states, and enterprises with fewer than 11 employees, including most farms in the United States (OSHA, 2012c). Expanding coverage and strengthening OSHA regulations and their enforcement is an important strategy for preventing soil-related occupational exposures.

13.3.6.3 Zoning

As a policy strategy, zoning addresses allowable land uses and permissions within each land use type. In order to generate zoning policy that supports soil health and long-term food security, it is important for zoning to consider land characteristics and land capability class (Table 13.4) for agricultural lands (Hockensmith, 1953). Additionally, zoning policy can improve farmland preservation. From 1982 to 2007 the United States lost 23 million acres of agricultural land to other land uses. Agricultural land lost to development is extremely difficult to recover (Dempsey, 2010; Lehman, 1992). Policies at state and local levels can reduce development on fertile agricultural land or restrict building to already developed areas (Nelson et al., 2009; Kruft, 2001). Zoning policy can also mandate measures to protect soil where development is occurring, such as by requiring plans for erosion control during construction or giving incentives to build structures that control stormwater runoff (Dempsey, 2010; Nelson et al., 2009). U.S. localities around the country have also been rewriting previously enacted zoning policies in efforts to encourage urban food production (Reinhardt, 2011; Abdul-Kareem et al., 2009).

Zoning and rezoning policy can be used to restrict industry locations, such as by segregating noxious land uses in industrial areas. While there are benefits to concentrating contaminated soil in certain locations, such strategies have limitations. First, in many urban areas, there are no "remote"

TABLE 13.4

USDA Land Capability Classes (LCC)

LCC	Definition
1	Soils with slight limitations that restrict their use.
2	Soils with moderate limitations that reduce the choice of plants or require moderate conservation practices.
3	Soils with severe limitations that reduce the choice of plants or require special conservation practices, or both.
4	Soils with very severe limitations that restrict the choice of plants or require very careful management, or both.
5	Soils with little or no hazard of erosion but that have other limitations that are impractical to remove and that limit their use mainly to pasture, range, forestland, or wildlife food and cover.
6	Soils with severe limitations that make them generally unsuited to cultivation and that limit their use mainly to pasture, range, forestland, or wildlife food and cover.
7	Soils with very severe limitations that make them unsuited to cultivation and that restrict their use mainly to grazing, forestland, or wildlife.
8	Soils and miscellaneous areas with limitations that preclude their use for commercial plant production and limit their use to recreation, wildlife, or water supply, or for esthetic purposes.

Source: NRCS, 2001, Land Capability Class, by State, 1997. http://www.nrcs.usda.gov/wps/portal/nrcs/detail/national/technical/nra/nri/?&cid = nrcs143_014040, verified 7 February 2012.

locations. One study found that 22% of all New York City residents lived in or adjacent to zoned industrial areas, with disparities according to minority status and income. Further, even if noxious uses are truly "away," chemicals will still travel from soils to water, and additionally there may be less public awareness and drive to minimize these hazards (Maantay, 2002).

13.3.6.4 Infrastructure Investment

A key direct intervention for building sustainable, healthy soils and food systems is recycling organic materials containing nitrogen, phosphorus, and other nutrients through composting, anaerobic digestion, or vermiculture. To enact this intervention on a broad scale, there is a need for public investment in the necessary physical infrastructure as well as ongoing funds for management, transportation, and education. There is also a critical need for investment in research and development on how to treat manure and biosolids adequately to protect human health.

Many farmers and, increasingly, gardeners use composted materials rather than synthetic fertilizers. There is a lost opportunity, however, in that an estimated 40% of food produced in the United States ends up in the waste stream and 97% of food waste goes to landfills (Levis et al., 2010; Hall et al., 2009). A few municipalities in the United States, including San Francisco and Seattle, have adopted municipal composting systems to gather and process such wastes and return the compost for public use (Gilmore, 2010; Olivares and Goldstein, 2008). Implementing such initiatives requires systems development, large-scale composting infrastructure, education, and outreach. Initially costly, such programs eventually yield financial and environmental savings (Morawski, 2008) plus benefits to surrounding farms and households.

13.3.7 MULTISECTORAL AND MULTISTAKEHOLDER STRATEGIES

After policy, the second set of strategies for enacting the direct interventions involves bringing together multiple sectors and stakeholders for action.

13.3.7.1 Stakeholder Engagement

Engaging with stakeholders is an important but sometimes neglected strategy for increasing the effectiveness with which the direct interventions are performed. Communication helps assure that competent plans and policies are developed with consideration of area conditions (social, political, economic, and ecological) and can avert unintended negative consequences. Additionally, participatory processes help create buy-in for these interventions. Stakeholders may include those directly affected, such as those living in an affected community or those whose livelihoods would be impacted by an intervention, community-based groups and coalitions, engaged organizations, government agencies, and professionals working on the ground. In particular, there is often a need to better engage populations who have lower incomes, minorities, and those with limited English proficiency, who may be significantly affected but are often excluded unintentionally.

For example, as engineers develop plans to remediate contaminated soil near current and former military sites, stakeholder engagement and multiagency work advance their direct intervention effort in important ways. Community members, whether formally engaged in Restoration Advisory Boards or not, play a critical role by raising concerns and helping to shape optimal site reuse plans. Engaging them also helps mitigate distrust and conflict that can hamper restoration efforts (EPA, 2011c; Office of the Secretary of Defense, 2007; Burke et al., 2003).

In many public planning processes, community engagement has become mandatory, however, the quality and level of that engagement varies, including by income and race (Maantay, 2002). Efforts are needed to "level the playing field" and assure that all voices are heard, including top-down commitment to make use of public input and to avoid concentrating unwanted land uses.

An additional strategy for advancing direct soil-related interventions is working with stakeholders in formal planning processes. There is a growing movement of city, state, and regional food policy councils working to promote local and regional agriculture and engaging in land use planning, zoning, and farmland preservation, thus indirectly addressing soil quality (Delaware Valley Regional Planning Commission, 2011; Harper et al., 2009; Winne, 2006). Few of these groups include a substantial focus on the ecological and environmental issues that affect soil, despite the potential impact on area food production capacity and economic development (Mark Winne, personal communication, 25 March 2011). Iowa notably has begun bringing soil into its planning conversation; for example, its 2011 Food System Blueprint includes 14 recommendations, several relevant for soil, including one directly on soil erosion (Tagtow and Roberts, 2011).

13.3.7.2 Governmental Interagency Activities

Soil issues are multidisciplinary in nature, and as such there is a great need for interagency governmental responses as a strategy for improving the effectiveness with which direct interventions are implemented. Continuing with the example of contaminated military sites, a small survey found that the Department of Defense often worked with departments of environment and the Army Corps of Engineers for 10 years or longer before health departments became engaged (Burke et al., 2003). Further, in the authors' experience, while health departments do sometimes develop relationships with environmental departments, they far more rarely work with agriculture departments. Engaging all stakeholders early, as partners, would be helpful in advancing direct soil interventions. Public health departments may help address potential health issues with education and screening, and may smooth relationships by communicating risk and health issues with community members. They can also provide expertise in risk assessment, surveillance, and monitoring to assure the interventions are performed in a health-protective way and achieve the desired outcomes.

13.3.8 Education, Communication, and Behavior Change

Education strategies are essential for motivating and shaping the direct interventions. Knowledge alone is rarely enough to motivate change, but it is a necessary precursor. Strategic behavior change interventions can also help move individuals and institutions from knowledge to action, and eventually from personal efforts to those with broader impact.

13.3.8.1 Safety Education

Individuals who interact directly with soil, both in personal and employment capacities, need access to readily understandable safety information in relevant languages and targeted to their needs. This includes information about soil hazards, protective strategies, personal protective equipment, safety and health regulations, and workers' rights. OSHA regulations require this information, however there are gaps in coverage and limitations in enforcement. For those with nonoccupational exposures there is generally no requirement for receiving such information.

13.3.8.2 Education about Direct Intervention Practices

Individuals, as well as management and governmental decision-makers, need clear information and technical assistance to increase the likelihood of direct intervention implementation. Across the many sectors, soil issues, and target audiences, it is difficult to generalize about the existing availability and accessibility of such technical guidance or its effectiveness. In the example of agricultural conservation practices, the Farm Bill is a central source of funds for cooperative extension services to provide such outreach (USDA Economic Research Service, 2008b). Some program funds are specifically directed to the needs of soil conservation and sustainable methods, but this allocation could be substantially increased. As for academic training, few college and university agriculture programs prioritize training in low-input agricultural methods of food production over mainstream production agriculture, although coursework has become widely available (Thompson, 2009; Parr et al., 2007).

13.3.8.3   Consumer and Institutional Food Purchaser Education

Consumers and institutional food purchasers need information about soil impacts of varying food and energy choices such as diets high in industrially produced red meat or use of corn ethanol. They also need information about the potential ecological and long-term food security benefits of shifting to foods produced with attention to soil health. For example, Smil (2002) estimates that if consumers in affluent countries ate 15%–35% less meat, there would be 5%–15% less phosphate fertilizer used to grow grain for animal feed. Few consumers are likely aware of, or concerned about, this benefit.

Even with knowledge and motivation, it can be difficult to identify "soil-friendly" products due to lack of labeling. Organics and grass-fed beef may be the best examples of labels that help consumers identify items produced with care for soil, but many sustainable farms are not covered by these labels. While some consumers might rely on farmers' market purchases or the designation of "local" as an indicator of soil stewardship, location of production or purchase alone is a weak indicator. Socially responsible purchasing for soil stewardship would be advanced with development of improved strategies for identifying and communicating to the public about recommended products.

13.3.9   RESEARCH

The final strategy for advancing the direct interventions is research. Evidence suggests that agricultural research and development can have significant benefits for soil conservation and thus public health (Heisey et al., 2010). To address the many critical research questions there is a need for increased funding designated for research on themes relevant to soil health through the USDA, National Institutes of Health, National Science Foundation, and other public and private funders. Key areas of emphasis include:

- Improving methods to promote soil health across the variety of local conditions and land uses, particularly in the context of climate change
- Improving methods to simultaneously increase soil health and yields
- Increasing understanding of health effects, exposure, plant uptake, and bioavailability of chemical and pathogen contaminants in soil
- Quantifying attributable risk or the portion of illnesses associated with soil versus other exposure media
- Improving methods for removing contaminants in biosolids and manure, as well as for assessing their safety
- Conducting social science and policy research on determinants of behavior change at the business, farm, and individual levels, and the relative effectiveness of various strategies, tools, incentive systems, barriers, and facilitators of change

Research training investments are needed to create the next generation of scientists with multidisciplinary skills and interest in these topics, and the commitment to seeking sustainable solutions. Collaborative research involving farmers, workers, community members, and others affected by the research can yield important insights. Overall, given the many gaps in knowledge, increased interdisciplinary research is one of the most needed strategies for advancing soil health and public health.

13.4   THE ROLE OF PUBLIC HEALTH

How can the public health field more effectively bring its strengths to bear in promoting the direct interventions and otherwise addressing soil concerns?

While public health budgets are perpetually limited, making it challenging to take on new obligations, many of the key soil-related actions are within the purview of existing programs and staff roles. Others can be addressed through partnerships. Public health professionals must work with and learn from farmers, soil scientists, engineers, cooperative extension agents, and others who already work with soil. All can be invaluable assets and resources towards the prevention of human morbidity and mortality. Further, developing these relationships helps assure that as soil-related policy and regulatory reform efforts develop, public health agencies and experts will be seated at the table, bringing health and equity lenses to bear.

Here, we discuss the Ten Essential Services of Public Health as modified by the Centers for Disease Control and Prevention (CDC) for environmental health (CDC, 2011), and highlight the soil context for each essential service. Some of these services are already in place, while others are in need of implementation and enforcement efforts.

13.4.1 ESSENTIAL SERVICE 1: MONITOR ENVIRONMENTAL AND HEALTH STATUS TO IDENTIFY AND SOLVE COMMUNITY ENVIRONMENTAL HEALTH PROBLEMS

This service includes public health surveillance on environmental hazards, exposures, and health outcomes including those from soil (Litt et al., 2004; Thacker et al., 1996). Screening for potential environmental hazards is often performed by environmental or agricultural agencies or by site managers for regulatory and reporting purposes. Public health agencies more typically become involved in tracking exposures and health outcomes and in linking such data with hazard information, including performing risk assessment to characterize the level of health risk from various hazard/exposure scenarios. For these interdisciplinary tasks, interagency collaboration is often needed, including creating active channels of communication, jointly developing priorities for monitoring, and designing shared protocols for action based on findings.

As an example of Essential Service 1, chemicals from World War I–era military chemical weapons research were discovered in soil in the Spring Valley neighborhood of Washington, DC. The U.S. Army Corps of Engineers examined hazards; their screening and assessment identified soil contamination with arsenic and other chemicals. From the health side, the District of Columbia Department of Health (DC DOH) and the federal agency for Toxic Substances and Disease Registry (ATSDR) assessed potential exposures among area residents and daycare center children and staff. ATSDR reported that no adverse health effects were anticipated based on the observed arsenic levels (ATSDR, 2005). The DC DOH contracted for a more complete health assessment, which concluded that the area's health profile was good and indicated challenges in assessing several health outcomes of concern. A human health risk assessment for exposure to arsenic and other chemicals was also performed. The findings supported ongoing remediation and surveillance for exposures and health outcomes (Fox et al., 2010). A follow-up health study is being funded by the Department of Environment, now engaged on health issues.

Beyond surveillance of potential environmental exposures, public health agencies and agricultural agencies should also be sharing monitoring data relevant to food security. For example, the Iowa "food system blueprint" *Cultivating Resilience* recommends that county boards of health (with boards of supervisors) establish regional food system planning councils that would, among other things, present periodic "report cards" evaluating food system health, including, as noted earlier, tracking erosion and other soil-relevant indicators (Tagtow and Roberts, 2011). While such coordination has been rare in the past, those engaged in planning to assure the public's health need to understand potential threats to long- and short-term regional and national food security.

13.4.2 ESSENTIAL SERVICE 2: DIAGNOSE AND INVESTIGATE ENVIRONMENTAL HEALTH PROBLEMS AND HEALTH HAZARDS IN THE COMMUNITY

This service includes diagnosis and epidemiologic investigation of soil-related health hazards and outcomes. There is a strong need for epidemiologic and other investigation to further clarify key soil-related risk factors of concern and the associations between soil exposures and health conditions. An important area of inquiry concerns attributable risk. For instance, in the case of understanding the public health risks of manure application to land, even when it is clear that manure contains *Escherichia coli* bacteria and human infections with *E. coli* are elevated in an area, to what extent are the soil exposures a concern? Additionally, there is a need to study the distribution of impacts, including environmental justice implications. Improving this understanding is critical for addressing risk factors and prioritizing action.

A second key area for public health epidemiology is investigating community concerns about disease clusters or potential health impacts from known or presumed soil contamination. In evaluating such cases it is usually challenging to find clear results due to small sample size, loss of historical exposure data, and other methodological issues (Juzych et al., 2007). Yet, health departments have an important role to play in working with soil scientists and others to seek further understanding and to respond to community concerns in sensitive and appropriate ways.

Finally, public health departments can supply public health interpretations for study findings. For example, the Ohio Department of Health and local health districts collaborated with others in a study to examine septic system malfunctioning. The study found that septic systems in soils with poor drainage characteristics were significantly more likely to malfunction than systems installed in soils having good drainage characteristics. Malfunctioning septic systems concerned the community because of contamination from bacteria and pathogens, nutrients (nitrogen and phosphorous), chloride, organic enrichment, and hazardous waste from improper disposal of household wastes. The local health department officials developed methods, reviewed the analysis, and considered how the findings could be used (CT Consultants, 2001).

13.4.3 ESSENTIAL SERVICE 3: INFORM, EDUCATE, AND EMPOWER PEOPLE ABOUT ENVIRONMENTAL HEALTH ISSUES

This service involves providing health education and communication as well as efforts to increase access to information and to assure cultural, literacy level, and linguistic appropriateness, and use of effective education methods. Public health policy organizations advocate for funds for education programs. Public health advocates have three main roles to play:

1. Providing health education for community members who could be exposed to hazards in soil

For example, environmental public health specialists at the EPA have sought to educate community gardeners about potential soil hazards, particularly at brownfield sites (EPA, 2009a), and to encourage steps to minimize exposure, including building raised beds. In another example, following discovery of arsenic in a public park in Baltimore, the city health department in conjunction with other local government officials and the federal ATSDR held public meetings, published educational materials, and provided an opportunity for individuals to ask questions (Baltimore City Health Department, 2007). Such education can help community members keep risk information in perspective and feel more empowered to take appropriate action. It can also help prevent harmful soil and human health exposures.

2. Promoting education regarding how best to address threats to soil health, mitigate soil damage, and reduce occupational and other exposures of human health concern, for those working with soil and hazardous substances that can enter soil

Such education is commonly implemented by work managers and in some cases by programs in community health cooperative extension, but public health educators and advocates can work to advance this education. They can also encourage managers to connect issues of worker protection with those of environmental sustainability, recognizing that often (though by no means always) interventions for one are beneficial for the other.

3. Providing public education about the impacts of varying food and energy choices on soil

A public health approach includes promoting the concept of co-benefits, meaning that some actions jointly benefit both health and the environment. Public health educators are beginning to address issues such as climate change and the multiple environmental harms of industrially produced animal products. An improved understanding in the field about the importance of healthy soil to sustaining human life could provide an additional motivation for environmentally friendly choices.

13.4.4 ESSENTIAL SERVICE 4: MOBILIZE COMMUNITY PARTNERSHIPS AND ACTION TO IDENTIFY AND SOLVE ENVIRONMENTAL HEALTH PROBLEMS

This essential service focuses on mobilizing and taking action using the above-described strategies and tools in order to advance the direct interventions. In particular, this service speaks to the public health role in advocacy for the policy strategies discussed in Section 13.3. The service is framed around the idea of working with key partners and stakeholders, incorporating their experiences and knowledge, and encouraging them to interact. Some groups and agencies may not initially seek public health alliances, so often the onus is on public health professionals and agencies to reach out to them.

Previously, with the exception of some policies and regulations relevant to soil contamination, the public health voice was largely absent from debates about policies that affect soils, including the Farm Bill and policies related to planning, zoning, and climate change. This is changing. For example, the American Public Health Association (APHA) mentions multiple concerns relevant to soil in its 2007 position paper "Toward a Healthy, Sustainable Food System" while the APHA, American Nurses Association, American Dietetic Association, and American Planning Association jointly expressed support for the principle that a healthy food system "Conserves, protects, and regenerates natural resources, landscapes, and biodiversity" (American Dietetic Association et al., 2010; APHA, 2007). In the lead-up to the 2012 Farm Bill, public health and sustainable agriculture advocates came together to identify shared priorities, noting in their introduction "Many of the sustainable agriculture movement's policy priorities—conservation assistance for improved soil health and pesticide use reduction, research into ecologically sustainable production systems, ... result in improved public health" (Klein et al., 2011). The public health community's ability to articulate the health importance of these soil-related policies is valuable in advancing action to benefit soil health.

13.4.5 ESSENTIAL SERVICE 5: DEVELOP POLICIES AND PLANS THAT SUPPORT INDIVIDUAL AND COMMUNITY ENVIRONMENTAL HEALTH EFFORTS

At all levels, federal to local, public health advocates, analysts, and agency staff play a direct role in policy development and planning relevant to soil. The field can strengthen such policies by providing data from surveillance, epidemiology, risk assessment, and health impact assessment as well as bringing a public health perspective to understanding policy issues. Public health advocates can also help ensure that these policies will serve the public good, including advocating for inclusion of perspectives from underserved communities and from stakeholders such as workers and family farmers. Public health has a role to play in using evidence and supporting all the policy strategies discussed in Section 13.3.

A more specific type of public health role in policy development involves creating and synthesizing the science base upon which EPA and other agency risk assessments are supported. By indicating levels of contamination relatively likely to result in health impacts, these risk assessments can have powerful influences on policy. Public health professionals contribute to these risk assessments in multiple ways. Environmental health scientists conduct the research on which EPA's risk assessments are performed, and environmental public health professionals and agency staff perform the literature reviews and risk assessments. Public health academic experts and advocates participate on review panels to provide input. Public health advocacy groups and academic centers provide public comment on draft policies. Health advocates and experts have also played a role in advocating for updated standards and new approaches to chemicals policy and risk assessment, such as demanding consideration of the cumulative risks of multiple chemicals acting in concert rather than each chemical separately.

Public health advocates also push for the revision or termination of existing policies that exacerbate health risk factors arising from soil and that encourage unsustainable practices, such as policies that promote high-input agricultural systems, or policies based on outdated risk data. For example, in 2007 the United States approved the fumigant pesticide methyl iodide for soil injection during strawberry production, despite scientific recognition that it is a particularly toxic chemical associated with cancer, neurotoxicity, and late-term miscarriages (Froines et al., 2010). To protect farmworkers and communities near strawberry fields, public health advocates pushed strongly for its ban, including mobilizing 200,000 public comments to EPA (Pesticide Action Network, 2011). In March 2012, the manufacturer voluntarily removed the fumigant from the U.S. market.

13.4.6   ESSENTIAL SERVICE 6: ENFORCE LAWS AND REGULATIONS THAT PROTECT HEALTH AND ENSURE SAFETY

In some cases, public health professionals and agencies are directly engaged in enforcement of laws relevant to soil and health (such as staff at OSHA); in others, they partner with enforcement agencies (such as staff at the EPA). Local city, county, and state environmental public health officials are often required to enforce regulations requiring soil evaluation prior to septic system construction and regulations in regard to the prevention of health hazards created by these systems. Limited resources and antiregulatory political forces can make it challenging to enforce health protective regulations (Hook et al., 2011; Cahill, 2010; McGarity et al., 2010). Efforts to increase enforcement can even backfire, leading to political challenges. Further, federal agencies have few enforcement staff relative to the size of the task: for instance, federal OSHA has enough inspectors to inspect each workplace under its jurisdiction only every 129 years (AFL-CIO, 2011). Some have alleged that the USDA may also have an internal disincentive for strong enforcement, given its dual role in promoting agriculture and enforcing rules (Mattera, 2004). As a result of these limitations, industry and agriculture may face relatively low incentive for compliance with regulations.

To advance the regulatory strategies described in Section 13.3, public health advocates raise awareness about gaps and increase pressure for effective enforcement as well as advocate for updated laws and meaningful penalties. Further, they can work strategically to leverage limited resources, such as by compliance monitoring or targeting enforcement based on level of risk.

13.4.7   ESSENTIAL SERVICE 7: LINK PEOPLE TO NEEDED PERSONAL HEALTH SERVICES AND ASSURE THE PROVISION OF HEALTH CARE WHEN OTHERWISE UNAVAILABLE

In the context of soil risk, this service includes assuring health care access for farmers, workers, and residents of communities exposed to contaminated soil and related risk factors. Many of these groups have challenges in affording care. At least as important, the public health field has a role in assuring that health care providers serving those potentially exposed to soil-related hazards are knowledgeable about relevant symptoms and risk reduction measures. Services such as environmental and occupational health clinics, migrant farmworker clinics, and other programs exist for this purpose. Some programs and materials exist to train health care workers who do not specialize in occupational medicine but there is a need for further expanded reach.

13.4.8 ESSENTIAL SERVICE 8: ASSURE A COMPETENT ENVIRONMENTAL HEALTH WORKFORCE

The public health workforce must become more educated about soil and its relationship to human health, including the areas of food security and ecological consequences. This essential service includes working with faculties of public health schools and other public health education providers (i.e., community colleges, undergraduate programs, cooperative extensions) to increase educational offerings, such as adding coursework about agricultural and ecological issues such as soil, climate change, and biodiversity. It is the authors' understanding that these themes are minimally addressed, if at all, in most public health and medical curricula; a more formal assessment is needed. By contrast, we expect that many environmental and occupational health sciences programs/industrial hygiene training programs already include content relevant to health threats from exposures to soil contamination and pathogens. Environmental public health professionals in county and state health departments also need training relevant to the soil concerns they address. For example, these professionals are often in charge of site and soil evaluation programs necessary to obtain permits for on-site sewage disposal systems. They need training in soil texture, soil morphology, and how soils can be used as media for wastewater treatment and recycling.

There is also a need for interdisciplinary professionals with knowledge of both health and soil issues. Multidisciplinary programs can provide seminars or continuing education for those outside the field; the extent to which such training currently exists is unclear.

13.4.9 ESSENTIAL SERVICE 9: EVALUATE EFFECTIVENESS, ACCESSIBILITY, AND QUALITY OF PERSONAL AND POPULATION-BASED ENVIRONMENTAL HEALTH SERVICES

Evaluation is core to a public health approach. Evaluation can help program managers, such as those running soil-related interventions, to improve program effectiveness and make the case for further funding. From a policy and planning perspective, evaluation assures that resources and efforts are properly allocated and targeted. Optimally, evaluation results should be disseminated to play a role in contributing to generalizable knowledge. Program evaluation plans should be included as a part of developing new soil programs, interventions, and communication campaigns relevant to soil. Policy plans should be examined for efficacy, effectiveness, efficiency, and equity.

13.4.10 ESSENTIAL SERVICE 10: RESEARCH FOR NEW INSIGHTS AND INNOVATIVE SOLUTIONS TO ENVIRONMENTAL HEALTH PROBLEMS AND ISSUES

This essential service involves conducting research to further refine understanding of the soil risk factors described here and of how best to address them while supporting long-term environmental sustainability. Public health research uses a broad range of tools, from microbiology to population-based epidemiology studies, policy research, economic research, qualitative methods, and community-based participatory research. Interdisciplinary research, providing opportunities for collaborators to initiate communication, is critical for addressing many of the most important questions relevant to soil. Individuals and organizations with direct experience in soil issues and related health concerns, including farmers and affected communities, are important partners. It will be critical to expand research exploring alternative approaches to longstanding challenges for soil health and human health.

Lastly, there is a significant need for research translation, for example, to bring into the public health literature evidence of how different soil conditions may relate to impacts on food security, climate change, or other ecologically driven outcomes. Health impact assessments of soil-related policies, as well as other methods of translating research into meaningful public health predictions, aid planning and may contribute to raising the level of public and policy-maker concern about soil health threats.

13.5 CONCLUSION

There is little awareness within the broad public health community about why soil is a public health concern. Those who focus on issues relevant to soil chemistry and biohazards generally approach their work from the perspective of the hazard, exposure, or health outcomes of concern rather than with a substantive focus on the medium. Even after the connections between soil and health are explained, most in the field question what, if anything, public health professionals and researchers can actually do to address these concerns. Further, the diverse soil issues are seen as separate sets of problems with segmented solutions, rather than intertwined.

This chapter has sought to draw together diverse strands of information to define a set of public health issues of concern, identify needed direct interventions, and define strategies and tools to advance the likelihood and extent of adopting the direct interventions. Lastly, we brought this information into the public health domain by recommending an expanded role for public health in advancing these tools based on already-recognized responsibilities.

Public health engagement on soil issues should not be delayed. Dramatic efforts to improve soil health, in the United States and globally, are an urgent part of the effort to stave off widespread hunger several decades hence. Human activities have resulted in widespread soil contamination with substances that can be toxic to humans, including those that will persist in the environment indefinitely. Remediation and exposure prevention are critical to prevent acute and chronic health effects. Negative ecological effects such as the release of greenhouse gases are also advanced with soil degradation and are already impacting the public's health.

Despite the magnitude and breadth of concerns, it is important to recognize that public health budgets are perennially tight, and public health agencies and professionals are often stretched to capacity with existing projects. How much additional effort is realistic? We emphasize that some of the recommended actions can occur within existing funded programs. Others should be prioritized due to their potential health impacts. Still others have public health co-benefits outside the arena of soil, including the co-benefits of building relationships with colleagues working in soil science, agriculture, planning, and advocacy. These co-benefits may provide additional impetus for funding and action.

We have not attempted to prioritize the recommendations, or to set a standard for the level of action needed by public health. Rather, we aim to present a framework that can be used to lay out the issues for prioritization. We recommend convening a high-level panel of soil scientists and experts from public health practice, policy, and research to further summarize evidence and grapple jointly with questions including which actions should be prioritized; what programs or activities are realistic, effective, and feasible; and how best to obtain buy-in and resources. It would be beneficial to maintain a forum for ongoing communication and advocacy.

Every handful of healthy soil holds billions of microorganisms. We have described how human actions contribute to the fates of those vast subterranean populations and their environs, and in turn how soil organisms, contaminants, structure, and quality can mediate above-ground processes relevant to public health outcomes. As agricultural expert Fred Kirschenmann has stated, "soil is, literally, the common ground on which we all stand" (Kirschenmann, 2010b). Multidisciplinary action and partnership, with strengthened public health engagement, can help assure it is a healthy commonality.

REFERENCES

Abdul-Kareem, M., D. Thornton, and L. Dunning. 2009. *Using Zoning to Create Healthy Food Environments in Baltimore City.* Harrison Institute for Public Law, Georgetown University Law Center, Washington, DC.

Abrahams, P.W. 2002. Soils: Their implications to human health. *Science of the Total Environment* 291(1–3): 1–32.

AFL-CIO. 2011. *Death on the Job: The Toll of Neglect*, 20th edn. AFL-CIO Safety and Health Department, Washington, DC.

American Dietetic Association, American Nurses Association, American Planning Association, and APHA. 2010. Principles of a healthy, sustainable food system. http://www.planning.org/nationalcenters/health/pdf/HealthySustainableFoodSystemsPrinciples.pdf (verified 6 February 2012).

APHA. 2007. Toward a healthy, sustainable food system. Policy number 200712. http://www.apha.org/advocacy/policy/policysearch/default.htm?id=1361 (verified 6 February 2012).

ATSDR. 2005. Health consultation: Spring Valley chemical munitions. Washington, District of Columbia. U.S. Department of Health and Human Services, Atlanta, GA. http://www.atsdr.cdc.gov/HAC/pha/SpringValleyChemical090705/SpringValleyHCFinal083005.pdf (verified 6 February 2012).

ATSDR. 2009. ATSDR—Hazardous Substances Emergency Events Surveillance (HSEES) home. http://www.atsdr.cdc.gov/HS/HSEES/ (verified 6 February 2012).

Baker, J.M., T.E. Ochsner, R.T. Venterea, and T.J. Griffis. 2007. Tillage and soil carbon sequestration-what do we really know? *Agriculture, Ecosystems and Environment* 118(1–4): 1–5.

Baltimore City Health Department. 2007. Swann Park updates from Baltimore City Health Department. http://swannpark.blogspot.com/ (verified 6 February 2012).

Bassil, K.L., C. Vakil, M. Sanborn, D.C. Cole, J.S. Kaur, and K.J. Kerr. 2007. Cancer health effects of pesticides. *Canadian Family Physician* 53: 1704–1711.

Batie, S.S. 2009. Green payments and the US farm bill: Information and policy challenges. *Frontiers in Ecology and the Environment* 7(7): 380–388.

Black, R.E., L.H. Allen, Z.A. Bhutta, L.E. Caulfield, M. de Onis, M. Ezzati, C. Mathers, and J. Rivera. 2008. Maternal and child undernutrition: Global and regional exposures and health consequences. *The Lancet* 371: 243–260.

Brevik, E.C. 2009a. Soil health and productivity. *In* W. Verheye (ed.), *Soils, Plant Growth and Crop Production.* Encyclopedia of Life Support Systems (EOLSS), Developed under the Auspices of the UNESCO, EOLSS Publishers, Oxford. http://www.eolss.net (verified 6 February 2012).

Brevik, E.C. 2009b. Soil, food security, and human health. *In* W. Verheye (ed.), *Soils, Plant Growth and Crop Production.* Encyclopedia of Life Support Systems (EOLSS), Developed under the Auspices of the UNESCO, EOLSS Publishers, Oxford. http://www.eolss.net (verified 3 January 2012).

Brown, P. 1995. Race, class, and environmental health: A review and systematization of the literature. *Environmental Research* 69(1): 15–30.

Burke, T.A., R. Neff, K. Chossek, and N. Tran. 2003. Strengthening the role of environmental public health at contaminated Air Force bases: Prevention, evaluation, and response. White paper, Johns Hopkins Bloomberg School of Public Health, Department of Health Policy & Management, Baltimore, MD.

Cahill, L.B. 2010. Achieving environmental, health, and safety compliance in the United States: Hard and getting harder. *Environmental Quality Management* 20(1): 1–7.

Carpenter, S.R. 2008. Phosphorus control is critical to mitigating eutrophication. *Proceedings of the National Academy of Sciences of the United States of America* 105(32): 11039–11040.

Carpenter, S.R., N.F. Caraco, D.L. Correll, R.W. Howarth, and A.N. Sharpley. 1998. Nonpoint pollution of surface waters with phosphorus and nitrogen. *Ecological Applications* 8: 559–568.

CDC. 2011. 10 Essential Environmental Public Health Services. http://www.cdc.gov/nceh/ehs/Home/HealthService.htm (verified 6 February 6, 2012).

CDC and NIOSH. 2010. Engineering controls. http://www.cdc.gov/niosh/topics/engcontrols/ (verified 6 February 2012).

Claassen, R. and M. Morehart. 2006. *Greening Income Support and Supporting Green.* Economic Research Service, USDA, Washington, DC.

Claassen, R., V. Breneman, S. Bucholtz, A. Cattaneo, R. Johansson, and M. Morehart. 2004. *Environmental Compliance in U.S. Agricultural Policy: Past Performance and Future Potential.* Economic Research Service, USDA, Washington, DC.

Clancy, K. and K. Ruhf. 2010. Is local enough? Some arguments for regional food systems. *Choices: The Magazine of Food, Farm, and Resource Issues* 25(1). http://www.choicesmagazine.org/magazine/article.php?article = 114 (verified 6 February 2012).

Clayton, C. 2011. Lucas opposes Ag budget cuts. *DTN/The Progressive Farmer.* http://www.dtnprogressivefarmer.com/dtnag/view/blog/getBlog.do;jsessionid = 951FBC9B9021A253318040161E23C174.agfreejvm1?blogHandle = policy&blogEntryId = 8a82c0bc301f591e01305602ca9d026e (verified 6 February 2012).

Cordell, D., J.O. Drangert, and S. White. 2009. The story of phosphorus: Global food security and food for thought. *Global Environmental Change* 19: 292–305.

Costello, A., M. Abbas, A. Allen, S. Ball, S. Bell, R. Bellamy, S. Friel, N. Groce, A. Johnson, and M. Kett. 2009. Managing the health effects of climate change. *The Lancet* 373(9676): 1693–1733.

Cox, C., A. Hug, and N. Bruzelius. 2011. *Losing Ground.* Environmental Working Group. http://static.ewg.org/reports/2010/losingground/pdf/losingground_report.pdf (verified 6 February 2012).

Cribb, J. 2010. *The Coming Famine: The Global Food Crisis and What We Can Do about It.* University of California Press, Berkeley, CA.

CT Consultants. 2001. Survey of northeast Ohio home sewage disposal systems and semi-public sewage disposal systems. Report prepared for the Northeast Ohio Areawide Coordinating Agency. http://www.noaca.org/Sewage.pdf (verified 7 February 2012).

Delaware Valley Regional Planning Commission. 2011. *Eating Here: Greater Philadelphia's Food System Plan.* Delaware Valley Regional Planning Commission, Philadelphia, PA.

Dembe, A.E., J.B. Erickson, and R. Delbos. 2004. Predictors of work-related injuries and illnesses: National survey findings. *Journal of Occupational and Environmental Hygiene* 1(8): 542–550.

Dempsey, J. 2010. 2007 NRI: Changes in land cover/use—Analysis. American Farmland Trust, Northampton, MA.

Department of Defense. 2011. DENIX—Defense environmental restoration program. http://www.denix.osd.mil/derp/ (verified 6 February 2012).

Doran, J. 2000. Soil health and sustainability: Managing the biotic component of soil quality. *Applied Soil Ecology* 15(1): 3–11.

EPA. 2002. *Onsite Wastewater Treatment Systems Manual.* EPA/625/R-00/008. http://www.epa.gov/nrmrl/pubs/625180012/625180012.htm (verified 6 February 2012).

EPA. 2006. Preventing septic system failure. http://cfpub.epa.gov/npdes/stormwater/menuofbmps/index.cfm?action = browse&Rbutton = detail&bmp = 25 (verified 6 February 2012).

EPA. 2008. *Septic Systems Fact Sheet.* EPA# 832-F-08-057. http://www.epa.gov/owm/septic/pubs/septic_systems_factsheet.pdf (verified 6 February 2012).

EPA. 2009a. *How Does Your Garden Grow: Brownfields Redevelopment and Local Agriculture.* EPA-560-F-09-024. http://www.epa.gov/brownfields/success/local_ag.pdf (verified 8 February 2012).

EPA. 2009b. *National Water Quality Inventory: Report to Congress. 2004 Reporting Cycle.* EPA 841-R-08-001. U.S. Environmental Protection Agency, Washington, DC.

EPA. 2010a. *Handbook for Managing Onsite and Clustered (Decentralized) Wastewater Treatment System.* EPA 832-B-05-001. U.S. Environmental Protection Agency National Service Center for Environmental Publications, Cincinnati, OH.

EPA. 2010b. *Concentrated Animal Feeding Operations (CAFO)—Final Rule.* http://cfpub.epa.gov/npdes/afo/cafofinalrule.cfm (verified 8 February 2012).

EPA. 2011a. *Inventory of U.S. Greenhouse Gas Emissions and Sinks: 1990–2009.* EPA 430-R-11-005. U.S. Environmental Protection Agency, Washington, DC.

EPA. 2011b. *Proposed NPDES CAFO Reporting Rule.* http://www.epa.gov/npdes/regulations/2011_npdes_cafo_proposedrule.pdf (verified 8 February 2012).

EPA. 2011c. *Superfund Community Involvement.* http://www.epa.gov/superfund/community/ (verified 8 February 2012).

EPA and OPP. 2011. Pesticides: Health and safety—Human health issues. http://www.epa.gov/opp00001/health/human.htm (verified 8 February 2012).

Eswaran, H., R. Lal, and P.F. Reich. 2001. Land degradation: An overview. *In* E.M. Bridges, I.D. Hannam, L.R. Oldeman, F.W.T.P. de Vries, S.J. Scherr, and S. Sompatpanit (eds), *Responses to Land Degradation. Proceedings of the 2nd International Conference on Land Degradation and Desertification,* Khon Kaen, Thailand. Oxford Press, New Delhi, India.

Eubanks, W. 2009. A rotten system: Subsidizing environmental degradation and poor public health with our nation's tax dollars. *Stanford Environmental Law* 28: 213–273.

FAO. 2010. *The State of Food Insecurity in the World.* Food and Agriculture Organization of the United Nations, Rome, Italy.

Feldman, J. 2011. *OSHA Inaction: Onerous Requirements Imposed on OSHA Prevent the Agency from Issuing Lifesaving Rules.* Public Citizen's Congress Watch, Washington, DC.

Fernandez-Cornejo, J. and M. Caswell. 2006. *The First Decade of Genetically Engineered Crops in the United States.* United States Department of Agriculture, Washington, DC.

Fox, M., F. Curriero, K. Kulbicki, B. Resnick, and T. Burke. 2010. Evaluating the community health legacy of WWI chemical weapons testing. *Journal of Community Health* 35(1): 93–103.

Francis, C.A., C.D. Flora, and L.D. King. 1990. *Sustainable Agriculture in Temperate Zones.* Wiley, New York, NY.

Frisvold, G.B. 2004. How federal farm programs affect water use, quality, and allocation among sectors. *Water Resources Research* 40(12): W12S05.

Froines, J.R., P. Blanc, K. Hammond, D. Hattis, E. Loechler, R. Melnick, T. McKone, and T. Slotkin. 2010. Report of the scientific review committee on methyl iodide to the department of pesticide regulation. Scientific Review Committee of the California Department of Pesticide Regulation. http://www.cdpr.ca.gov/docs/risk/mei/peer_review_report.pdf (verified 6 February 2012).

Gilmore, S. 2010. New Seattle law will cook restaurant waste into compost. *The Seattle Times*, 30 June. http://seattletimes.nwsource.com/html/localnews/2012249781_cedar01m.html (verified 6 February 2012).

Godfray, H.C., J.R. Beddington, I.R. Crute, L. Haddad, D. Lawrence, J.F. Muir, J. Pretty, S. Robinson, S.M. Thomas, and C. Toulmin. 2010. Food security: The challenge of feeding 9 billion people. *Science* 327(5967): 812–818.

Graham, J.P. and K.E. Nachman. 2010. Managing waste from confined animal feeding operations in the United States: The need for sanitary reform. *Journal of Water & Health* 8(4): 646–670.

Hall, K.D., J. Guo, M. Dore, and C.C. Chow. 2009. The progressive increase of food waste in America and its environmental impact. *PLoS One* 4(11): e7940. doi:10.1371/journal.pone.0007940.

Handschumacher, P. and D. Schwartz. 2010. Do pedo-epidemiological systems exist? *In* E.R. Landa and C. Feller (eds), *Soil and Culture*, pp. 355–368. Springer, New York, NY.

Harper, A., A. Shattuck, E. Holt-Giménez, A. Wolf, M. Workman, P. Clare-Roth, A. El-Khoury, S. Turrell, and D. Strong. 2009. *Food Policy Councils: Lessons Learned*. Institute for Food and Development Policy, Oakland, CA.

Harter, R.D. 1983. Effect of soil pH on adsorption of lead, copper, zinc, and nickel. *Soil Science Society of America Journal* 47(1): 47–51.

Haynes, R.J. and R. Naidu. 1998. Influence of lime, fertilizer and manure applications on soil organic matter content and soil physical conditions: A review. *Nutrient Cycling in Agroecosystems* 51(2): 123–137.

Heisey, P.W., J.L. King, K.D. Rubenstein, D.A. Bucks, and R. Welsh. 2010. *Assessing the Benefits of Public Research within an Economic Framework: The Case of USDA's Agricultural Research Service*. Economic Research Report No. (ERR-95), U.S. Department of Agriculture, Washington, DC.

Heisler, J., P.M. Gilbert, J.M. Burkholder, D.M. Anderson, W. Cochlan, W.C. Dennison, Q. Dortch, et al. 2008. Eutrophication and harmful algal blooms: A scientific consensus. *Harmful Algae* 8(1): 3–13.

Hellerstein, D. and S. Malcolm. 2011. *The Influence of Rising Commodity Prices on the Conservation Reserve Program*. Economic Research Report No. 110, U.S. Department of Agriculture, Washington, DC.

Herrero, M., P.K. Thornton, A.M. Notenbaert, S. Wood, S. Msangi, H.A. Freeman, D. Bossio, et al. 2010. Smart investments in sustainable food production: Revisiting mixed crop-livestock systems. *Science* 327(5967)(02): 822–825.

Hockensmith, R. 1953. Soil classification—Using soil survey information for farm planning. *Soil Science Society of America Journal* 17(3): 285–287.

Hook, J., N. Bendavid, and S. Power. 2011. GOP wins deep cuts in environment spending. The *Wall Street Journal*, April 13. http://online.wsj.com/article/SB10001424052748703385404576258550820756980.html (verified 7 February, 2012).

Horowitz, J., R. Ebel, and K. Ueda. 2010. *"No-till" Farming Is a Growing Practice*. Economic Information Bulletin No. 70, Economic Research Service, USDA, Washington, DC.

Howard, J. 2010. Seven challenges for the future of occupational safety and health. *Journal of Occupational and Environmental Hygiene* 7(4): D11–D18.

Hussain, S., T. Siddique, M. Saleem, M. Arshad, and A. Khalid. 2009. Impact of pesticides on soil microbial diversity, enzymes, and biochemical reactions. *Advances in Agronomy* 102: 159–200.

Institute of Medicine. 1988. *The Future of Public Health*. National Academies of Science, Washington, DC.

Jackson-Smith, D. 2010. *Toward Sustainable Agricultural Systems in the 21st Century*. National Academy Press, Washington, DC.

Johnson, J.M.F., A.J. Franzluebbers, S.L. Weyers, and D.C. Reicosky. 2007. Agricultural opportunities to mitigate greenhouse gas emissions. *Environmental Pollution* 150(1): 107–124.

Juzych, N.S., B. Resnick, R. Streeter, J. Herbstman, J. Zablotsky, M. Fox, and T.A. Burke. 2007. Adequacy of state capacity to address noncommunicable disease clusters in the era of environmental public health tracking. *American Journal of Public Health* 97(Suppl. 1): S163–S169.

Karlen, D. 2003. Soil quality: Why and how? *Geoderma* 114(3–4): 145–156.

Karlen, D. 2004. Soil quality as an indicator of sustainable tillage practices. *Soil and Tillage Research* 78(2): 129–130.

Kemp, D.R. and D.L. Michalk. 2007. Towards sustainable grassland and livestock management. *Journal of Agricultural Science* 145(6): 543–564.

Kirschenmann, F. 2010a. Alternative agriculture in an energy- and resource-depleting future. *Renewable Agriculture and Food Systems* 25: 85–89.

Kirschenmann, F. 2010b. On becoming lovers of the soil. *In* C.L. Falk (ed.), *Essays from a Farmer Philosopher*, pp. 284–289. The University Press of Kentucky, Lexington, KY.

Klein, R., K. Fitzgerald, F. Hoefner, K. Clancy, R. Neff, D. Wallinga, and M. Muller. 2011. Mapping a sustainable agriculture and public health alliance for the 2012 farm bill. http://www.healthyeatingresearch.com/images/stories/meetings/MappingASustainableAgriculturePublicHealthAllianceForThe2012FarmBill.pdf (verified 9 February 2012).

Kloppenburg, J., J. Hendrickson, and G.W. Stevenson. 1996. Coming in to the foodshed. *Agriculture and Human Values* 13(3): 33–42.

Konen, M.E. 1999. Human impacts on soils and geomorphic processes on the Des Moines Lobe, Iowa. Unpublished PhD dissertation, Iowa State University, Ames, IA.

Kruft, D. 2001. Agricultural zoning. The Agricultural Law Resource and Reference Center, The Dickinson School of Law of The Pennsylvania State University. http://law.psu.edu/_file/aglaw/Agricultural_Zoning.pdf (verified 7 February 2012).

Lal, R. 2007. Carbon management in agricultural soils. *Mitigation and Adaptation Strategies for Global Change* 12: 303–322.

Lehman, T. 1992. Public values, private lands: Origins and ironies of farmland preservation in Congress. *Agricultural History* 66(2): 257–272.

Levis, J.W., M.A. Barlaz, N.J. Themelis, and P. Ulloa. 2010. Assessment of the state of food waste treatment in the United States and Canada. *Waste Management* 30(8–9): 1486–1494.

Litt, J., N. Tran, and T.A. Burke. 2002. Examining urban brownfields through the public health "macroscope". *Environmental Health* 110: 183–193.

Litt, J., N. Tran, K. Malecki, R. Neff, B. Resnick, and T.A. Burke. 2004. Identifying priority health conditions, environmental data, and infrastructure needs: A synopsis of the Pew environmental health tracking project. *Environmental Health Perspectives* 112(14): 1414–1418.

Looker, D. 2011. Feed grain end users want easier opt-out for CRP. *Agriculture Online*. http://www.agriculture.com/news/policy/feed-grain-end-users-wt-easier-optout-f_4-ar16498 (verified 7 February 2012).

Lubowski, R., M. Vesterby, S. Bucholtz, A. Baez, and M.J. Roberts. 2006. *Major Uses of Land in the United States, 2002*. Economic Information Bulletin No. 14, Economic Research Service, USDA, Washington, DC.

Maantay, J. 2002. Zoning law, health, and environmental justice: What's the connection? *The Journal of Law, Medicine & Ethics* 30(4): 572–593.

Manna, M.C., A. Swarup, R.H. Wanjari, B. Mishra, and D.K. Shahi. 2007. Long-term fertilization, manure and liming effects on soil organic matter and crop yields. *Soil and Tillage Research* 94: 397–409.

Martinez, C.E. and H.L. Motte. 2000. Solubility of lead, zinc and copper added to mineral soils. *Environmental Pollution* 107(1): 153–158.

Martuzzi, M., F. Mitis, and F. Forastiere. 2010. Inequalities, inequities, environmental justice in waste management and health. *European Journal of Public Health* 20(1): 21–26.

Mattera, P. 2004. *USDA, Inc.: How Agribusiness Has Hijacked Regulatory Policy at the U.S. Department of Agriculture*. Good Jobs First, Washington, DC.

McGarity, T., R. Steinzor, S. Shapiro, and M. Shudtz. 2010. Workers at risk: Regulatory dysfunction at OSHA. Center for Progressive Reform White Paper #1003. http://www.progressivereform.org/articles/OSHA_1003.pdf (verified 7 February 2012).

Melero, S., E. Madejon, J.C. Ruiz, and J.F. Herencia. 2007. Chemical and biochemical properties of a clay soil under dryland agriculture system as affected by organic fertilization. *European Journal of Agronomy* 26(3): 327–334.

Merriam-Webster. 2008. *Merriam-Webster's Collegiate Dictionary*, 11th edn. Merriam-Webster, Springfield, MA.

Monke, J. 2010. Previewing the next farm bill: Unfunded and early-expiring provisions. Congressional Research Service Report for Congress 7-5700. Washington, DC.

Morawski, C. 2008. Composting—Best bang for MSW management buck. *Biocycle* 49(10): 23–27.

Multiple Authors. 2011. Letter to The Honorable Deborah Stabenow, Chairwoman of the Senate Committee on Agriculture, Nutrition and Forestry, and The Honorable Frank Lucas, Chairman of the House Committee on Agriculture (Letter from 72 organizations regarding Conservation Reserve Program contracts). http://www.ngfa.org/files//misc/CRPLetter.pdf (verified 9 February 2012).

Murray, L.R. 2003. Sick and tired of being sick and tired: Scientific evidence, methods, and research implications for racial and ethnic disparities in occupational health. *American Journal of Public Health* 93(2): 221–226.

Nachman, K.E., J.P. Graham, L.B. Price, and E.K. Silbergeld. 2005. Arsenic: A roadblock to potential animal waste management solutions. *Environmental Health Perspectives* 113(9): 1123–1124.

Nachman, K.E., J.N. Mihalic, T.A. Burke, and A.S. Geyh. 2008. Comparison of arsenic content in pelletized poultry house waste and biosolids fertilizer. *Chemosphere* 71(3): 500–506.

National Coalition for Agricultural Safety. 1989. *A Report to the Nation: Agricultural Occupational and Environmental Health: Policy Strategies for the Future.* Institute of Agricultural Medicine and Occupational Health, Iowa City, IA.

National Environmental Services Center. 2004. Septic systems—A practical alternative for small communities. *Pipeline* 15(3): 1–7.

National Sanitation Foundation. 2011. NSF/ANSI standard 245. http://www.nsf.org/Certified/Wastewater/Listings.asp?Standard = 245& (verified 7 February 2012).

National Sustainable Agriculture Coalition. 2010. Groups urge ag funding fairness. http://sustainableagriculture.net/blog/groups-urge-ag-funding-fairness/ (verified 7 February 2012).

Natural Resources Defense Council. 2010. EPA, environmental groups reach settlement on factory farm pollution lawsuit. http://www.nrdc.org/media/2010/100526.asp (verified 7 February 2012).

Neff, R.A. and L.R. Goldman. 2005. Regulatory parallels to Daubert: Stakeholder influence, "sound science," and the delayed adoption of health-protective standards. *American Journal of Public Health* 95(Suppl. 1): S81–S91.

Neff, R.A., C.I. Parker, F.L. Kirschemann, J. Tinch, and R.S. Lawrence. 2011. Peak oil, food systems and public health. *American Journal of Public Health* 101(9): 1587–1597.

Nelson, K., A. Doll, W. Schroeer, J. Charlier, V. Dover, M. Flippen, C. Duerksen, L. Einsweiler, D. Farr, L. Oberholtzer, and R. Williams. 2009. *Essential Smart Growth Fixes for Urban and Suburban Zoning Codes.* U.S. Environmental Protection Agency, Washington, DC.

Neumann, W. and A. Pollack. 2010. U.S. farmers cope with roundup-resistant weeds. *The New York Times,* 3 May.

NOAA. 2009. State of the science fact sheet: Harmful algal blooms. http://nrc.noaa.gov/plans_docs/2009/NOAAHABFactSheetOct2009.pdf (verified 7 February 2012).

Nord, M., A. Coleman-Jensen, M. Andrews, and S. Carlson. 2010. *Household Food Security in the United States, 2009.* Economic Research Report No. 108, Economic Research Service, USDA, Washington, DC.

NRCS. 2001. Land Capability Class, by State, 1997. http://www.nrcs.usda.gov/wps/portal/nrcs/detail/national/technical/nra/nri/?&cid = nrcs143_014040 (verified 7 February 2012).

NRCS. 2006. Nutrient management. ftp://ftp-fc.sc.egov.usda.gov/NHQ/practice-standards/standards/590.pdf (verified 7 February 2012).

NRCS. 2010a. 2007 National resources inventory: Soil erosion on cropland. http://www.nrcs.usda.gov/Internet/FSE_DOCUMENTS/nrcs143_012269.pdf (verified 8 February 2012).

NRCS. 2010b. EQIP 2009 contracts and funding data. http://www.nrcs.usda.gov/wps/portal/nrcs/detailfull/national/programs/financial/eqip/?&cid = nrcs143_008294 (verified 8 February 2012).

NRCS. 2011. Highly erodible land conservation compliance provisions. http://www.nrcs.usda.gov/wps/portal/nrcs/detail/national/programs/alphabetical/camr/?&cid = nrcs143_008440 (verified 7 February 2012).

Office of Congresswoman Louise Slaughter. 2011. FDA reports to Slaughter: Over 70 percent of antibiotics administered to animals in feed. http://www.louise.house.gov/index.php?option = com_content&view = article&id = 2481:fda-reports-to-slaughter-over-70-percent-of-antibiotics-administered-to-animals-in-feed-&catid = 95:2011-press-releases&Itemid = 55 (verified 7 February 2012).

Office of the Secretary of Defense. 2007. Restoration Advisory Board rule handbook. http://www.saw.usace.army.mil/campbutner/RAB%20Rule%20Handbook_23%20Jan%202007-v2.pdf (verified 7 February 2012).

Olivares, C. and N. Goldstein. 2008. Food composting infrastructure. *BioCycle* 49(12): 30–31.

OSHA. 2012a. Occupational safety and health standards for agriculture. http://www.osha.gov/pls/oshaweb/owastand.display_standard_group?p_toc_level = 1&p_part_number = 1928. Accessed 2 February 2012 (verified 8 February 2012).

OSHA. 2012b. Hazardous waste operations and emergency response. 29 CFR 1910.120. http://www.osha.gov/pls/oshaweb/owadisp.show_document?p_table = standards&p_id = 9765. Accessed 2 February 2012 (verified 8 February 2012).

OSHA. 2012c. OSHA enforcement. http://www.osha.gov/dep/index.html (verified 8 February 2012).

Parr, D., C. Trexler, N. Khanna, and B. Battisti. 2007. Designing sustainable agriculture education: Academics' suggestions for an undergraduate curriculum at a land grant university. *Agriculture and Human Values* 24(4): 523–533.

Pepper, I.L., C.P. Gerba, D.T. Newby, and C.W. Rice. 2009. Soil: A public health threat or savior? *Critical Reviews in Environmental Science and Technology* 39: 416–432.

Perez, M. 2007. *Trouble Downstream: Upgrading Conservation Compliance.* Environmental Working Group, Washington, DC.

Pesticide Action Network. 2011. Safe strawberries. http://www.panna.org/cancer-free-strawberries (verified 8 February 2012).

Peters, C.J., J.L. Wilkins, and G.W. Fick. 2007. Testing a complete-diet model for estimating the land resource requirements of food consumption and agricultural carrying capacity: The New York State example. *Renewable Agriculture and Food Systems* 22: 145–153.

Peters, C.J., N.L. Bills, J.L. Wilkins, and G.W. Fick. 2008. Foodshed analysis and its relevance to sustainability. *Renewable Agriculture and Food Systems* 24(1): 1–7.

Pew Commission on Industrial Farm Animal Production. 2008. Putting meat on the table: Industrial farm animal production in America. http://www.livablefutureblog.com/pdf/Putting_Meat_on_Table_FULL.pdf (verified 8 February 2012).

Pimentel, D. 2005. Environmental and economic costs of the application of pesticides primarily in the United States. *Environment, Development and Sustainability* 7(2): 229–252.

Pimentel, D. 2006. Soil erosion: A food and environmental threat. *Environment, Development and Sustainability* 8: 119–137.

Powles, S.B. 2008. Evolved glyphosate-resistant weeds around the world: Lessons to be learnt. *Pest Management Science* 64(4): 360–365.

Public Health Functions, Steering Committee. 1994. Public health in America. http://www.health.gov/phfunctions/public.htm (verified 9 February 2012).

Rabalais, N.N., E.R. Turner, and D. Scavia. 2002. Beyond science into policy: Gulf of Mexico hypoxia and the Mississippi River. *Bioscience* 52(2): 129–142.

Rapport, D.J., R. Costanza, and A.J. McMichael. 1998. Assessing ecosystem health. *Trends in Ecology & Evolution* 13(10): 397–402.

Ray, D.E., D.G. De La Torre Ugarte, and K.J. Tiller. 2003. *Rethinking U.S. Agricultural Policy: Changing Course to Secure Farmer Livelihoods Worldwide.* Agricultural Policy Analysis Center, University of Tennessee, Knoxville, TN.

Reinhardt, J. 2011. GitC launches interactive urban ag zoning map. http://growninthecity.com/2011/03/gitc-launches-interactive-urban-ag-zoning-map/ (verified 8 February 2012).

Reynolds, K.A., K.D. Mena, and C.P. Gerba. 2008. Risk of waterborne illness via drinking water in the United States. *Reviews of Environmental Contamination and Toxicology* 192: 117–158.

Rezg, R., B. Mornagui, S. El-Fazaa, and N. Gharbi. 2010. Organophosphorus pesticides as food chain contaminants and type 2 diabetes: A review. *Trends in Food Science & Technology* 21(7): 345–357.

Ribaudo, M.O., N.R. Gollehon, and J. Agapoff. 2003. Land application of manure by animal feeding operations: Is more land needed? *Journal of Soil and Water Conservation* 58(1): 30–38.

Robinson, J.C. 1989. Exposure to occupational hazards among Hispanics, blacks, and non-Hispanic whites in California. *American Journal of Public Health* 79(5): 629–630.

Sanborn, M., K.J. Kerr, L.H. Sanin, and D.C. Cole. 2007. Non-cancer health effects of pesticides. *Canadian Family* 53: 1712–1720.

Silbergeld, E.K. and K. Nachman. 2008. The environmental and public health risks associated with arsenical use in animal feeds. *Annals of the New York Academy of Sciences* 1140(10): 346–357.

Silbergeld, E.K., J. Graham, and L.B. Price. 2008. Industrial food animal production, antimicrobial resistance, and human health. *Annual Review of Public Health* 29: 151–169.

Silverstein, M. 2008. Getting home safe and sound: Occupational Safety and Health Administration at 38. *American Journal of Public Health* 98(3): 416–423.

Sing, C.F. and D.B. Sing. 2008. Soil and human health. *In* R. Lal (ed.), *Encyclopedia of Soil Science*, pp. 838–841. Taylor and Francis Group, Boca Raton, FL.

Smil, V. 2002. Phosphorus: Global transfers. *In* I. Douglas (ed.), *Encyclopedia of Global Environmental Change Volume 3: Causes and Consequences of Global Environmental Change*, pp. 536–542. Wiley, Chichester.

Stingone, J.A. and S. Wing. 2011. Poultry litter incineration as a source of energy: Reviewing the potential for impacts on environmental health and justice. *New Solutions* 21(1): 27–42.

Streitfeld, D. 2008. As prices rise, farmers spurn conservation program. *The New York Times*, 9 April.

Tagtow, A.M. and S.L. Roberts. 2011. *Cultivating Resilience: A Food System Blueprint That Advances the Health of Iowans, Farms and Communities.* Iowa Food Systems Council, Elkhart, IA.

Thacker, S.B., D.F. Stroup, R.G. Parrish, and H.A. Anderson. 1996. Surveillance in environmental public health: Issues, systems, and sources. *American Journal of Public Health* 86(5): 633–638.

Thompson, B. 2009. *Educational and Training Opportunities in Sustainable Agriculture*, 19th edn. Agricultural Research Service, USDA, Washington, DC.

Tisdale, S.L., W.L. Nelson, and J.D. Beaton. 1985. *Soil Fertility and Fertilizers*, 4th edn. Macmillian Publishing Company, New York, NY.

Tyler, E.T., R. Laak, E. McCoy, and S.S. Sandhu. 1977. The soil as a treatment system. *In Home Sewage Treatment, Proceedings of the Second National Home Sewage Treatment Symposium*, pp. 22–37. ASAE Publication 5-77. American Society of Agricultural Engineers, St. Joseph, MI.

U.S. Census Bureau. 2009. Projections of the population and components of change for the United States: 2010 to 2050. http://www.census.gov/population/www/projections/summarytables.html (verified 8 February 2012).

U.S. Census Bureau. 2010. International data base—Total midyear population for the world: 1950–2050. http://www.census.gov/population/international/data/idb/worldpoptotal.php (verified 8 February 2012).

USDA. 2009. *Summary Report: 2007 Natural Resources Inventory*. Natural Resources Conservation Service and Center for Survey Statistics and Methodology, Iowa State University, Ames, IA.

USDA Agricultural Research Service. 2005. Salinity in agriculture. http://www.ars.usda.gov/Aboutus/docs. htm?docid = 10201&page = 1 (verified 8 February 2012).

USDA Economic Research Service. 2008a. 2008 farm bill side-by-side—Title II: Conservation. http://www. ers.usda.gov/farmbill/2008/titles/titleIIConservation.htm (verified 8 February 2012).

USDA Economic Research Service. 2008b. 2008 Farm bill side-by-side—Title VII: Research and related matters. http://www.ers.usda.gov/FarmBill/2008/Titles/TitleVIIResearch.htm (verified 8 February 2012).

USDL. 2004. *The National Agricultural Workers Survey*. U.S. Department of Labor, Washington, DC.

Veenstra, J.J. 2010. Fifty years of agricultural soil change in Iowa. Unpublished PhD dissertation, Iowa State University, Ames, IA.

Victora, C.G., L. Adair, C. Fall, P.C. Hallal, R. Martorell, L. Richter, and H.S. Sachdev. 2008. Maternal and child undernutrition: Consequences for adult health and human capital. *The Lancet* 371(9609): 340–357.

Walker, P., P. Rhubart-Berg, S. McKenzie, K. Kelling, and R.S. Lawrence. 2007. Public health implications of meat production and consumption. *Public Health Nutrition* 8(4): 348–356.

Wallinga, D. 2009. Today's food system: How healthy is it? *Journal of Hunger & Environmental Nutrition* 4(3): 251–281.

Weichenthal, S., C. Moase, and P. Chan. 2010. A review of pesticide exposure and cancer incidence in the agricultural health study cohort. *Environmental Health Perspectives* 118(8): 1117–1125.

Wilson, D. 2001. *Fateful Harvest: The True Story of a Small Town, a Global Industry and a Toxic Secret*. Harper Collins, New York, NY.

Winne, M. 2006. Policy: Issue contacts. North American Food Policy Council. http://www.foodsecurity.org/ FPC/policy-issue.html (verified 8 February 2012).

Woods, J., A. Williams, J. Hughes, M. Black, and R. Murphy. 2010. Energy and the food system. *Philosophical Transactions of the Royal Society B: Biological Sciences* 365(1554): 2991–3006.

WHO. 2011. Environmental and occupational cancers. Fact sheet N°350. http://www.who.int/mediacentre/ factsheets/fs350/en/index.html (verified 8 February 2012).

Xue, X. and A.E. Landis. 2010. Eutrophication potential of food consumption patterns. *Environmental Science & Technology* 44(16): 6450–6456.

Section IV

Food Security and Climate Change

14 Soils and Food Security

Winfried E.H. Blum and Stephen Nortcliff

CONTENTS

14.1 COMPLEX ISSUE OF FOOD SECURITY AND THE FUNCTIONS OF SOIL

14.1.1 COMPLEX ISSUE OF FOOD SECURITY

The 1996 World Summit agreed that food security represents "a situation that exists when all people, at all times, have physical, social and economic access to sufficient safe and nutritious food that meets their dietary needs and food preferences for an active and healthy life" (Barrett, 2010). This definition comprises complex targets, because food security means not only to have food available through sustainable production but also to have access to food under different conditions. Table 14.1 explains the complex issue of food security, distinguishing between availability/sustainability and accessibility.

Availability (production) encompasses environmental, social, and economic as well as technological opportunities and threats, for example, the availability of natural resources such as land

TABLE 14.1
Complex Issue of Food Security

Focus Area	Topics	Subjects
Availability/sustainability (= sustainable food production)	1. Environmental opportunities/ threats	Natural resources: land surface, topography, soils, water, biota (animals, plants), climatic conditions, etc.
	2. Social/economic opportunities/threats	Human resources, capital resources, farming systems, land ownership, food policy systems, etc.
	3. Technological opportunities/ threats	Available tools (e.g., machinery), agrochemicals, etc.
Accessibility	1. Physical accessibility	Storage, transport, market conditions, quality/safety control, etc.
	2. Economic/social accessibility	Food policy, pricing systems, income generation, etc.
	3. Cultural accessibility	Food habits, world views, myths, religions, etc.
	4. Physiological accessibility	Genetic constitutions, health conditions, for example, allergies, etc.

surface, topography, soils, water, biota, climatic conditions, and others. Moreover, human and capital resources, as well as farming systems, land ownership, and food policy systems, in addition to the availability of tools, for example, agricultural machinery as well as agrochemicals (plant protection products, fertilizers, etc.), are important.

Food availability is very different from food accessibility by individuals or groups of individuals, because there are a number of constraints, such as physical accessibility (e.g., through storage, transport, market conditions, quality and safety control, and others) and economic and social accessibility, which are mainly determined by food policies at national and international levels, pricing systems, and income generation, because most people need to purchase food. Cultural accessibility is most significant and very diverse at a worldwide level, because of different food habits based on worldviews, myths, religions, and cultural perspectives in general. Finally, because of different genetic constitutions, the physiological accessibility to certain types of food differs worldwide, depending also on the health conditions of individuals. The increase in allergies worldwide is an example.

For the discussion of food security in relation to soils, the focus is mainly on the availability aspect, which means the aspect of sustainable food production. This depends on the availability of natural resources, summarized under "environmental opportunities and threats," such as available land surface, topography, soil distribution and soil quality, and water availability; availability of biota, such as husbandry animals and plants (e.g., seed material); and climatic conditions. The complexity of this approach is shown in detail by Mueller et al. (2010).

In the following text, we will therefore consider mainly the relationship between global soil resources and sustainable food production.

14.1.2 MAIN FUNCTIONS OF SOIL

Figure 14.1 depicts the goods and services provided by land and soil. These goods and services are based on the capacity of land and soil to perform specific functions, each of which is important for human well-being and the environment (Blum, 2005):

1. Production of biomass through agriculture and forestry
2. Protection of groundwater and the food chain against contamination, and maintenance of biodiversity by filtering, buffering, and transformation activities

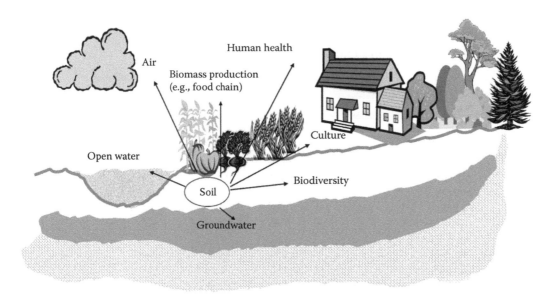

FIGURE 14.1 Goods and services provided by land and soil. (From Van Camp, L., Bujarrabal, B., Gentile, A.R., Jones, R.J.A., Montanarella, L., Olazabal, C., and Selvaradjou, A.-K., *Reports of the Technical Working Groups Established under the Thematic Strategy for Soil Protection, Vol. IV: Contamination and Land Management* (EUR 21319 EN/4). Joint Research Centre, European Commission, Brussels, Belgium, 2004. http://ec.europa.eu/environment/soil/pdf/vol4.pdf, verified 3 January 2012.)

3. Preservation of the gene reserve, which is by far the largest on Earth and three to four times larger within the soil than that above ground, thus providing a very significant habitat
4. Provision of the physical basis for infrastructural development, such as housing, industrial production, transport, dumping of refuse, sports, recreation, and others
5. Provision of raw materials such as gravel, sand, clay, and others (e.g., for infrastructural development)
6. Preservation of geogenic and cultural heritage by concealing and protecting archaeological and paleontological remains

In the context of food security, the function of producing biomass is the most relevant to sustaining food production and responding to cultural practices conducive to sustainable agricultural land management. Therefore, in the following section we will discuss food security from the perspective of sustainable production. We will focus on environmental opportunities and threats, especially those concerning natural resources, such as land surface, topography, and soils. The aspects of water and climatic conditions will also be included (see Table 14.1).

14.2 GLOBAL SOIL RESOURCES AND FOOD PRODUCTION

14.2.1 Global Soil Resources: Their Quality and Distribution

Blum and Eswaran (2004) presented data derived from earlier work by Eswaran (1989) illustrating the global distribution of soil and land resources (the principal categories being the soil orders of soil taxonomy [Soil Survey Staff, 1999]) and corresponding populations. The population data have been adjusted to reflect the 2010 estimate of global population (6.9 billion). Table 14.2 presents the land area occupied by each soil order or land class and an estimate of the number of people living on each. Ultisols, Alfisols, Inceptisols, and Entisols have high populations, together supporting over 70% of the global population. This group of soils present favorable conditions for agriculture

TABLE 14.2
Global Soil and Land Quality Classes and Population

	Land		2010 Population	
Soil and Land Quality Classes	**Area 10^6 km^2**	**%**	**Population 10^6**	**%**
1.Total ice-free land/population	128.57	100	6900	100
2. Kinds of soils				
Alfisols	11.52	8.96	1166	16.9
Andisols	0.84	0.65	117	1.7
Aridisols	14.34	11.15	380	5.5
Entisols	19.30	15.01	1097	15.9
Gelisols	10.29	8.00	28	0.4
Histosols	1.40	1.09	35	0.5
Inceptisols	11.73	9.13	1352	19.6
Mollisols	8.23	6.40	455	6.6
Oxisols	8.96	6.97	269	3.9
Spodosols	3.06	2.38	117	1.7
Ultisols	10.08	7.84	1221	17.7
Vertisols	2.89	2.25	380	5.5
Shifting sands	4.86	3.78	90	1.3
Rocky land	11.93	9.28	186	2.7
Glaciers, water bodies	9.14	7.11	7	0.1

but represent only 44% of the land area. In addition, it is important to note that, historically, communities first settled alluvial plains and low-relief undulating land that required low traction for management. Technological developments have enabled Vertisols and Aridisols to be more intensively used. In temperate parts of the world, Alfisols and Mollisols have high concentrations of people. The Mollisols occupy about 6.4% of the land surface and have about 6.6% of the global population. These two soil orders are some of the most productive soils of the world but are mostly found in temperate regions. In the tropics, a high proportion of the population is associated with river terraces (Entisols and Inceptisols) and Ultisols. Ultisols and Oxisols have presented major problems for sustained low-input agricultural production, but with increased understanding of the nature of these soils and careful management more sustainable production is increasingly possible (see also Eswaran, 1989).

The Gelisols of the boreal zone have the lowest population density of approximately 2 persons/km^2, while the Andisols (developed on volcanic pyroclastic materials) have the highest, with more than 129 persons/km^2. Some of the highest population densities in the world are on Andisols in Central Africa (Rwanda, Burundi, and parts of western Zaire). The Ultisols and Vertisols, which are extensively used for agricultural production in the tropics, have population densities in excess of 100 and 120 persons/km^2, respectively. Fragile systems such as those with Histosols and Aridisols have densities of 21 and 24 persons/km^2, respectively, and although these are low in comparison to other soil orders, in many regions these densities are threatening the sustainability of the systems. In many parts of the tropics, the rapidly increasing population has forced the landless poor to move into often marginal fragile ecosystems or degrade the better resources of their countries by overuse or inappropriate use.

Land quality is a measure of the ability of land to perform specific functions. It is normal for land to be placed in classes according to the ability to perform the functions outlined earlier, although in most systems the focus of the quality evaluation is in terms of the performance with respect to biomass production and frequently focusing further on the arable agricultural component of biomass production (see also Mueller et al., 2010). Beinroth et al. (2001) produced a nine-class land classification based on grain production. Table 14.3 presents a summary of the structure of the classification.

The respective areas of Earth occupied by these classes and the population associated with these classes are presented in Table 14.4. Class I lands, which have ideal soils occurring in ideal climates for crop production and are characterized by high productivity, high response to management, and minimal limitations, occupy only 2.38% of the global land surface but contribute over 40% of global food production. The 9.53% of global land resources in Classes II and III have minor limitations that are easily corrected and do not pose permanent restrictions to the use of land. Most of these lands are in the temperate regions of the world where the climate is moderate, with rare extremes of rainfall or temperature. These soils respond well to management and the positive effects of appropriate management persist for long periods.

Land quality Classes IV, V, and VI cover about 34% of the global land surface, largely in the tropics, and support over 50% of the population. These soils and the environmental contexts in which they are found have a range of constraints from high ambient temperatures that reduce germination

TABLE 14.3
Properties of the Land Quality Classes for Grain Production

Land Quality Class	Properties
I	This is prime land. Soils are highly productive, with few management-related constraints. Soil temperature and moisture conditions are ideal for annual crops. Soil management consists largely of sensible conservation practices to minimize erosion, appropriate fertilization, and use of best available plant materials. Risk for sustainable grain crop production is generally <20%.
II and III	These soils are good and have few problems for sustainable production. For Class II soils, however, care must be taken to reduce potential degradation. The lower resilience characteristics of Class II soils make them more risky, particularly for low-input grain crop production. Their productivity is however generally very high and consequently response to management is high. Risk for sustainable grain crop production is generally 20%–40% but can be reduced with good conservation practices.
IV, V, and VI	If there is a choice these soils must not be used for grain crop production, particularly soils belonging to Class IV. All three classes of soils require important inputs of conservation management if grain production is planned. Lack of plant nutrients is a major constraint, and careful fertilizer management is essential. Soil degradation must be carefully monitored and actions taken to prevent it. Productivity is not high and so low-input farmers must either be discouraged from using these soils or provided with substantial support. Land may be better used for national parks or biodiversity zones and in semiarid areas for rangeland. Risk for sustainable crop production is 40%–60%.
VII	These soils may only be used for grain crop production if there is no better quality land available. They require high levels of management and are not suitable for low-input agricultural production. Their low resilience makes them very prone to degradation. As in Classes V and VI, biodiversity management is crucial in these areas. Risk for sustainable crop production is 60%–80%.
VIII and IX	These soils belong to very fragile ecosystems or are very uneconomical for grain crop production. They should be retained under their natural state. Some areas may be used for recreational purposes but under very controlled conditions. In Class VIII, which is largely confined to tundra and boreal areas, timber harvesting must be done very carefully to prevent ecosystem damage. Class IX is mainly in deserts where biomass production is very low. Risk for sustainable crop production is >80%.

Source: After Beinroth, F.H., Eswaran, H., and Reich, P.F., *Sustaining the Global Farm*, pp. 569–574. Selected papers from the 10th International Soil Conservation Organisation Meeting, 24–29 May 1999, West Lafayette, IN, 2001.

TABLE 14.4
Land Area in Land Quality Classes with Estimated Population in Each Class

Land Quality Class	Area		Population	
	Million km²	%	Millions	%
I	3.06	2.38	405	5.9
II	6.40	4.98	949	13.8
III	5.85	4.55	320	4.6
IV	5.08	3.95	787	11.4
V	21.23	16.51	1985	28.8
VI	17.13	13.32	812	11.8
VII	11.58	9.01	768	11.1
VIII	21.46	16.69	124	1.8
IX	36.78	28.61	751	10.9

rates to low nutrient availability that limits biomass production of annual crops. Because much of this class of land lies within the tropics where production is often predominantly low-input, population is high, and growth rates are higher than in other regions, these lands must be carefully managed and are very vulnerable to degradation if poorly managed.

Class VII land, occupying a little over 9% of the land surface, includes shallow soils, those with high salt concentrations, and those with high organic matter levels. Shallow soils are normally considered unsuitable for agriculture, and saline soils may be used with specific adaptive crops and cropping practices. The high organic matter peatlands are a group of soils particularly vulnerable to degradation and permanent loss through drainage. Class VIII lands occupy almost 17% of the land surface; they have low temperatures and/or steep slopes and are generally considered unsuitable for agriculture. Class IX lands occupy over 28% of the land surface and are comprised of soils with inadequate moisture to support annual crop production.

The worldwide distribution of these nine land quality classes is shown in Figure 14.2. In Table 14.5, the percentage of land area in the main biomes as a function of land quality (Blum and Eswaran, 2004) is shown, revealing that only about 35% of the highly productive soils (Classes I–III) occur in the tropics, whereas 65% occur in regions with boreal, temperate, and Mediterranean climates, mostly in the northern hemisphere.

Based on Buringh (1985), FAO (1995), and our own calculations, about 12% of the world's land surfaces are suitable for food and fiber production, 24% can be used for grazing, 31% produce forests, and 33% are unsuitable for any kind of sustainable use, mainly because of climatic constraints.

Summarizing, it can be stated that food security depends essentially on the 12% of the land surfaces with soil quality Classes I–III, where about 25% of the world population lives and all traded food and fiber for the world market is produced.

14.2.2 Threats to Soil That May Compromise Food Production

While soils develop in response to a range of environmental conditions, there is concern that the natural functions of soils are increasingly threatened by changes in the environmental context (Scheffer et al., 2001). These changes are frequently human-induced or human-influenced (Foley et al., 2005). Agricultural activities have a clear impact on global environmental change (Tilman et al., 2001). A number of recent national and international approaches to soil protection have highlighted a list of possible threats to soil's capacity to perform its functions. For example, in presenting the case for a European Soil Protection Strategy, the Commission of the European Communities (2002) outlined

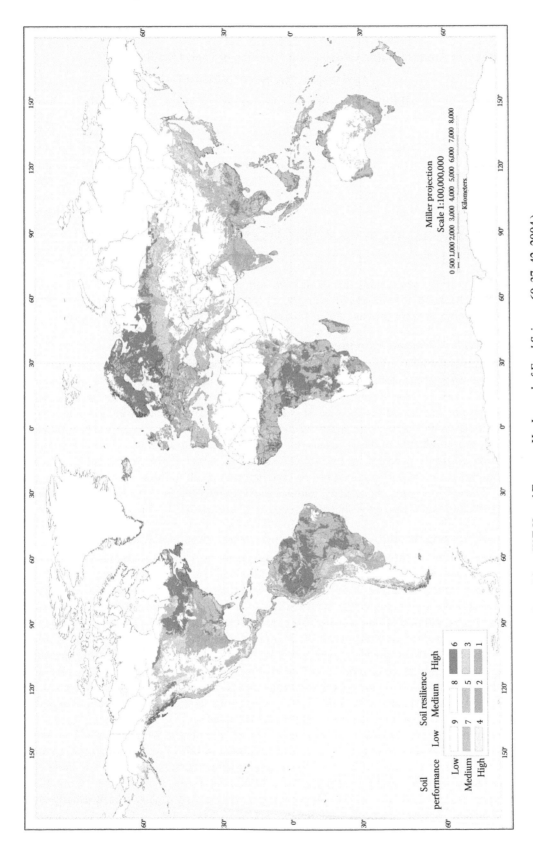

FIGURE 14.2 Global map of land quality. (From Blum, W.E.H. and Eswaran, H., *Journal of Food Science*, 69, 37–42, 2004.)

TABLE 14.5

Percent of Land Area in Major Biomes as a Function of Land Quality

Biomes	\multicolumn{10}{c}{Land Quality Class (percent of ice-free land surface)}
	I	II	III	IV	V	VI	VII	VIII	IX	Total
Tundra								15.62		15.62
Boreal			2.03	0.67	0.50	3.05	2.63	1.08	0.09	10.02
Temperate	2.14	2.55	0.70	1.31	4.76	1.66	2.01		0.15	15.29
Mediterranean			0.30	0.15	1.35	0.08	0.65		0.03	2.56
Desert							1.42		28.19	29.61
Tropical	0.25	2.43	1.51	1.83	9.90	8.53	2.31		0.16	26.90
Total	2.38	4.98	4.55	3.95	16.51	13.32	9.01	16.69	28.61	100.00

Source: Blum, W.E.H. and Eswaran, H., *Journal of Food Science*, 69, 37–42, 2004.

a series of threats to the sustainable use of soil: (a) soil sealing, (b) erosion, (c) decline in organic matter, (d) contamination, (e) loss of biodiversity, (f) compaction, (g) salinization, and (h) flooding and landslides. Some of these threats related to human activities are summarized in Figure 14.3.

14.2.2.1   Impact of Human Activities on Soil

While the Commission of the European Communities highlighted these threats in the context of Western Europe, they occur globally; their relative importance depends upon location and land use. To a degree these threats arise because we expect the soil to perform a range of functions, in some cases many functions at the same time (see Section 14.1.2). By steadily increasing the demands on the soil from these functions, we have often created an unstable system where the soil becomes less resilient and more vulnerable to these threats and often the consequence is that there is a dramatic reduction in its ability to perform its normal functions (Lal, 2007, 2009; Pretty, 2008). In some cases, overemphasis on the performance of one function may result in threats to the ability to perform others. These threats are increasingly seen as particularly relevant to the biomass production function of soils and hence impact global food security considerably.

14.2.2.2   Soil Sealing through Urbanization, Industrialization, and Transport

The covering of soil for the infrastructure of modern life, housing, roads, or other land developments is known as soil sealing. When land is sealed, the soil is unable to perform many of its functions including the absorption of rainwater for infiltration, and filtering is also not possible. In addition, sealed areas may have a great impact on surrounding soils by changing water flow patterns and by increasing the fragmentation of biodiversity. Burghardt (2006) noted that for the period 1997–2001 in Germany, the mean daily rate of growth of settlements and traffic areas, with the resultant loss of land open directly to the atmosphere, was 1.29 km^2. Soil sealing is almost irreversible, and there is increasing concern among governments and environmental regulators about this permanent loss of soil and the associated loss of ecosystem functions. Many national authorities are now considering guidance to land developers to encourage reduction in the total amount of land sealed and to increase the proportion of permeable land in future developments.

A novel way to observe the extent of soil sealing is to examine the view of Earth from space at night with the lights associated with urban development visible. Figure 14.4 shows a recent example of the nighttime view of Earth. While not an exact correspondence, it is clear that much of the higher-quality land is associated with the areas of the highest levels of urbanization (compare also Foley et al., 2005). There is clear evidence of high levels of urbanization in North America, Western Europe, and Japan (The Earth Institute, 2005), areas frequently strongly associated with good-quality land (see Classes I–III in Figure 14.2). But this image also

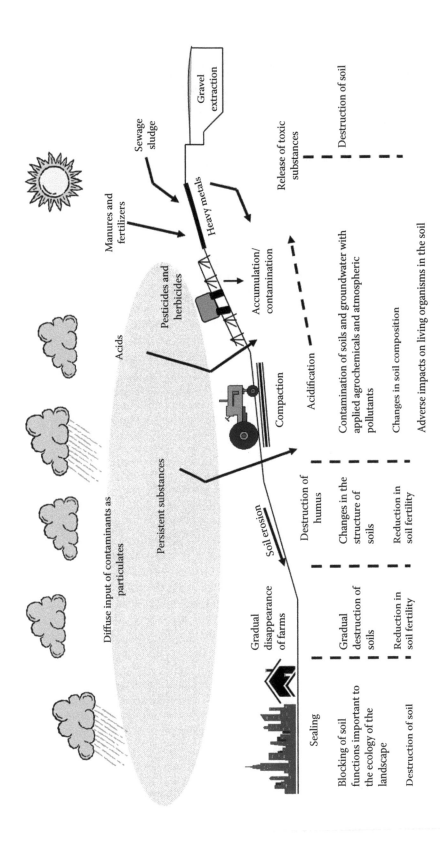

FIGURE 14.3 The impact of human activities on soil. (From Blum, W.E.H., *Threats to Soil Quality in Europe*, pp. 5–10, JRC Scientific and Technical Reports EUR 23438 EN, Ispra, Italy, 2008.)

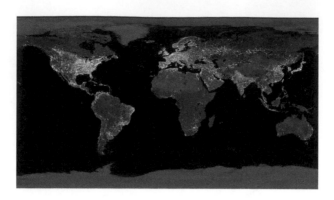

FIGURE 14.4   Nighttime view of Earth. (Courtesy of NASA.)

highlights some recent changes such as the increasing urbanization in the Indian subcontinent and China. This urbanization and the associated sealing are important because sealing permanently removes land from the function of biomass production. In areas such as China and India, this has a double impact: first, the loss of land for agricultural production, and second, the shift from rural to urban lifestyles of significant proportions of the population means that they do not contribute to biomass production as in the past (see also Section 14.3.2). The combination of the loss of land from sealing and the increasing proportion of the population living in urban environments has a significant impact on global food security. An estimation of current daily losses of soil through urbanization, industrialization, and transport in the European Union (total surface 4,324,782 km²) amounts to about 1,200 ha/day, corresponding to 12 km². A very rough estimation of daily soil losses at the global scale amounts to about 25,000–30,000 ha/day, corresponding to 250–300 km².

14.2.2.3   Erosion

Erosion is a natural geological phenomenon where soil particles are removed by water or wind, transporting them elsewhere. Sedimentary rocks, found widely on Earth's surface, are a product of erosion and deposition. While erosion is a natural process, some activities of humankind may result in a dramatic increase in erosion rates, especially unsustainable agricultural land use (Lal, 2001). As soil is an essentially nonrenewable resource, when erosion is serious it is generally irreversible and the soil is lost forever. Erosion occurs because of a combination of conditions such as steep slopes, climate (e.g., long dry periods followed by intense rainfall may enhance the potential for erosion by water), inappropriate land use, poor vegetation cover, and sometimes ecological disasters such as tectonic activity, which disturbs the environmental balance, or forest fires, which remove the protective cover of vegetation (Bai et al., 2008). Soils are not equally vulnerable to erosion; some intrinsic features of a soil can make it more prone to erosion (e.g., a thin layer of topsoil, silty texture, or low organic matter content) (Boardman and Poesen, 2006).

Soil erosion results in the loss of soil functions and in extreme cases the loss of the whole soil, exposing unproductive parent material at the surface. Soil erosion principally occurs at the surface and the loss of the productive topsoil, which is enriched with soil carbon and frequently of higher fertility than underlying layers, has a significant effect on the capacity of the soil to sustain biomass production (Den Biggelar et al., 2003). A loss of a few centimeters of topsoil, though it may be a small proportion of the total depth of the soil, will often have a disproportionately large impact on the soil's ability to sustain high levels of biomass production. Lal (2003, 2005) also highlighted the close link between the preferential erosional loss of organically enriched topsoil and the impact this may have on the global carbon budget. While the loss of the soil and the associated soil carbon loss are major concerns, frequently there are additional environmental impacts because the soil is transported to watercourses, adding contaminants and increasing the turbidity of the water, and its deposition downstream may cause further environmental damage.

14.2.2.4   Decline in Soil Organic Matter

Soil organic matter is composed of organic material (plant root remains, leaves and other plant material, dead fauna and excrement), living organisms (bacteria, fungi, earthworms, and other soil flora and fauna), and humus. In natural systems organic material is constantly added to the soil and decomposed. During this decomposition, plant nutrients are released and often rapidly taken up by the plants, forming an almost closed cycle. Additionally, carbon is released to the atmosphere as CO_2, which is then recaptured through the process of photosynthesis.

Organic matter plays a central role in maintaining many key soil functions and is a major determinant of a soil's resistance to erosion and underlying soil fertility (Lal, 2002). It contributes to the aggregation of soil particles and the chemical and physical buffering capacity of soil, thus contributing to limit the diffusion of pollution from soil to water. It is widely recognized that farming and forestry practices have an important impact on soil organic matter, and despite the recognized importance of maintaining the organic matter content of soil, there is evidence that with a shift in the last half century toward greater specialization and cereal monoculture, particularly in temperate regions, losses of soil organic matter through decomposition are often not completely replaced. Specialization in farming has led to the separation of livestock from arable production so that rotational practices, which were important in the past in maintaining soil organic matter content, are often no longer a feature of the farming system. There has also been concern that recent shifts in climate may have further contributed to changes in soil organic matter levels (Bellamy et al., 2005; Section 14.3.3).

The buildup of organic matter in soils is a slow process (much slower than the decline in organic matter observed as a result of poor management); however, these declines have been halted and in some cases reversed with the adoption of land management practices such as conservation tillage, including no-tillage cropping techniques, organic farming, permanent grassland, cover crops, and mulching and manuring with green legumes, farmyard manure, and compost. A number of authors have shown significant accumulations of carbon in the soil following the adoption of conservation tillage practices. Fuentes et al. (2010) compared soil organic carbon contents following 16 years of conventional tillage and zero tillage in the central highlands of Mexico and found levels significantly higher in the zero-tillage sites; they also noted that the retention of organic residues on site strongly influenced the soil carbon levels. Sombrero and de Benito (2010), reporting a study from Castile-León in Spain, noted increases in soil carbon levels and structural stability under minimum or no-tillage systems. They also noted that the levels of soil carbon were also affected by soil type, crop type, crop rotation, and the quality and quantity of crop residues on the soil surface. These techniques, in addition to addressing the issue of soil organic matter, have also been shown to be effective in preventing erosion, increasing fertility, and enhancing soil biodiversity.

In many smallholder farming systems in the tropics and subtropics, soil organic matter plays a significant role in the provision of nutrients for crop production. Traditional practices frequently involve periods of fallow when no crop production takes place and soil amendments with crop and plant residues and manures, practices which increase or at least maintain the levels of soil organic matter. Increasing pressure on land through increasing populations and losses of land through sealing and erosion has often reduced the fallow period and resulted in lower levels of organic matter recycling. Reductions in the levels of soil organic matter often result in soils with lower capacity to produce biomass and increased vulnerability to erosion and compaction, which will also reduce the biomass production capacity of the soil.

Carbon is a major component of soil organic matter, which in turn plays a major role in the global carbon cycle. Recent studies (e.g., Post and Kwon, 2000; Guo and Gifford, 2002) have emphasized the important role of the soil carbon pool in the context of global carbon fluxes and have increased awareness of the role of land use management practices in both maintaining soil carbon stocks and potentially increasing the sequestration of carbon in the soil pool.

14.2.2.5 Soil Contamination

The introduction of contaminants in the soil may result in damage to or loss of individual or several functions of soils and the possible contamination of water. The occurrence of contaminants in soils above certain levels entails multiple negative consequences for the food chain and thus for human health, and for all types of ecosystems and other natural resources. To assess the potential impact of soil contaminants, not only their concentrations but also their environmental behavior and exposure mechanisms for human and ecosystem health have to be taken into account. A distinction is often made between soil contamination originating from clearly confined sources (local or point source contamination) and that caused by diffuse sources. The report of the Technical Working Group on soil contamination (Van Camp et al., 2004) established under the European Commission's Thematic Strategy for Soil Protection provides a comprehensive summary of the sources and distribution of contaminated soils in Europe.

14.2.2.5.1 Local Soil Contamination

Local (or point source) contamination is generally associated with mining, industrial sites, waste landfills, and other facilities, both during their operation and after closure. These activities can pose risks to both soil and water. In mining, the risk is associated with the storage or disposal of tailings, acid mine drainage, and the use of certain chemical reagents. Industrial facilities can also be a major source of local contamination. In some cases the contamination may persist for many years.

14.2.2.5.2 Diffuse Soil Contamination

Diffuse pollution is generally associated with atmospheric deposition, certain farming practices, and inadequate waste and wastewater recycling and treatment. Contaminants released as emissions from industry, traffic, and agriculture are the major sources of atmospheric pollution. Deposition of airborne pollutants releases acidifying contaminants (e.g., SO_2, NO_x), metals (e.g., cadmium, lead, arsenic, mercury), and several organic compounds (e.g., dioxins, PCBs, polycyclic aromatic hydrocarbons) into soils. Bourennane et al. (2010) noted marked patterns for topsoil contamination with Cd, Zn, and Pb in the Nord-Pas de Calais region of France. Topsoil Cd contamination was widespread in the region and was linked not only to urban and industrial activities but also to intensive agriculture where urban sludge, manures, and Cd-enriched phosphate fertilizers had resulted in widespread contamination. The Zn and Pb contamination in the area was more restricted and linked to current or former industrial activity. In addition to the soil–atmosphere links, there are also strong links between the soil and the hydrosphere. In a study from northern France, Gutierrez and Baran (2009) noted the importance of considering the impact of diffuse pollution, which may arise from the application of agrochemicals, on water resources. Because of the often complex nature of the linkages between the soil and underlying hydrological systems, the consequences of this pollution may not be reflected in the quality of the groundwater for some time. It is important therefore to consider the nature of these linkages when attempts at management to prevent further contamination and when remediation of contaminated soils and water resources are being planned.

Figure 14.5 illustrates the contamination that can occur through excessive use of fossil energy and raw materials taken from inert positions in the inner part of the globe and deposited on the land surfaces through the atmosphere pathway, water pathway, and terrestrial transport (e.g., in the case of plant protection products, fertilizers, biosolids, composts, and other materials). Soil contamination can impact food production in two ways. The first direct impact is the reduction of the productive capacity of the soil because the contaminants inhibit growth. A second impact on food production is that the food may be unfit for human consumption because of uptake of contaminants during growth. The increased urbanization mentioned previously, if allowed to proceed without

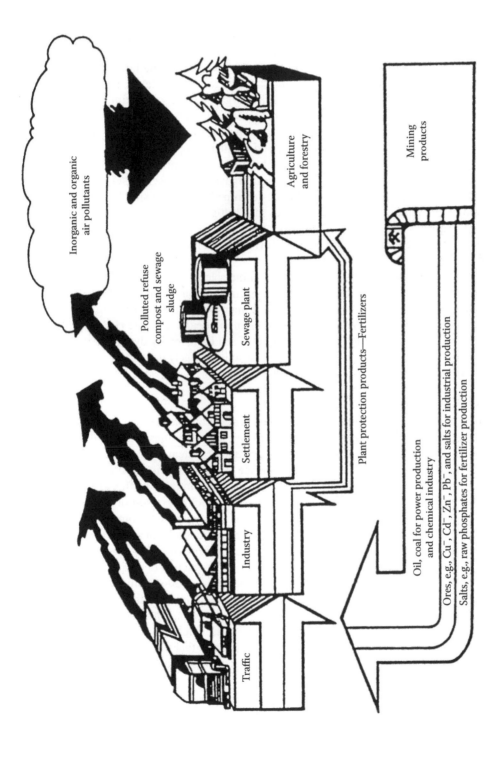

FIGURE 14.5 Soil contamination through excessive use of fossil energy and raw materials. (From Blum, W.E.H., *Problems of Soil Conservation*, Nature and environment series No. 39, Council of Europe, Strasbourg, France, 1988.)

careful planning and control, has the potential to produce soil contamination and further impact global food production (Blum, 1998).

14.2.2.6   Decline in Soil Biodiversity

Soil is a habitat for a huge variety of living organisms, and because of the interactions of the soil-forming factors there is considerable diversity in the nature of the soil and the soil organisms (Bardgett et al., 2005; Gardi and Jefferey, 2009). Our knowledge of the larger organisms found in the soil system is incomplete. We have some knowledge of the relative magnitude but very sparse knowledge of the nature and function of microorganisms (e.g., see Pimentel et al., 2006). Soil bacteria, fungi, protozoa, and other microorganisms play an essential role in maintaining the physical and biochemical properties needed for soil fertility (Barrios, 2007; Brussaard et al., 2007). Larger organisms such as worms, snails, and small arthropods contribute to reducing the size of organic matter, which is further degraded by microorganisms, and carry it to deeper layers of soil, where it is more stable. Furthermore, soil organisms themselves serve as reservoirs of nutrients, suppress external pathogens, and break down pollutants into simpler, often less harmful, components (Decaëns et al., 2006; Turbé et al., 2010). The interrelationships and interdependence among species are complex. The loss of a single species may have a cascading effect because of this interdependence, resulting in a further loss of species and associated functions (Pimentel et al., 2006).

Reductions in soil biodiversity make soils more vulnerable to other degradation processes and frequently reduce their ability to perform many ecosystem functions (Hunt and Wall, 2002; Matson et al., 1997). Because soil biodiversity interacts with many soil and broader environmental functions, it is often used as an overall indicator of the state of soil health (e.g., see Harris and Bezdicek, 1994; Chapin et al., 2000). Although the complexity of soil biodiversity dynamics is not yet fully understood, there is evidence that biological activity in soils is largely dependent on the occurrence of appropriate levels and quality of organic matter.

14.2.2.7   Soil Compaction

Soil compaction occurs on agricultural land when soil is subject to mechanical pressure through the use of heavy machinery or overgrazing, especially in wet soil conditions (Horn and Peth, 2011). It has long been recognized as a potential problem in agricultural systems, and in the United Kingdom a special report was commissioned following a sequence of poor yields in the 1960s (MAFF, 1970). In addition to recognizing the need to avoid compaction by minimizing traffic when soils are wet and more vulnerable to damage, the report also highlighted declines in soil organic matter levels, which were a possible contributory factor to the increased vulnerability to compaction. In sensitive areas, walking tourism (hiking) and skiing also contribute to the problem and in construction sites compaction is a common occurrence. Compaction reduces the pore space between soil particles and the soil partially or fully loses its capacity to absorb water. Compaction of deeper soil layers is very difficult to reverse (Horn et al., 2000). The overall deterioration in soil structure caused by compaction restricts root growth, water storage capacity, biological activity and stability and significantly reduces fertility and food production (Horn et al., 2006; Clarke et al., 2008). Moreover, when heavy rainfall occurs, water can no longer easily infiltrate the soil, which may generate conditions conducive to soil erosion. Possibly more significantly though, water that runs off will not replenish the soil and near-surface storage of water. Near-surface stored water is essential in most agricultural systems to sustain plant growth during rainless periods.

14.2.2.8   Salinization

Salinization is the accumulation of soluble salts of principally sodium, magnesium, and calcium in soils to the extent that crop production is severely reduced. This process is often associated with irrigation practices as irrigation water will contain variable amounts of salts, in particular in regions where low rainfall, high evapotranspiration rates, or soil textural characteristics impede the washing out of the salts, which subsequently build up in the soil surface layers (Singh, 2009). Irrigation with waters with a high salt content dramatically worsens the problem. In coastal areas, salinization can also be associated with groundwater overexploitation (caused by the demands of growing urbanization, industry, and agriculture) leading to a lower water table and triggering the intrusion of saline marine water. With the increasing demand for food from a growing global population and changes in the diet associated with increasing wealth, marginal land is frequently brought into agricultural production. Where insufficient stored water is the reason for the marginalization of the land, irrigation is often the immediate option for bringing the land into food production. Without very careful management of both the water resource used for irrigation and the soil being irrigated, these lands may have a short productive life. This is particularly the case where scarce high-quality water resources and the conflicting demands from human consumption result in lower-quality, often saline, water being the only resource available for irrigating crops (e.g., see Galvani, 2007).

14.2.2.9 Floods and Landslides

Floods and landslides are mainly natural hazards intimately related to soil and land management practices, although their impact is often exacerbated by unusual environmental conditions. Landslides have a predominantly local impact on food production although they may temporarily impact food distribution through the disruption of communication networks. Floods may be both local, impacting a few hectares, or in extreme cases nationwide, impacting thousands of square kilometers. Flooding may cause soil erosion, with the loss of soil, seed, and, in extreme cases, crops, and pollution with sediments. Often in addition to the damage to soil and the natural environment, there are also major impacts on human activities and human lives, damage to buildings and infrastructures, and loss of agricultural land.

Floods and landslides are not a threat to soils in the same manner as the threats already listed. Floods can however, in some cases, result in part from soil not performing its role of controlling the water cycle due to compaction or sealing. Floods and landslides may also be favored by erosion of the soil cover, often caused by deforestation or by abandonment of land.

14.2.2.10 Soil Nutrient Mining

While not mentioned as a threat in the European context, or indeed in any of the other soil protection policies recently developed, soil nutrient mining is possibly one of the most significant threats to food production in large parts of the tropics. Agricultural production in much of Africa is threatened by nutrient mining (Hartemink, 1997; De Jager et al., 2001). The context of agricultural production in much of the continent is one of fragile ecosystems, low inherent soil fertility, and less use of modern inputs such as mineral fertilizers and improved crop varieties. The traditional practice in Africa, and in particular sub-Saharan Africa, is one of fallow systems, where soil is left uncultivated to allow "recovery." Increasing pressure on land from both rising population and, in some countries, exclusion of indigenous populations from parts of the landscape through landgrabbing has resulted in a reduction in the length of fallow periods and in some cases their removal. This trend, particularly where no or at best very low levels of fertilizer are applied, has often resulted in the depletion or "mining" of the soil nutrient store. With reduced nutrient levels the soils are less productive, and to maintain overall levels of food production the populations seek to exploit marginal lands often with lower inherent fertility and other constraints on crop production; many may have significant ecological value. These declining fertility levels in African soils as a result of nutrient mining have resulted in decreased crop yields and per capita food production with a reduction in food security and in the mid to long term are likely to be a key initiator of broader land degradation (Henao and Baanante, 2006).

Nutrient balances, which consider the inputs and outputs from the system, have been used to estimate the magnitude and extent of nutrient mining. Inputs are fertilizers, organic residues, manures, nitrogen fixation, and sedimentation. Outputs are through crop uptake and export, erosion, leaching, and volatilization. During the period 2002–2004, 85% of African agricultural land (1.85 million km^2) had annual nutrient mining rates of over 35 kg (N, P, and K)/ha and 40% had annual rates greater than 60 kg/ha. There are of course wide variations in the observed rates across the continent, with an annual rate of 8 kg/ha in Egypt and 88 kg/ha in Somalia.

Unless policies are introduced to limit this "mining" through support for agroinputs together with associated conservation practices and improvements in the competitive environments for African farmers so that they are able to embrace an element of forward planning with indicative prices for their produce, this degradation process is likely to continue. In some areas, a key change is to establish for farmers tenure of the land that they farm. Where there is no tenure, there is no incentive to manage the soil for the long term. Short-term management will encourage nutrient mining.

14.2.2.11 Desertification

Desertification is a complex process of land degradation through natural and human-induced impacts (e.g., as a result of environmental responses to climate change). It is expressed in terms of increased periods of droughts and overuse of natural resources, especially vegetation cover, by grazing or fuel wood collection, with subsequent soil degradation and losses, including salinization (Anjum et al., 2010). An example for South Africa is given by Stringer and Reed (2007).

Land prone to desertification was normally not used as cropland for food production but was always an important part of the agriculturally productive areas and often used as pastures, for example, by nomads. The loss of these areas by desertification increases the pressure on still-productive land and soils for food production and may even cause social conflicts (Blum, 2009). Dregne (1998) estimated that 3.592 billion ha of land had been affected by desertification. Eswaran et al. (2001) estimated that a "desertification tension zone" affects a total land area of about 4.23 billion ha, of which 1.17 billion ha is in areas with high population density (<41 persons/km^2). Bai et al. (2008) estimated land degradation affected 3.5 billion ha or approximately 24% of Earth's land surface. Reynolds et al. (2007) discuss more sustainable approaches in the use of drylands at a global level in response to these pressures.

14.3 GLOBAL CHANGE, THREATENING FOOD SECURITY

14.3.1 Global Change and Food Security: An Introduction

Besides the threats to soil directly endangering food production through soil losses (e.g., sealing and erosion), as well as through degradation processes, such as loss of organic matter, compaction, desertification, salinization, and nutrient mining, there are further global processes threatening food security, indirectly linked to soil and food production, such as population increase, migration from rural to urban areas, and change in food habits, which increase the demand for food and at the same time threaten the basis of food production. Moreover, through intensive use of energy and the destruction of natural resources for the extension of food production, the existing problems of climate change due to the use of fossil energy are exacerbated. The tendency to increase production of biofuels, which competes for space and energy with food production, is a further threat to food security, in addition to the existing economic crisis and emerging procedures in food production and marketing, such as landgrabbing and hedging. These developments and processes at a worldwide level will be discussed under global change.

14.3.2 World Population Increase, Migration, and Changes in Food Habits

The world's population increases by about 80 million per year, with very unequal distribution worldwide (IIASA, 2007; Lutz et al., 2008). About 70% occurs in sub-Saharan Africa, South and Southeast Asia, and South America.

Even with an increase in food production of more than 12% since 1990, the global undernourished population increased by about 9% (Barrett, 2010). Lal (2010) noted that the number of food-insecure people rose from 854 million in 2007 to about 1020 million in 2009, also due to changes in market conditions and food pricing (see also FAO, 2008; IFPRI, 2008). According to FAOSTAT (2011), the number of people suffering hunger declined in 2010 to about 926 million. All these figures include only people who are protein/calorie malnourished and do not include those who are iron malnourished (about 2 billion) and iodine malnourished (about 750 million) or show other specific nutritional deficiencies (WHO, 2000). Pimentel estimates that in total about 60% of the world population is malnourished (personal communication, based on FAO and WHO data).

In the second half of the twentieth century, increases in food production of nearly 250% matched increases in population, from about 2.5 billion in 1950 to about 6.0 billion in 2000 (Blum and Eswaran, 2004; Godfray et al., 2010), mainly through the success of the "green revolution" (Borlaug, 2007). However, in the final years of the twentieth century, population growth began to outstrip productivity increases. Therefore, the per capita food production showed a slight decrease over the time period. Moreover, in some areas (e.g., Africa), the green revolution had relatively little impact (Sanchez and Swaminathan, 2005), in part because of nutrient-poor soils and limited access to fertilizers and insufficient agricultural tools (Scherr, 1999).

St. Clair and Lynch (2010) report that the global demand for food is expected to increase between twofold and fivefold in the period 1990–2030. At the same time, the per capita area of land dedicated to food production, currently about 0.27 ha, continues to decrease because of population growth, as well as increased urbanization and soil degradation.

Lele (2010) noted that the main focus for poverty reduction is Africa and Asia, where 97% of the world's food-insecure are found (see also Rosegrant and Cline, 2003). The food price hike in 2008 exacerbated the problem of food shortages, with many in Africa and Asia not having the buying power to purchase food. Bai et al. (2008) note that, at the same time, the growth in cereal yields seen for much of the second half of the twentieth century has slowed and that the climatic shifts predicted by the Intergovernmental Panel on Climate Change may result in a 20%–40% drop in cereal yields, mostly in Africa and Asia. In addition, by 2050, there will be about 2.3 billion more people, which means that removing this poverty stress and feeding the additional population will require a 70% increase in cereal yields and a doubling or more of the output of developing countries. Even under an increase of more than 100% in grain production during the last 50 years, the land devoted to arable agriculture has increased by less than 10% (Godfray et al., 2010). It is likely that, by 2050, we shall need 70%–100% more food to feed the estimated global population of 9 billion (Royal Society of London, 2009).

Will there be new land for agricultural production? The competition for land from other uses, especially biofuel production, makes this increasingly unlikely. The recent awareness of the need to protect biodiversity and to retain the goods and services provided by natural ecosystems makes substantial increases in the area of land devoted to agriculture very unlikely.

Besides the increase in world population, the migration of about 80–100 million per year from rural to urban areas (United Nations, 2004) is driving urbanization and results in further sealing and degradation of productive land, as can be seen in many slum areas around urban centers in Africa, Asia, and Latin America.

Moreover, the worldwide increasing demand for animal products, especially protein in the form of meat, but also eggs and milk (FAOSTAT, 2011; Godfray et al., 2010), is threatening food security because of the consumption of grain in animal husbandry. For the production of 1 kg of chicken meat, about 2–3 kg of grain; for 1 kg of pork, about 4–6 kg of grain; and for 1 kg of beef, about 7–10 kg of grain are needed (Zhou et al., 2008). This means that large quantities of vegetable food are lost for human consumption, as they are used as animal feed without consideration of the additional environmental impacts (De Vries and de Boer, 2010; Xiong et al., 2008).

14.3.3 Climate Change Threatening Food Security

Climate change is threatening food security in different ways. More frequent extreme weather events cause water and wind erosion, floods and landslides, desertification, and salinization, which in turn increase soil losses and degradation (Alley et al., 2003; Haron and Dragovich, 2010). Also, there is an increasing lack of fresh water for irrigation and competition for water used in biofuel production, industry, and households (Rosegrant and Cai, 2002). Schmidhuber and Tubiello (2007) discussed the complexity of climate change impact on food security at the global level (see also IPCC, 2007). Lavalle et al. (2009) investigated the impact of climate change on agriculture and forestry in Europe. A case study for Mali was published by Butt et al. (2005).

Moreover, food and biofuel production are exacerbating climate change through increasing deforestation for agricultural land use, urbanization, industrialization, and transport, thus reducing carbon sequestration through soils. Intensive agricultural production of food and biofuels increases the output of greenhouse gases such as methane (CH_4) and nitrous oxide (N_2O), which have a much greater effect on climate change than CO_2, from soils and animals.

The actual increase in water-deficient regions, for example, in the Near East, the Mediterranean, and other world regions, is already playing a decisive role in food production through irrigation (IWMI, 2007).

On the other hand, the increase in temperature, especially in the northern regions of America, Asia, and Europe that have sufficient freshwater reserves, may partly compensate for the losses of food production in water-insecure areas and those with extreme soil losses and soil degradation.

14.3.4 Food or Biofuel: A Competition for Space, Energy, and Water

Due to the steadily increasing prices for fossil energy in the form of oil and gas, the production of biofuel in gaseous form as biogas, in liquid form as biodiesel or ethanol, and in solid form such as straw and wood, has become economically interesting, especially in cases where financial subsidies are granted in view of the contribution to the mitigation of climate change by the use of biofuels. Goldemberg and Guardabassi (2009) discussed the advantages and disadvantages of biofuel products from carbohydrates (see also Demibras, 2007; OECD, 2007; Pimentel and Pimentel, 2007). Pimentel (2010) discussed the environmental impacts of cellulosic ethanol production.

In many parts of the world, especially in Asia and Latin America, natural vegetation cover (e.g., forests) is destroyed in order to gain new surfaces for biofuel production (Fargione et al., 2008), mainly sugar cane and oil-producing plants. The trilemma between food security, energy security, and environmental security was highlighted by Popp (2010).

In view of the constantly decreasing surface for food production, due to urbanization, industrialization, and transport, and soil losses and degradation through erosion and other forms, biofuel production is in strong competition with food production and therefore with food security, at least in the medium to long run (Pimentel et al., 2009; Tilman et al., 2009). Moreover, there are clear indications that agricultural biofuel production is endangering soil quality, because of the complete removal of organic matter and the return of insufficient organic matter residues to the soil for maintaining the organic matter status of soils and biodiversity (Cruse et al., 2010; Lal and Stewart, 2010). Further forms of possible soil degradation can be distinguished (Blum et al., 2010).

Further competition exists for water. In many regions of the globe (e.g., in the Mediterranean basin), there is already insufficient water for food production (Bernades, 2008).

14.3.5 World Economy and Emerging Economic Trends in Food Production and Marketing

In the year 2008, during the world economic crisis, new economic concepts in food production and marketing emerged or were increasingly realized (Piesse and Thirtle, 2009). Speculation on the food market, for example, through hedging (Garcia and Leuthold, 2004; Williams, 2001), became an important issue. Landgrabbing (i.e., the purchase or rent of large territories, especially in countries in development) for food or biofuel production also increased considerably. More than

3 million ha of land were grabbed in Africa, Asia, and Latin America in the years 2006–2009 for food supplies only (von Braun and Meinzen-Dick, 2009).

Whereas hedging is a form of economic speculation with agricultural products, landgrabbing can totally change the agricultural and rural landscapes of countries, as described by Adesina (2010) for Africa. Moreover, direct problems with soil fertility are to be expected, especially when food and biofuel production is maximized to make maximum profit out of purchased or rented areas (Robertson and Pinstrup-Andersen, 2010). Landgrabbing is a clear sign that many industrialized countries and international enterprises have recognized that land cannot be increased and due to constant losses of productive land in their own environment and increasing demands for food and energy at the local and global levels, secured land in foreign countries and regions for sustaining food and biofuel production has become an important asset.

14.4 SUMMARY AND CONCLUSIONS

The large increase in food production in the second half of the last century, mainly through the "green revolution," and further positive developments could not prevent the situation that today nearly 1 billion people live with hunger, mostly because of lack of access to food. In total, about 4 billion people can be considered malnourished, due to further specific nutritive deficiencies.

The sustainable production (availability) of food is also increasingly threatened by impacts deriving from human activities, especially changing forms of land use at local and global scales. Most critical are soil losses through sealing by urbanization, industrialization, and transport, probably the most important threat to food security of all, but also erosion by water and wind, and further severe forms of soil degradation, such as loss of organic matter, contamination, loss of soil biodiversity, compaction, salinization, flooding, nutrient mining, and desertification.

Moreover, the decrease in productive agricultural area and the degradation of the remaining productive surfaces is not the only threat to food security. There is an annual increase in the world population of about 80 million and a migration of about 80–100 million per year from rural to urban areas, where people depend on food markets without being able to produce their own food.

Moreover, climate change is threatening food security directly with increasing losses and degradation of soil, mainly through extreme events, and in many regions a decrease of water resources is threatening rain-fed and irrigation agriculture. Meanwhile, there is serious competition for space, energy, and water emerging from biofuel production and a concomitant increase in demand for food and fiber on local and world food markets.

New economic concepts and speculation on food markets (e.g., in the form of hedging) can hamper access to food through high fluctuations in prices. Moreover, the decreasing availability of productive agricultural land has led to landgrabbing, mostly in developing countries of Africa, Asia, and Latin America, threatening land and soil quality. Without new approaches in food production at local and worldwide levels, it has to be expected that within one or two decades food shortage will severely further threaten millions of people and increase hunger, especially in countries in development.

REFERENCES

Adesina, A.A. 2010. Conditioning trends shaping the agricultural and rural landscape in Africa. *Agricultural Economics* 41(S1): 73–82.

Alley, R.B., J. Marotzke, W.D. Nordhaus, J.T. Overpeck, D.M. Peteet, R.A. Pielke Jr, R.T. Pierrehumbert, et al. 2003. Abrupt climate change. *Science* 299: 2005–2010.

Anjum, S.A., L.C. Wang, L.L. Xue, M.F. Saleem, G.X. Wang, and C.M Zou. 2010. Desertification in Pakistan: Causes, impacts and management. *Journal of Food, Agriculture and Environment* 8: 1203–1208.

Bai, Z.G., D.L. Dent, L. Olsson, and M.E. Shaepman. 2008. Proxy global assessment of land degradation. *Soil Use and Management* 24: 223–234.

Bardgett, R.D., M.B. Usher, and D. Hopkins. 2005. *Biological Diversity and Function in Soils*. Cambridge University Press, Cambridge.

Barrett, C.B. 2010. Measuring food security. *Science* 327: 825–828.

Barrios, E. 2007. Soil biota, ecosystem services and land productivity. *Ecological Economics* 64: 269–285.

Beinroth, F.H., H. Eswaran, and P.F. Reich. 2001. Global assessment of land quality. *In* D.E. Stott, R.H. Mohtar, and G.C. Steinhard (eds), *Sustaining the Global Farm*, pp. 569–574. Selected papers from the 10th International Soil Conservation Organisation Meeting, 24–29 May 1999, West Lafayette, IN.

Bellamy, P.H., P.J. Loveland, R.I. Bradley, M.R. Lark, and G.J.D. Kirk. 2005 Carbon losses from soils across England and Wales 1978–2003. *Nature* 437: 245–248.

Bernades, G. 2008. Future biomass energy supply: The consumptive water use perspective. *International Journal of Water Resources Development* 24: 235–245.

Blum, W.E.H. 1988. *Problems of Soil Conservation*. Nature and environment series No. 39, Council of Europe, Strasbourg, France.

Blum, W.E.H. 1998. Soil degradation caused by industrialisation and urbanisation. *In* H.-P. Blume, H. Eger, E. Fleischhauer, A. Hebel, C. Reij, and K.G. Steiner (eds), *Towards Sustainable Land Use*, Vol. I, pp. 755–766. Advances in Geoecology 31. Catena Verlag, Reiskirchen, Germany.

Blum, W.E.H. 2005. Functions of soil for society and the environment. *Reviews in Environmental Science and Biotechnology* 4: 75–79.

Blum, W.E.H. 2008. Characterisation of soil degradation risk: An overview. *In* G. Tóth, L. Montanarella, and E. Rusco (eds), *Threats to Soil Quality in Europe*, pp. 5–10. JRC Scientific and Technical Reports EUR 23438 EN, Ispra, Italy.

Blum, W.E.H. 2009. Reviewing land use and security linkages in the Mediterranean region. *In* J. Rubio, U. Safriel, R. Daussa, W. Blum, and F. Pedrazzini (eds), *Water Scarcity, Land Degradation and Desertification in the Mediterranean Region*, pp. 101–117. Springer, Dordrecht, The Netherlands.

Blum, W.E.H. and H. Eswaran. 2004. Soils for sustaining global food production. *Journal of Food Science* 69(2): 37–42.

Blum, W.E.H., M.H. Gerzabek, K. Hackländer, R. Horn, G. Rampazzo-Todorovic, F. Reimoser, W. Winiwarter, S. Zechmeister-Boltenstern, and F. Zehetner. 2010. Ecological consequences of biofuels. *In* R. Lal and B.A. Stewart (eds), *Soil Quality and Biofuel Production*, pp. 63–92. Advances in Soil Science. CRC Press, Boca Raton, FL.

Boardman, J. and J. Poesen. 2006. *Soil Erosion in Europe*. Wiley, Chichester.

Borlaug, N.E. 2007. Feeding a hungry world. *Science* 318: 359.

Bourennane, H., F. Douay, T. Sterckeman, E. Villanneau, H. Cieselski, D. King, and D. Baize. 2010. Mapping of anthropogenic trace element inputs in agricultural topsoil from Northern France using enrichment factors. *Geoderma* 157: 165–174.

Brussaard, L., P.C. de Ruiter, and G.G. Brown. 2007. Soil biodiversity for agricultural sustainability. *Agriculture, Ecosystems and Environment* 121: 233–244.

Burghardt, W. 2006. Soil sealing and soil properties related to sealing. *In* E. Frossard, W.E.H. Blum, and B.P. Warkentin (eds), *Functions of Soils for Human Societies and the Environment*, pp. 117–124. Special Publication 266, The Geological Society, London.

Buringh, P. 1985. The land resource for agriculture. *Philosophical Transactions of the Royal Society B* 310(1144): 151–159.

Butt, T.A., B.A. McCarl, J. Angerer, P.T. Dyke, and J.W. Stuth. 2005. The economic and food security implications of climate change in Mali. *Climate Change* 68: 355–378.

Chapin III, F.S., E.S. Zavaleta, V.T. Eviner, R.L. Naylor, P.M. Vitousek, H.L. Reynolds, D.U. Hooper, et al. 2000. Consequences of changing biodiversity. *Nature* 405: 234–242.

Clarke, M.A., R.E. Creamer, L.K. Deeks, D.J.G. Gowing, I.P. Holman, R.J.A. Jones, R.C. Palmer, et al. 2008. Scoping study to assess soil compaction affecting upland and lowland grassland in England and Wales. Defra Project Code BD2304 Final Project Report. Available at http://randd.defra.gov.uk/Default.aspx?Menu=Menu&Module=More&Location=None&Completed=0&ProjectID=14699 (verified 3 January 2012).

Commission of the European Communities. 2002. *Towards a Thematic Strategy for Soil Protection*. COM (2002) 179 final. European Commission, Brussels, Belgium.

Cruse, R.M., M.J. Cruse, and D.C. Reicosky. 2010. Soil quality impacts of residue removal for biofuel feedstock. *In* R. Lal and B.A. Stewart (eds), *Soil Quality and Biofuel Production*, pp. 45–62. Advances in Soil Science. CRC Press, Boca Raton, FL.

Decaëns, T., J.J. Jiménez, C. Gioia, J. Measey, and P. Lavelle. 2006. The value of soil animals for conservation biology. *European Journal of Soil Biology* 42(Suppl. 1): S23–S38.

De Vries, M. and I.J.M. de Boer. 2010. Comparing environmental impacts for livestock products: A review of life cycle assessments. *Livestock Science* 128: 1–11.

De Jager, A., D. Onduru, M.S. van Wijk, J. Vlaming, and G.N. Gachini. 2001. Assessing sustainability of low-external-input farm management systems with the nutrient monitoring approach: A case study in Kenya. *Agricultural Systems* 69: 99–118.

Demibras, A. 2007. Progress and recent trends in biofuels. *Progress in Energy and Combustion Science* 33: 1–18.

Den Biggelar, C., R. Lal, K. Wiebe, H. Eswaran, and V. Breneman. 2003. The global impact of soil erosion on productivity II: Effects on crop yields and production over time. *Advances in Agronomy* 81: 49–95.

Dregne, H.E. 1998. Desertification assessment. *In* R. Lal, W.E.H. Blum, C. Valentine, and B. Stewart (eds), *Methods for Assessment of Soil Degradation*, pp. 441–458. CRC Press, Boca Raton, FL.

Eswaran, H. 1989. Soil taxonomy and agrotechnology transfer. *In* O. Van Cleemput (ed.), *Soils for Development*, pp. 9–28. Publication Series No. 1. ITC Ghent, Belgium.

Eswaran, H., P. Reich, and F. Beinroth. 2001. Global desertification tension zones. *In* D.E. Stott, R.H. Mohtar, and G.C. Steinhardt (eds), *Sustaining the Global Farm*, pp. 24–28. Selected papers from the 10th International Soil Conservation Organisation Meeting, 24–29 May 1999, West Lafayette, IN.

Fargione, J., J. Hill, D. Tilman, S. Polasky, and P. Hawthorne. 2008. Land clearing and the biofuel carbon debt. *Science* 319: 1235–1238.

FAO. 1995. *Food Safety and Trade*. Food, nutrition and agriculture review 15. Food and Agriculture Organization of the United Nations, Rome, Italy.

FAO. 2008. *The State of Food Insecurity in the World*. Food and Agriculture Organization of the United Nations, Rome, Italy.

FAOSTAT. 2011. FAOSTAT Database on Agriculture. Available at http://faostat.fao.org/ (verified 3 January 2012).

Foley, J.A., R. Defries, G.P. Asner, C. Barford, G. Bonan, S.R. Carpenter, F.S. Chapin, et al. 2005. Global consequences of land use. *Science* 309: 570–574.

Fuentes, M., B. Govaerts, C. Hidalgo, J. Etchevers, I. Gonzalez-Martin, J. Miguel Hernandez-Hierro, K.D. Sayre, and L. Dendooven. 2010. Organic carbon and stable ^{13}C isotope in conservation agriculture and conventional systems. *Soil Biology and Biochemistry* 42(4): 551–557.

Galvani, A. 2007. The challenge of the food sufficiency through salt tolerant crops. *Reviews in Environmental Science and Biotechnology* 6: 3–16.

Garcia, P. and R.M. Leuthold. 2004. A selected review of agricultural commodity futures and options markets. *European Review of Agricultural Economics* 31: 235–272.

Gardi, C. and S. Jefferey. 2009. *Soil Biodiversity*. JRC Scientific and Technical Research Series. Available at http://eusoils.jrc.ec.europa.eu/ESDB_Archive/eusoils_docs/other/EUR23759.pdf (verified 3 January 2012).

Godfray, H.C.J., J.R. Beddington, I.R. Crute, L. Haddad, D. Lawrence, J.F. Muir, J. Prett, S. Robinson, S.M. Thomas, and C. Toulmin. 2010. Food security: The challenge of feeding 9 billion people. *Science* 327: 812–817.

Goldemberg, J. and P. Guardabassi. 2009. Are biofuels a feasible option? *Energy Policy* 37(1): 10–14.

Guo, L.B. and R.M. Gifford. 2002. Soil carbon stocks and land use change: A meta analysis. *Global Change Biology* 8: 345–360.

Gutierrez, A. and M. Baran. 2009. Long-term transfer of diffuse pollution at catchment scale: Respective roles of soil and the saturated and unsaturated zones (Brévilles, France). *Journal of Hydrology* 369: 381–391.

Harris, R.F. and D.F. Bezdicek. 1994. Descriptive aspects of soil quality/health. *In* J.W. Doran, D.C. Coleman, D.F. Bezdicek, and B.A. Stewart (eds), *Defining Soil Quality for a Sustainable Environment*, pp. 23–35. SSSA Special Publication no. 35, Madison, WI.

Haron, M. and D. Dragovich. 2010. Climatic influences on dryland salinity in central west New South Wales, Australia. *Journal of Arid Environments* 74: 1216–1224.

Hartemink, A.E. 1997. Soil fertility decline in some major soil groupings under permanent cropping in Tanga region, Tanzania. *Geoderma* 75: 215–229.

Henao, J. and C. Baanante. 2006. *Agricultural Production and Soil Nutrient Mining in Africa: Implication for Resource Conservation and Policy Development*. IFDC Technical Bulletin, International Fertilizer Development Center, Muscle Shoals, AL.

Horn, R. and S. Peth 2011. Mechanics of unsaturated soils for agricultural applications. *In* P.M. Huang, Y. Li, and M.E. Sumner (eds), *Handbook of Soil Science*, 2nd edn, pp. 3-1–3-30. CRC Press, Boca Raton, FL.

Horn, R., H. Fleige, S. Peth, and X. Peng. 2006. *Soil Management for Sustainability*. Advances in geoecology 38. Catena Verlag, Reiskirchen, Germany.

Horn, R., J.J.H. van den Akker, and J. Arvidsson. 2000. *Subsoil Compaction—Distribution, Processes and Consequences*. Advances in geoecology 32. Catena Verlag, Reiskirchen, Germany.

Hunt, A.W. and D.H. Wall. 2002. Modelling the effects of loss of soil biodiversity on ecosystem function. *Global Change Biology* 8: 33–50.

IFPRI. 2008. *International Food Prices: The What, Who, and How of Proposed Policy Action.* International Food Policy Research Institute, Washington, DC.

IIASA. 2007. A world of simultaneous population growth and shrinking unified by accelerating aging. *Population Network Newsletter* 39: 1–3.

IPCC. 2007. *Climate Change 2007: Mitigation of Climate Change.* Cambridge University Press, Cambridge. Available at http://www.ipcc.ch/publications_and_data/ar4/wg3/en/contents.html (verified 3 January 2012).

IWMI. 2007. *Water for Food, Water for Life: A Comprehensive Assessment of Water Management in Agriculture.* International Water Management Institute, London.

Lal, R. 2001. Soil degradation by erosion. *Land Degradation & Development* 12: 519–539.

Lal, R. 2002. Soil carbon dynamics in cropland and rangeland. *Environmental Pollution* 116(3): 353–362.

Lal, R. 2003. Soil erosion in the global carbon budget. *Environmental Interaction* 1036: 1–14.

Lal, R. 2005. Soil erosion and carbon dynamics. *Soil & Tillage Research* 81: 137–142.

Lal, R. 2007. Anthropogenic influences on world soils and implications to global food security. *Advances in Agronomy* 93: 69–93.

Lal, R. 2009. Soils and food sufficiency: A review. *Agronomy for Sustainable Development* 29: 113–133.

Lal, R. 2010. Managing soils and ecosystems for mitigating anthropogenic carbon and advancing global food security. *BioScience* 60(9): 708–721.

Lal, R. and B.A. Stewart (eds). 2010. *Soil Quality and Biofuel Production.* Advances in Soil Science. CRC Press, Boca Raton, FL.

Lavalle, C., F. Micale, H.T. Durrant, A. Camia, R. Hiederer, C. Lazar, C. Conte, G. Amatulli, and G. Genovese. 2009. Climate change in Europe - 3: Impact on agriculture and forestry—A review. *Agronomy for Sustainable Development* 29: 433–446.

Lele, U. 2010. Food security for a billion poor. *Science* 326: 1554.

Lutz, W., W. Sandersen, and S. Scherbov. 2008. The coming acceleration of global population ageing. *Nature* 451: 716–719.

MAFF (Ministry of Agriculture, Fisheries and Food). 1970. *Modern Farming and the Soil.* Her Majesty's Stationery Office, London.

Matson, P.A., W.J. Parton, A.G. Power, and M.J. Swift. 1997. Agricultural intensification and ecosystem properties. *Science* 277: 504–509.

Mueller, L., U. Schindler, W. Mirschel, T.G. Shepherd, B.C. Ball, K. Helming, J. Rogasik, F. Eulenstein, and H. Wiggering. 2010. Assessing the productivity function of soils: A review. *Agronomy for Sustainable Development* 30: 601–614.

OECD. 2007. *Biofuels: Is the Cure Worse than the Disease?* Organization for Economic Cooperation and Development, Paris, France.

Pimentel, D. 2010. Corn and cellulosic ethanol problems and soil erosion. *In* R. Lal and B.A. Stewart (eds), *Soil Quality and Biofuel Production,* pp. 119–136. Advances in Soil Science. CRC Press, Boca Raton, FL.

Pimentel, D. and M. Pimentel. 2007. *Food, Energy and Society.* Taylor & Francis, Boca Raton, FL.

Pimentel, D., T. Petrova, M. Riley, J. Jacquet, V. Ng, J. Honigman, and E. Valero. 2006. Conservation of biological diversity in agricultural, forestry and marine systems. *In* A.R. Burk (ed.), *Focus on Ecology Research.* Nova Science, Hauppauge, NY.

Pimentel, D., A. Marklein, M.A. Toth, M.N. Karpoff, G.S. Paul, R. McCormack J. Kyriazis and T. Krueger. 2009. Food versus biofuels: Environmental and economic costs. *Human Ecology* 37: 1–12.

Piesse, J. and C. Thirtle. 2009. Three bubbles and a panic: An explanatory review of recent food commodity price events. *Food Policy* 34: 119–129.

Popp, J. 2010. Economic balance—Competition between food production and biofuels expansion. *In* R. Lal and B.A. Stewart (eds), *Soil Quality and Biofuel Production,* pp. 151–182. Advances in Soil Science. CRC Press, Boca Raton, FL.

Post, W.M. and K.C. Kwon. 2000. Soil carbon sequestration and land use change: Processes and potential. *Global Change Biology* 6: 317–327.

Pretty, J. 2008. Agricultural sustainability: Concepts, principles and evidence. *Philosophical Transactions of the Royal Society B* 363: 447–465.

Reynolds, J.F., D.M. Stafford Smith, E.F. Lambin, B.L. Turner II, M. Mortimore, S.P.J. Batterbury, T.E. Downing, et al. 2007. Global desertification: Building a science for dryland development. *Science* 316: 847–851.

Robertson, D. and P. Pinstrup-Andersen. 2010. Global land acquisition: Neo-colonialism or development opportunity? *Food Security* 2: 271–283.

Rosegrant, M.W. and X. Cai. 2002. Global water demand and supply projections. Part 2: Results and prospects to 2025. *Water International* 27: 170–182.

Rosegrant, M.W. and S.A. Cline. 2003. Global food security: Challenges and policies. *Science* 302: 1917–1919.

Royal Society of London. 2009. *Reaping the Benefits: Science and the Sustainable Intensification of Global Agriculture.* The Royal Society, London.

Sanchez, P.A. and M.S. Swaminathan. 2005. Hunger in Africa: The link between unhealthy people and unhealthy soils. *Lancet* 365: 442–444.

Scheffer, M., S. Carpenter, J.A. Foley, C. Falke, and B. Walker. 2001. Catastrophic shifts in ecosystems. *Nature* 413: 591–596.

Scherr, S.J. 1999. *Soil Degradation. A Threat to Developing-Country Food Security by 2010?* Food, Agriculture and the Environment Discussion Paper 27. International Food Policy Research Institute, Washington, DC.

Schmidhuber, J. and F.N. Tubiello. 2007. Global food security under climate change. *Proceedings of the National Academy of Sciences of the United States of America* 104: 19703–19708.

Singh, G. 2009. Salinity-related desertification and management strategies: Indian experience. *Land Degradation and Development* 20: 367–385.

Soil Survey Staff. 1999. *Soil Taxonomy: A Basic System of Soil Classification for Making and Interpreting Soil Surveys,* 2nd edn. U.S. Department of Agriculture Handbook 436. U.S. Government Printing Office, Washington, DC.

Sombrero, A. and A. de Benito. 2010. Carbon accumulation in soil. Ten-year study of conservation tillage and crop rotation in a semi-arid area of Castile-Leon, Spain. *Soil & Tillage Research* 107(2): 64–70.

St. Clair, S.B. and J.P. Lynch. 2010. The opening of Pandora's Box: Climate change impacts on soil fertility and crop nutrition in developing countries. *Plant and Soil* 335: 110–115.

Stringer, L.C. and M.S. Reed. 2007. Land degradation assessment in southern Africa: Integrating local and scientific knowledge bases. *Land Degradation and Development* 18: 99–116.

The Earth Institute. 2005. *The Growing Urbanisation of the World.* Columbia University, New York, NY.

Tilman, D., J. Fargione, B. Wolff, C. D'Antonio, A. Dobson, R. Howarth, D. Schindler, W. Schlesinger, D. Simberloff, and D. Swackhamer. 2001. Forecasting agriculturally driven global environmental change. *Science* 292: 281–284.

Tilman, D., R. Socolow, J.A. Foley, J. Hill, E. Larson, L. Lynd, S. Pacala, et al. 2009. Beneficial biofuels—The food, energy and environment trilemma. *Science* 325: 270–271.

Turbé, A., A. De Toni, P. Benito, P. Lavelle, N. Ruiz, W.H. van der Putten, E. Labouze, and S. Mudgal. 2010. *Soil Biodiversity: Functions, Threats and Tools for Policy Makers.* Bio Intelligence Service Technical Report 2010-049, European Commission, Paris, France.

United Nations. 2004. *World Population to 2300* (and associated database). Department of Economic and Social Affairs, United Nations, New York, NY.

Van Camp, L., B. Bujarrabal, A.R. Gentile, R.J.A. Jones, L. Montanarella, C. Olazabal, and A.-K. Selvaradjou. 2004. *Reports of the Technical Working Groups Established under the Thematic Strategy for Soil Protection, Vol. IV: Contamination and Land Management* (EUR 21319 EN/4). Joint Research Centre, European Commission, Brussels, Belgium. Also available at http://ec.europa.eu/environment/soil/pdf/vol4.pdf (verified 3 January 2012).

Von Braun, J. and R. Meinzen-Dick. 2009. "Land grabbing" by foreign investors in developing countries: Risks and opportunities. IFPRI Policy Brief 13, http://www.ifpri.org/sites/default/publications/bp013all.pdf. Accessed December 2009 (verified 3 January 2012).

Williams, J.C. 2001. Commodity futures and options. *In* B.L. Gardner and G.C. Rausser (eds), *Handbook of Agricultural Economics, Vol. 1B: Marketing, Distribution and Consumers,* pp. 745–816. Elsevier Science B.V., Amsterdam, The Netherlands.

World Health Organization (WHO). 2000. *Nutrition for Health and Development—A Global Agenda for Combating Malnutrition.* World Health Organization, Geneva, Switzerland.

Xiong, Z.Q., J.R. Freney, A.R. Mosier, Z.L. Zhu, Y. Lee, and K. Yagi. 2008. Impacts of population growth, changing food preferences and agricultural practices on the nitrogen cycle in East Asia. *Nutrient Cycling in Agroecosystems* 80: 189–198.

Zhou, Z.Y., W.M. Tian, and B. Malcolm. 2008. Supply and demand estimates for feed grains in China. *Agricultural Economics* 39: 111–122.

15 Soils, Climate, and Human Adaptability
Review of History in the Holocene

Miroslav Kutílek and Donald R. Nielsen

CONTENTS

15.1 ADAPTATION OF *HOMO SAPIENS* TO CLIMATE OSCILLATION AS THEIR MOST IMPORTANT EVOLUTIONARY FEATURE

The relationship between environmental alteration and ancient cultures includes their rise, endurance, and decline. To properly understand human evolution, we must first understand the effects specific climate changes have on soil and vegetation. An enormous transformation of the eastern part of Africa began as a result of continental drift, which was realized in this region by the movement of three tectonic plates. Two of them collided and the third one started the first phase of separation into two parts. This process began some time ago, during the second half of the Miocene. A large depression, the Great East African Rift, formed, and a range of high mountains and volcanoes was squeezed out on the east coast. The depression now has a length of about 5000 km and a depth of several hundred meters. With the high rising mountains disturbing the atmospheric circulation, the majority of the monsoon rains were kept close to the east coast, leaving the East African Rift in the rain shadow. Before the East African Rift Valley and the high mountains were formed, the entire region was a tropical humid forest in a continuous belt from the west to the east coast of Africa. With its new climate, the newly appearing Rift manifested a change of vegetation. With the slow regression of tropical forest and the invasion of savanna beginning 10 myr BP (millions of years before present) in East Africa, a strong vegetative change took place between 6.3 and 6 myr BP (Bonnefille, 2010). The transition was not smooth and the whole period of drying was interrupted by several climatic oscillations. The earlier environmental conditions before the Rift formation are still visibly marked at many localities by subsoils containing plinthite (Showers, 2009) or as plinthic spots in the soil profile and iron coatings of particles. The soils of the northern part of the Rift were influenced by desertification processes during the dry periods. At the foot of the mountains, the uplift yielded a staircase of terraces (Veldkamp et al., 2007).

This principal climatic change was the basis for a small number of secondary changes of short, alternating periods of extreme wetness and aridity. Prior to 2.7 myr BP, the wet phases appeared every 400 kyr (thousand years), whereas after 2.7 myr BP, the wet phases appeared every 800 kyr, with periods of precessional-forced extreme climate variability at 2.7–2.5 myr, 1.9–1.7 myr, and 1.1–0.9 myr BP (Trauth et al., 2007). High latitude forcing in both hemispheres compressed the Intertropical Convergence Zone so that East Africa became locally sensitive to precessional forcing, resulting in the mentioned periods of climate extremes.

Miocene apes had difficulties adapting to the new environmental conditions. They were used to the rich food resources of the tropical forest in which they also found secure refuge from their predators. Missing the advantages of the dense forest vegetation in the new tall grass, they were not always able to see approaching predators. It was far beyond their faculties to adapt to a new environment. Bipedality, stone use, and Pleistocene brain enlargement would enable ways of adapting to environmental novelty and its dynamic properties (Potts, 1998). With the first signs of bipedality appearing about 6 myr BP, this new quality was subsequently developed by Australopithecus and more generally in hominids. Several branches of evolution ended before the genus *Homo* finally appeared. The extreme climate variability had consequences for migration and eventually ended the existence of species among hominids who were not adaptable to new climate periods. Thus, the evolution was not a smooth process but was characterized by many blind alleys and extinction events (Trauth et al., 2007; Kutilek and Nielsen, 2010). *Homo sapiens*, the final link of this evolutionary chain, had the greatest adaptation ability, which enabled the species to rapidly colonize Earth. They appeared for the first time as a new species in Eastern Africa about 200 kyr BP, emigrated from Africa about 70 kyr BP, and spread across Europe and Asia by about 40 kyr BP. They had to adapt to the most severe period of the last Würm (Wisconsinan) glaciation, especially when they used the "bridge" formed by the glacier sheet between Siberia and America—today's Bering Strait. At that time, sea level was about 100 m below its recent level (Roberts, 2004).

H. sapiens needed less than 200,000 years to colonize the whole planet Earth. This is a much shorter period than the spreading of *Homo erectus*, who needed about 1 million years, and other human species, who took more than half a million years—none of them colonized the whole Earth. They primarily missed America. These rates of colonization demonstrate the much higher adaptability of *H. sapiens* to a variety of ecological conditions compared with that of other earlier human species. We can characterize the whole evolutionary *Homo* process with the words of Oliver (1991): "Archaeological evidence shows the high grasslands of eastern and central Africa to have been the 'cradle of humanity'."

H. sapiens lived in small nomadic groups; they were hunter–gatherers when the warming period started about 14 kyr BP in two waves interrupted by two Dryas cooling periods. Then, 11.5 kyr BP our warm interglacial, the Holocene, finally started. The warm climate brought more rains in the eastern part of the Mediterranean, mainly in the wintertime, and the moist soil of the spring was favorable to the growth of grasses and increased production of seeds. Groups of nomads learned to store the grains of seeds for the rest of the year, and the style of wandering became modified by some groups. When some of the stored "grains" dropped into the soil, the nomads discovered the advantage of sowing the seeds and profited from the increased "yield." Other groups of nomads imitated them. From this practice came the idea of clearing a specific area free of competing weeds and sowing seeds of their favorite plants. After moving on to hunt and gather elsewhere, they found, upon their return, desirable plants ready for a harvest. Other groups tried to rob the harvested seeds, but eventually the sowing practice was extended and many gangs of nomads discovered one of the principles of agriculture. With the cultivation of these first crops not yet leading to settled regular agriculture, the newly discovered practice was simply added to existing practices of hunting and foraging. Keeping of animals first started as an additional tool against hunger during late summer, in wintertime, and in years when nature offered insufficient edible products. Sheep and goats were kept primarily as a source of food, and after about two millennia, cattle were also intensively kept. The greatest invention of *H. sapiens*—the agricultural revolution or Neolithic Revolution—was about to begin.

The transition from glacial to interglacial was not a smooth process and two cool waves, the Older and Younger Dryas, interrupted the warming. Because the cooling was accompanied by dryness, the half-settled nomad groups suffered more than the groups practicing a lifestyle of hunting, fishing, and gathering. With the majority of the primitive settlements being destroyed, only after the arrival of the next warming phase were the simple principles of the agricultural revolution reinvented. The break by the Younger Dryas was much more severe than the Older Dryas and the urgent need for nourishment contributed to the accelerated domestication of animals. The nomads kept animals with them during their periodic migrations. Eventually, the animals provided the foundation of nutrition when groups of nomads once again established settlements after the end of Younger Dryas when the Holocene began. The beginning of the Holocene differed by hundreds of years from one region to another, so we assign a conventional date of 11,500 years BP.

The transition from the glacial/interglacial and the early Holocene has sometimes been thought of as populations living with an abundance of resources without working. The reality was a hard-working population of active learners who finally survived the Older and Younger Dryas' extremely cool climate oscillations.

15.2 ADAPTATION OF ANCIENT CULTURES TO CLIMATE CHANGE

15.2.1 SAHARA

The Sahara was a desert during all of the Pleistocene glacial epochs and during some of them it expanded several hundreds of kilometers farther south than its recent southern boundary. After the start of the Holocene time, climate warming was accompanied by an increase in precipitation that transformed the majority of the Sahara into a savanna (Ritchie et al., 1985). The permanent warming, which lasted for thousands of years, caused a rise in atmospheric CO_2 concentration from about 180 ppm (parts per million) at the end of the glacial up to roughly 270 ppm early in the Holocene. This abrupt rise of atmospheric CO_2 had a fertilizing effect and plant growth was stimulated to increase by up to 40%. The increase in CO_2 and the more efficient photosynthesis were accompanied by a relative reduction in transpiration (Kimball et al., 2002; Kirkham, 2005). This effect encouraged crop planting and augmented the success of the agricultural revolution (Cunniff et al., 2008). The richness of vegetation attracted game and the whole Sahara region was occupied by animals typical of savanna regions (Goudie, 1983; Roberts, 2004).

The development of a paleoclimate is reflected by remnants of paleosols. Beneath the recent aridic arenosols and leptosols (aridisols) we find a paleosol horizon with evidence of processes influenced by water availability and warm temperature, such as rubification, clay translocation, and calcite redistribution. The paleosol remnants that have been identified date mainly to pluvial periods since the Early/Middle Pleistocene. They are interrupted by several erosional and aeolian arid phases. An example of a published study of Sahara paleosols is Zerboni et al. (2010).

Nomads, already the first to have domesticated animals, were attracted by early Holocene rich savanna vegetation and game. They practiced primitive agriculture on land close to lakes and rivers including the Nile. The social system of cooperation within simple villages differed from the habits of nomads. That difference deepened during dry periods when the nomads lacked food. They attacked the settlers who stored food supplies and other necessities in their villages. To protect themselves, the settlers established principles of cooperation between neighboring communities against the raids of robbers. Such agreements of cooperation were the first social and organizational human structures.

Significant cooling 5.9 kyr BP marked a permanent change in the climate in North Africa, leading to aridization of the whole Sahara region (Bond et al., 1997). In less than 1000 years, the region was first transformed into semidesert and then into a vast desert of primarily sand dunes associated with rocky uplands, ergs, and hammada or desert pavement. Lakes also dried up as their tributaries no longer conducted water and were transformed into dry waddies. The lack of nutrition drove the nomads against settlers around oases; many primitive villages were repeatedly plundered and finally destroyed.

15.2.2 ANCIENT EGYPT

Many oases were disappearing and their original inhabitants either returned to the nomadic style of life or tried to move to the greatest oasis of the Sahara, the valley of the river Nile, after the aridization started. The regular floods of the river arrived in June each year from the monsoon-fed Blue Nile in the Ethiopian mountains. The floods rose up until September/October when they reached their peak and then subsided by the end of December. The Nile is one of the most predictable rivers in the world with its floods averaging more than a hundred days compared to the short-lasting floods of other rivers. The regular periodicity of its floods was attractive for the increased population of the valley, who readily learned how to benefit from the use of temporarily flooded terrain.

By learning how to conduct the excess water away after each flood retreated, the inhabitants brought about the invention of appropriate management tools and increased the level of general knowledge. Since the moist soil held enough water in its entire profile, the planted crops never suffered from lack of water and the land yielded well year after year. Because the flood water brought fresh products of soil erosion from the upper part of the river, the soil of the valley was not only moistened but also fertilized by fine suspensions. Hence, a special type of fluvisol, mainly calcaric fluvisol known as black mud, was formed. Since the river water was virtually free of salts, there was no danger of salinization, even over centuries. Although techniques for conducting excess water away were primitive, the inhabitants gradually found out how to move the water to the terrain of the valley not directly encroached by flood water: check basin irrigation was invented. The indispensable cooperation between the neighboring villages led to the gradual formation of small power centers that also mutually cooperated. This cooperation was not always spontaneous and on occasion it was interrupted by acts of violence. The protection against the raids and robberies of nomads was more efficient when these early power centers cooperated, so the process of unification continued, bringing positive results in both agricultural production and defense. The protodynastic period of Upper Egypt (3100–3000 BC) was the first step of unification, ending with King Narmer, who according to some Egyptologists united the two lands of Upper and Lower Egypt and contributed to the founding of the Egyptian state. According to the historian Manetho, the name of the first king and unifier was Menes.

The ruling of such a large state required a complicated structure of governance with the appropriate types of cults and ceremonies resulting in the unification of various social layers and of different tribal habits. Without a strict bureaucracy headed by the members of the king's (the title in Egyptian: *nesut-bity*) broad family, a united rule would have been impossible. The unification also required the buildup of a strong military force to protect the state from outside attacks. Just the feeding of the soldiers and the countless workers of the pyramids, temples, and palaces was an extraordinary task. All management was conditioned by the use of script.

The deterioration of the climate in the fifth millennium BP depressed nomadic activity and caused the demise of many local settlements around oases. If the fate of the original population of the Sahara region was the sole consideration, the term "catastrophe" would aptly fit the devastating scenario described. However the change in climate also contributed to the formation of a new, prosperous, and advanced culture. In the Nile valley, *H. sapiens* were not only able to adapt to the new environmental conditions but also began to modify natural processes to their advantage.

The richness of the kingdom was due principally to the skillful use of regular floods and fertile soil. Today, we say alluvial soil or fluvisol, while in old Egypt it was *kemet*, meaning the black fertile mud. Because of its tremendous importance, the whole of ancient Egypt was also called Kemet. *Deshret*, the name for red desert land, was known as the seat of death and evil spirits. The ancient Egyptians did not consider the desert to be part of their land, but in spite of its reputation, the kingdom's army completed many expeditions there, partly for economic reasons and partly to demonstrate the powers of the pharaoh. Thousands of years ago Egyptian soils were already classified according to their quality: *nemhura* soils were three times more costly than *sheta-teni* soils (Krupenikov, 1981 cited in Krasilnikov and Tabor, 2003, according to Showers, 2009).

The kingdom flourished under dynastic rule up to the Fifth Dynasty, when the initial reforms gradually weakened the central rule of pharaoh and government. The decline continued into the next dynasty, especially under the long reign of Pepi II, during which more power was relegated to local administrations. With regional rulers practicing their own politics without coordination, the original richness of the kingdom crumbled and was wasted on local ineffective mandates. The rulings of subsequent dynasties were characterized by chaos during the First Intermediate Period.

When the chaotic long-lasting deficiency of central rule was faced with severe climate change, probably caused by an asteroid impact about 4200 years BP (Dalfes et al., 1997), the Old Kingdom came to an end. Regional cooling was accompanied by a drastic reduction in precipitation, and the regular, periodic flooding of the Blue Nile was reduced to such an extent that the river was easily crossed by foot in some years. Without the annual recharge of the Nile, crop yields declined substantially and, in some years, there were no yields at all. This critical situation with long-lasting famine lasted for decades. Robbery, plundering, and looting were common during this Intermediate Period. Many temples and edifices were ruined, and in the centers of local governments, the artifacts produced decreased in quality. The chaos lasted for two centuries. When the climate reverted to its earlier favorable conditions, the strong centralism of the Middle Kingdom was reestablished. This part of Ancient Egyptian history provides evidence that if society is losing its strength and effectiveness, climate oscillation and environmental change may well cause the collapse of the entire system because the adaptability of human society is reduced or even lost (Kutilek and Nielsen, 2010).

15.2.3 Mesopotamia

A similar situation to that in Egypt occurred in the ancient kingdom of Sumer, now southern Iraq, in Mesopotamia. Mesopotamia has low rainfall and is supplied with surface water by only two major rivers, the Tigris and Euphrates. Both rivers have high spring floods from snowmelt in the highlands of Anatolia. The plains of Mesopotamia are very flat and poorly drained, so that the region has always had problems with drought, severe flooding, waterlogging, silting, and soil salinity. From time to time catastrophic floods have inundated the whole region. At Ur, a clay layer 1.5 m thick separates two layers of pottery. The layer of clay without any archeological debris most probably originated as a result of a major flood and is likely the basis for the flood story in the Sumerian Epic of Gilgamesh and much later the biblical story of the Flood.

On the western side of the flat plain of Mesopotamia is the Euphrates, with its riverbed higher than that of the Tigris to the east. Sometimes floods from the Euphrates have found their way across the slightly sloping landscape into the Tigris. Sumerians used this gradient as the driving force for irrigation waters from the Euphrates while the Tigris served as a drain.

The first inhabitants practicing simple agriculture started arriving in lower Mesopotamia about 8000 years BP from the hilly country in the north. They were driven by climate change. The Global Thermal Optimum existing from about 9000 years BP was interrupted by four centuries of intensive cooling 8200 BP. The cooling may have been caused by the abrupt breaking of natural dams of the large North American glacial lake, Lake Agassiz, which was due to melting of the continental glacier. The dams were moraines formed by unconsolidated glacial debris of soil, rock, and large pieces of ice. When the ice melted, the dams were no longer stable and collapsed. The enormous amount of fresh water flowing into the Atlantic Ocean disrupted the thermohaline circulation for centuries. The warming of the Northern Hemisphere was interrupted and the cooling was accompanied by aridization (Kutilek and Nielsen, 2010). Precipitation in West Asia probably decreased more than 40% and the people reacted by migrating from the hilly regions of Anatolia to lower Mesopotamia, where they found a warmer climate and water in the rivers (Staubwasser and Weiss, 2006).

Even earlier, the Sumerians may have invented irrigation (Brevik and Hartemink, 2010). They likely found that yields were higher near blind branches of the river or on the closed meanders when the excess water drained away. This discovery inspired them to dig artificial river branches, which served as irrigation canals. Oates and Oates (1976) describe the archeological evidence of irrigation canals as early as 8200 years BP (quoted according to Roberts, 2004) and it was on the slopes above the Mesopotamian floodplains where the first empirical steps of irrigation started. The new knowledge was brought about 1000 years later to alluvial plains, first to small tributaries, where the technology was gradually improved. "The traditions of the ancient irrigators in the cradles of civilization were based on generations of trial and error" (Gardner, 1987).

The technologic knowledge of irrigation developed further in lower Mesopotamia along irrigation canals and close to rivers where large, socially stratified cities along with temples were found. The richness of cities and the blossoming of arts and sciences, including astronomy and mathematics (using the sexadecimal system based on 60), earned Mesopotamia a distinction as a "cradle of culture." The Uruk civilization lasted a little more than one millennium, ending with an arid period about 4900 years BP. The arid period coincided with the end of the Holocene climate optimum. Aridity was also responsible for the intensification of salinization and ultimately the decline of Uruk.

While they dwelt in the mountainous regions in the north, the inhabitants placed their gods on inaccessible peaks. After their migration to the plains of lower Mesopotamia, where there were no naturally occurring isolated high-level seats appropriate to locate their gods, the Sumerians built tower-like temples known as ziggurats. They seated their gods on top of the ziggurats where only priests had access.

Surrounding each ziggurat was the community meeting place. However, the gods did not protect the inhabitants from the most serious enemy, the salinization of their soil.

The irrigation system, denoted here as protobasin irrigation, differed from the techniques applied in Ancient Egypt where the floodwater irrigation was closely related to what we today commonly call "leaching irrigation." Although the Mesopotamian system enabled the soil profile to become water saturated, no water percolated completely through and drained out the bottom of the soil profile. Because the groundwater level was relatively close to the surface, inadequate and inefficient drainage was a continual problem. The soils of Mesopotamia belonged to the slightly developed Reference Soil Group of Solonchak, where salts accumulated due to the rise of water from the relatively shallow groundwater as water intensively evaporated from the soil surface. A horizon rich with salts always developed in the soil profile, sometimes even reaching the surface. Additional irrigation water did not move the salts downward. Any saline horizon originally in the profile was shifted upwards toward the soil surface owing to the intensification of evaporation associated with the additional irrigation. Eventually, due to the irrigation of these poorly drained soils, salts accumulated in the surface layers of the A horizon and even covered the soil surface as a blanket of whitish, crystalline dust. Yields were depressed and the less resistant crops did not grow at all. The progress of salinization was evident from the historical succession of crops. Wheat and barley were grown in a ratio of 50/50 about 5500 BP, but about one century later the more salt-tolerant barley represented 80% of the crop, and finally wheat could not be grown because of the salts that had accumulated excessively in the soil and groundwater (Hadan, 1971).

The inhabitants responded in four ways as they contended with the reduced yields caused by salinization. The simplest was to abandon the salted fields and to begin irrigating somewhere else on soils not so strongly affected by salinity. Sometimes this meant abandoning the nearby town and moving to a new locality. Another way of getting rid of the infertile top horizon of soil was to "peel off" the salty top horizon and dump the waste material on nearby areas. Areas receiving deposits of over-salinized topsoil eventually formed broad regions of low, gently rolling hills which have attracted the curiosity of modern-day archeologists. After spending decades digging in vain trying to discover new archeological sites, nothing has been found except soil and a distinct salty collar on the foot of each slope. Through chemical analyses and discussions with soil scientists, the likely solution to the puzzle has been found.

The third response, a more advanced technique of irrigation agriculture on salt-affected soils, was a system of alternate year fallowing. After the harvest on the irrigated area the groundwater level was usually at a depth of about 1 m or even less, and soil salinization was in progress. The soil was left fallow for the next year without irrigation, cultivation, or any form of management. Phreatophytic weeds with deep roots invaded the soil and the roots reaching down to the water table absorbed water for photosynthesis and transpiration. With this "pumping" of water lowering the water table well below 1 m, the migration of salts to the topographical surface was interrupted, allowing the salinity of soils to be slightly reduced by leaching rains and irrigation water applied at the beginning of the next season. Nevertheless, without adequate drainage, the fallow system was only marginally effective because the process of soil salinization was merely delayed until the salty groundwater once again moved to the soil surface.

The fourth response, washing the salts out by excessive irrigation, was a much more advanced procedure that was practiced only as the population learned how to drain the soil and remove the salty drain water from the vicinity of the cultivated field.

This description helps us understand how the population proceeded from adaptation to modification of the unsuitable environment. The first step imitated the natural process. Never being completely satisfied with the initial results, mankind subsequently tried repeatedly to adapt specific parts of the environment to their beneficial use. However, these sequential periods of adaptation were not connected in a simple, smooth process.

During the earliest culture of the Ubaid period (7900–6400 BP) in Southern Mesopotamia, the tendencies toward urbanization and animal husbandry of the first settlements were widely practiced in sedentary communities. The Uruk period from 6400 to 5400 BP ended with a collapse caused most likely by a short climatic oscillation, soil salinization, and a substantial decrease in soil fertility. Keeping the original region of Uruk, the Early Sumerian Dynasty (4900–4350 BP) extended its power more to the north. Sumer was divided into about 12 relatively independent city states that practiced agriculture with monocropping, land cultivation, and intensive irrigation. As the density of the population increased, both the social structure and labor organization were extended.

With continued development and accrued knowledge, the Akkadian Empire with regular kings reached its peak during the rule of Sargon the Great (about 4270–4215 BP). Unfortunately, its new social structures, cults, handicrafts, and developments in general technology, all linked to knowledge of script and to advancements in the arts, suffered a shock about 4200 years BP. The main cause was abrupt climate change with cooling, decreasing precipitation, reductions in river discharge, and the virtual absence of floods. As noted previously regarding ancient Egypt, the principal factor causing the climate change was likely an asteroid impact (Weiss et al., 1993). The increased aridity was reflected strongly in the reduction or absence of drought-susceptible crop species (Riehl, 2009). Due to salinization, overuse of the agricultural soils without fallow, and massive increase in population, the situation was no longer sustainable. Frequent dust storms suffocated crops and yields dropped to only 10%–20% of earlier yields. The wind transport of salts together with dust to the agronomic soils cannot be excluded (Yaalon and Ganor, 1973). Because of economic decline, many state-controlled cities could no longer pay taxes. Additionally, in contrast to its normally positive attributes, urbanization made the human social system less flexible and less adaptable to the environmental change (van der Leeuw, 2008). Upon its collapse, the Akkad Empire was replaced by the Neo-Sumerian Dynasty and, later on, the Babylonian and Assyrian Kingdoms followed. The famous first Babylonian Dynasty collapsed when climate change due to the Santorini eruption (about 3650–3620 BP) brought dryness and an increased rate of salinization, reducing yields for several decades.

15.2.4   Environmental Resilience

The environmental resilience of Mesopotamian cultures had two common elements for the whole of ancient Mesopotamia. First, its resilience was lowered if we define "resilience" as the ability of a system to remain functionally stable in the face of stress and to recover following a disturbance. Each collapse brought about a new type of urban state in Mesopotamia, even though the principal features of the culture were retained and further developed. Second, with each new type of state, social and cultural development was shifted northward, probably to districts with soils less harmed by salinity, while the infertile soils with high salt content and sometimes with solonetz development were abandoned. Generally, the old urban states disappeared and perished, being replaced by new urban states, usually in a new place in Mesopotamia. On the other hand, after each of the three Intermediate Periods in Ancient Egypt, the same type of state structure and culture continued regardless of the Intermediate Period, even though many detailed aspects were further developed. Comparing Mesopotamia's environmental resilience with that of Ancient Egypt, we find them nonidentical. The main difference in the environments of the two cultural centers was the salinity of soils in Mesopotamia. We hypothesize that soil salinization was one of the main factors causing the lower environmental resilience in Mesopotamia when compared to Ancient Egypt. Developing the hypothesis further, the quality and stable fertility of soil are two of the main factors that maintain a stable environmental resilience.

15.3   RECENT CLIMATE OSCILLATION AND THEORIES

15.3.1   Recent Global Warming

The recent climate change started in about 1850. Indeed, the estimation of the year is a matter of general agreement and not strictly subject to local or regional observations. Climate change is usually characterized by global temperature, in recent studies, whereas other climate characteristics are either mentioned in passing or ignored. The temperature increase in the twentieth century was estimated as 0.74°C and the rate of its rise is increasing according to the Fourth Assessment of the Intergovernmental Panel for Climate Change (IPCC, 2007). Eleven of the twelve years between 1995 and 2006 rank among the 12 warmest years in the instrumental record of global surface temperature kept since 1850. The warmest years in the instrumental record of global surface temperatures are 1998 and 2005 for the time period ending in 2007 (Trenberth et al., 2007). However, after the initiative of McIntyre in 2010, the Goddard Institute for Space Studies acknowledged an error and 1934 is now considered to be the warmest year on record in the United

States, not 1998.[*] Temperature measurement of the global ocean has increased to depths of at least 3000 m in the last 40 years. The energy balance shows that the ocean absorbs more than 80% of the heat added to the climate system. Global warming caused a rate of rise of seawater level by 1.8 ± 0.3 mm/year during the second half of the twentieth century (White et al., 2005). The expansion was mainly due to heating of the water.

Although the rise in global temperature has an average value of 0.74°C, the amount of warming has been greater over land than that over the oceans, owing to the smaller thermal capacity of the land. Further, because the continents occupy a larger portion of the northern hemisphere than the southern, temperatures have tended to rise more in the northern than in the southern hemisphere (Trenberth et al., 2007). Although the rise is minor in the equatorial regions and increases with the latitudes in both hemispheres, latitude dependence is only one of many factors that play a role. The geographical position on a continent is important—temperature values differ for inland climates compared to those on lands close to the oceans. In central Europe, the average temperature increased by between 1.1°C and 1.3°C in 100 years (Czech Hydro-Meteorological Institute, 2007), while in Northern Ireland, at the Armagh Observatory, the increase was only 1.2°C in 200 years, that is, from 1796 to 2002. Within the Arctic Circle at latitudes between 70° and 90°, the average temperature rise was 2.1°C in the period 1880–2004 (Trenberth et al., 2007, from Figure 3.7). "No single location follows the global average, and the only way to monitor the globe with any confidence is to include observations from as many diverse places as possible" (Trenberth et al., 2007, 250). The temperature rise was not universal—it rose more in the period 1920–1940 than its current rise and there was even a period of temperature decrease in roughly the next 10-year period. The estimation of the global temperature from surface meteorological station data has been difficult. During the last 28 years, instrumentation in satellites has been applied to measure Earth's global surface temperature. These methods of measurement are considered by meteorologists to be the most objective proof of global warming. Accepting their appreciation of the reliability of such measurements, we emphasize that the global temperature has no longer been rising during the last decade (Solomon et al., 2007; Easterling and Wehner, 2009). See also Figure 15.1, where the temperature in the last 30 years (1979–2009), taken from satellite data, is evaluated.

Most climatologists in meteorological state services subscribe to the hypothesis that the climate warming during the last two centuries was caused dominantly by the greenhouse effect of CO_2 released by burning fossil fuels. Before the industrial revolution, which started by using the energy from coal, oil, and gas, the concentration of CO_2 in the atmosphere ranged from 250 to 280 ppm. Today's value is close to 390 ppm. A simple correlation performed several decades ago showed a high correlation coefficient between the global temperature and CO_2 atmospheric concentration. More recently, physical models of global warming have been presented with dominant greenhouse gas (GHG) forcing combined with solar, aerosol, and ozone forcing. The models were tested against either the last 150 years or shorter periods. Since global warming in the twentieth century occurred in two periods, the first from 1900 to 1940 and the second from the late 1960s to the end of the century, the measured data of those periods were frequently used to verify the models (e.g., Meehl et al., 2003 and literature quoted there). GHG forcing was ascertained as the dominant factor in the majority of cases, or at least it was dominant for the second period. However, the hypothesis has to be generally valid for each and every case. As shown in Figure 15.1, despite the still rising atmospheric content of CO_2, the temperature has had a decreasing tendency for the entire past decade and the correlation between temperature and CO_2 content is inconsistent and weak for the three decades of temperature measurement from the satellite data. It would be revealing to falsify the hypothesis

---

[*] After McIntyre, Goddard (a NASA group) considered 1934 the warmest year on record in the United States, but not globally. At this time NASA considers 2010 and 2005 to be the warmest years on record globally, with 9 of the 10 warmest years occurring since 2000. The only year in NASA's top 10 outside of the 2000s is 1998. The year 1934 is not considered by NASA to be in the top 10 warmest years globally. http://www.nasa.gov/home/hqnews/2012/jan/HQ_12-020_2011_Global_Temp.html.

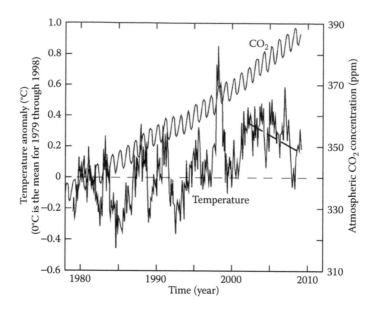

FIGURE 15.1 The temperature change in the lower troposphere from 1979 to 2009 determined by satellite measurements and evaluated with the remote sensing system. The slightly oscillating and regularly rising curve represents the increase in CO_2 concentration as measured at Mauna Loa Hawaii. (From Kutílek, M. and Nielsen, D.R., *Essays in GeoEcology*, Catena Verlag, Reiskirchen, Germany, 2010, Figure 2.)

about the influence of CO_2 on every climate oscillation during the entirety of the Holocene; see Karl Popper's principles of epistemology.

15.3.2 EXPERIMENTS PERFORMED BY NATURE

The best way to test the hypothesis is to repeat the experiment with slightly modified boundary conditions. When we transpose the problem to the case of recent global warming, the falsification of the hypothesis requires the release and neglect of one of the principal factors of climate change, as we learn by studying the geologic history of planet Earth. There are at least eight such factors:

1. Milankovitch cycles include three parameters that change the solar radiation reaching the planet Earth due to the regular periodic repetition of the change of Earth's orbital geometry: (a) The geometry of Earth's orbit around the Sun changes from a nearly circular to an elliptical path, with an eccentricity of 0.058. One complete cycle of the change lasts 100,000 years. (b) The inclination or tilt of Earth's axis in relation to its plane of orbit around the Sun ranges from 21.5° to 24.5° and has a periodicity of 41,000 years. (c) The inclination of Earth's axis due to its slow wobble (precession) has a periodicity of 21,000 years. Collectively, these cause glacials and interglacials.
2. Solar activity is not constant but manifests cycles of various lengths and fluctuates periodically around a value of 1365 W/m². Accompanied by the existence of sunspots, it is correlated with climatic oscillations. Periodicities of solar activity and climate change, observed by Lane et al. (1994), have been physically explained by the Svensmark hypothesis (Svensmark et al., 2007).
3. Continental drift is the shift of plates carrying the continents, other land masses, and the bottoms of oceans. The plates are sliding very slowly on the asthenosphere (viscoelastic material) creating new continents and supercontinents as well as breaking up existing continents. The drift is also associated with geological events such as earthquakes, volcanoes,

uplift of mountains, and mid-ocean rifts. All these phenomena influence the climate of continents.

4. The greenhouse gas (GHG) effect plays an important role in absorbing the long-wave radiation emanating from Earth's surface as it is heated by solar short-wave radiation. Asymmetric molecules of gases are effective absorbers of the long-wave radiation. Among the naturally occurring gases in the atmosphere, water vapor is the most efficient absorber, followed by methane, N_2O, and CO_2. The authors of the Assessments of the IPCC have stated that the increase in atmospheric CO_2 concentration from preindustrial values of 280 ppm to recent values near 380 ppm is the main driver of the recent global warming. It is the opinion of the authors of this chapter that CO_2 contributed much less, at least one order of magnitude less, to the absorption of long-wavelength radiation than water vapor and that contributions by methane and N_2O were even smaller.

5. Thermohaline circulation by the main oceanic streams is caused by temperature gradients and differences in seawater density, owing to differences in water salinity. Thermohaline circulation patterns are important thermoregulators of our planet, and whenever they weaken they cause great climate changes.

6. Aerosols in the troposphere, due to eruptions of volcanoes or impacts of asteroids, reduce the intensity of the incoming solar radiation and reduce Earth's temperature, depending upon the increase in solid microparticle and nanoparticle concentrations in the atmosphere. The reflection of solar radiation depends primarily upon the particle size distribution of the aerosols. Incoming solar radiation, with maximum energy at wavelengths between 0.4 and 1 mm, is reflected back into outer space by particles having sizes that range from 0.1 to 2 μm. The reflection is increased if the solid particles are covered with a liquid water film.

7. Vegetation cover. Incoming solar radiation is absorbed and partially reflected in different ways as the vegetation changes naturally or due to human activities. Whenever the vegetation is altered, changes in the albedo change the entire radiation balance. The most drastic change occurs when original forests are logged and the soil surface is left without vegetation. Not only does the radiation balance change, but the soil water balance also changes due to the evaporative loss of soil water. Together with the ensuing change of evapotranspiration, the thermal regime of the region is changed. Deforestation and agricultural use of soils across large regions cause relative aridization and, in the long term, cooling of the climate.

8. Earth's magnetic field is not constant and neither are the positions of its magnetic poles. Their reversal has occurred several times during geologic history when the magnetic field was extremely weak. Their recent average values in the Holocene remain high compared to earlier values. The last reversal was 780,000 years ago. Moreover, the magnetic pole positions are also not constant—they move as much as 15 km/year. As Earth's magnetic field changes, two important phenomena develop. Instabilities in the ozone layer occur in both vertical and horizontal directions, leading to modification of atmospheric circulation and temperature gradients. Also, the solar wind and cosmic ray flux are modulated. Hence, as the geomagnetic field changes its shape and intensity, distinct changes in the climate result (Knudsen and Riisager, 2009).

Experimentation with Earth and its atmosphere is not possible. However, we are not helpless in testing the hypothesis that global warming is caused by an anthropogenic increase in CO_2 in the atmosphere. Nature itself offers information on past climate changes and on the concentration of atmospheric CO_2. We restrict our discussion to the late Pleistocene and Holocene—the time period of *H. sapiens*. Reliable and indisputable data are available from the deep drilling of Antarctic glaciers on the stations Vostok and EPICA. Based on isotope ratios of $^2H/^1H$ and $^{18}O/^{16}O$ in thin ice layers, the temperature has been estimated with high precision. Accurately measured contents of CO_2 in the air enclosed in the pores of ice are also available. We have obtained temperature and

CO_2 concentration levels from recent times to approximately 400,000 years BP on the Vostok polar station and about 800,000 years BP on the EPICA station. The values of temperature change, taking the recent average value as reference, and the CO_2 content in parts per million for Vostok are shown in the bottom graph of Figure 15.2. The oldest values are on the right of the graph, in the late Mindel, the third glacial before present. The interglacial Mindel/Riss (pre-Illinoian) follows in the period between about 240,000 and –225,000 years BP. Next is the Riss (Illinoian) glacial, after which comes the Riss/Würm (Eem or Sangamon) interglacial, which was significantly warmer than our recent period. Next, the Würm (Wisconsinan) glacial lasted until 11,500 years BP, the time since our present Holocene interglacial began. These relatively regular changes back and forth from glacials to interglacials are caused dominantly by the Milankovitch eccentricity of Earth's orbit. The concentration of CO_2 appears to be correlated well with the temperature, and from a detailed analysis we know that the CO_2 peaks show a delay of about 300–800 years behind the temperature peaks. That is, the CO_2 reacted after the temperature had already changed. Under the title *Relative Trends between Temperature and CO₂* in the top part of Figure 15.2, we present a simplified illustration of the changing magnitudes of the two variables. There were five distinct time intervals when the relations between temperature and CO_2 were not only different from their expected correlation but acted counter to the GHG effect hypothesis. In two cases, the temperature was, on average, constant while the CO_2 concentration was decreasing with time; in another case, the temperature was constant while CO_2 was increasing; in still another case, the temperature was decreasing while CO_2 concentration remained constant; and in the fifth case, the temperature was rising while the CO_2 concentration remained constant. This natural experiment from the late Pleistocene documents that there must be factors influencing the trends of temperature that are counter to the role of the GHG effect.

The Northern Hemisphere climate is closely related to temperature estimations derived from isotope ratios of ^{18}O measured in the Greenland glaciers. Assuming that the Holocene climate was decisive for the cultural and technological evolution of mankind, we use the evaluation of data of Alley (2004) by Koutsoyiannis (2008) for a more detailed discussion on climate oscillation during this period. Our modification of the Koutsoyiannis graph is shown in Figure 15.3. Using 20-year

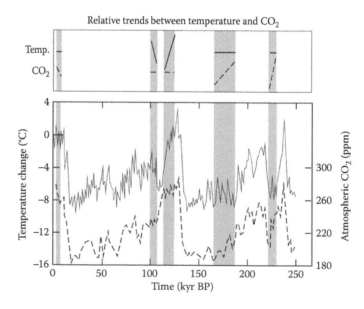

FIGURE 15.2 Evaluation of the Vostok ice core drilling data by Klemeš (2009, personal communication). Time periods denoted in gray are those during which the temperature change or temperature constancy does not correspond to the assumed greenhouse effect of CO_2 concentration. CO_2 concentration is shown by the dashed line and temperature by the solid line. (Top part of the figure from Kutílek, M. and Nielsen, D.R., *Essays in GeoEcology*, Catena Verlag, Reiskirchen, Germany, 2010, Figure 21.)

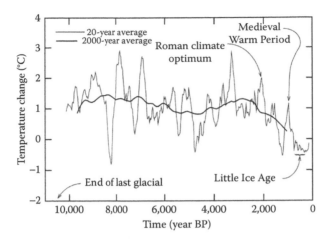

FIGURE 15.3 Temperatures in the Holocene according to ice core drilling data from the Greenland glacier. The measured data were "smoothed" by the method of moving averages. The recent average temperature is denoted as 0°C on the vertical timescale. We are still below the 2000-year averages of the Holocene and below all the warm peaks of the 20-year averages. (From Koutsoyiannis, D., Lecture at EUREAU Workshop on Climate Change Impact on Water Resources with Emphasis on Potable Water, Chania, Crete, Greece, 2008. http://www.itia.ntua.gr/en/docinfo/865, verified 8 January 2012; modified from Alley, R.B., IGBP PAGES/World Data Center for Paleoclimatology Data Contribution Series #2004-013, NOAA/NGDC Paleoclimatology Program, Boulder, CO, 2004.)

moving averages, the measured data offer a full explanation of warm and cool climate periods related to the scale of our observation of recent climate. Neglecting the secondary, smaller oscillations in this discussion, we can identify nine main peaks (warm periods) with temperatures at least 1°C higher than the recent average temperature. Among the nine peaks were three peaks having temperatures 2.6°C–2.8°C higher than the recent average temperature. The atmospheric CO_2 concentration derived from Vostok was in ranges from about 255 ppm for the temperature peak around 7800 years BP to 275 ppm for the temperature peak around 3300 years BP. In the absence of any indication of CO_2 change close in time to the warm peaks, it seems logical to conclude that the GHG concentration gradient did not cause any significant change in temperature. Except for its recent rise, the atmospheric CO_2 concentration during the entire Holocene ranged from 250 to 285 ppm. Its minima, observed and discussed in preceding paragraphs, are also clearly distinct in Figure 15.3. Except for a few scenarios made for the last millenium, no attempt has been made to test the otherwise broadly used climatic models against the preponderance of Holocene data.

An evaluation of measured data using 2000-year moving averages provides information on the scale of the last geologic epoch. The Holocene temperature optimum lasted to about 7000 years BP and then the decrease can be correlated to the drying phase observed in Egypt and Mesopotamia. The secondary temperature maximum was reached around 3200 years BP and a rapid decrease in temperature followed. A possible cause was the expansion of agriculture and the change of native vegetation, causing a change of albedo and other microclimatically important characteristics.

The climate in the last millennium is characterized by two main oscillations—the Medieval Warm Period (about AD 800–1300) and the Little Ice Age beginning in the fourteenth through seventeenth century AD and lasting until AD 1850 when temperatures increased sharply. In 1850, the atmospheric CO_2 concentration was below 285 ppm and GHG forcing was absent. Figure 15.4 documents that the average temperature during the Medieval Warm Period was 0.5°C–1°C above the recent global value according to published proxies. The shape of the 20-year averages in Figure 15.3 suggests that the recent warming is actually a gradual return to temperature values typical for the Holocene, although still below those levels of the relatively mild Medieval Warm Period (Akasofu, 2010).

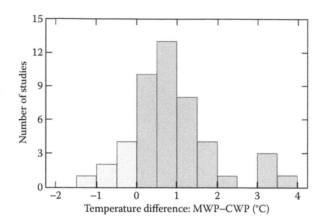

FIGURE 15.4 Distribution of a number of studies on the Medieval Warm Period (MWP) (*vertical axis*) and the temperature estimated as proxy and plotted on the *horizontal axis* as MWP-CWP, where CWP is the Contemporary Warm Period. In the most frequent instances, the temperature was higher than the recent temperature by 0.5°C–1°C. (From Kutílek, M. and Nielsen, D.R., *Essays in GeoEcology*, Catena Verlag, Reiskirchen, Germany, 2010, Figure 8.)

15.3.3 FALSE THEORY

All global climate oscillations that have occurred in both the late Pleistocene and Holocene times have been without the active benefit of greenhouse-effect forcing. The models developed on the assumption that GHGs have a dominating role upon climate change (or recent global warming) are not applicable for the Pleistocene or Holocene and GHGs have not affected other major geological climate changes (e.g., see Zeebe et al., 2009).

We compare the modeled global surface temperatures during the past two millennia in Figure 15.5 (Mann and Jones, 2003) with temperature trends drawn in Figure 15.3, which shows a distinct temperature peak in the Middle Ages and a deep sink thereafter in the Little Ice Age. We emphasize that the results shown in Figure 15.4 are from 47 studies published in peer-reviewed journals on estimates of medieval temperatures, and they manifest an approximate Gaussian distribution with a mean between 0.5°C and 1.0°C above the recent global temperature. On the other hand, the numerical model in Figure 15.5 indicates neither a medieval temperature maximum nor a distinct Little Ice Age minimum. Excluding the time period between AD 1850 and AD 2000, the estimated temperatures from their model oscillate between a minimum value of −0.4°C and a maximum of −0.05°C from the recent global temperature. This published result illustrates the failure of the model that is based on the hypothesis of a dominant greenhouse effect influence.

There is no better example of the improper functioning of global temperature models than those illustrated in Figure 15.6, where modeled temperatures (Hansen et al., 1988) are compared with observed data. The model results matched the observations relatively well during the 30-year period between 1958 and 1988 (time of publication), with the exception of two cool oscillations: one occurring between 1973 and 1976 and the other around 1983. Hansen claimed that the real world had followed a course closer to that estimated by Scenario B. Predictions from each model made in 1988 into the future, shown in Figure 15.6, were easily compared with data measured up till 2010. In 1992, Scenario A started to deviate from observations, and in 2006, Scenario B began to fail. In both cases, predicted temperatures were much too high. In 2010, the temperature predicted by Scenario B was 0.3°C higher than the actual temperature and that by Scenario A exceeded measured values by 0.5°C.

Douglas et al. (2008) compared predictions of 22 widely used climate models with actual readings by surface stations, weather balloons, and orbiting satellites over the past three decades.

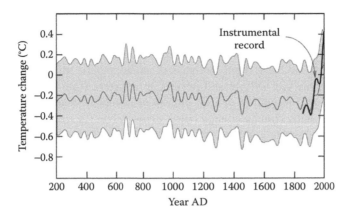

FIGURE 15.5 Global temperature changes in the last 2000 years by numerical model with uncertainties indicated by the gray area. (Modified from Mann, M.E. and Jones, P.D., *Geophysical Research Letters* 30, 2003; Jones, P.D. and Mann, M.E., *Reviews of Geophysics* 42, RG2002, 2004. http://www.ncdc.noaa.gov/paleo/pubs/jones2004/fig5.jpg.)

Although they found that most of the models predicted the middle and upper parts of the troposphere warming drastically over the past 30 years, actual observations showed very little warming over those regions of the atmosphere. From their findings and from other examples in the literature (e.g., Kutílek and Nielsen, 2010), it is evident that results from computer models do not agree with observations of reality. Being based merely on the assumptions of their creators, such hypothetical scenarios were never successfully tested against long periods of the past climate change. Their lack of validity is not only because of the underestimation of other forcing factors but also because feedbacks diagnosed from relatively short-term satellite data sets may have nothing to do with primary feedbacks involved in long-term processes, regardless of the indisputable accuracy of these data sets.

When we summarize the information in the previous subsection and that on models based upon the GHG hypothesis (not comprehensively cited here for brevity), we find that the assumption of the dominant GHG effect fails and is not acceptable when we deal with the recent global warming.

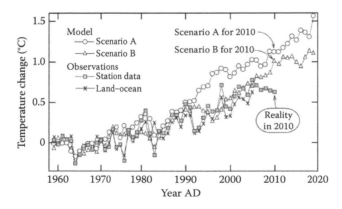

FIGURE 15.6 The model of Hansen et al. (1988) was tested for the period between 1958 and 1984. The predicted temperatures are after 1984. Increase in GHG in Scenario A: CH_4, 0.5%; N_2O, 0.25%; CO_2, 3% developing, 1% developed. In Scenario B: CH_4, 0.25%; N_2O, 0.25%; CO_2, 2% developing, 0% developed. Modeled predictions are compared to temperature observations. (From Real Climate, Hansen's 1988 projections, 2007. http://www.realclimate.org/index.php/archives/2007/05/hansens-1988-projections/.)

15.4 ADAPTATION OR FIGHTING CLIMATE CHANGE

15.4.1 EMPTY THREATS

When it has been demonstrated that the hypothesis of anthropogenic influence upon global warming caused by huge amounts of CO_2 released into the atmosphere from the burning of fossil fuels is not valid, it follows that a reduction of CO_2 emissions cannot decrease the rate of warming or ultimately stop it. Hence, with climate change being unavoidable, the unsolved remaining question is if the warming continues, might we expect a cooling interruption of the unrelenting warming process. If cooling were to bring temperatures back to those of the 1940s, there would be nothing new to solve. If, on the other hand, the warming process continues long term, even with short-term oscillations, we are obliged to explore new, developing situations and eventually define global and regional threats. Let us define and briefly discuss them.

15.4.1.1 Disastrous Hurricanes Due to Continued Global Warming

Hurricanes are natural disasters and the physics of their evolution belongs to atmospheric sciences. Because tropical cyclone activity seems to rise and fall in cycles lasting for many decades, interpretation of their long-term trends is difficult. The question of whether global warming is affecting tropical cyclones was given a nonconclusive reply by those attending a World Meteorologic Organization workshop in 2006.

In spite of such ambiguity, climatologists belonging to the group working for IPCC ascribe the alleged increasing strength and frequency of tropical cyclones to global warming. They found a positive correlation between sea surface temperature and Atlantic tropical cyclone frequency from 1871 to 2005 and modeled hurricane occurrence in a similar way to the modeling of global warming for the last 2000 years (Mann et al., 2009). Their results are dubious for three reasons. First, the structure of the models was identical to that of climate change models and we have shown that the latter have not been successfully tested against temperature data of the past. Such modeling of cyclones carries the burden that the results have yet to be confirmed with reality. Second, Pielke et al. (2008), using normalized damage as the scale of hurricane force for more than a century (1900–2005), found multidecadal variability but no tendency of a continuously increased force of hurricanes with the increase in global temperature. A third complicating consideration is that changes in climate can also change the potential paths of tropical cyclones—they could preferentially either remain over oceans or strike land.

Even so, there is no argument against the general rule that the occurrence of hurricanes is strongly reduced or probably even absent during the long-lasting cool periods of glaciation. However, the strength of winds or wind storms on continents was much higher during cool periods compared with that of recent experience. This statement is also supported by the evidence of dust storms in the last glacial (Harrison et al., 2001) and by the thick layers of loess that originated during glaciation. The probable absence of hurricanes in glacial periods is not proof that the rising temperature in warm periods of the Holocene would bring an increased occurrence of hurricanes.

15.4.1.2 Floods

According to the IPCC authors, river flood regimes are related to man-made climate change. The long-term data on river discharge and flood frequency indicate that there is no relation between floods and climate change on a centenarian scale. However, hydrologists have found a tendency of clustering floods (Brázdil et al., 2005), and paleorecords show that floods due to excessive rainfall or snowmelt are quite sensitive to modest changes in climate—smaller than changes expected from IPCC-predicted future global warming in the twenty-first century. Frequencies of large floods have increased along with increases in the number of air waves and their amplitudes in the middle and upper tropospheric circumpolar circulation (Knox, 2000). Nevertheless, owing to changing air circulation systems, flash floods restricted to small regions could appear more frequently, without the opportunity to predict them well ahead of their occurrence.

15.4.1.3 Sea Level Rise

With the warming and thawing of inland ice, the volume of seawater increases. With warming, the volume of seawater increases further because its thermal expansivity is temperature-dependent; hence, both processes cause sea level to rise. Although the IPCC predicts a rise between 38 cm and 1 m, some prognoses parallel conditions during the last Eem interglacial when the seawater level was 4–6 m above the recent level. The latter assertion is surely invalid because the time for melting is not comparable. The Eem temperature was 3°C–5°C higher than our recent temperature, and it lasted for thousands of years, compared with our recent perturbation, which has thus far lasted only two centuries. With the rate of recent rise of sea level ranging from 1.5 to 2 mm/year, the most extreme rise by 2100 will be less than 20 cm, assuming that the rate of rise does not increase. Although potentially troublesome, protecting coastal areas from such rises is not much more demanding than that already commonly achieved today against tides frequently associated with local, short-lasting storms.

15.4.1.4 Increase in Aridity and Desertification

The IPCC states that due to global warming and drying, the menace of drought increases, and further decreased precipitation is the dominant factor in the aridization process. Studies of the geological near past, that is, of the last two interglacials that had average temperatures that were 3°C–5°C higher than now, provide unambiguous results pertaining to increased precipitation and the disappearance of great deserts, or at least the substantial restriction of their areas. Even during the first quarter of the Holocene, the majority of the recent area of the Sahara desert was covered by savanna during a time that was warmer than now (see Figure 15.3) and also relatively humid. On the other hand, the climate cooling during glacial periods caused an increase in deserts. For example, in the Sahara the southern boundary of sand dunes shifted about 400 km to the south compared to the recent situation in our Holocene (Roberts, 2004). Similar expansions occurred in other desert zones. Regions in the middle latitudes suffered extreme aridity, with strong dust storms contributing to aridity in the cool climate of glacials, whereas the climate in the interglacials was generally warmer and more humid, judging from the features of paleosols (Smolíková, 1990; Kutílek and Nielsen, 2010). Therefore, if desertification is detected in some regions, it cannot be a consequence of global warming—the only perpetrator is human society with its inability to harmlessly coexist with arid and semiarid environments without transforming them into super-arid zones by extreme exploitation of inadequate local natural resources.

15.4.1.5 Menace of Famines

According to publications of the IPCC and some climatologists within its sphere of influence, increased aridity increases the menace of famines. They ignore the multitude of experiments performed since the end of the nineteenth century, verifying the positive influence of CO_2 upon photosynthesis and upon crop yields. The first ideas on "atmospheric fertilization" appeared in several of 20 volumes on Agricultural Physics edited by E. Wollny (*Forschungen an dem Gebiete der Agrikultur-Physik*) starting in 1878. The first monograph on this theme written in 1926 was published by Reinau (1927). Since that time, the understanding of photosynthesis has greatly advanced, together with the closely related topic of the plant–water regime. The experiments were shifted outside of greenhouses where photosynthesis, plant–water regime, and transpiration were studied simultaneously in the field with atmospheric CO_2 concentrations controlled at various levels (Kimball et al., 2002; Kirkham, 2005, 2011). Transpiration is influenced by an increase in atmospheric CO_2 in several ways. First, the number of stomata openings decreases mainly in C_3 plants if they grow in elevated concentrations of atmospheric CO_2. Through them the physical loss of water by evaporation from the plant is realized and reduced under the higher CO_2 concentration. Moreover, the diffusional resistance of the stomata increases with the increase in CO_2 concentration since they change their shape more closely to that of a crescent, which further reduces transpiration

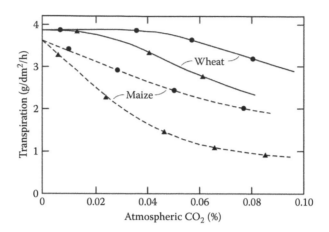

FIGURE 15.7 Effect of CO_2 concentration on transpiration of wheat and maize grown under different light intensities. The closed, dark circles are for plants grown at 0.400 ly/min (langleys per minute). The closed triangles are for plants grown at 0.015 ly/min. (From Akita, S. and Moss, D.N., *Crop Science* 12, 789–793, 1972, Figure 1, quoted according to Kirkham [2011].)

(the aperture gets smaller as the CO_2 concentration in the atmosphere increases). Owing primarily to these two mechanisms, the amount of water needed for the production of one unit of mass of organic material is reduced when CO_2 concentration increases (Figure 15.7). In other words, plant yield increases when CO_2 concentration increases, while water consumption is essentially not influenced (or the increase in water consumption is small and not proportional to the increased yield). These same relations are even more pronounced if the temperature is increased. Generally, the rise in temperature and the rise in atmospheric CO_2 concentration act positively upon food production and upon efficient soil water management.

15.4.2 Adaptation to Global Warming: The Optimal in Healthy Society

We purposely omit any of the countless economic comparisons of the two basic approaches to the contemporary, overexaggerated societal problem of global warming, that is, fighting or adapting to climate change. Although many messages have recently appeared in the specialized literature (e.g., Lomborg, 2007; Klaus, 2007), we have dealt principally with those aspects involving naturally occurring processes and with their eventual consequences for human cultures and societies.

"Fighting" the climate calls for everyone in all societies to follow scenarios embracing too many uncertainties not proven by scientific methodology although the popular media enthusiastically advocates them. Consequently, even if the reduction of CO_2 emissions is realized, with or without the politics of the recently introduced CO_2 capture and storage scenario being universally promoted, global warming will not necessarily stop.

Adaptation to new phenomena caused by the continuity of global climate change in specific regions is inevitable, with or without the modification of proven technological procedures. Methods of adaptation must be fitted to local and regional needs using procedures selected from technological steps already known or from those empirically developed without waiting for rigorous scientific methodology. Additional study of the history of several human cultures convincingly reveals that the adaptation of such procedures proved to be the most efficient process only when human culture and technology was further developed. The essential role of soil, its properties, and its theoretical or empirical management in providing locally available food resources in variable climates, regardless of their occurrence and sustainability, is abundantly evident.

REFERENCES

Akasofu, S.-I. 2010. On the recovery from the Little Ice Age. *Natural Science* 2: 1211–1224.

Akita, S. and D.N. Moss. 1972. Differential stomatal response between C3 and C4 species to atmospheric CO_2 concentration and light. *Crop Science* 12: 789–793. Quoted according to Kirkham (2011).

Alley, R.B. 2004. GISP2 ice core temperature and accumulation data. IGBP PAGES/World Data Center for Paleoclimatology Data Contribution Series #2004-013. NOAA/NGDC Paleoclimatology Program, Boulder, CO.

Brázdil, R., P. Dobrovolný, L. Elleder, V. Kakos, O. Kotyza, V. Květoň, J. Macková, et al. 2005. *Historical and Recent Floods in the Czech Republic.* Masaryk University and the Czech Hydro-Meteorological Institute, Brno, Czech Republic.

Bond, G., W. Showers, M. Cheseby, R. Lotti, P. Almasi, P. deMenocal, P. Priore, H. Cullen, I. Hajdas and G. Bonani. 1997. A pervasive millennial-scale cycle in North Atlantic Holocene and glacial climates. *Science* 278(5341): 1257–1266.

Bonnefille, R. 2010. Cenozoic vegetation, climate changes and hominid evolution in tropical Africa. *Global and Planetary Change* 72: 390–411.

Brevik, E.C. and A.E. Hartemink. 2010. Early soil knowledge and the birth and development of soil science. *Catena* 83: 23–33.

Cunniff, J., C.P. Osborne, B.S. Ripley, M. Charles, and G. Jones. 2008. Response of wild C4 crop progenitors to subambient CO_2 highlights a possible role in the origin of agriculture. *Global Change Biology* 14: 576–587.

Czech Hydro-Meteorological Institution. 2007. Climate Change at the start of 21st century and its further evolution, 21 February 2007. Prague, Czech Republic.

Dalfes, H.N., G. Kukla, and H. Weiss. 1997. *Third Millennium BC Climate Change and Old World Collapse.* NATO ASI Series I: Global Environmental Change, Vol. 49. Springer, Berlin, Germany.

Douglas, D.H., J.R. Christy, B.D. Pearson, and S.F. Singer. 2008. A comparison of tropical temperature trends with model predictions. *International Journal of Climatology* 28: 1693–1701.

Easterling, D.R. and M.F. Wehner. 2009. Is the climate warming or cooling? *Geophysical Research Letters* 36: L08706.

Gardner, W.R. 1987. Irrigation, past and present. *In Reports Prepared for a Conference on Agricultural and Water Resources Development in the Arab World*, pp. 30–33. Iowa State University, Ames, IA.

Goudie, A.S. 1983. The arid Earth. *In* R. Gardner and H. Scoging (eds), *Megageomorphology*, pp. 152–171. Clarendon Press, Oxford.

Hadan, A. 1971. Archaeological methods for dating of soil salinity in the Mesopotamian Plain. *In* D.A. Yaalon (ed.), *Paleopedology: Origin, Nature and Dating of Paleosols*, pp. 181–187. International Soil Science Society and Israel Universities Press, Jerusalem, Israel.

Hansen, J., I. Fung, A. Lacis, D. Rind, S. Lebedeff, R. Ruedy, G. Russell, and P. Stone. 1988. Global climate changes as forecast by Goddard Institute for Space Studies three-dimensional model. *Journal of Geophysical Research* 93(D8): 9341–9364.

Harrison, S.P., E.K. Kohfeld, C. Roelandt, and T. Claquin. 2001. The role of dust in climate changes today, at the last glacial maximum and in the future. *Earth-Science Reviews* 54: 43–80.

IPCC. 2007. Summary for policymakers. *In* S. Solomon, D. Qin, M. Manning, Z. Chen, M. Marquis, K.B. Averyt, M. Tignor, and H.L. Miller (eds), *Climate Change 2007: The Physical Science Basis*, pp. 1–18. Contribution of Working Group I to the Fourth Assessment Report of the Intergovernmental Panel on Climate Change, Cambridge University Press, Cambridge.

Jones, P.D. and M.E. Mann. 2004. Climate over past millennia. *Reviews of Geophysics* 42. RG2002, doi:10.1029/2003RG000143.

Kimball, B.A., K. Kobayashi, and M. Bindi. 2002. Responses of agricultural crops to free-air CO_2 enrichment (FACE). *Advances in Agronomy* 77: 293–368.

Kirkham, M.B. 2005. *Principles of Soil and Plant Water Relations.* Elsevier, Amsterdam, The Netherlands.

Kirkham, M.B. 2011. *Elevated Carbon Dioxide: Impacts on Soil and Plant Water Relations.* CRC Press, Boca Raton, FL.

Klaus, V. 2007. *Blue Planet in Green Shackles. What Is Endangered: Climate or Freedom?* Competitive Enterprise Institute, Mininova, Utrecht, The Netherlands.

Knox, J.C. 2000. Sensitivity of modern and Holocene floods to climate change. *Quaternary Science Reviews* 19: 439–457.

Knudsen M.F. and P. Riisager. 2009. Is there a link between Earth's magnetic field and low-latitude precipitation? *Geology* 37: 71–74.

Koutsoyiannis, D. 2008. Climate change as a scapegoat in water science, technology and management. Lecture at EUREAU Workshop on Climate Change Impact on Water Resources with Emphasis on Potable Water, Chania, Crete, Greece, 26 June 2008. http://www.itia.ntua.gr/en/docinfo/865 (verified 8 January 2012).

Kutílek, M. and D.R. Nielsen. 2010. Facts about global warming. Rational or emotional issue? *Essays in GeoEcology*. Catena Verlag, Reiskirchen, Germany.

Lane, L.J., M.H. Nichols, and H.B. Osborn. 1994. Time series analyses of global change data. *Environmental Pollution*. 83: 63–68.

Lomborg, B. 2007. *Cool It: The Skeptical Environmentalist's Guide to Global Warming*. Knopf, New York, NY.

Mann, M.E. and P.D. Jones. 2003. Global surface temperatures over the past two millennia. *Geophysical Research Letters* 30. doi:10.1029/2003GL017814.

Mann, M.E., J.D. Woodruff, J.P. Donnelli, and Z. Zhang. 2009. Atlantic hurricanes and climate over the past 1,500 years. *Nature* 460: 880–883.

Meehl, G.A., W.M. Washington, T.M.L. Wigley, J.M. Arblaster, and A. Dai. 2003. Solar and greenhouse gas forcing and climate response in the twentieth century. *Journal of Climate* 16: 426–444.

Oliver, R. 1991. *The African Experience: Major Themes in African History from Earliest Times to the Present*. Icon Editions, New York, NY.

Pielke Jr, R.A., J. Gratz, C.W. Landsea, D. Collins, M.A. Saunders, and R. Musulin. 2008. Normalized hurricane damage in the United States: 1900–2005. *Natural Hazards Review* 9: 29–42.

Potts, R. 1998. Environmental hypotheses of hominin evolution. *Yearbook of Physical Anthropology* 41: 93–136.

Real Climate. 2007. Hansen's 1988 projections. http://www.realclimate.org/index.php/archives/2007/05/hansens-1988-projections/, 15 May 2007.

Reinau, E. 1927. *Praktische Kohlensäuredüngung in Gärtnerei und Landwirtschaft*. Verlag von Julius Springer, Berlin, Germany.

Riehl, S. 2009. Archaeobotanical evidence for the interrelationship of agricultural decision-making and climate change in the ancient Near East. *Quaternary International* 197: 93–114.

Ritchie, J.C., C.H. Eyeles, and C.V. Haynes. 1985. Sediment and pollen evidence of an early to mid Holocene humid period in the eastern Sahara. *Nature* 314: 352–355.

Roberts, N. 2004. *The Holocene*. Blackwell, Oxford.

Showers, K.B. 2006. A history of African soil: Perceptions, use and abuse. *In* J.R. McNeill and V. Winiwarter (eds), *Soils and Societies: Perspectives from Environmental History*, pp. 118–176. White Horse Press, Isle of Harris.

Smolíková, L. 1990. Regionální paleopedologie. Zákonitosti půdního vývoje v kvartéru (In Czech: Regional paleopedology. Soil evolution principles in Quartenary). *In* J. Němeček, L. Smolíková, and M. Kutílek (eds), *Pedologie a paleopedologie (Pedology and Paleopedology)*, pp. 405–508. Academia, Prague, Czech Republic.

Solomon, S., D. Qin, M. Manning, R.B. Alley, T. Berntsen, N.L. Bindoff, Z. Chen, et al. 2007. Technical summary. *In* S. Solomon, D. Qin, M. Manning, Z. Chen, M. Marquis, K.B. Averyt, M. Tignor, and H.L. Miller (eds), *Climate Change 2007: The Physical Science Basis*, pp. 19–91. Contribution of Working Group I to the Fourth Assessment Report of the Intergovernmental Panel on Climate Change, Cambridge University Press, Cambridge.

Staubwasser, M. and H. Weiss. 2006. Holocene climate and cultural evolution in late prehistoric-early historic West Asia. *Quaternary Research* 66: 372–387.

Svensmark, H., J.O.P. Pedersen, N.D. Marsh, M.B. Enghoff, and U.I. Uggerhoj. 2007. Experimental evidence for the role of ions in particle nucleation under atmospheric conditions. *Proceedings Royal Society A* 463: 385–396.

Trauth, M.H., M.A. Maslin, A.L. Deino, M.R. Strecker, A.G.N. Bergner, and M. Duhnforth. 2007. High- and low-latitude forcing of Plio-Pleistocene East African climate and human evolution. *Geophysical Research Abstracts* 9: 05299.

Trenberth, K.E., P.D. Jones, P. Ambenje, R. Bojariu, D. Easterling, A. Klein Tank, D. Parker, et al. 2007. Observations: Surface and atmospheric climate change. *In* S. Solomon, D. Qin, M. Manning, Z. Chen, M. Marquis, K.B. Averyt, M. Tignor, and H.L. Miller (eds), *Climate Change 2007: The Physical Science Basis*, pp. 235–336. Contribution of Working Group I to the Fourth Assessment Report of the Intergovernmental Panel on Climate Change, Cambridge University Press, Cambridge.

van der Leeuw, S.E. 2008. Climate and society: Lessons from the past 10 000 years. *Journal of the Human Environment* 37: 476–482.

Veldkamp, A., E. Buis, J.R. Wijbrans, D.O. Olago, E.H. Boshoven, M. Marée, and R.P. van den Berg van Saparoea. 2007. Late Cenozoic fluvial dynamics of the River Tana, Kenya, an uplift dominated record. *Quaternary Science Reviews* 26: 2897–2812.

Weiss, H., M.A. Courty, W. Wetterstrom, F. Guichard, L. Senior, R. Meadow, and A. Carnow. 1993. The genesis and collapse of third millennium North Mesopotamian civilization. *Science* 261: 995–1004.

White, N.J., J.A. Church, and J.M. Gregory. 2005. Coastal and global averaged sea level rise for 1950 to 2000. *Geophysical Research Letters* 32: L011601.

Yaalon, D.H. and E. Ganor. 1973. The influence of dust on soils during the Quaternary. *Soil Science* 116: 146–155.

Zeebe, R.E., J.C. Zachos, and G.R. Dickens. 2009. Carbon dioxide forcing alone insufficient to explain palaeocene–eocene thermal maximum warming. *Nature Geoscience* 2: 576–580.

Zerboni, A., L. Trombino, and M. Cremaschi. 2010. Micromorphological approach to polycyclic pedogenesis on the Messak Settafet plateau (central Sahara): Formative processes and paleoenvironmental significance. *Geomorphology* 125: 319–335.

16 Climate Change, Soils, and Human Health

Eric C. Brevik

CONTENTS

16.1 INTRODUCTION

Global climate change is a phenomenon that has been known to scientists for over 70 years, but it has only been at the forefront of the public consciousness over the last two decades (Corfee-Morlot et al., 2007). The most recent report of the Intergovernmental Panel on Climate Change (IPCC) indicates that the average global temperature will probably rise between 1.1°C and 6.4°C by 2090–2099 as compared with 1980–1999 temperatures, with the most likely rise being between 1.8°C and 4.0°C (IPCC, 2007a). The IPCC also concluded that "Most of the observed increase in global average temperatures since the mid-20th century is *very likely* due to the observed increase in anthropogenic greenhouse gas concentrations" (IPCC, 2007a). The "very likely" in this statement means the IPCC scientists agreed there was a greater than 90% probability that anthropogenic greenhouse gas emissions have caused most of the warming witnessed since the mid-twentieth century (Solomon et al., 2007).

It seems that one common rallying cry for those who choose not to believe that climate change is occurring is to opine that climate change is scientifically controversial or is not proven (Cooney, 2010; Corfee-Morlot et al., 2007; Lahsen, 2005). The truth is there is little controversy; the idea that Earth's climate is changing due to anthropogenic activity is now almost universally accepted in the scientific community (Cooney, 2010; Corfee-Morlot et al., 2007). It is possible to find some scientists researching climate change who question whether the driving force behind that change is natural or anthropogenic (e.g., Kutílek and Nielsen, Chapter 15, this volume; Kutílek, 2011; Carter, 2007; Davis and Bohling, 2001; Bluemle et al., 1999; Idso, 1998), but those scientists are a distinct minority (Corfee-Morlot et al., 2007; Lahsen, 2005). Furthermore, many of those scientists do not dispute that climate change does occur and is occurring (e.g., Kutílek and Nielsen, Chapter 15, this volume; Kutílek, 2011; Carter, 2007; Davis and Bohling, 2001; Bluemle et al., 1999). Therefore, even if we cannot agree on *why* climate change is happening, it should be possible to agree that it *is* happening. With climate change happening, there will be effects on the environment, including on the organisms (such as humans) that live there.

Studies into the effects climate change will have on soils are in their early stages, so there is much more to be learned in this area. However, through the results of the studies that have been done and our understanding of soil processes and properties, it is possible to provide some insight into the expected effects of climate change. For example, we know that changing climates will influence important cycles such as the carbon and nitrogen cycles, which will in turn affect soil processes and fertility (Wan et al., 2011; Davidson and Janssens, 2006; Gorissen et al., 2004; Hungate et al., 2003; Kirschbaum, 1995). Climate change will also influence soil moisture levels (Kirkham, 2011; Backlund et al., 2008; Chiew et al., 1995). Soil fertility and moisture are important components of crop production, which means climate change has the potential to influence our ability to feed the human population by impacting food security (St. Clair and Lynch, 2010; Huntingford et al., 2005; Parry et al., 2005; Rosenzweig and Parry, 1994). Soil erosion is also expected to increase as climate changes (Maeda et al., 2010; Ravi et al., 2010; Favis-Mortlock and Boardman, 1995), leading to reductions in crop production and increased food security concerns (Li et al., 2011; Zhang et al., 2004). Aeolian erosion of soils is expected to increase in dryland regions (Ravi et al., 2010); the soil-generated dust may lead to respiratory disease (Rask-Andersen, 2001; Schenker, 2000), providing another link between climate change, soils, and their effects on human health. This brief discussion serves as an introduction to the study of the effects of climate change on soils, an appetizer if you will, and how those effects could influence human health. This chapter will assess what we currently know about gas fluxes in soil related to climate change. It will also discuss the potential effects of climate change on (1) human health, (2) soil processes and properties, and (3) how changes in soil processes and properties could potentially impact human health in the future.

16.2   GAS FLUXES AND SOILS

In 2004 carbon dioxide (CO_2), methane (CH_4), and nitrous oxide (N_2O) made up most of the anthropogenic greenhouse gas emissions (IPCC, 2007b; Le Treut et al., 2007). These three gases are also the most important of the long-lived greenhouse gases (Hansen et al., 2007). More specifically, CO_2 emissions in 2004 accounted for 77% of potential greenhouse warming by anthropogenic gases, with CH_4 and N_2O accounting for about 22% of potential warming (Figure 16.1) (IPCC, 2007b). These gases are a part of the global carbon and nitrogen cycles (Figures 16.2 and 16.3) (Richter et al., 2007). Prior to the industrial revolution, the global carbon and nitrogen cycles were in a balance, with inputs approximately equaling outputs; burning of fossil fuels, tilling of soil, and other human activities have altered the natural balance such that we are now releasing more carbon and nitrogen into the atmosphere each year than is taken up by global sinks (Pierzynski et al., 2009).

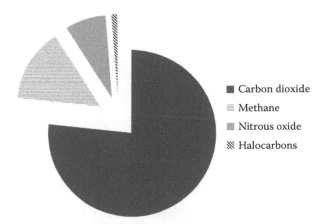

FIGURE 16.1 Relative contribution to potential greenhouse warming by anthropogenic gas emissions in 2004. (Data from IPCC, *Working Group Contributions to the Fourth Assessment Report of the Intergovernmental Panel on Climate Change,* Cambridge University Press, Cambridge, 2007b.)

Agriculture currently contributes about 20% of total annual anthropogenic greenhouse gas emissions (Lokupitiya and Paustian, 2006).

Because soils are part of the carbon and nitrogen cycles, it is possible to influence atmospheric levels of carbon- and nitrogen-based greenhouse gases through soil management (Wagner-Riddle and Weersink, 2011; Hobbs and Govaerts, 2010; Lal, 2007). A fourth group of greenhouse gases,

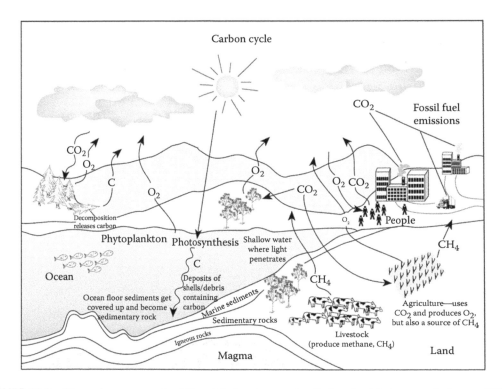

FIGURE 16.2 The main components of the global carbon cycle. Carbon is able to move between different pools within the cycle. For example, burning of fossil fuels or decomposition of soil organic matter sends carbon gases into the atmosphere, while photosynthesis locks atmospheric carbon up in plant tissues and deposition of organic-rich sediments on the ocean floor locks carbon up in geologic rocks and sediments. (Courtesy of NASA.)

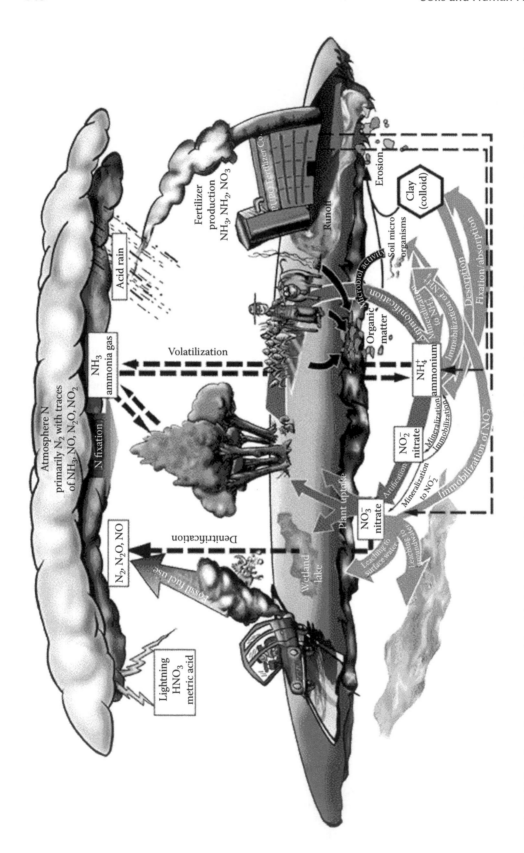

FIGURE 16.3 The main components of the global nitrogen cycle. As with the carbon cycle, nitrogen is able to move between pools in its cycle, including the soil pool. Some processes put gases such as N₂O into the atmosphere, while other processes remove those gases from the atmosphere and transfer them into other pools. (Courtesy of NASA.)

the halocarbons, will not be discussed in this chapter. However, halocarbons can be created naturally and halocarbon formation has been documented in the soil (Gribble, 2003; Hoekstra et al., 1999; Keene et al., 1999). It should also be noted that the discussions on CO_2, CH_4, and N_2O presented here are brief and are only meant to demonstrate that ties exist between these gases, the soil system, and the atmosphere, with some examples of how human management can influence those relationships. In this section and throughout the chapter, readers are referred to the references cited for more complete discussions on the topics covered.

16.2.1 SOILS AND CO_2

The largest active terrestrial carbon pool is in soil, which contains an estimated 2500 Pg of carbon, compared with 620 Pg of carbon in terrestrial biota and detritus and 780 Pg of carbon in the atmosphere (Figure 16.4) (Lal, 2010). In addition to these pools, there is approximately 90,000,000 Pg of carbon in the geological formations of Earth's crust, 38,000 Pg of carbon in the ocean as dissolved carbonates, 10,000 Pg of carbon sequestered as gas hydrates, and 4,000 Pg of carbon in fossil fuels (Rustad et al., 2000). While Earth's crust, the oceans, and the gas hydrates are much larger carbon pools than soils, humans are not able to easily manipulate conditions that influence carbon exchange in these pools. We could reduce carbon emissions sharply by ceasing the use of fossil fuels, but this requires the development of alternative fuel sources. While conservation of fossil fuels and development of alternative energy sources are being investigated, we are still searching for other ways to manage ever-growing levels of CO_2 in our atmosphere.

What is a Pg?

The discussion of global carbon pools involves very large measures of mass relative to the masses that most people are typically accustomed to talking about. A petagram (Pg) is 1,000,000,000,000,000 (1 quadrillion) g, or 1,000,000,000,000 (1 trillion) kg. So when calculations indicate there are 2,500 Pg of carbon in soil, that is the same as saying 2,500 trillion (or 2.5 quadrillion) kg of carbon.

One of the potential ways that is readily available to mitigate climate change is carbon sequestration by soils using the soil–plant system. Plants remove CO_2 from the atmosphere during

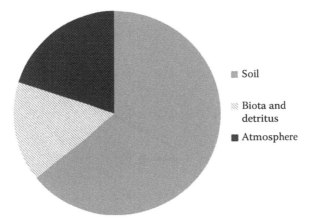

FIGURE 16.4 Relative size of the active terrestrial carbon pools. The size of the soil carbon pool relative to the biological and atmospheric pools demonstrates the importance of soils in the carbon cycle. (Data from Lal, R., *BioScience*, 60, 708–721, 2010.)

photosynthesis and create carbohydrates, some of which are incorporated into plant tissues. As plants or plant parts die, some of the plant tissues are incorporated into the soil as soil organic matter (Lal et al., 1998). Given the proper conditions some soils can become net carbon sinks, effectively removing CO_2 from the atmosphere (Figure 16.5) (Mosier, 1998). Because of this capability of the soil–plant system, carbon sequestration by soils as a means of mitigating global warming has received a considerable amount of research interest.

Carbon can be sequestered in any soil, but mankind has the greatest potential control over sequestration in intensively managed systems such as agricultural and agroforestry soils. Soil management techniques such as reduced tillage or no-till systems often result in lower CO_2 emissions from the soil and greater carbon sequestration in the soil as compared to management systems based on intensive tillage (Figure 16.6) (Farina et al., 2011; Hobbs and Govaerts, 2010; Lokupitiya and Paustian, 2006; Steinback and Alvarez, 2006; Post et al., 2004; West and Post, 2002), as do changes such as using cover crops, crop rotations instead of monocropping, and reducing or eliminating fallow periods (Álvaro-Fuentes and Paustian, 2011; Lokupitiya and Paustian, 2006; Post et al., 2004; West and Post, 2002). The use of reduced or no-till systems has the added benefit of using less fuel for working the soil, which reduces CO_2 emissions by agricultural machinery (Wagner-Riddle and Weersink, 2011; Hobbs and Govaerts, 2010; Schneider and Smith, 2009; Wolf and Snyder, 2003); fuel savings of around 32.7 L/ha (3.5 gal per acre) have been estimated for no-till versus conventional tillage systems in cotton farming (Wolf and Snyder, 2003). Returning land from agricultural use to native forest or grassland can also lead to significant carbon sequestration in soils (Post and Kwon, 2000; Silver et al., 2000). Sequestration of carbon tends to be rapid initially, with declining rates over time (Figure 16.6) (Dixon-Coppage et al., 2005; Silver et al., 2000; Neill et al., 1998; Chadwick et al., 1994). Maximizing carbon sequestration in soils requires adequate nitrogen to allow carbon accumulation. Hungate et al. (2003) question whether or not there will be enough nitrogen available to maximize carbon sequestration as climate change occurs.

Management decisions can restrict the ability of a soil to sequester carbon as well. For example, the decision to plow native soils has led to significant releases of CO_2 (Figure 16.6) (Lokupitiya and Paustian, 2006; Rustad et al., 2000). As another example, the extensive use of heavy equipment in modern production agriculture has made soil compaction a major problem (DeNeve and Hofman, 2000), which has been shown to limit carbon sequestration (Dixon-Coppage et al., 2005; Brevik et al., 2002; Brevik, 2000). Organic soils can be a particular carbon management challenge as they typically form in wet conditions and have to be drained for agricultural uses. This drainage changes

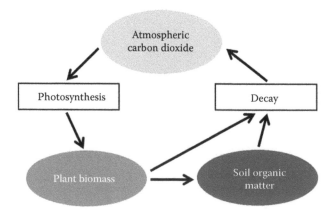

FIGURE 16.5 The concept of carbon sequestration by soils. Atmospheric CO_2 is utilized during photosynthesis and transformed into plant biomass. As the biomass enters the soil and decays, some of it is transferred into soil organic matter and some returns to the atmosphere as CO_2. Soil organic matter also decays and releases CO_2 to the atmosphere. If more plant biomass is added to the soil than decays, the total amount of soil organic matter increases, resulting in carbon sequestration.

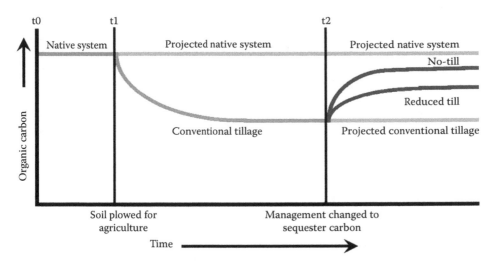

FIGURE 16.6 Typical changes in soil organic carbon over time under different soil management. Time t0 to t1 represents the soils under a native ecosystem. At t1, the soil was broken for agricultural production using conventional tillage, leading to a decline in soil organic matter. The "Projected native system" is the expected soil organic matter content if the native ecosystem had not been disturbed. At t2, management was changed to sequester carbon in the soil. Two possible paths are shown: conversion to reduced till or conversion to no-till. Both increase soil organic carbon, no-till more so than reduced till, but neither returns soil carbon to native system levels. The "Projected conventional tillage" is the expected soil organic matter content if conventional tillage had been maintained. Note that changes in organic carbon content are rapid immediately following management changes but rates of change decrease and then stop as the soil reaches carbon equilibrium.

the soil environment from anaerobic to aerobic, which speeds decomposition of the organic matter in the soil and releases greenhouse gases into the atmosphere (Troeh et al., 2004; Wolf and Snyder, 2003). A study in Finland that looked at the effect of crops on greenhouse gas fluxes from soils showed that the organic soils in their study were a net source of CO_2 for all cropping systems studied (Martikainen et al., 2002).

Most agricultural soils only sequester carbon for about 50–150 years following management changes before they reach carbon saturation (Figure 16.6) (Lal, 2010; Mosier, 1998), putting a limit on the ultimate effectiveness of soils in mitigating CO_2 additions to the atmosphere. The IPCC estimates that 0.4–0.8 PgC/year (the equivalent of 1.4–2.9 Pg of CO_2/year) could be sequestered globally in agricultural soils, with soil carbon saturation occurring after 50–100 years (Smith et al., 2007). Estimated anthropogenic CO_2 emissions to the atmosphere in 2004 totaled about 38 PgC/year (IPCC, 2007b) and natural carbon sinks have taken about 45% of anthropogenic CO_2 emissions out of the atmosphere since 1959 (Denman et al., 2007), meaning about 21 Pg of anthropogenic C/year remained in the atmosphere long-term in 2004. Using the aforementioned numbers, carbon sequestration by agricultural soils would be able to remove about 2%–4% of the annual anthropogenic additions of carbon to the atmosphere for the next 50–100 years.

Intensely managed soils have received the most attention in carbon sequestration research, but there are other soils that have potential for significant carbon sequestration as well. Coastal wetland soils have the ability to sequester carbon at higher rates than most agricultural soils (Jesperson and Osher, 2007; Johnson et al., 2007; Brevik and Homburg, 2004; Hussein et al., 2004). Coastal wetland soils can sequester carbon over hundreds or thousands of years rather than the decades possible with agricultural soils. In addition, they can sequester carbon to depths of several meters as opposed to the typical single-meter depth measured for agricultural soils, making coastal wetland soils more efficient at sequestering carbon than most agricultural soils on a per unit area basis (Johnson et al., 2007; Brevik and Homburg, 2004; Hussein et al., 2004). Coastal wetlands also release much lower levels of CH_4 and N_2O than freshwater wetlands (Bartlett and Harris, 1993; DeLaune et al., 1990), a distinct advantage when considering carbon sequestration as a means of mitigating greenhouse gas-driven climate change. Findings such as these indicate that the conservation and restoration of coastal wetlands would be well advised (Zedler and Kercher, 2005; Brevik and Homburg, 2004). However, the current global trend is a rapid reduction of coastal wetlands from natural activities such as erosion due to rising sea levels. Human activities also lead to the loss of wetlands as they are drained and filled in or otherwise modified for uses such as agriculture, urban development, petroleum extraction, and salt production (White and Morton, 1997; Tsihrintzis et al., 1996; Steyer and Stewart, 1992; Titus, 1991). Significant carbon sequestration is also possible in alluvial soils (Grossman et al., 1998) and in some abandoned mine and quarry sites (Sperow, 2006; Ussiri et al., 2006; Dixon-Coppage et al., 2005; Akala and Lal, 2001).

While we have the ability to sequester carbon in the soil, management decisions can also release carbon from the soil, making those soils a net source of greenhouse gases. Plowing native soils for agricultural production (Figure 16.6), introducing more aggressive forms of tillage to an agricultural management system, and draining wetlands are examples of management changes that increase CO_2 emissions from soils. It is also true that carbon that has been sequestered can be returned to the atmosphere at a future time if the management changes that lead to its sequestration are altered. In short, managed soils can be either net sinks or net sources of CO_2 depending on their management (Mosier, 1998; Schlesinger, 1995).

Arctic soils are of particular concern regarding the release of carbon to the atmosphere. Due to the cold conditions under which they form, microbial activity and decomposition rates tend to be low in Arctic soils, thus soil organic carbon reaches high levels (Barber et al., 2008). However, warming these soils can switch them from a carbon sink to a carbon source (Bliss and Maursetter, 2010; Welker et al., 1999; Oechel and Vourlitis, 1995), with well-drained soils releasing CO_2 to the atmosphere (Barber et al., 2008; Sjögersten et al., 2006). This is of particular concern because Arctic soils contain about 30% of the world's soil carbon (Chapin et al., 2004; Oechel and Vourlitis, 1995) and thus have the potential to release large quantities of greenhouse gases into the atmosphere as they thaw (Chapin et al., 2004).

16.2.2 Soils and CH_4

Methane (CH_4) concentrations in the atmosphere did not increase between 1998 and 2005, but were more than 2.5 times higher than in 1800 (Forster et al., 2007). Agriculture accounts for about 47% of annual global anthropogenic emissions of CH_4 (Smith et al., 2007). Production of CH_4 in the soil is associated with the anaerobic decomposition of organic matter. Because of this, the main anthropogenic source of soil-derived methane is rice production, while natural soil-derived methane comes primarily from wetlands (Figure 16.7) (Stępniewski et al., 2011; Heilig, 1994). Termites are also a major natural source of methane (Heilig, 1994). A significant portion of the CH_4 produced in soil is oxidized by soil microorganisms aerobically (Stępniewski et al., 2011; Mosier, 1998; Schütz et al., 1990) into products including CO_2 (Figure 16.7) (Heilig, 1994). Increasing soil temperatures lead to increased CH_4 production in rice paddy soils and wetlands, which is a concern with rising global temperatures (Stępniewski et al., 2011; Schütz et al., 1990). The melting of soils that have been permanently frozen (permafrost) (Figure 16.8) is also becoming a major source of atmospheric CH_4 (Barber et al., 2008).

Management makes a difference in CH_4 fluxes in soil. The presence of ammonium ions in the soil from nitrogen fertilization has been shown to inhibit the ability of agricultural soils to serve as a CH_4 sink (Stępniewski et al., 2011). Different vegetation growing on the same soil will also cause differences in CH_4 emission or consumption. In a study by Hu et al. (2001), a soil under forest vegetation acted as a net sink of CH_4 while the same soil in a nearby field planted with maize was essentially CH_4 neutral and a third field of the same soil planted with grass cover was a net source of CH_4 to the atmosphere.

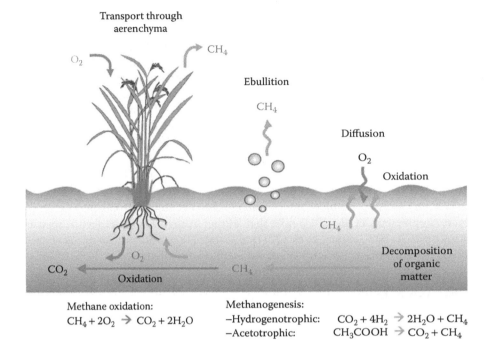

Methane oxidation:
$CH_4 + 2O_2 \rightarrow CO_2 + 2H_2O$

Methanogenesis:
−Hydrogenotrophic: $CO_2 + 4H_2 \rightarrow 2H_2O + CH_4$
−Acetotrophic: $CH_3COOH \rightarrow CO_2 + CH_4$

FIGURE 16.7 Generation and emission of methane from wet soils. (Courtesy of Josef Zeyer, ETH Zurich, Switzerland.)

⁓ Modern southern permafrost boundary

▨ Permafrost area likely to thaw by 2100

▨ Permafrost area projected to be under different
   stages of degradation

FIGURE 16.8 The projected shift of the permafrost boundary in northern Asia by 2100 due to climate change. (From Cruz, R.V., Harasawa, H., Lal, M., Wu, S., Anokhin, Y., Punsalmaa, B., Honda, Y., Jafari, M., Li, C., and Huu Ninh, N., *Climate Change 2007: Impacts, Adaptation and Vulnerability*, Contribution of Working Group II to the Fourth Assessment Report of the Intergovernmental Panel on Climate Change, Cambridge University Press, Cambridge, 2007, Figure 10.5.)

Rice production management has the greatest potential to reduce anthropogenic additions of soil-derived CH_4 (Neue, 1992). Dryland tillage and dry seeding or other means of reducing the period of soil saturation lead to less CH_4 production (Stępniewski et al., 2011; Neue, 1992). However, the production of CH_4 and N_2O are inversely related in rice soils; managing soil moisture levels to prevent the generation of one tends to encourage generation of the other (Neue, 1992). Since both are greenhouse gases, the balance between them must be carefully assessed. Adding organic amendments such as manure to flooded soils as a nutrient source increases CH_4 emissions (Stępniewski et al., 2011; Zhang et al., 2011; Wassmann et al., 1993), while high percolation rates in wetland rice soils reduce methane emissions (Bouwman, 1991). Fertilizer experiments have produced some mixed results (Neue, 1992). Wassmann et al. (1993) found no mineral fertilizer effect on methane generation in rice paddy soils when adding potassium fertilizers, but Lu et al. (1999) found that phosphorus fertilizers decreased CH_4 emissions. Lu et al. (1999) attributed this to increased root exudates in phosphorus-deficient soils as the plant tried to manipulate the soil environment to increase phosphorus uptake. Stępniewski et al. (2011) note that adding oxidizing mineral fertilizers can reduce CH_4 emissions by 20%–70%. Zhang et al. (2011) also noted that a mixed management system that incorporated ducks into the rice system decreased methane emissions.

16.2.3 Soils and N_2O

Agriculture accounts for about 58% of anthropogenic nitrous oxide (N_2O) emissions (Smith et al., 2007). From a soil perspective, N_2O is created when soil water contents approach field capacity, or a state of near-saturation, and biological reactions in the soil convert NO_3^- to NO^-, N_2O, or N_2 (Mullen, 2011). Enhanced microbial production in expanding agricultural lands that are amended with fertilizers and manure is believed to be the primary driver behind increased atmospheric N_2O levels (Forster et al., 2007; Lokupitiya and Paustian, 2006). Well over half of the global emissions of N_2O appear to come from the tropics (equator to 23.5° N and S) (Forster et al., 2007), with 13%–37% of global N_2O emissions coming from tropical forest soils (Melillo et al., 2001). N_2O emissions are expected to increase with increasing soil clay content when other factors are held constant (Chatskikh et al., 2005). There are also some indications that warming of cold-region soils could lead to increased N_2O emissions from those soils (Williams et al., 1998; Brooks et al., 1997).

Agricultural management is a major factor in N_2O emissions. As nitrogen fertilizer applications increase, denitrification and the generation of N_2O in the soil also increase (Figure 16.9) (Mullen, 2011; Stępniewski et al., 2011; Grant et al., 2006; Chatskikh et al., 2005). Emissions of N_2O are usually lower in organic farming systems than in conventional systems (Stępniewski et al., 2011). Some studies have found higher N_2O emissions from tilled soils than from no-till soils (Stępniewski et al., 2011; Wagner-Riddle and Weersink, 2011; Steinback and Alvarez, 2006), but this is not true in all cases (Grandy et al., 2006). The conversion of tropical forest to pasture led to an initial increase in N_2O emissions followed by a decline in emissions relative to the original forest in a Brazilian study (Melillo et al., 2001), however, conversion of tropical forest to fertilized crop production on Borneo led to an order of magnitude increase in N_2O emissions from the agricultural soils as compared with the forest soils (Hall et al., 2004).

16.2.4 Gas Fluxes and Soils: Summary

In summary, human management can have a profound impact on processes that emit or consume CO_2, CH_4, and N_2O in the soil. Current soil carbon estimates for soils of the world are given in Table 16.1. Any land management changes that lead to reduced production (i.e., the use of oxidizing mineral fertilizers or decreased flood times to reduce CH_4 emissions from rice fields) or increased sequestration (i.e., converting to a no-till system with cover crops to sequester carbon) of greenhouse gases in the soil system have the overall effect of reducing climate change-related issues. However, soils can also serve as a source of greenhouse gases if not managed properly.

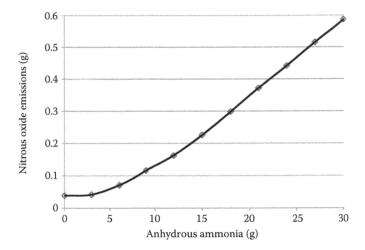

FIGURE 16.9 Modeled nitrous oxide emissions per meter squared at various application rates of anhydrous ammonia fertilizer. (Data from Grant, R.F., Pattey, E., Goddard, T.W., Kryzanowski, L.M., and Puurveen, H., *Soil Science Society of America Journal*, 70, 235–248, 2006.)

For example, converting a no-till field using cover crops to an intensively tilled field with periods of fallow would lead to increased greenhouse gas emissions from that field. In fact, soils are a major source of anthropogenic non-CO_2 emissions from agriculture. N_2O emissions from soils constituted 38% and CH_4 emissions from rice production 11% of the total non-CO_2 greenhouse gas emissions from agriculture in 2005 (Smith et al., 2007). Additionally, increasing soil temperatures have been

TABLE 16.1
Organic and Total Carbon for Soils of the World

Order[a]	Organic C (Pg)	Total C (Pg)	Ice-Free Land Surface (km²)	Organic C (kg/m²)	Total C (kg/m²)
Alfisols	136	236	12,620,000	10.8	18.7
Andisols	69	70	912,000	75.7	76.8
Aridisols	110	1,154	15,700,000	7.0	73.5
Entisols	106	223	23,390,000	4.5	9.5
Histosols	390	390	3,780,000	103.2	103.2
Inceptisols	267	552	15,110,000	17.7	36.5
Mollisols	72	211	11,260,000	6.4	18.7
Oxisols	150	150	9,810,000	15.3	15.3
Spodosols	98	98	5,600,000	17.5	17.5
Ultisols	101	101	11,050,000	9.1	9.1
Vertisols	38	63	3,160,000	12.0	19.9
Misc. land[b]	18	18	18,400,000	1.0	1.0

Source: Organic and total carbon data for each order are from Eswaran et al. (1995) and total area of ice-free land is from Blum and Eswaran (2004). The average organic and total carbon per meter squared for each order are calculated using the data from Eswaran et al. (1995) and Blum and Eswaran (2004).

[a] Eswaran et al. (1995) was published before the Gelisols order was established in 1998. Prior to 1998, most of the soils currently classified as Gelisols were classified in the Entisols, Inceptisols, Histosols, Mollisols, and Spodosols orders (Buol et al., 2003). For the purposes of this table, the Gelisols area has been split equally among these soil orders.

[b] Ice-free land without soil cover.

shown to lead to increased CH_4 production in rice paddies and wetlands (Stępniewski et al., 2011; Schütz et al., 1990). However, soils are not currently considered to be a major net source of CO_2. While agricultural soils produce large quantities of CO_2 each year, they also take up large quantities, such that agricultural soils are estimated to contribute less than 1% of net global anthropogenic CO_2 emissions (Smith et al., 2007). Therefore, soils influence climate change to the extent that they contribute to the presence of these greenhouse gases in the atmosphere and also contribute proportionally to any human health influences related to climate change caused by these gases. Some of the health issues that might arise due to climate change are discussed in the next section.

16.3   CLIMATE CHANGE IMPACTS ON HUMAN HEALTH

It is important to note that many of the ways climate change might impact human health are not fully understood and more research is needed to quantitatively define health challenges (Ebi and McGregor, 2008; Campbell-Lendrum and Woodruff, 2006; Ebi et al., 2006; Kovats et al., 2001). However, climate change is expected to have several impacts on human health and we can make educated inferences regarding those impacts. The IPCC identifies five main types of empirical studies that provide insight into the current sensitivity of human health to weather and climate: (1) health impacts of individual extreme events; (2) spatial studies where climate is an explanatory variable in the distribution of a disease or disease vector; (3) temporal studies assessing the health effects of climate variability; (4) experimental laboratory and field studies of vector, pathogen, or allergen biology; and (5) intervention studies that investigate the effectiveness of public health measures (Confalonieri et al., 2007). Such studies indicate there are a wide array of impacts climate change may have on human health, including but not limited to temperature extremes, food safety, air-quality issues, vectorborne and other infectious diseases, and occupational health issues (Confalonieri et al., 2007).

16.3.1   TEMPERATURE EXTREMES

In the average year about 65% of weather-related human deaths in the United States are caused by extreme heat or cold (Figure 16.10) (Aguado and Burt, 2007). Large numbers of deaths are also associated with temperature extremes globally, with problems associated with cold most common in,

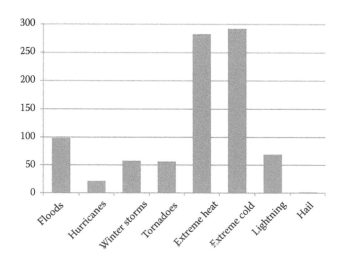

FIGURE 16.10   Average annual deaths due to weather events in the United States during the 1990s. (Data from Aguado, E. and Burt, J.E., *Understanding Weather and Climate*, Pearson-Prentice Hall, Upper Saddle River, NJ, 2007.)

but not completely confined to, northern latitudes (Confalonieri et al., 2007). Hansen et al. (2008) have also documented an association between heat waves and hospital admissions for mental and behavioral disorders. Heat-related deaths are expected to increase with climate change as the frequency and intensity of heat waves increase (Ebi et al., 2008).

Many factors other than temperature help determine the impact of temperature extremes on human health. For example, people with chronic obstructive pulmonary disease (COPD) have a greater risk than the general population of dying on cold days (Schwartz, 2005). Reid et al. (2009) identified four factors that explained over 75% of vulnerability to death during a heat wave in the United States: (1) social and/or environmental vulnerability: a grouping that included factors such as education, income status (defined as being above or below the poverty line), race, and proximity to green space. (2) Social isolation: whether or not people lived alone. (3) Air conditioning prevalence: a factor that looked at whether or not individuals had air conditioning and, for those who did, whether or not it was central air conditioning. (4) Proportion of the elderly and those with diabetes, which looked at people who were 65 years old or older and/or those who had diabetes, which have been shown to be risk factors for death during heat waves (Schwartz, 2005). In a study in England and Wales, Hajat et al. (2007) found the elderly to be at particular risk to temperature extremes, but found little difference between social classes, with the exception of a slightly increased risk from cold for low-income people in rural areas. Hajat et al. (2007) also found stronger heat effects in females than males and in urban areas as opposed to rural areas. City design may play a role as well, with sprawling cities being more susceptible to mortality from extreme heat events than more compact cities (Stone et al., 2010). It may not be just extreme heat we have to worry about, however. Some studies suggest that even moderate heat can cause increases in death rates; therefore, the expected rise in global temperatures over the next century makes heat-related health issues a primary concern (Kalkstein and Greene, 1997).

16.3.2 FOOD SAFETY

Food safety issues are primarily concerned with the effects of increasing temperatures on incidences of food poisoning (Confalonieri et al., 2007). While it does not hold true for all types of food poisoning (Lake et al., 2009; Louis et al., 2005), some studies have found a relationship between increasing ambient air temperature and reported cases of some types of food poisoning, such as salmonellosis (Figure 16.11) (Lake et al., 2009; D'Souza et al., 2004; Kovats et al., 2004). If the relationship between salmonellosis cases and temperature holds true, this represents another health

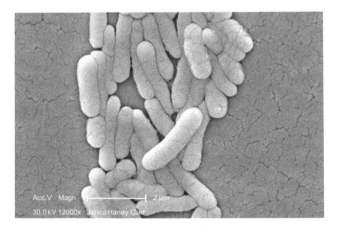

FIGURE 16.11 A scanning electron micrograph of *Salmonella typhimurium* bacteria grown in a pure culture and magnified ×12,000. (Courtesy of Centers for Disease Control and Prevention, image #10970.)

concern with rising global temperatures. More information is also needed concerning the transport of microbial pathogens that can end up on agricultural products, particularly produce, through water transmission (Rose et al., 2001).

16.3.3 AIR QUALITY

Air-quality issues related to human health include ground-level ozone, fine particulate matter, and long-range transport of air pollutants (Confalonieri et al., 2007). Changes in conditions such as atmospheric chemical reaction rates and boundary layer heights could lead to increased concentrations of tropospheric ozone, particularly in developed countries (English et al., 2009; Ebi and McGregor, 2008; Pyle et al., 2007). Tropospheric ozone can cause health problems including chest pain, coughing, throat irritation, congestion, worsening of airway problems such as bronchitis, emphysema, and asthma, and inflammation of the lining of the lungs (Baird and Cann, 2005). Warming temperatures will likely lead to more dust generation during agricultural operations, introducing increased levels of particulate matter, bacteria, fungal and bacterial spores, and agricultural chemicals into the atmosphere (Boxall et al., 2009). As global temperatures have risen, asthma incidence has increased as well (Casimiro et al., 2006; Beggs and Bambrick, 2005; Hales et al., 1998). There are concerns that increased temperatures are linked to increased pollen exposure, with pollen being an important trigger for some types of asthma (Figure 16.12) (Beggs and Bambrick, 2005; Hales et al., 1998).

16.3.4 VECTORBORNE AND OTHER DISEASES

Vectorborne diseases (VBDs) are spread by arthropods such as mosquitos, ticks, sandflies, and blackflies (Figure 16.13) and are among the best studied of the diseases associated with climate change because they are widespread and sensitive to climatic factors (Confalonieri et al., 2007). Studies carried out in many parts of the world have concluded that the risk of VBDs could increase given climate change and, in particular, global warming. The diseases most commonly cited as potentially increasing due to climate change are dengue fever and malaria (Figure 16.14) (Chan et al., 1999), but they are not the only diseases of concern. A study in Portugal determined that the risks of contracting West Nile virus, leishmaniasis, Lyme disease, and Mediterranean spotted fever could increase if Portugal warms as climate models predict (Figure 16.15) (Casimiro

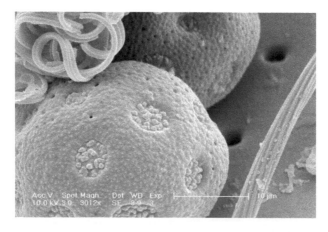

FIGURE 16.12   A scanning electron micrograph of particulate matter consisting primarily of pollen granules and disintegrated plant tissues magnified ×3012. (Courtesy of Centers for Disease Control and Prevention, image #7982.)

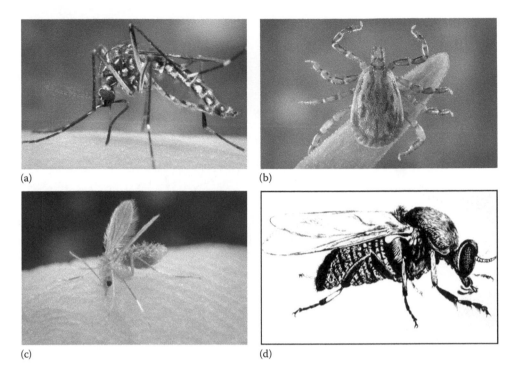

FIGURE 16.13   Some common disease vectors. (a) A female *Aedes aegypti* mosquito in the process of feeding, (b) dorsal view of a male yellow dog tick, *Amblyomma aureolatum*, (c) a *Phlebotomus papatasi* sandfly in the process of feeding, and (d) an illustration of a female *Simulium* sp., more commonly known as the blackfly. (Courtesy of Centers for Disease Control and Prevention, images #9261, #11070, #10275, and #4639.)

et al., 2006). In northern Canada, climate change is expected to increase the range and activity of vectors such as biting flies, which could introduce new VBDs (Furgal and Seguin, 2006). Warmer temperatures have already increased the northern range of ticks in Sweden (Lindgren et al., 2000). Pascual et al. (2006) note a correlation between increased mosquito-borne malaria and warming temperatures in the East African highlands. Bultó et al. (2006) concluded that more frequent epidemic outbreaks of dengue fever were likely in Cuba, given climate change projections, and in the United States it is expected that climate change could increase incidences of Lyme disease (Ebi et al., 2006).

However, each of these studies also indicated the need for additional study of climate change and VBDs. Climate change may or may not increase incidences of VBDs; there are many other factors that must be considered and we need a better understanding of the complex interactions taking place (English et al., 2009; Ebi et al., 2006; Pascual et al., 2006). Climate change may also influence the occurrence of other diseases such as coccidioidomycosis (valley fever), human hantavirus cardiopulmonary syndrome (English et al., 2009), and plague (Stenseth et al., 2006), but these studies are subject to the same types of uncertainties as the previously discussed VBDs (English et al., 2009).

16.3.5   OCCUPATIONAL HEALTH ISSUES

Occupational health deals with workplace conditions. The primary occupational health concern that relates to climate change is heat exhaustion or heat stress caused by increasing average global temperatures. While both indoor and outdoor workers are at risk, the greatest risk is taken by those

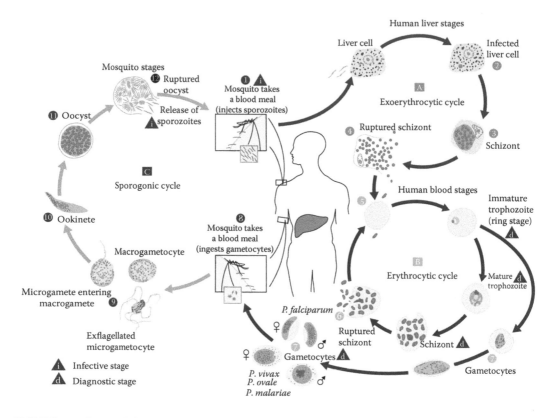

FIGURE 16.14   The life cycle of the parasites of the genus *Plasmodium* that are causal agents of malaria. (Courtesy of Centers for Disease Control and Prevention, image #3405.)

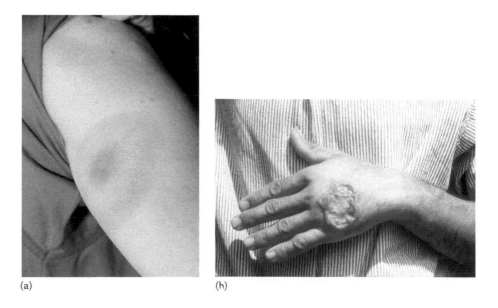

(a)   (h)

FIGURE 16.15   (a) The characteristic Lyme disease "bull's-eye" rash called *erythema migrans* and (b) a skin ulcer due to leishmaniasis. (Courtesy of Centers for Disease Control and Prevention, images #9873 and #352.)

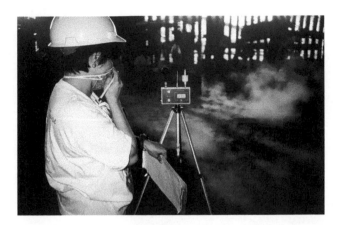

FIGURE 16.16   An inspector taking heat-stress measurements in an industrial setting. (Courtesy of Centers for Disease Control and Prevention, image #6800.)

who work outdoors or in otherwise non-air-conditioned work spaces (Figure 16.16) (Confalonieri et al., 2007).

16.3.6   Climate Change Impacts on Human Health: Summary

In each of the areas discussed in previous sections, soils themselves are not directly responsible for any adverse effects to human health that may be brought about by climate change. However, as potential sources or sinks of greenhouse gases to the atmosphere, soils are indirectly linked to such health impacts due to their influence on atmospheric greenhouse gas concentrations. Therefore, there is a link between soils and human health as it relates to temperature extremes, food safety, air quality, VBDs, and occupational health insofar as there are links between these areas and climate change. In short, the link is the impact of soils on global climate change.

16.4   SOIL HEALTH

We humans expect a lot from our soils. For example, soils grow much of our food and fiber needs, purify wastes for us, and provide aesthetic appeal through the growth of flower gardens and lush lawns (Figure 16.17). In order to perform all these various tasks, the soil itself needs to be healthy. A healthy soil is said to be one that has the capacity to function, within ecosystem and land use boundaries, to sustain biological productivity, maintain environmental quality, and promote plant and animal health (Doran et al., 1994). There are several physical, chemical, and biological properties a soil needs to have in order to be considered healthy. While the exact specifications differ from soil to soil depending on the environment the soil is in and the use it is being put to, a healthy soil does need adequate organic matter, good structure, and a diverse mixture of microorganisms and macroorganisms (Brevik, 2009a). This is an oversimplification of a fairly complex idea, but the overriding theme is the establishment of the soil as a medium that can do a good job of whatever function it is being put to. In natural systems soils often reach this point by achieving equilibrium with the surrounding environment; in a human-managed system it is up to the caretaker to provide the inputs needed to achieve a state of soil health (Wolf and Snyder, 2003). One of the major concerns surrounding climate change and soils is that climate change will alter the processes that take place in soils such that those soils will no longer be in a healthy condition to do things such as provide food and fiber for human needs (Lal, 2010).

FIGURE 16.17 (a) Soils are used to raise food such as these strawberries, (b) fiber such as cotton, and (c) ornamentals such as flower gardens. ([a, c] Photos by Lynn Betts, USDA NRCS and [b] photo by Gary Kramer, USDA NRCS.)

16.5 CLIMATE CHANGE AND SOIL PROCESSES

Climate change is expected to have several effects on the soil system. Changes in atmospheric concentrations of CO_2, temperature, and precipitation amounts and patterns will modify the soil–plant system and influence decomposition rates, which will have impacts on soil organic carbon levels. Organic carbon in turn has a significant influence on soil structure, soil fertility, microbial processes and populations in the soil, and other important soil properties. The challenge in figuring out how climate change will influence soil properties and processes is in working out the complex interactions that take place as conditions are changed in the atmosphere and soil.

16.5.1 Carbon

Early expectations were that increased atmospheric CO_2 would lead to increased plant productivity coupled with increased carbon sequestration by soil, meaning increased plant growth and the soil–plant system would help offset increasing atmospheric CO_2 levels (Hättenschwiler et al., 2002; Idso, 1998; Coughenour and Chen, 1997). This increase in plant growth is known as the CO_2 fertilization effect. However, recent studies indicate the CO_2 fertilization effect may not be as large as originally thought (Jarvis et al., 2010; Zaehle et al., 2010; Körner, 2006; Long et al., 2005; Poorter and Navas, 2003; Zavaleta et al., 2003). Increasing levels of ozone may actually counteract the CO_2 fertilization effect, leading to reduced plant growth under elevated CO_2 (Long et al., 2005), and the negative effects of increased temperatures on plant growth may also cancel out any CO_2 fertilization effect that does take place (Jarvis et al., 2010). A long-term experiment in a forest ecosystem indicated that nitrogen was transferred from the soil to plants at an accelerated rate under increased atmospheric CO_2 (Lichter et al., 2008), a condition that eventually could lead to nitrogen limitation negatively affecting plant growth (Hungate et al., 2003), and modeling of carbon dynamics as influenced by nitrogen indicates less carbon sequestration by soil than originally expected given CO_2 fertilization (Zaehle et al., 2010). Another long-term elevated CO_2 experiment in a grasslands ecosystem indicated that nitrogen and phosphorus became limiting after a time, again limiting plant biomass response to elevated CO_2 (Niklaus and Körner, 2004). Niklaus and Körner (2004) concluded that the increases in plant productivity they did see were due primarily to soil moisture status as opposed to a CO_2 fertilization effect. Experiments looking at the decomposition of plant tissues grown under elevated atmospheric CO_2 also indicate that increased levels of CO_2 are emitted during that decomposition (Kirkham, 2011). Therefore, elevated CO_2 levels will not necessarily lead to increased soil carbon sequestration, but will likely result in more carbon turnover (Eglin et al., 2011). This conclusion is supported by research done by Carney et al. (2007), who observed soil organic carbon levels declining due to increased microbial activity.

Increased temperature is likely to have a negative effect on carbon allocation to the soil, leading to reductions in soil organic carbon and creating a positive feedback in the global carbon cycle as global temperatures rise (Wan et al., 2011; Gorissen et al., 2004; Kirschbaum, 1995). Link et al. (2003) observed that soil warming and drying led to a 32% reduction in soil carbon over a 5-year time period. Modeling of carbon responses to climate change in Canada predicted small increases in aboveground biomass in forest and tundra ecosystems but larger decreases in soil and litter pools, for an overall increase in atmospheric carbon (Price et al., 1999). Another modeling study predicted decreases in soil organic carbon of 2.0%–11.5% in the north central United States (Grace et al., 2006). Niklińska et al. (1999) measured humus respiration rates under increased temperatures in samples from European Scots pine stands and concluded that the ecosystems studied would switch from net sinks to net sources of atmospheric carbon with global warming.

What this all means from a soils perspective is that soils cannot necessarily be expected to become massive carbon sinks as atmospheric CO_2 levels rise. The actual impact of elevated atmospheric CO_2 on carbon storage in soils is very difficult to predict. However, if the results of the aforementioned studies are representative of what does occur, soils may actually lose organic matter as atmospheric CO_2 levels and global temperatures increase, creating a positive feedback system that could push temperatures even higher. If too much organic matter is lost, that will also have negative impacts on soil physical, chemical, and biological properties, which will ultimately lead to negative impacts on crop production (Brevik, 2009a; Wolf and Snyder, 2003) and thus food security and human health.

16.5.2 Nitrogen

When CO_2 enrichment increases the soil C:N ratio, decomposing organisms in the soil need more nitrogen, which can reduce nitrogen mineralization (Reich et al., 2006a; Hungate et al., 2003; Gill et al., 2002). Mineralization is an essential step in supplying nitrogen to plants (Mullen, 2011; Reich et al., 2006b; Havlin et al., 2005; Pierzynski et al., 2009). Therefore, if nitrogen mineralization is reduced, it would be expected that plant-available nitrogen levels in the soil would also be reduced and plant productivity would be negatively affected. Holland (2011) reports that nitrogen limitation of CO_2-fertilized plants is consistent with the results reported by Hungate et al. (2003), but that increased temperatures stimulate nitrogen availability in the soil, enhancing terrestrial carbon uptake relative to the results of Hungate et al. (2003). However, the stimulated carbon uptake is not enough to offset the nitrogen limitation, and the net result is still an increase in atmospheric CO_2 and an overall reduction in soil carbon levels (Holland, 2011).

It should be noted that nitrogen supplements, that is, fertilizers, can alter these results (Reich et al., 2006a). Fertilization occurs much more often on agricultural soils than on forest or grassland soils. However, Mulvaney et al. (2009) have reported that adding synthetic nitrogen fertilizers in excess of crop needs has the long-term effect of decreasing both soil organic carbon and total soil nitrogen, negatively impacting soil productivity and agronomic efficiency. Therefore, nitrogen fertilization needs to be used carefully.

Some researchers have reported that increasing temperatures increase nitrogen mineralization (Reich et al., 2006b; Joshi et al., 2005; Norby and Luo, 2004), which could have a positive effect on plant growth. However, a warming study by An et al. (2005) showed that nitrogen mineralization was stimulated in the first year but depressed afterward.

Szukics et al. (2010) studied the effects of increasing temperature (5°C–25°C) and soil water (30%, 55%, and 70% water-filled pore space) on the activity of microorganisms responsible for nitrification and denitrification. Nitrification is the process by which soil organisms convert NH_4^+ to NO_3^-; denitrification is the process of converting NO_3^- into nitrogen gases, including N_2O. They found that increasing soil temperature from 5°C to 25°C induced a rapid stimulation of nitrogen cycling rates. The nitrification rate and NO_3^- concentration increased most rapidly at the 55% water content. In the 70% water content soils, the NO_3^- pool was increasingly depleted as soil temperature increased and was almost completely depleted at 25°C. The depletion in hot, wet soils was attributed to complete denitrification and the release of nitrogen gases into the atmosphere. NO was the primary nitrogen gas released from the 30% water content soils and N_2O emissions were highest from the 55% water content soils. This research demonstrates that increased emissions of N gases into the atmosphere from soils are possible as global temperatures rise.

Symbiotic biological N_2 fixation often increases with elevated CO_2 levels but usually only in experiments where phosphorus, potassium, and/or other nonnitrogen nutrients were added (Reich et al., 2006b). In experiments where increases in N_2 fixation were observed without the addition of nonnitrogen nutrients, they tended to be short-term responses to elevated CO_2 levels that declined with time (Reich et al., 2006b). Free-living and associated N_2-fixing organisms appear to be unresponsive to elevated CO_2 levels in the limited long-term field experiments that have been conducted (Reich et al., 2006b). Therefore, it is unlikely that we can rely on increased rates of atmospheric nitrogen fixation to ensure that nitrogen does not become a limiting factor to carbon sequestration by soils in a warmer world.

The relationships between climate change and soil nitrogen, and how those relationships will affect carbon sequestration by soils, is one of the more controversial issues being addressed right now in the study of soil science and climate change, and more study is needed to resolve it (Holland, 2011; Reich et al., 2006b; Luo et al., 2004). Understanding these carbon–nitrogen interactions is critical in order to determine whether soil organic matter levels will increase or decline under elevated CO_2 levels.

16.5.3 SOIL WATER

Soil water content in soils of semiarid grassland systems is expected to be higher under elevated atmospheric CO_2, a condition attributed to reduced transpiration due to increased stomatal resistance

(Kirkham, 2011). An experiment in a desert scrub ecosystem did not find increased soil water content under elevated CO_2 levels (Nowak et al., 2004), presumably because the stomata of desert plants already act to reduce transpiration. However, another study in an irrigated desert agroecosystem showed a trend towards higher soil water contents under elevated CO_2 as compared to ambient CO_2 (Kirkham, 2011). In short, different parts of the world will be impacted differently in terms of soil water (Kang et al., 2009).

Doubling atmospheric CO_2 has been shown to reduce seasonal evapotranspiration by 8% in wheat and cotton and by 9% in soybeans grown under day/night temperatures of 28°C/18°C (Hatfield, 2011). However, the reduction in transpiration by soybeans was eliminated if the plants were grown under temperatures of 40°C/30°C (Hatfield, 2011). In a study on rice, doubling CO_2 decreased evapotranspiration by 15% at 26°C but increased evapotranspiration at 29.5°C (Hatfield, 2011). These results show that the reducing effect of CO_2 on evapotranspiration is temperature dependent, and if temperatures get high enough the benefits of elevated CO_2 are lost.

Elevated CO_2 levels increase the water use efficiency and decrease evapotranspiration rates of many plants, potentially leaving them better able to survive climate change. This could influence crop production in the future. However, evapotranspiration rates appear to be temperature dependent, meaning the water benefits of increased atmospheric CO_2 could be reduced or lost in areas where temperatures rise too high.

16.5.4 Erosion

Through climate change and anthropogenic activities, many of our world's soils have or will become more susceptible to erosion by wind and/or water (Sivakumar, 2011; Ravi et al., 2010; Zhang et al., 2004). In Australia, simulations showed that increased rainfall due to climate change could lead to significant increases in runoff, with amplification greater in arid areas (up to five times more runoff than the percentage increase in rainfall) than in wet and temperate areas (twice as much runoff as the percentage change in rainfall) (Chiew et al., 1995). Barring adequate increased vegetative cover or enhanced soil infiltration rates, greater runoff would be expected to cause increased erosion (Zhang et al., 2004; Favis-Mortlock and Boardman, 1995). Water erosion models in the United Kingdom predicted that a 10% increase in winter rainfall could increase annual soil erosion by as much as 150% during wet years, but that long-term averages of soil erosion would show a modest increase over current conditions (Favis-Mortlock and Boardman, 1995). Li et al. (2011) predicted changes in water erosion of –5% to 195% for conventional tillage and 26%–77% for conservation tillage in China's Loess Plateau region, while Zhang et al. (2004) predicted increased erosion in Oklahoma of 19% and 40% under conservation tillage and conventional tillage, respectively.

The negative effects of soil erosion on crop yields and food production are well established (Figure 16.18) (Bakker et al., 2007; Montgomery, 2007; Pimentel, 2006; Sparovek and Schnug, 2001; Poudel et al., 1999; Evans, 1997; Tengberg et al., 1997). During their study of a semiarid Mediterranean ecosystem in Spain, García-Fayos and Bochet (2009) found strong correlations between the interaction between climate change and soil erosion and negative impacts on aggregate stability, bulk density, water-holding capacity, pH, organic matter content, total nitrogen content, and soluble phosphorus content of the soil. All of the soil properties listed are important from a soil health perspective (Brevik, 2009a; Magdoff and van Es, 2009; Wolf and Snyder, 2003). Therefore, it can be stated that if climate change increases soil erosion, it will also damage soil properties that are important in the production of food and fiber resources needed by humans.

16.5.5 Soil Organisms

Soil organisms perform many functions essential to the creation of healthy soil (Magdoff and van Es, 2009; Wolf and Snyder, 2003). Some soil organisms break down fresh organic matter added to the soil, releasing nutrients that can be used by plants and cycling nutrients through the soil system.

FIGURE 16.18 Erosion of topsoil from this hilltop has led to reduced crop production in the eroded areas. (Photo by Gene Alexander, USDA NRCS.)

Other soil organisms fix atmospheric nitrogen, making it available to plants. During organic matter decomposition soil organisms create organic "glues" that help arrange individual sand, silt, and clay particles in the soil into groupings called peds (Figure 16.19). Soils with well-formed peds have good structure, or aggregation. Spaces (pores) between the peds serve as pathways for the movement of water, air, and roots through the soil; pores within the peds act to store water in between rain events. The role of soil organisms in decomposing organic matter means they are an integral part of the global carbon and nitrogen cycles, which influence the abundance of greenhouse gases in the atmosphere.

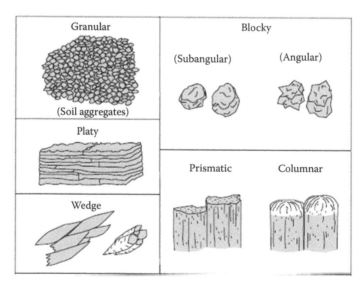

FIGURE 16.19 Common ped shapes. (Modified from Schoeneberger, P.J., Wysocki, D.A., Benham, E.C., and Broderson, W.D. (eds), *Field Book for Describing and Sampling Soils, Version 2.0,* Natural Resources Conservation Service, National Soil Survey Center, Lincoln, NE, 2002.)

The effect of climate change on soil organisms is not easy to determine. Soil organisms respond to a wide array of soil conditions including temperature, water content, pH, nutrient levels, oxygen status, and the presence or absence of other soil organisms (Brady and Weil, 2008). It is very difficult if not impossible to predict how all these variables will change at any single location given changes in global climate. Therefore, studies in this area tend to look at how changes in one or two variables, temperature and/or rainfall, for example, would influence soil organisms at a given location. The results of some of those studies are summarized here.

Briones et al. (1997) conducted an experiment that took intact soil cores and raised the average annual temperature of those cores by 2.5°C. They tracked Enchytraeidae, Tardigrada, and Diptera responses and determined that some species were tolerant to the new, higher temperatures and would increase their numbers, some species would migrate to deeper layers of the soil, and some would go dormant or extinct. Briones et al. (1997) concluded that predicted global temperature changes would have significant effects on the soil ecosystem that would have important implications for organic matter decomposition and nutrient cycling.

Lipson et al. (2006) studied the effects of three elevated CO_2 levels (400, 800, and 1200 ppm CO_2) on microbial biomass, activity, and community structure. The experiments were carried out in the artificial ecosystem of the Biosphere 2 Laboratory in Arizona, United States, which allowed good control of experimental conditions. The plots on the south end of the study received more light than other parts of the plots. Chao 1 estimates of microbial diversity indicated the study soils had a similar number of unique ribotypes to agricultural soils. The south plots had lower Chao 1 estimates than the other plots, indicating a possible interaction between CO_2 and light in these plots favored some species over others at the expense of species diversity. Total heterotrophic microbial biomass and activity increased with increasing CO_2 levels. Several other tests run during the experiment did not produce consistent results. Lipson et al. (2006) concluded that a basic understanding of the soil microbial community is necessary to predict how that community might react to global change.

Kardol et al. (2011) looked at the influence of CO_2 level, temperature, and precipitation on microarthropods. They found that the community composition shifted in response to the treatments, with most of the composition shift attributed to the effects of precipitation and temperature and how those two variables affected soil water content. They concluded that climate change can affect the structure of soil microarthropod communities, which could in turn impact ecosystem functions such as soil organic matter decomposition.

Plant roots are important contributors to the soil ecosystem. Roots secrete a number of organic compounds into the soil around them in a zone called the rhizosphere (Figure 16.20). Microbial numbers in the rhizosphere are typically 2–10 times greater than microbial numbers in the bulk soil because many of the root organic secretions provide an inviting environment for soil microbes (Brady and Weil, 2008). Drennan and Nobel (1996) carried out a study to look at the effects of elevated CO_2 concentrations and temperature on the root systems of three desert plants, *Encelia farinosa* (a C_3 plant), *Pleuraphis rigida* (C_4), and *Agava deserti* (CAM). They found that *A. deserti* increased its average daily root elongation under elevated CO_2, but there was no root elongation effect on the other two species. They also found that shading of the soil reduced daily variations in soil temperature and altered root distribution and elongation patterns.

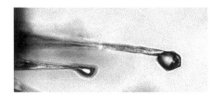

FIGURE 16.20 Sorghum root hairs with droplets of oily organic secretions exuding from the tips. (Image by Franck Dayan, USDA ARS.)

While this brief discussion of climate change effects on soil organisms does not definitively answer how soil organisms will change in response to climate change, it does provide valuable information. It tells us that soil ecosystems will change as a result of climate change and it tells us that some very important processes involving soil organisms, like organic matter decomposition and nutrient cycling, will also likely change. This conclusion differs substantially from the assumption by some early in the study of soil organisms and climate change that predicted temperature changes would have little effect on the soil ecosystem except in response to shifts in vegetation (Whitford, 1992).

16.5.6　Summary

There is much more to learn about climate change and its effects on soil processes and properties. However, there are some general trends we can look at based on the studies that have been completed. Soil processes, and therefore soil properties, will change as the climate changes, with the exact direction and magnitude of those changes being dependent on the amount of change in atmospheric CO_2, temperature, and precipitation amounts and patterns. The response of plants to climate change will also be critical, as plant residues are a major contributor to soil organic matter. Recent studies give reason to believe soils may become net sources of atmospheric carbon as temperatures rise; this is particularly true of high-latitude regions with permanently frozen soils. Soil erosion by both wind and water is also likely to increase. All of these changes have the potential to cause reductions in crop production.

There are still many things we need to know more about. Much research is needed to better understand soil water–CO_2 level–temperature relationships, and how climate change will affect the nitrogen cycle and, in turn, how the nitrogen cycle will affect carbon sequestration in soils. There is a critical need for deeper knowledge of the response of plants to elevated atmospheric CO_2, given limitations in nutrients such as nitrogen and phosphorus, and how that affects soil organic matter dynamics. There is also a great need for a better understanding of how soil organisms will respond to climate change because those organisms are incredibly important in a number of soil processes, including the carbon and nitrogen cycles.

16.6　SOILS, CLIMATE CHANGE, AND HUMAN HEALTH

Soils have an indirect influence on human health as it relates to climate change through their place in the carbon and nitrogen cycles, as discussed earlier in this chapter. There are also more direct influences that soils can have on human health. Through nutrient cycling, soils supply many of the nutrients essential to healthy plant growth (Brevik, 2009a; Magdoff and van Es, 2009; Wolf and Snyder, 2003). These same nutrients are important to healthy human growth as we obtain those soil nutrients through consumption of plants grown in the soil or meat products produced from animals that consumed the plant material (Brevik, 2009b; Combs, 2005). Human pathogens are found in the soil, as are a number of organic chemicals (Boxall et al., 2009; Brevik, 2009b; Bultman et al., 2005), all of which will be influenced by climate change (Boxall et al., 2009). Airborne dust is another serious health implication associated with soils (Brevik, 2009b; Derbyshire, 2005; Leikauf, 2002) and increased wind erosion is a possible result of climate change, particularly in semiarid and arid environments (Ravi et al., 2010).

16.6.1　Food Security

"Food security [is] a situation that exists when all people, at all times, have physical, social and economic access to sufficient, safe and nutritious food that meets their dietary needs and food preferences for an active and healthy life" (FAO, 2003). Over 97.5% of human food needs, as measured by calories consumed, come from the land while less than 2.5% comes from aquatic systems

(Table 16.2). This means soils are critical to food security. Climate change has already caused and will continue to cause changes in global temperature and precipitation patterns (Meehl et al., 2007; Trenberth et al., 2007) as well as changes in soil processes and properties, as previously discussed. This has led to considerable concern that climate change could compromise food security (Sauer and Nelson, 2011; Lal, 2010; Kang et al., 2009; Park et al., 2009; Funk et al., 2008; Paeth et al., 2008; Schmidhuber and Tubiello, 2007; Fischer et al., 2005; Gregory et al., 2005; Long et al., 2005; Parry et al., 2005; Rosegrant and Cline, 2003), which would lead to an overall decline in human health.

There were at least 1 billion people living in a state of food insecurity in 2010; eliminating this insecurity and feeding the additional 2.3 billion people expected by 2050 will require global cereal production, for example, to be increased by 70% (Lele, 2010; St. Clair and Lynch, 2010). The IPCC expects that climate change will impact all four dimensions of food security: (1) food availability, (2) stability of food supplies, (3) access to food, and (4) food utilization (Easterling et al., 2007). Global food production is projected to increase with a rise in local average temperature over a range of 1°C–3°C but to decrease if the temperature surge is beyond 3°C (Easterling et al., 2007). The distribution of this increased food production is not uniform, however. As a general trend, food production is expected to increase at mid- to high latitudes and to decrease near the equator (Figures 16.21 and 16.22) (Parry et al., 2005; Olesen and Bindi, 2002; Rosenzweig and Parry, 1994).

TABLE 16.2
Daily Per Capita Food Intake as a Worldwide Average, 2001–2003

Food Source	Calories[a]	Percent of Calories
Rice	557	25.5
Wheat	521	23.9
Maize	147	6.7
Sorghum	33	1.5
Potatoes	60	2.7
Cassava	42	1.9
Sugar	202	9.3
Soybean oil	87	4.0
Palm oil	50	2.3
Milk	122	5.6
Animal fats (raw and butter)	62	2.8
Eggs	33	1.5
Meat (pig)	117	5.4
Meat (poultry)	46	2.1
Meat (bovine)	40	1.8
Meat (sheep and goats)	11	0.5
Fish and other aquatic products[b]	52	2.4
Total	2182	

Source: Brevik, E.C., *Soils, Plant Growth and Crop Production*, Encyclopedia of Life Support Systems (EOLSS), Developed under the Auspices of the UNESCO, EOLSS Publishers, Oxford, 2009. http://www.eolss.net (verified 2 January 2012); based on information from http://faostat.fao.org/Portals/_Faostat/documents/pdf/world.pdf and http://www.fao.org/fishery/statistics/global-consumption/en.

[a] Aquatic products data from 2003. All other data from 2001–2003.

[b] Includes both marine and freshwater products.

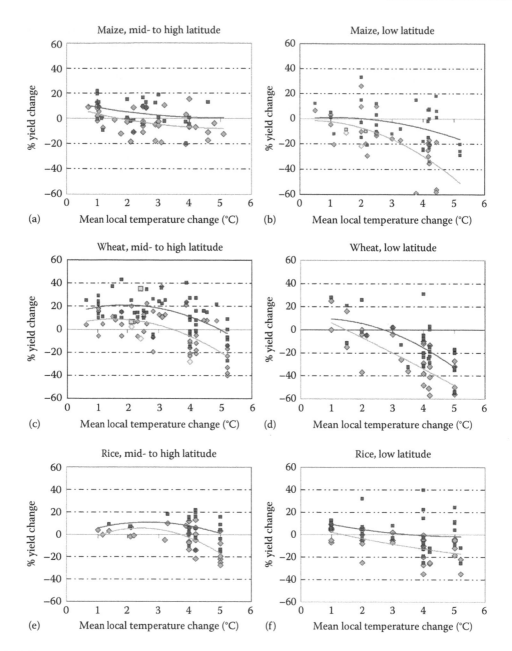

FIGURE 16.21 (a–f) Expected sensitivity of cereal yields to climate change. The upper trend lines are for yields with adaptations to address climate change; the lower trend lines are for yields without adaptations to address climate change. The boxes are individual data points for yields with adaptations, the diamonds for yields without adaptations. (From Easterling, W.E., Aggarwal, P.K., Batima, P., Brander, K.M., Erda, L., Howden, S.M., Kirilenko, A., et al., *Climate Change 2007: Impacts, Adaptation and Vulnerability*, Contribution of Working Group II to the Fourth Assessment Report of the Intergovernmental Panel on Climate Change, Cambridge University Press, Cambridge, 2007, Figure 5.2.)

Low soil fertility is currently a food security problem in many developing countries, particularly in Africa and South Asia (St. Clair and Lynch, 2010; Sanchez and Swaminathan, 2005; Lal, 2004). Africa and South Asia are also among the regions most at risk of food insecurity (Lele, 2010; Huntingford et al., 2005; Sanchez and Swaminathan, 2005) and deteriorating soil health due to climate change (Tan et al., 2010). Proper soil management has the potential to drastically reduce food

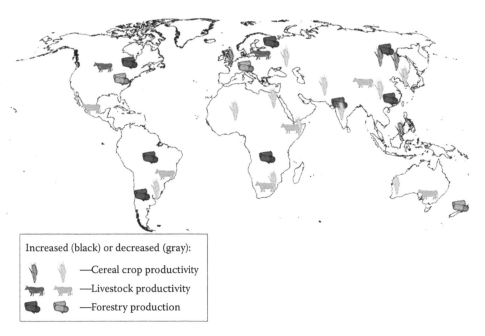

FIGURE 16.22 Expected major impacts of climate change on crop and livestock yields and forestry production by 2050, shown by region. (From Easterling, W.E., Aggarwal, P.K., Batima, P., Brander, K.M., Erda, L., Howden, S.M., Kirilenko, A., et al., *Climate Change 2007: Impacts, Adaptation and Vulnerability*, Contribution of Working Group II to the Fourth Assessment Report of the Intergovernmental Panel on Climate Change, Cambridge University Press, Cambridge, 2007, Figure 5.4.)

security issues in these regions. For example, maize and bean yields in Kenya were increased four times when crop residues were left on the fields and fertilizer and manure were applied, increasing soil organic carbon (Lal, 2004); and Tan et al. (2010) concluded that soil organic carbon losses in West Africa could be stopped with the addition of 30–60 kg of nitrogen fertilizer per hectare. Increases in yields have also been realized in the United States, Argentina, Thailand, and Nigeria when carbon was sequestered in the soil due to improvements in soil health (Lal, 2004). Increasing soil organic carbon in degraded soils would be a major step forward in enhancing food security in developing countries (Lal, 2010; St. Clair and Lynch, 2010). However, if carbon sequestration practices are started in these countries, accelerated decomposition of soil organic matter due to global warming could potentially thwart these efforts and have a negative impact on food security. If carbon sequestration practices are not started, these already degraded soils will be even more vulnerable to the effects of climate change than relatively healthy soils in developed countries (Tan et al., 2010). Therefore, climate change has the potential to exacerbate food security issues through its potential effects on soil health.

Nutrients are passed from soil to plants to humans or, in the case of livestock, from soil to plants to livestock to humans (Tables 16.3 and 16.4) (Brevik, 2009b). However, if the nutrient is not present in the soil, or if it is not plant-available due to being tied up in the soil or antagonistic affects from other ions, plants cannot access the nutrient and pass it up the food chain (Brevik, 2009b). Unhealthy soils tend to have a lower overall nutrient status. Low-nutrient status in agricultural soils not only reduces the amount of food available for human consumption, it also makes those who rely on the low-nutrient soils more susceptible to disease (Sanchez and Swaminathan, 2005). If problems from VBDs, for example, become more pronounced with changes in climate, low-nutrient status soils will simply make those who rely on them even more prone to problems (Pimentel et al., 2007). In this case, soils could have both direct and indirect effects on human health due to climate change.

TABLE 16.3
Some Important Plant-Tissue Sources of Elements Essential to Human Life

Element	Important Sources
Ca	Kale, collards, mustard greens, broccoli
Cu	Beans, peas, lentils, whole grains, nuts, peanuts, mushrooms, chocolate
I	Vegetables, cereals, fruit
K	Fruits, cereals, vegetables, beans, peas, lentils
Mg	Seeds, nuts, beans, peas, lentils, whole grains, dark green vegetables
Mn	Whole grains, beans, peas, lentils, nuts, tea
Mo	Beans, peas, lentils, dark green leafy vegetables
P	Nuts, beans, peas, lentils, grains
Se	Grain products, nuts, garlic, broccoli (if grown on high-Se soils)
Zn	Nuts, whole grains, beans, peas, lentils

Source: Brevik, E.C., *Soils, Plant Growth and Crop Production*, Encyclopedia of Life Support Systems (EOLSS), Developed under the Auspices of the UNESCO, EOLSS Publishers, Oxford, 2009. http://www.eolss.net (verified 2 January 2012).

Toxicity issues can also be passed on from soil to humans through the food we consume. Li et al. (2010) found that elevated atmospheric CO_2 concentrations may lead to increased uptake of Cd in rice. Other studies have indicated that elevated CO_2 may increase uptake of Cu and Cs by some crops (Wu et al., 2009; Tang et al., 2003). In contrast, several researchers have reported reductions in Cu uptake under elevated atmospheric CO_2 and other research has indicated no significant variation in the uptake of various metals under elevated CO_2 (Li et al., 2010). Whether or not elevated CO_2 causes increased heavy metal uptake, which metals are or are not a problem, and which crop

TABLE 16.4
Some Important Animal-Product Sources of Elements Essential to Human Life

Element	Important Sources
Ca	Dairy products
Cl	Dairy products, meats, eggs
Cu	Organ meats
Fe	Meats, especially red meat
K	Dairy products, meats
Mo	Organ meats
Na	Dairy products, meats, eggs
P	Meats, eggs, dairy products
Se	Meats from Se-fed livestock
Zn	Meats, organ meats

Source: Brevik, E.C., *Soils, Plant Growth and Crop Production*, Encyclopedia of Life Support Systems (EOLSS), Developed under the Auspices of the UNESCO, EOLSS Publishers, Oxford, 2009. http://www.eolss.net (verified 2 January 2012).

varieties have or do not have such metal uptake problem are key questions that need to be addressed regarding climate change and food security.

16.6.2 AIRBORNE DUST

Soils are a major source of atmospheric dust (Sing and Sing, 2010; Brevik, 2009b; Washington et al., 2003). Evidence from Antarctic ice cores indicates that dust levels in the atmosphere increased during the twentieth century (McConnell et al., 2007), likely due to overgrazing, deforestation/desertification, and global warming (Sing and Sing, 2010; McConnell et al., 2007). Ravi et al. (2010) predict that climate change could lead to increased soil erosion by aeolian processes in dry environments, and Boxall et al. (2009) state that changes in temperature and precipitation could lead to increased aerial inputs of dust. The main factor influencing dust transport is climate (Garrison et al., 2003). Prosperol and Lamb (2003) have demonstrated an inverse relationship between dust generation off the Sahara and precipitation, while Meehl et al. (2007) state that rainfall is expected to decrease by 6%–10% over the 30°N and 30°S latitude regions, an area that includes the Sahara. Given that the Sahara is one of the dominant sources of dust globally (Sing and Sing, 2010; Washington et al., 2003), there could be significant increases in atmospheric dust load. If atmospheric dust levels increase, this could have a negative impact on human health. Mineral dust exposure can cause chronic bronchitis, pulmonary fibrosis, COPD, and interstitial fibrosis depending on the exposure levels (Schenker, 2000), as well as silicosis and pulmonary tuberculosis (Sing and Sing, 2010). Dust can also lead to chemical and pathogen exposures that can cause negative health effects (Boxall et al., 2009; Brevik, 2009b). Viable bacterial and fungal spores have been transported great distances by dust, such as from the Sahara/Sahel to the U.S. Virgin Islands and Mali and from the Gobi/Taklamakan to Korea (Sing and Sing, 2010).

16.6.3 SOILS, CLIMATE CHANGE, AND HUMAN HEALTH: SUMMARY

Very little work has been done on the direct effects of climate change on soils and human health. The majority of such work has been done on crop yields under changing climates, but even that work tends to focus more on the crop–atmosphere interaction than on a crop–soil–atmosphere approach. The health of the soils used to produce food products will make a big difference in human health as we go forward. The important thing for us to discern is how we can keep our soils healthy even as climate change is altering the systems that determine processes and process rates in our soils. More work has been done on the indirect effects soils and climate change have on human health, although it is not, and rightfully so, typically presented as a situation where soils have an impact on human health. That being said, it is important to recognize that soils influence climate, which in turn influences human health in a number of ways.

16.7 CONCLUSIONS

Earth's climate system is changing; of that we are certain. We can also say with a high level of confidence that anthropogenic activities are significantly contributing to those changes. Beyond those two statements, most of what is covered in this chapter is less certain. In discussions of how sinks and sources of greenhouse gases will behave as atmospheric levels of those gases increase, the impact of climate change on human health, or the effects of climate change on soil processes, more information is needed. All of these questions involve highly complex and interconnected systems that make it difficult to isolate a single variable, such as temperature or precipitation patterns, and reach meaningful conclusions about how a change in that single variable affects the system being studied. However, we do know that there is the potential for some highly undesirable things to occur as a result of climate change. There is the possibility that soils could contribute increasing amounts of greenhouse gases to the atmosphere, losing their ability to act as a sink for carbon as global temperatures increase. There is the chance of increased risk of some diseases and a chance that we will see negative impacts on the physical and chemical properties of our soils that are essential for the production of food and fiber products. Therefore, it is critical that we support continued research into these areas, with the particular goal of understanding the complex interactions that take place in the natural environment.

It is also important to note that there is an imbalance in terms of who will suffer most from climate change. In all cases, it is the inhabitants of developing countries who are least prepared to cope with a changing climate and with soil degradation due to climate change; developed countries are better equipped to mitigate those changes and cope than are developing countries.

REFERENCES

Aguado, E. and J.E. Burt. 2007. *Understanding Weather and Climate*. Pearson-Prentice Hall, Upper Saddle River, NJ.

Akala, V.A. and R. Lal. 2001. Soil organic carbon pools and sequestration rates in reclaimed minesoils in Ohio. *Journal of Environmental Quality* 30: 2098–2104.

Álvaro-Fuentes, J. and K. Paustian. 2011. Potential soil carbon sequestration in a semiarid Mediterranean agroecosystem under climate change: Quantifying management and climate effects. *Plant and Soil* 338: 261–272.

An, Y., S. Wan, X. Zhou, A.A. Subedar, L.A. Wallace, and Y. Luo. 2005. Plant nitrogen concentration, use efficiency, and contents in a tallgrass prairie ecosystem under experimental warming. *Global Change Biology* 11: 1733–1744.

Backlund, P., A. Janetos, D.S. Schimel, J. Hatfield, M.G. Ryan, S.R. Archer, and D. Lettenmaier. 2008. Executive summary. *In The Effects of Climate Change on Agriculture, Land Resources, Water Resources, and Biodiversity in the United States*, pp. 1–10. A Report by the U.S. Climate Change Science Program and the Subcommittee on Global Change Research, Washington, DC.

Baird, C. and M. Cann. 2005. *Environmental Chemistry*, 3rd edn. W.H. Freeman and Company, New York, NY.

Bakker, M.M., G. Govers, R.A. Jones, and M.D.A. Rounsevell. 2007. The effect of soil erosion on Europe's crop yields. *Ecosystems* 10(7): 1209–1219.

Barber, D.G., J.V. Lukovich, J. Keogak, S. Baryluk, L. Fortier, and G.H.R. Henry. 2008. The changing climate of the Arctic. *Arctic* 61(1): 7–26.

Bartlett, K.H. and R.C. Harris. 1993. Review and assessment of methane emissions from wetlands. *Chemosphere* 26: 261–320.

Beggs, P.J. and H.J. Bambrick. 2005. Is the global rise of asthma an early impact of anthropogenic climate change? *Environmental Health Perspectives* 113(8): 915–919.

Bliss, N.B. and J. Maursetter. 2010. Soil organic carbon stocks in Alaska estimated with spatial and pedon data. *Soil Science Society of America Journal* 74: 565–579.

Bluemle, J.P., J.M. Sabel, and W. Karlén. 1999. Rate and magnitude of past global climate changes. *Environmental Geosciences* 6: 63–75.

Blum, W.E.H. and H. Eswaran. 2004. Soils for sustaining global food production. *Journal of Food Science* 69(2): 37–42.

Bouwman, A.F. 1991. Agronomic aspects of wetland rice cultivation and associated methane emissions. *Biogeochemistry* 15(2): 65–88.

Boxall, A.B.A., A. Hardy, S. Beulke, T. Boucard, L. Burgin, P.D. Falloon, P.M. Haygarth, et al. 2009. Impacts of climate change on indirect human exposure to pathogens and chemicals from agriculture. *Environmental Health Perspectives* 117(4): 508–514.

Brady, N.C. and R.R. Weil. 2008. *The Nature and Properties of Soils*, 14th edn. Pearson Prentice Hall, Upper Saddle River, NJ.

Brevik, E.C. 2000. A comparison of soil properties in compacted versus non-compacted Bryant soil series twenty-five years after compaction ceased. *Soil Survey Horizons* 41(2): 52–58.

Brevik, E.C. 2009a. Soil health and productivity. *In* W. Verheye (ed.), *Soils, Plant Growth and Crop Production*. Encyclopedia of Life Support Systems (EOLSS), Developed under the Auspices of the UNESCO, EOLSS Publishers, Oxford. http://www.eolss.net (verified 2 January 2012).

Brevik, E.C. 2009b. Soil, food security, and human health. *In* W. Verheye (ed.), *Soils, Plant Growth and Crop Production*. Encyclopedia of Life Support Systems (EOLSS), Developed under the Auspices of the UNESCO, EOLSS Publishers, Oxford. http://www.eolss.net (verified 2 January 2012).

Brevik, E.C. and J.A. Homburg. 2004. A 5000 year record of carbon sequestration from a coastal lagoon and wetland complex, Southern California, USA. *Catena* 57: 221–232.

Brevik, E.C., T. Fenton, and L. Moran. 2002. Effect of soil compaction on organic carbon amounts and distribution, South-Central Iowa. *Environmental Pollution* 116: S137–S141.

Briones, M.J.I., P. Ineson, and T.G. Piearce. 1997. Effects of climate change on soil fauna; responses of enchytraeids, Diptera larvae, and tardigards in a transplant experiment. *Applied Soil Ecology* 6: 117–134.

Brooks, P.D., S.K. Schmidt, and M.W. Williams. 1997. Winter production of CO_2 and N_2O from alpine tundra: Environmental controls and relationship to inter-system C and N fluxes. *Oecologia* 110(3): 403–413.

Bultman, M.W., F.S. Fisher, and D. Pappagianis. 2005. The ecology of soil-borne human pathogens. *In* O. Selinus, B. Alloway, J.A. Centeno, R.B. Finkelman, R. Fuge, U. Lindh, and P. Smedley (eds), *Essentials of Medical Geology: Impacts of the Natural Environment on Human Health*, pp. 481–511. Elsevier Academic Press, Amsterdam, The Netherlands.

Bultó, P.L.O., A.P. Rodríguez, A.R. Valencia, N.L. Vega, M.D. Gonzalez, and A.P. Carrera. 2006. Assessment of human health vulnerability to climate variability and change in Cuba. *Environmental Health Perspectives* 114(12): 1942–1949.

Buol, S.W., R.J. Southard, R.C. Graham, and P.A. McDaniel. 2003. *Soil Genesis and Classification*, 5th edn. Iowa State University Press, Ames, IA.

Campbell-Lendrum, D. and R. Woodruff. 2006. Comparative risk assessment of the burden of disease from climate change. *Environmental Health Perspectives* 114(12): 1935–1941.

Carney, K.M., B.A. Hungate, B.G. Drake, and J.P. Megonigal. 2007. Altered soil microbial community at elevated CO_2 leads to loss of soil carbon. *Proceedings of the National Academy of Sciences of the United States of America* 104(12): 4990–4995.

Carter, R.M. 2007. The myth of dangerous human-caused climate change. *Proceedings of the Australasian Institute of Mining & Metallurgy New Leaders Conference*, Brisbane, 2–3 May 2007, pp. 61–74.

Casimiro, E., J. Calheiros, F.D. Santos, and S. Kovats. 2006. National assessment of human health effects of climate change in Portugal: Approach and key findings. *Environmental Health Perspectives* 114(12): 1950–1956.

Chadwick, O.A., E.F. Kelly, D.M. Merritts, and R.G. Amundson. 1994. Carbon dioxide consumption during soil development. *Biogeochemistry* 24(3): 115–127.

Chan, N.Y., K.L. Ebi, F. Smith, T.F. Wilson, and A.E. Smith. 1999. An integrated assessment framework for climate change and infectious diseases. *Environmental Health Perspectives* 107(5): 329–337.

Chapin III, F.S., G. Peterson, F. Berkes, T.V. Callaghan, P. Angelstam, M. Apps, C. Beier, et al. 2004. Resilience and vulnerability of northern regions to social and environmental change. *Ambio* 33(6): 344–349.

Chatskikh, D., J.E. Olesen, J. Berntsen, K. Regina, and S. Yamulki. 2005. Simulation of effects of soils, climate, and management on N_2O emission from grasslands. *Biogeochemistry* 76: 395–419.

Chiew, F.H.S., P.H. Whetton, T.A. McMahon, and A.B. Pittock. 1995. Simulation of the impacts of climate change on runoff and soil moisture in Australian catchments. *Journal of Hydrology* 167: 121–147.

Combs Jr, G.F. 2005. Geological impacts on nutrition. *In* O. Selinus, B. Alloway, J.A. Centeno, R.B. Finkelman, R. Fuge, U. Lindh, and P. Smedley (eds), *Essentials of Medical Geology: Impacts of the Natural Environment on Human Health*, pp. 161–177. Elsevier Academic Press, Amsterdam, The Netherlands.

Confalonieri, U., B. Menne, R. Akhtar, K.L. Ebi, M. Hauengue, R.S. Kovats, B. Revich and A. Woodward. 2007. Human health. *In* M.L. Parry, O.F. Canziani, J.P. Palutikof, P.J. van der Linden, and C.E. Hanson (eds), *Climate Change 2007: Impacts, Adaptation and Vulnerability*, pp. 391–431. Contribution of Working Group II to the Fourth Assessment Report of the Intergovernmental Panel on Climate Change, Cambridge University Press, Cambridge.

Cooney, C.M. 2010. The perception factor: Climate change gets personal. *Environmental Health Perspectives* 118(11): A484–A489.

Corfee-Morlot, J., M. Maslin, and J. Burgess. 2007. Global warming in the public sphere. *Philosophical Transactions of the Royal Society A* 365: 2741–2776.

Coughenour, M.B. and D.-X. Chen. 1997. Assessment of grassland ecosystem responses to atmospheric change using linked plant–soil process models. *Ecological Applications* 7(3): 802–827.

Cruz, R.V., H. Harasawa, M. Lal, S. Wu, Y. Anokhin, B. Punsalmaa, Y. Honda, M. Jafari, C. Li and N. Huu Ninh. 2007. Asia. *In* M.L. Parry, O.F. Canziani, J.P. Palutikof, P.J. van der Linden, and C.E. Hanson (eds), *Climate Change 2007: Impacts, Adaptation and Vulnerability*, pp. 469–506. Contribution of Working Group II to the Fourth Assessment Report of the Intergovernmental Panel on Climate Change, Cambridge University Press, Cambridge.

Davidson, E.A. and I.A. Janssens. 2006. Temperature sensitivity of soil carbon decomposition and feedbacks to climate change. *Nature* 440: 165–173.

Davis, J.C. and G.C. Bohling. 2001. The search for patterns in ice-core temperature curves. *In* L.C. Gerhard, W.E. Harrison, and B.M. Hanson (eds), *Geological Perspectives of Global Climate Change*, pp. 213–229. AAPG, Tulsa, OK.

DeLaune, R.D., W.H. Patrick, C.W. Lindau, and C.J Smith. 1990. Nitrous oxide and methane emission from Gulf Coast wetlands. *In* A.F. Bouwman (ed.), *Soils and the Greenhouse Effect*, pp. 497–502. Wiley, New York, NY.

DeNeve, S. and G. Hofman. 2000. Influence of soil compaction on carbon and nitrogen mineralization of soil organic matter and crop residues. *Biology and Fertility of Soils* 30: 544–549.

Denman, K.L., G. Brasseur, A. Chidthaisong, P. Ciais, P.M. Cox, R.E. Dickinson, D. Hauglustaine, et al. 2007. Couplings between changes in the climate system and biogeochemistry. *In* S. Solomon, D. Qin, M. Manning, Z. Chen, M. Marquis, K.B. Averyt, M. Tignor, and H.L. Miller (eds), *Climate Change 2007: The Physical Science Bbasis*, pp. 499–587. Contribution of Working Group I to the Fourth Assessment Report of the Intergovernmental Panel on Climate Change, Cambridge University Press, Cambridge.

Derbyshire, E. 2005. Natural aerosolic mineral dusts and human health. *In* O. Selinus, B. Alloway, J.A. Centeno, R.B. Finkelman, R. Fuge, U. Lindh, and P. Smedley (eds), *Essentials of Medical Geology: Impacts of the Natural Environment on Human Health*, pp. 459–480. Elsevier Academic Press, Amsterdam, The Netherlands.

Dixon-Coppage, T.L., G.L. Davis, T. Couch, E.C. Brevik, C.I. Barineau, and P.C. Vincent. 2005. A forty-year record of carbon sequestration in an abandoned borrow-pit, Lowndes County, GA. *Soil and Crop Science Society of Florida Proceedings* 64: 8–15.

Doran, J.W., D.C. Coleman, D.F. Bezdicek, and B.A. Stewart (eds). 1994. *Defining Soil Quality for a Sustainable Environment*. Special Publication #34, American Society of Agronomy, Madison, WI.

Drennan, P.M. and P.S. Nobel. 1996. Temperature influences on root growth for *Encelia farinosa* (Asteraceae), *Pleuraphis rigida* (Poaceae), and *Agave deserti* (Agavaceae) under current and doubled CO_2 concentrations. *American Journal of Botany* 83(2): 133–139.

D'Souza, R.M., N.G. Becker, G. Hall, and K.B.A. Moodie. 2004. Does ambient temperature affect foodborne disease? *Epidemiology* 15(1): 86–92.

Easterling, W.E., P.K. Aggarwal, P. Batima, K.M. Brander, L. Erda, S.M. Howden, A. Kirilenko, et al. 2007. Food, fibre and forest products. *In* M.L. Parry, O.F. Canziani, J.P. Palutikof, P.J. van der Linden, and C.E. Hanson (eds), *Climate Change 2007: Impacts, Adaptation and Vulnerability*, pp. 273–313. Contribution of Working Group II to the Fourth Assessment Report of the Intergovernmental Panel on Climate Change, Cambridge University Press, Cambridge.

Ebi, K.L. and G. McGregor. 2008. Climate change, tropospheric ozone and particulate matter, and health impacts. *Environmental Health Perspectives* 116(11): 1449–1455.

Ebi, K.L., D.M. Mills, J.B. Smith, and A. Grambsch. 2006. Climate change and human health impacts in the United States: An update on the results of the U.S. national assessment. *Environmental Health Perspectives* 114(9): 1318–1324.

Ebi, K.L., M. Helmer, and J. Vainio. 2008. The health impacts of climate change: Getting started on a new theme. *Prehospital and Disaster Medicine* 23(2): S60–S64.

Eglin, T., P. Ciasis, S.L. Piao, P. Barré, V. Belassen, P. Cadule, C. Chenu, T. Gasser, M. Reichstein, and P. Smith. 2011. Overview on response of global soil carbon pools to climate and land-use changes. *In* T.J. Sauer, J.M. Norman, and M.V.K. Sivakumar (eds), *Sustaining Soil Productivity in Response to Global Climate Change: Science, Policy, and Ethics*, pp. 183–199. Wiley, Oxford.

English, P.B., A.H. Sinclair, Z. Ross, H. Anderson, V. Boothe, C. Davis, K. Ebi, et al. 2009. Environmental health indicators of climate change for the United States: Findings from the state environmental health indicator collaborative. *Environmental Health Perspectives* 117(11): 1673–1681.

Eswaran, H., E. Van den Berg, P. Reich, and J. Kimble. 1995. Global soil carbon resources. *In* R. Lal, J.M. Kimble, R.F. Follett, and B.A. Stewart (eds), *Soil Processes and the Carbon Cycle*, pp. 27–43. CRC Press, Boca Raton, FL.

Evans, R. 1997. Accelerated soil erosion in Britain. *Geography* 82(2): 149–162.

FAO. 2003. *Trade Reforms and Food Security: Conceptualizing the Linkages*. Food and Agriculture Organization of the United Nations, Rome, Italy.

Farina, R., G. Seddaiu, R. Orsini, E. Steglich, P.P. Roggero, and R. Francaviglia. 2011. Soil carbon dynamics and crop productivity as influenced by climate change in a rainfed cereal system under contrasting tillage using EPIC. *Soil and Tillage Research* 112: 36–46.

Favis-Mortlock, D. and J. Boardman. 1995. Nonlinear responses of soil erosion to climate change: A modeling study on the UK South Downs. *Catena* 25: 365–387.

Fischer, G., M. Shah, F.N. Tubiello, and H. van Velhuizen. 2005. Socio-economic and climate change impacts on agriculture: An integrated assessment, 1990–2080. *Philosophical Transactions of the Royal Society B* 360(1463): 2067–2083.

Forster, P., V. Ramaswamy, P. Artaxo, T. Berntsen, R. Betts, D.W. Fahey, J. Haywood, et al. 2007. Changes in atmospheric constituents and in radiative forcing. *In* S. Solomon, D. Qin, M. Manning, Z. Chen, M. Marquis, K.B. Averyt, M. Tignor, and H.L. Miller (eds), *Climate Change 2007: The Physical Science Basis*, pp. 129–234. Contribution of Working Group I to the Fourth Assessment Report of the Intergovernmental Panel on Climate Change, Cambridge University Press, Cambridge.

Funk, C., M.D. Dettinger, J.C. Michaelsen, J.P. Verdin, M.E. Brown, M. Barlow, and A. Hoell. 2008. Warming of the Indian Ocean threatens eastern and southern African food security but could be mitigated by agricultural development. *Proceedings of the National Academy of Sciences of the United States of America* 105(32): 11081–11086.

Furgal, C. and J. Seguin. 2006. Climate change, health, and vulnerability in Canadian northern aboriginal communities. *Environmental Health Perspectives* 114(12): 1964–1970.

García-Fayos, P. and E. Bochet. 2009. Indication of antagonistic interaction between climate change and erosion on plant species richness and soil properties in semiarid Mediterranean ecosystems. *Global Change Biology* 15: 306–318.

Garrison, V.H., E.A. Shinn, W.T. Foreman, D.W. Griffin, C.W. Holmes, C.A. Kellogg, M.S. Majewski, L.L. Richardson, K.B. Ritchie, and G.W. Smith. 2003. African and Asian dust: From desert soils to coral reefs. *BioScience* 53(5): 469–480.

Gill, R.A., H.W. Polley, H.B. Johnson, L.J. Anderson, H. Maherali, and R.B. Jackson. 2002. Nonlinear grassland responses to past and future atmospheric CO_2. *Nature* 417: 279–282.

Gorissen, A., A. Tietema, N.N. Joosten, M. Estiarte, J. Peñuelas, A. Sowerby, B.A. Emmett, and C. Beier. 2004. Climate change affects carbon allocation to the soil in shrublands. *Ecosystems* 7: 650–661.

Grace, P.R., M. Colunga-Garcia, S.H. Gage, G.P. Robertson, and G.R. Safir. 2006. The potential impact of agricultural management and climate change on soil organic carbon of the north central region of the United States. *Ecosystems* 9: 816–827.

Grandy, A.S., T.D. Loecke, S. Parr, and G.P. Robertson. 2006. Long-term trends in nitrous oxide emissions, soil nitrogen, and crop yields of till and no-till cropping systems. *Journal of Environmental Quality* 35: 1487–1495.

Grant, R.F., E. Pattey, T.W. Goddard, L.M. Kryzanowski, and H. Puurveen. 2006. Modeling the effects of fertilizer application rate on nitrous oxide emissions. *Soil Science Society of America Journal* 70: 235–248.

Gregory, P.J., J.S.I. Ingram, and M. Brklacich. 2005. Climate change and food security. *Philosophical Transactions of the Royal Society B* 360(1463): 2139–2148.

Gribble, G.W. 2003. The diversity of naturally produced organohalogens. *Chemosphere* 52: 289–297.

Grossman, R.B., D.S. Harms, M.S. Kuzila, S.A. Glaum, S.L. Hartung, and J.R. Fortner. 1998. Organic carbon in deep alluvium in southeast Nebraska and northeast Kansas. *In* R. Lal, J.M. Kimble, R.F. Follett, and B.A. Stewart (eds), *Soil Processes and the Carbon Cycle*, pp. 45–55. CRC Press, Boca Raton, FL.

Hajat, S., R.S. Kovats, and K. Lachowycz. 2007. Heat-related and cold-related deaths in England and Wales: Who is at risk? *Occupational and Environmental Medicine* 64(2): 93–100.

Hales, S., S. Lewis, T. Slater, J. Crane, and N. Pearce. 1998. Prevalence of adult asthma symptoms in relation to climate in New Zealand. *Environmental Health Perspectives* 106(9): 607–610.

Hall, S.J., G.P. Asner, and K. Kitayama. 2004. Substrate, climate, and land use controls over soil N dynamics and N-oxide emissions in Borneo. *Biogeochemistry* 70: 27–58.

Hansen, J., M. Sato, P. Kharecha, G. Russell, D.W. Lea, and M. Siddall. 2007. Climate change and trace gases. *Philosophical Transactions of the Royal Society A* 365: 1925–1954.

Hansen, A., P. Bi, M. Nitschke, P. Ryan, D. Pisaniello, and G. Tucker. 2008. The effect of heat waves on mental health in a temperate Australian city. *Environmental Health Perspectives* 116(10): 1369–1375.

Hatfield, J.L. 2011. Soil management for increasing water use efficiency in field crops under changing climates. *In* J.L. Hatfield and T.J. Sauer (eds), *Soil Management: Building a Stable Base for Agriculture*, pp. 161–173. Soil Science Society of America, Madison, WI.

Hättenschwiler, S., I.T. Handa, L. Egli, R. Asshoff, W. Ammann, and C. Körner. 2002. Atmospheric CO_2 enrichment of alpine treeline conifers. *New Phytologist* 156(3): 363–375.

Havlin, J.L., J.D. Beaton, S.L. Tisdale, and W.L. Nelson. 2005. *Soil Fertility and Fertilizers: An Introduction to Nutrient Management*, 7th edn. Pearson Prentice Hall, Upper Saddle River, NJ.

Heilig, G.K. 1994. The greenhouse gas methane (CH_4): Sources and sinks, the impact of population growth, possible interventions. *Population and Environment* 16(2): 109–137.

Hobbs, P.R. and B. Govaerts. 2010. How conservation agriculture can contribute to buffering climate change. *In* M.P. Reynolds (ed.), *Climate Change and Crop Production*, pp. 177–199. CPI Antony Rowe, Chippenham.

Hoekstra, E.J., H. de Weerd, E.W.B. de Leer, and U.A.Th. Brinkman. 1999. Natural formation of chlorinated phenols, dibenzo-p-dioxins, and dibenzofurans in soil of a Douglas fir forest. *Environmental Science and Technology* 33(15): 2543–2549.

Holland, E.A. 2011. The role of soils and biogeochemistry in the climate and Earth system. *In* T.J. Sauer, J.M. Norman, and M.V.K. Sivakumar (eds), *Sustaining Soil Productivity in Response to Global Climate Change: Science, Policy, and Ethics*, pp. 155–168. Wiley, Oxford.

Hu, R., K. Kusa, and R. Hatano. 2001. Soil respiration and methane flux in adjacent forest, grassland, and cornfield soils in Hokkaido, Japan. *Soil Science and Plant Nutrition* 47(3): 621–627.

Hungate, B.A., J.S. Dukes, M.R. Shaw, Y. Luo, and C.B. Field. 2003. Nitrogen and climate change. *Science* 302: 1512–1513.

Huntingford, C., F.H. Lambert, J.H.C. Gash, C.M. Taylor, and A.J. Challinor. 2005. Aspects of climate change prediction relevant to crop productivity. *Philosophical Transactions of the Royal Society B* 360: 1999–2009.

Hussein, A.H., M.C. Rabenhorst, and M.L. Tucker. 2004. Modeling of carbon sequestration in coastal marsh soils. *Soil Science Society of America Journal* 68: 1786–1795.

Idso, S.B. 1998. Carbon-dioxide-induced global warming: A skeptic's view of potential climate change. *Climate Research* 10: 69–82.

IPCC. 2007a. Summary for policymakers. *In* S. Solomon, D. Qin, M. Manning, Z. Chen, M. Marquis, K.B. Averyt, M. Tignor, and H.L. Miller (eds), *Climate Change 2007: The Physical Science Basis*, pp. 1–18. Contribution of Working Group I to the Fourth Assessment Report of the Intergovernmental Panel on Climate Change, Cambridge University Press, Cambridge.

IPCC. 2007b. Climate change 2007 synthesis report. *In* R.K. Pachauri and A. Reisinger (eds), *Working Group Contributions to the Fourth Assessment Report of the Intergovernmental Panel on Climate Change*, pp. 23–73. Cambridge University Press, Cambridge.

Jarvis, A., J. Ramirez, B. Anderson, C. Leibing, and P. Aggarwal. 2010. Scenarios of climate change within the context of agriculture. *In* M.P. Reynolds (ed.), *Climate Change and Crop Production*, pp. 9–37. CPI Antony Rowe, Chippenham.

Jesperson, J.L. and L.J. Osher. 2007. Carbon storage in the soils of a mesotidal Gulf of Maine estuary. *Soil Science Society of America Journal* 71: 372–379.

Johnson, B.J., K.A. Moore, C. Lehmann, C. Bohlen, and T.A. Brown. 2007. Middle to Late Holocene fluctuations of C3 and C4 vegetation in a Northern New England Salt Marsh, Sprague Marsh, Phippsburg Maine. *Organic Geochemistry* 38: 394–403.

Joshi, A.B., D.R. Vann, and A.H. Johnson. 2005. Litter quality and climate decouple nitrogen mineralization and productivity in Chilean temperate rainforests. *Soil Science Society of America Journal* 70: 153–162.

Kalkstein, L.S. and J.S. Greene. 1997. An evaluation of climate/mortality relationships in large U.S. cities and the possible impacts of climate change. *Environmental Health Perspectives* 105(1): 84–93.

Kang, Y., S. Khan, and X. Ma. 2009. Climate change impacts on crop yield, crop water productivity, and food security—A review. *Progress in Natural Science* 19: 1665–1674.

Kardol, P., W.N. Reynolds, R.J. Norby, and A.T. Classen. 2011. Climate change effects on soil microarthropod abundance and community. *Applied Soil Ecology* 47: 37–44.

Keene, W.C., M. Aslam, K. Khalil, D.J. Erickson III, A. McCulloch, T.E. Graedel, J.M. Lobert, et al. 1999. Composite global emissions of reactive chlorine from anthropogenic and natural sources: Reactive chlorine emissions inventory. *Journal of Geophysical Research* 104(D7): 8429–8440.

Kirkham, M.B. 2011. *Elevated Carbon Dioxide*. CRC Press, Boca Raton, FL.

Kirschbaum, M.U.F. 1995. The temperature dependence of soil organic matter decomposition and the effect of global warming on soil organic C storage. *Soil Biology and Biochemistry* 27(6): 753–760.

Körner, C. 2006. Plant CO_2 responses: An issue of definition, time and resource supply. *New Phytologist* 172(3): 393–411.

Kovats, R.S., D.H. Campbell-Lendrum, A.J. McMichael, A. Woodward, and J. St H. Cox. 2001. Early effects of climate change: Do they include changes in vector-borne disease? *Philosophical Transactions of the Royal Society B* 356: 1057–1068.

Kovats, R.S., S.J. Edwards, S. Hajat, B.G. Armstrong, K.L. Ebi, and B. Menne. 2004. The effect of temperature on food poisoning: A time-series analysis of salmonellosis in ten European countries. *Epidemiology and Infection* 132(3): 443–453.

Kutílek, M. 2011. Soils and climate change. *Soil and Tillage Research* 117: 1–7.

Lahsen, M. 2005. Technocracy, democracy, and U.S. climate politics: The need for demarcations. *Science, Technology, & Human Values* 30(1): 137–169.

Lake, I.R., I.A. Gillespie, G. Bentham, G.L. Nichols, C. Lane, G.K. Adak, and E.J. Threlfall. 2009. A re-evaluation of the impact of temperature and climate change on foodborne illness. *Epidemiology and Infection* 137: 1538–1547.

Lal, R. 2004. Soil carbon sequestration impacts on global climate change and food security. *Science* 304: 1623–1627.

Lal, R. 2007. Soil science and the carbon civilization. *Soil Science Society of America Journal* 71: 1425–1437.

Lal, R. 2010. Managing soils and ecosystems for mitigating anthropogenic carbon emissions and advancing global food security. *BioScience* 60(9): 708–721.

Lal, R., J. Kimble, and R.F. Follett. 1998. Pedospheric processes and the carbon cycle. *In* R. Lal, J.M Kimble, R.F. Follett, and B.A. Stewart (eds), *Soil Processes and the Carbon Cycle*, pp. 1–8. CRC Press, Boca Raton, FL.

Le Treut, H., R. Somerville, U. Cubasch, Y. Ding, C. Mauritzen, A. Mokssit, T. Peterson and M. Prather. 2007. Historical overview of climate change. *In* S. Solomon, D. Qin, M. Manning, Z. Chen, M. Marquis, K.B. Averyt, M. Tignor, and H.L. Miller (eds), *Climate Change 2007: The Physical Science Basis*, pp. 93–127. Contribution of Working Group I to the Fourth Assessment Report of the Intergovernmental Panel on Climate Change, Cambridge University Press, Cambridge.

Leikauf, G.D. 2002. Hazardous air pollutants and asthma. *Environmental Health Perspectives* 110(4): 505–526.

Lele, U. 2010. Food security for a billion poor. *Science* 326: 1554.

Li, Z., S. Tang, X. Deng, R. Wang, and Z. Song. 2010. Contrasting effects of elevated CO_2 on Cu and Cd uptake by different rice varieties grown on contrasting soils with two levels of metals: Implication for phytoextraction and food security. *Journal of Hazardous Materials* 177: 352–361.

Li, Z., W.-Z. Lui, X.-C. Zhang, and F.-L. Zheng. 2011. Assessing the site-specific impacts of climate change on hydrology, soil erosion, and crop yields in the Loess Plateau of China. *Climatic Change* 105: 223–242.

Lichter, J., S.A. Billings, S. Ziegler, D. Gaindh, R. Ryals, A.C. Finzi, R.B. Jackson, E.A. Stemmler, and W.H. Schlesinger. 2008. Soil carbon sequestration in a pine forest after 9 years of atmospheric CO_2 enrichment. *Global Change Biology* 14: 1–13.

Lindgren, E., L. Tälleklint, and T. Polfeldt. 2000. Impact of climatic change on the northern latitude limit and population density of the disease-transmitting European tick *Ixodes ricinus*. *Environmental Health Perspectives* 108(2): 119–123.

Link, S.O., J.L. Smith, J.J. Halverson, and H. Bolton Jr. 2003. A reciprocal transplant experiment within a climatic gradient in a semiarid shrub-steppe ecosystem: Effects on bunchgrass growth and reproduction, soil carbon, and soil nitrogen. *Global Change Biology* 9: 1097–1105.

Lipson, D.A., M. Blair, G. Barron-Gafford, K. Grieve, and R. Murthy. 2006. Relationships between microbial community structure and soil processes under elevated atmospheric carbon dioxide. *Microbial Ecology* 51: 302–314.

Lokupitiya, E. and K. Paustian. 2006. Agricultural soil greenhouse gas emissions: A review of national inventory methods. *Journal of Environmental Quality* 35: 1413–1427.

Long, S.P., E.A. Ainsworth, A.D.B. Leakey, and P.B. Morgan. 2005. Global food insecurity. Treatment of major food crops with elevated carbon dioxide or ozone under large-scale fully open-air conditions suggests recent models may have overestimated future yields. *Philosophical Transactions of the Royal Society B* 360(1463): 2011–2020.

Louis, V.R., I.A. Gillespie, S. J. O'Brien, E. Russek-Cohen, A.D. Pearson, and R.R. Colwell. 2005. Temperature-driven Campylobacter seasonality in England and Wales. *Applied and Environmental Microbiology* 71(1): 85–92.

Lu, Y., R. Wassmann, H.-U. Neue, and C. Huang. 1999. Impact of phosphorus supply on root exudation, aerenchyma formation and methane emission of rice plants. *Biogeochemistry* 47(2): 203–218.

Luo, Y., B. Su, W.S. Currie, J.S. Dukes, A. Finzi, U. Hartwig, B. Hungate, et al. 2004. Progressive nitrogen limitation of ecosystem responses to rising atmospheric carbon dioxide. *BioScience* 54(8): 731–739.

Maeda, E.E., P.K.E. Pellikka, M. Siljander, and B.J.F. Clark. 2010. Potential impacts of agricultural expansion and climate change on soil erosion in the Eastern Arc Mountains of Kenya. *Geomorphology* 123: 279–289.

Magdoff, F. and H. van Es. 2009. *Building Soils for Better Crops: Sustainable Soil Management*, 3rd edn. Sustainable Agriculture Network Handbook Series #10. Sustainable Agriculture Publications, Waldorf, MD.

Martikainen P.J., K. Regina, E. Syväsalo, T. Laurila, A. Lohila, M. Aurela, J. Silvola, et al. 2002. Agricultural soils as a sink and source of greenhouse gases: A research consortium (AGROGAS). *In* J. Käyhkö and L. Talve (eds), *Understanding the Global System, the Finnish Perspective*, pp. 55–68. Finnish Global Change Research Programme FIGARE, Turku, Finland.

McConnell, J.R., A.J. Aristarain, J.R. Banta, P. Edwards, and J.C. Simões. 2007. 20th-century doubling in dust archived in an Antarctic Peninsula ice core parallels climate change and desertification in South America. *Proceedings of the National Academy of Sciences of the United States of America* 104(14): 5743–5748.

Meehl, G.A., T.F. Stocker, W.D. Collins, P. Friedlingstein, A.T. Gaye, J.M. Gregory, A. Kitoh, et al. 2007. Global climate projections. *In* S. Solomon, D. Qin, M. Manning, Z. Chen, M. Marquis, K.B. Averyt, M. Tignor, and H.L. Miller (eds), *Climate Change 2007: The Physical Science Basis*, pp. 747–845. Contribution of Working Group I to the Fourth Assessment Report of the Intergovernmental Panel on Climate Change, Cambridge University Press, Cambridge.

Melillo, J.M., P.A. Steudler, B.J. Feigl, C. Neill, D. Garcia, M.C. Piccolo, C.C. Cerri, and H. Tian. 2001. Nitrous oxide emissions from forests and pastures of various ages in the Brazilian Amazon. *Journal of Geophysical Research* 106(D24): 34179–34188.

Montgomery, D.R. 2007. *Dirt: The Erosion of Civilizations*. University of California Press, Berkley, CA.

Mosier, A.R. 1998. Soil processes and global change. *Biology and Fertility of Soils* 27: 221–229.

Mullen, R.W. 2011. Nutrient cycling in soils: Nitrogen. *In* J.L. Hatfield and T.J. Sauer (eds), *Soil Management: Building a Stable Base for Agriculture*, pp. 67–78. Soil Science Society of America, Madison, WI.

Mulvaney, R.L., S.A. Khan, and T.R. Ellsworth. 2009. Synthetic nitrogen fertilizers deplete soil nitrogen: A global dilemma for sustainable cereal production. *Journal of Environmental Quality* 38: 2295–2314.

Neill, C., C.C. Cern, J.M. Melillo, B.J. Feigl, P.A. Steudler, J.F.L. Moraes, and M.C. Piccolo. 1998. Stocks and dynamics of soil carbon following deforestation for pasture in Rondonia. *In* R Lal, J.M. Kimble, R.F. Follett, and B.A. Stewart (eds), *Soil Processes and the Carbon Cycle*, pp. 9–28. CRC Press, Boca Raton, FL.

Neue, H.-U. 1992. Agronomic practices affecting methane fluxes from rice cultivation. *Ecological Bulletins* 42: 174–182.

Niklaus, P.A. and C. Körner. 2004. Synthesis of a six-year study of calcareous grassland responses to *in situ* CO_2 enrichment. *Ecological Monographs* 74(3): 491–511.

Niklińska, M., M. Maryański, and R. Laskowski. 1999. Effect of temperature on humus respiration rate and nitrogen mineralization: Implications for global climate change. *Biogeochemistry* 44(3): 239–257.

Norby, R.J. and Y. Luo. 2004. Evaluating ecosystem responses to rising atmospheric CO_2 and global warming in a multi-factor world. *New Phytologist* 162: 281–293.

Nowak, R.S., S.F. Zitzer, D. Babcock, V. Smith-Longozo, T.N. Charlet, J.S. Coleman, J.R. Seemann, and S.D. Smith. 2004. Elevated atmospheric CO_2 does not conserve soil water in the Mojave Desert. *Ecology* 85(1): 93–99.

Oechel, W.C. and G.L. Vourlitis. 1995. Effects of global change on carbon storage in cold soils. *In* R. Lal, J. Kimble, E. Levine, and B.A. Stewart (eds), *Soils and Global Change*, pp. 117–129. CRC Press, Boca Raton, FL.

Olesen, J.E. and M. Bindi. 2002. Consequences of climate change for European agricultural productivity, land use, and policy. *European Journal of Agronomy* 16: 239–262.

Paeth, H., A. Capo-Chichi, and W. Endlicher. 2008. Climate change and food security in tropical West Africa—A dynamic-statistical modeling approach. *Erdkunde* 62(2): 101–115.

Park, S.E., S.M. Howden, S.J. Crimp, D.S. Gaydon, S.J. Attwood, and P.N. Kokic. 2009. More than eco-efficiency is required to improve food security. *Crop Science* 50: S132–S141.

Parry, M., C. Rosenzweig, and M. Livermore. 2005. Climate change, global food supply, and risk of hunger. *Philosophical Transactions of the Royal Society B* 360: 2125–2138.

Pascual, M., J.A. Ahumada, L.F. Chaves, X. Rodó, and M. Bouma. 2006. Malaria resurgence in the East African Highlands: Temperature trends revisited. *Proceedings of the National Academy of Sciences of the United States of America* 103(15): 5829–5834.

Pierzynski, G.M., J.T. Sims, and G.F. Vance. 2009. *Soils and Environmental Quality*, 3rd edn. CRC Press, Boca Raton, FL.

Pimentel, D. 2006. Soil erosion: A food and environmental threat. *Environment, Development, and Sustainability* 8: 119–137.

Pimentel, D., S. Cooperstein, H. Randell, D. Filiberto, S. Sorrentino, B. Kaye, C. Nicklin, et al. 2007. Ecology of increasing diseases: Population growth and environmental degradation. *Human Ecology* 35(6): 653–668.

Poorter, H. and M.-L. Navas. 2003. Plant growth and competition at elevated CO_2: On winners, losers and functional groups. *New Phytologist* 157(2): 175–198.

Post, W.M. and K.C. Kwon. 2000. Soil carbon sequestration and land-use change: Processes and potential. *Global Change Biology* 6: 317–327.

Post, W.M., R.C. Izaurralde, J.D. Jastrow, B.A. McCarl, J.E. Amonette, V.L. Bailey, P.M. Jardine, T.O. West, and J. Zhou. 2004. Enhancement of carbon sequestration in US soils. *BioScience* 54(10): 895–908.

Poudel, D.D., D.J. Midmore, and L.T. West. 1999. Erosion and productivity of vegetable systems on sloping volcanic ash-derived Philippine soils. *Soil Science Society of America Journal* 63: 1366–1376.

Price, D.T., C.H. Peng, M.J. Apps, and D.H. Halliwell. 1999. Simulating effects of climate change on boreal ecosystem carbon pools in central Canada. *Journal of Biogeography* 26: 1237–1248.

Prosperol, J.M. and P.J. Lamb. 2003. African droughts and dust transport to the Caribbean: Climate change implications. *Science* 302: 1024–1027.

Pyle, J.A., N. Warwick, X. Yang, P.J. Young, and G. Zeng. 2007. Climate/chemistry feedbacks and biogenic emissions. *Philosophical Transactions of the Royal Society A* 365: 1727–1740.

Rask-Andersen, A. 2001. Asthma increase among farmers: A 12-year follow-up. *Uppsala Journal of Medical Sciences* 116: 60–71.

Ravi, S., D.D. Breshears, T.E. Huxman, and P. D'Odorico. 2010. Land degradation in drylands: Interactions among hydraulic-aeolian erosion and vegetation dynamics. *Geomorphology* 116: 236–245.

Reich, P.B., S.E. Hobbie, T. Lee, D.S. Ellsworth, J.B. West, D. Tilman, J.M. Knops, S. Naeem, and J. Trost. 2006a. Nitrogen limitation constrains sustainability of ecosystem response to CO_2. *Nature* 440: 922–925.

Reich, P.B., B.A. Hungate, and Y. Luo. 2006b. Carbon–nitrogen interactions in terrestrial ecosystems in response to rising atmospheric carbon dioxide. *Annual Review of Ecology, Evolution, and Systematics* 37: 611–636.

Reid, C.E., M.S. O'Neill, C.J. Gronlund, S.J. Brines, D.G. Brown, A.V. Diez-Roux, and J. Schwartz. 2009. Mapping community determinants of heat vulnerability. *Environmental Health Perspectives* 117(11): 1730–1736.

Richter, D., M. Hofmockel, M.A. Callaham, D.S. Powlson, and P. Smith. 2007. Long-term soil experiments: Keys to managing Earth's rapidly changing ecosystems. *Soil Science Society of America Journal* 71: 266–279.

Rose, J.B., P.R. Epstein, E.K. Lipp, B.H. Sherman, S.M. Bernard, and J.A. Patz. 2001. Climate variability and change in the United States: Potential impacts on water- and foodborne diseases caused by microbiologic agents. *Environmental Health Perspectives* 109(2): 211–221.

Rosegrant, M.W. and S.A. Cline. 2003. Global food security: Challenges and policies. *Science* 302(5652): 1917–1919.

Rosenzweig, C. and M. Parry. 1994. Potential impact of climate change on world food supply. *Nature* 367: 133–138.

Rustad, L.E., T.G. Huntington, and R.D. Boone. 2000. Controls on soil respiration: Implications for climate change. *Biogeochemistry* 48: 1–6.

Sanchez, P.A. and M.S. Swaminathan. 2005. Hunger in Africa: The link between unhealthy people and unhealthy soils. *Lancet* 365: 442–444.

Sauer, T.J., and M.P. Nelson. 2011. Science, ethics, and the historical roots of our ecological crisis. Was White right? *In* T.J. Sauer, J.M. Norman, and M.V.K. Sivakumar (eds), *Sustaining Soil Productivity in Response to Global Climate Change: Science, Policy, and Ethics*, pp. 3–16. Wiley, Oxford.

Schenker, M. 2000. Exposures and health effects from inorganic agricultural dusts. *Environmental Health Perspectives* 108(4): 661–664.

Schlesinger, W.H. 1995. An overview of the carbon cycle. *In* R. Lal, J. Kimble, E. Levine, and B.A. Stewart (eds), *Soils and Global Change*, pp. 9–25. CRC Press, Boca Raton, FL.

Schmidhuber, J. and F.N. Tubiello. 2007. Global food security under climate change. *Proceedings of the National Academy of Sciences of the United States of America* 104(50): 19703–19708.

Schneider, U.A. and P. Smith. 2009. Energy intensities and greenhouse gas emission mitigation in global agriculture. *Energy Efficiency* 2: 195–206.

Schoeneberger, P.J., D.A. Wysocki, E.C. Benham, and W.D. Broderson (eds). 2002. *Field Book for Describing and Sampling Soils, Version 2.0.* Natural Resources Conservation Service, National Soil Survey Center, Lincoln, NE.

Schütz, H., W. Seiler, and R. Conrad. 1990. Influence of soil temperature on methane emission from rice paddy fields. *Biogeochemistry* 11(2): 77–95.

Schwartz, J. 2005. Who is sensitive to extremes of temperature? A case-only analysis. *Epidemiology* 16(1): 67–72.

Silver, W.L., R. Osterlag, and A.E. Lugo. 2000. The potential for carbon sequestration through reforestation of abandoned tropical agricultural and pasture lands. *Restoration Ecology* 8: 394–407.

Sing, D. and C.F. Sing. 2010. Impact of direct soil exposures from airborne dust and geophagy on human health. *International Journal of Environmental Research and Public Health* 7: 1205–1223.

Sivakumar, M.V.K. 2011. Climate and land degradation. *In* T.J. Sauer, J.M. Norman, and M.V.K. Sivakumar (eds), *Sustaining Soil Productivity in Response to Global Climate Change: Science, Policy, and Ethics*, pp. 141–154. Wiley, Oxford.

Sjögersten, S., R. van der Wal, and S.J. Woodin. 2006. Small-scale hydrological variation determines landscape CO_2 fluxes in the high Arctic. *Biogeochemistry* 80(3): 205–216.

Smith, P., D. Martino, Z. Cai, D. Gwary, H. Janzen, P. Kumar, B. McCarl, et al. 2007. Agriculture. *In* B. Metz, O.R. Davidson, P.R. Bosch, R. Dave, and L.A. Meyer (eds), *Climate Change 2007: Mitigation*, pp. 497–540. Contribution of Working Group III to the Fourth Assessment Report of the Intergovernmental Panel on Climate Change, Cambridge University Press, Cambridge.

Solomon, S., D. Qin, M. Manning, R.B. Alley, T. Berntsen, N.L. Bindoff, Z. Chen, et al. 2007. Technical summary. *In* S. Solomon, D. Qin, M. Manning, Z. Chen, M. Marquis, K.B. Averyt, M. Tignor, and H.L. Miller (eds), *Climate Change 2007: The Physical Science Basis*, pp. 19–91. Contribution of Working Group I to the Fourth Assessment Report of the Intergovernmental Panel on Climate Change, Cambridge University Press, Cambridge.

Sparovek, G. and E. Schnug. 2001. Temporal erosion-induced soil degradation and yield loss. *Soil Science Society of America Journal* 65: 1479–1486.

Sperow, M. 2006. Carbon sequestration potential in reclaimed mine sites in seven east-central States. *Journal of Environmental Quality* 35: 1428–1438.

St. Clair, S.B. and J.P. Lynch. 2010. The opening of Pandora's Box: Climate change impacts on soil fertility and crop nutrition in developing countries. *Plant and Soil* 335: 101–115.

Steinback, H.S. and R. Alvarez. 2006. Changes in soil organic carbon contents and nitrous oxide emissions after introduction of no-till in Pampean agroecosystems. *Journal of Environmental Quality* 35: 3–13.

Stenseth, N.C., N.I. Samia, H. Viljugrein, K.L. Kausrud, M. Begon, S. Davis, H. Leirs, et al. 2006. Plague dynamics are driven by climate variation. *Proceedings of the National Academy of Sciences of the United States of America* 103(35): 13110–13115.

Stępniewski, W., Z. Stępniewski, and A. Rożej. 2011. Gas exchange in soils. *In* J.L. Hatfield and T.J. Sauer (eds), *Soil Management: Building a Stable Base for Agriculture*, pp. 117–144. Soil Science Society of America, Madison, WI.

Steyer, G.D. and R.E. Stewart Jr. 1992. *Monitoring Program for Coastal Wetlands Planning, Protection, and Restoration Act Projects*. National Wetlands Research Center Open File Report 93-01, U.S. Fish and Wildlife Service.

Stone, B., J.J. Hess, and H. Frumkin. 2010. Urban form and extreme heat events: Are sprawling cities more vulnerable to climate change than compact cities? *Environmental Health Perspectives* 118(10): 1425–1428.

Szukics, U., G.C.J. Abell, V. Hödl, B. Mitter, A. Sessitsch, E. Hackl, and S. Zechmeister-Boltenstern. 2010. Nitrifiers and denitrifiers respond rapidly to changed moisture and increasing temperature in a pristine forest soil. *FEMS Microbiology Ecology* 72: 395–406.

Tan, Z., L.L. Tieszen, S. Liu, and E. Tachie-Obeng. 2010. Modeling to evaluate the response of savanna-derived cropland to warming-drying stress and nitrogen fertilizers. *Climatic Change* 100: 703–715.

Tang, S.R., L. Xi, X.M. Zhang, and H.Y. Li. 2003. Response to elevated CO_2 of Indian mustard and sunflower growing on copper contaminated soil. *Bulletin of Environmental Contamination and Toxicology* 71: 988–997.

Tengberg, A., M. Stocking, and S.C.F. Dechen. 1997. The impact of erosion on soil productivity. An experimental design applied in São Paulo state, Brazil. *Geografiska Annaler. Series A, Physical Geography* 79(1/2): 95–107.

Titus, J.G. 1991. Greenhouse effect and coastal wetland policy: How Americans could abandon an area the size of Massachusetts at minimum cost. *Environmental Management* 15: 39–58.

Trenberth, K.E., P.D. Jones, P. Ambenje, R. Bojariu, D. Easterling, A. Klein Tank, D. Parker, et al. 2007. Observations: Surface and atmospheric climate change. *In* S. Solomon, D. Qin, M. Manning, Z. Chen, M. Marquis, K.B. Averyt, M. Tignor, and H.L. Miller (eds), *Climate Change 2007: The Physical Science Basis*, pp. 235–336. Contribution of Working Group I to the Fourth Assessment Report of the Intergovernmental Panel on Climate Change, Cambridge University Press, Cambridge.

Troeh, F.R., J.A. Hobbs, and R.L. Donahue. 2004. *Soil and Water Conservation for Productivity and Environmental Protection*, 4th edn. Prentice Hall, Upper Saddle River, NJ.

Tsihrintzis, V.A., G.M. Vasarhelyi, and J. Lipa. 1996. Ballona wetland: A multi-objective salt marsh restoration plan. *Proceedings of the Institution of Civil Engineers: Water, Maritime, and Energy* 118: 131–144.

Ussiri, D.A.N., R. Lal, and P. A. Jacinthe. 2006. Soil properties and carbon sequestration of afforested pastures in reclaimed minesoils of Ohio. *Soil Science Society of America Journal* 70: 1797–1806.

Wagner-Riddle, C. and A. Weersink. 2011. Net agricultural greenhouse gases: Mitigation strategies and implications. *In* T.J. Sauer, J.M. Norman, and M.V.K. Sivakumar (eds), *Sustaining Soil Productivity in Response to Global Climate Change: Science, Policy, and Ethics*, pp. 169–182. Wiley, Oxford.

Wan, Y., E. Lin, W. Xiong, Y. Li, and L. Guo. 2011. Modeling the impact of climate change on soil organic carbon stock in upland soils in the 21st century in China. *Agriculture, Ecosystems, and Environment* 141: 23–31.

Washington, R., M. Todd, N.J. Middleton, and A.S. Goudie. 2003. Dust-storm source areas determined by the total ozone monitoring spectrometer and surface observations. *Annals of the Association of American Geographers* 93(2): 297–313.

Wassmann, R., H. Schütz, H. Papen, H. Rennenberg, W. Seiler, D. Aiguo, S. Renxing, S. Xingjian, and W. Mingxing. 1993. Quantification of methane emissions from Chinese rice fields (Zhejiang province) as influenced by fertilizer treatment. *Biogeochemistry* 20(2): 83–101.

Welker, J.M., K.B. Brown, and J.T. Fahnestock. 1999. CO_2 flux in Arctic and alpine dry tundra: Comparative field responses under ambient and experimentally warmed conditions. *Arctic, Antarctic, and Alpine Research* 31(3): 272–277.

West, T.O. and W.M. Post. 2002. Soil organic carbon sequestration rates by tillage and crop rotation: A global data analysis. *Soil Science Society of America Journal* 66: 1930–1946.

White, W.A. and R.A. Morton. 1997. Wetland losses related to fault movement and hydrocarbon production, southeastern Texas coast. *Journal of Coastal Research* 13: 1305–1320.

Whitford, W.G. 1992. Effects of climate change on soil biotic communities and soil processes. *In* R.L. Peters and T.E. Lovejoy (eds), *Global Warming and Biological Diversity*, pp. 124–136. Yale University Press, New York, NY.

Williams, M.W., P.D. Brooks, and T. Seastedt. 1998. Nitrogen and carbon soil dynamics in response to climate change in a high-elevation ecosystem in the Rocky Mountains, USA. *Arctic and Alpine Research* 30(1): 26–30.

Wolf, B. and G.H. Snyder. 2003. *Sustainable Soils: The Place of Organic Matter in Sustaining Soils and Their Productivity*. Haworth Press, New York, NY.

Wu, H.B., S.R. Tang, X.M. Zhang, J.K. Guo, Z.G. Song, S. Tian, and D. Smith. 2009. Using elevated CO_2 to increase the biomass of a *Sorghum vulgare x Sorghum vulgare var. sudanense* hybrid and *Trifolium pretense* L. and to trigger hyperaccumulation of cesium. *Journal of Hazardous Materials* 170: 861–870.

Zaehle, S., P. Friedlingstein, and A.D. Friend. 2010. Terrestrial nitrogen feedbacks may accelerate future climate change. *Geophysical Research Letters* 37: L01401.

Zavaleta, E.S., M.R. Shaw, N.R. Chiariello, B.D. Thomas, E.E. Cleland, C.B. Field, and H.A. Mooney. 2003. Grassland responses to three years of elevated temperature, CO_2, precipitation, and N deposition. *Ecological Monographs* 73(4): 585–604.

Zedler, J.B. and S. Kercher. 2005. Wetland resources: Status, trends, ecosystem services, and restorability. *Annual Reviews of Environment and Resources* 30: 39–74.

Zhang, J.-E., Y. Ouyang, Z.-X. Huang, and G.-M. Quan. 2011. Dynamic emission of CH_4 from a rice-duck farming ecosystem. *Journal of Environmental Protection* 2: 537–544.

Zhang, X.C., M.A. Nearing, J.D. Garbrecht, and J.L. Steiner. 2004. Downscaling monthly forecasts to simulate impacts of climate change on soil erosion and wheat production. *Soil Science Society of America Journal* 68: 1376–1385.

Index

9 781138 199316